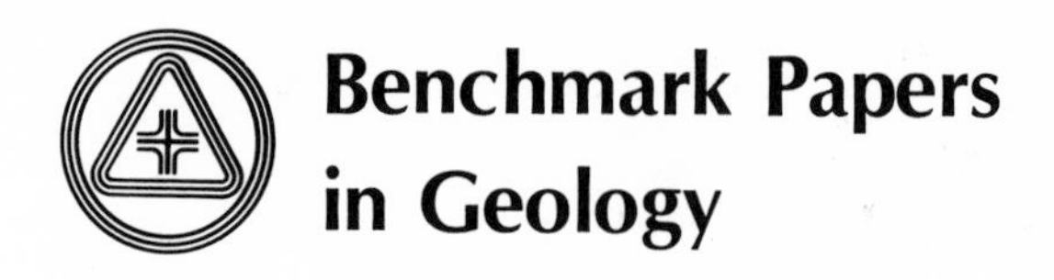

Series Editor: Rhodes W. Fairbridge
Columbia University

PUBLISHED VOLUMES

RIVER MORPHOLOGY / *Stanley A. Schumm*
SLOPE MORPHOLOGY / *Stanley A. Schumm and M. Paul Mosley*
SPITS AND BARS / *Maurice L. Schwartz*
BARRIER ISLANDS / *Maurice L. Schwartz*
ENVIRONMENTAL GEOMORPHOLOGY AND LANDSCAPE CONSERVATION, VOLUME I: Prior to 1900 / VOLUME II: Urban Areas / VOLUME III: Non-Urban Regions / *Donald R. Coates*
TEKTITES / *Virgil E. Barnes and Mildred A. Barnes*
GEOCHRONOLOGY: Radiometric Dating of Rocks and Minerals / *C. T. Harper*
MARINE EVAPORITES: Origin, Diagenesis, and Geochemistry / *Douglas W. Kirkland and Robert Evans*
GLACIAL ISOSTASY / *John T. Andrews*
GLACIAL DEPOSITS / *Richard P. Goldthwait*
PHILOSOPHY OF GEOHISTORY: 1785–1970 / *Claude C. Albritton, Jr.*
GEOCHEMISTRY OF GERMANIUM / *Jon N. Weber*
GEOCHEMISTRY AND THE ORIGIN OF LIFE / *Keith A. Kvenvolden*
GEOCHEMISTRY OF WATER / *Yasushi Kitano*
GEOCHEMISTRY OF IRON / *Henry Lepp*
SEDIMENTARY ROCKS: Concepts and History / *Albert V. Carozzi*
METAMORPHISM AND PLATE TECTONIC REGIMES / *W. G. Ernst*
SUBDUCTION ZONE METAMORPHISM / *W. G. Ernst*
PLAYAS AND DRIED LAKES: Occurrence and Development / *James T. Neal*
GEOCHEMISTRY OF BORON / *C. T. Walker*
PLANATION SURFACES: Peneplains, Pediplains, and Etchplains / *George Adams*
SUBMARINE CANYONS AND DEEP-SEA FANS: Modern and Ancient / *J. H. McD. Whitaker*
ENVIRONMENTAL GEOLOGY / *Frederick Betz, Jr.*

Additional volumes in preparation

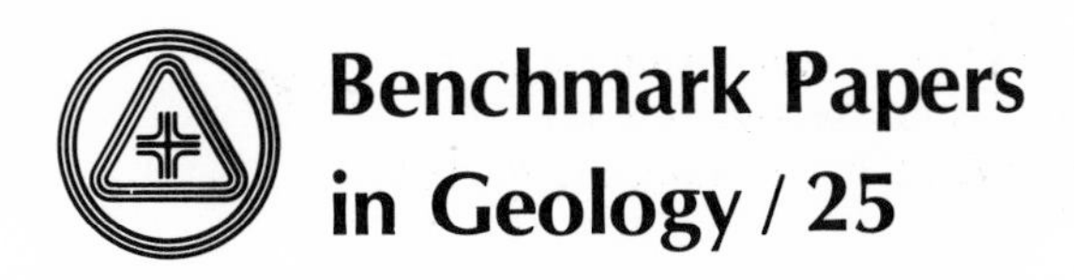

A BENCHMARK® Books Series

ENVIRONMENTAL GEOLOGY

Edited by
FREDERICK BETZ, JR.

Distributed by
HALSTED PRESS *A Division of John Wiley & Sons, Inc.*

Meant for Betty
1921–1974

Benchmark Papers in Geology, Volume 25
Library of Congress Catalog Number: 75-29646
ISBN: 0-470-07170-2

77 76 75 1 2 3 4 5
Manufactured in the United States of America.

LIBRARY OF CONGRESS CATALOGING IN PUBLICATION DATA

Main entry under title:

Enviromental geology.

(Benchmark papers in geology ; 25)
Includes bibliographies and indexes.
1. Engineering geology--Addresses, essays, lectures.
2. Environmental protection--Addresses, essays, lectures.
3. Man--Influence of environment--Addresses, essays, lectures. 4. Man--Influence on nature--Addresses, esaays, lectures. I. Betz, Frederick, 1915-
TA705.E64 301.31 75-29646
ISBN 0 470-07170-2

Exclusive Distributor: **Halsted Press**
A Division of John Wiley & Sons, Inc.

ACKNOWLEDGMENTS AND PERMISSIONS

ACKNOWLEDGMENTS

APPALACHIAN REGIONAL COMMISSION, WASHINGTON, D.C.—*Appalachia*
Environmental Analysis in Local Development Planning

ASSOCIATION OF ENGINEERING GEOLOGISTS—*Geological and Geographical Problems of Areas of High Population Density*
Geological and Geographical Factors in the Siting of New Towns in Great Britain

BRITISH ASSOCIATION FOR THE ADVANCEMENT OF SCIENCE—*The Advancement of Science*
The Influence of Geology on Military Operations in North-west Europe

DEPARTMENT OF GEOLOGY–GEOGRAPHY, UNIVERSITY OF MISSOURI–KANSAS CITY—*University of Missouri–Kansas City Geographical Publication*
Occupance and Use of Underground Space in the Greater Kansas City Area

GEOLOGICAL SURVEY OF ALABAMA—*Sinkhole Problems Along Proposed Route of Interstate Highway 459 near Greenwood, Alabama*
Excerpts

THE GEOLOGICAL SOCIETY OF AMERICA—*Military Geology*
Excerpts

SOILTEST, INC.—*The Testing World*
The Earthquakes that Rearranged Mid-America

24TH INTERNATIONAL GEOLOGICAL CONGRESS, CANADA—*24th International Geological Congress, Proceedings*
Land Subsidence and Population Growth
Urban and Environmental Geology of Hamburg (Germany)

U.S. GEOLOGICAL SURVEY
Geologic and Seismologic Aspects of the Managua, Nicaragua, Earthquakes of December 23, 1972
Excerpts
Hydrologic Implications of Solid-Waste Disposal
Mineral Resources off the Northeastern Coast of the United States
Excerpts

Acknowledgments and Permissions

PERMISSIONS

The following papers have been reprinted with the permission of the authors and copyright holders.

AMERICAN ASSOCIATION FOR THE ADVANCEMENT OF SCIENCE—*Science*
Hydrological and Ecological Problems of Karst Regions

AMERICAN GEOLOGICAL INSTITUTE—*Environmental Geology: Short Course Lecture Notes*
An Approach to Environmental Geology

AMERICAN MINING CONGRESS—*Mining Congress Journal*
Resources and Environment—Quest for Balance

AMERICAN WATER WORKS ASSOCIATION, INC.—*Journal American Water Works Association*
System for Evaluation of Contamination Potential of Some Waste Disposal Sites

ASSOCIATION OF ENGINEERING GEOLOGISTS—*Bulletin of the Association of Engineering Geologists*
Missouri's Approach to Engineering Geology in Urban Areas

THE COLORADO SCHOOL OF MINES—*Mineral Industries Bulletin*
Geothermal Energy: Geology, Exploration, and Developments, Parts 1 and 2

FRIEDRICH VIEWEG & SOHN GMBH, WEST GERMANY—*Geoforum*
Heavy Metal Accumulation in River Sediments: A Response to Environmental Pollution

GEOLOGICAL SOCIETY OF CZECHOSLOVAKIA—*23rd International Geological Congress, Proceedings*
Geology As Applied to Urban Planning: An Example from the Greater Anchorage Area Borough, Alaska

HARPER & ROW, PUBLISHERS, INC. (for THE NATIONAL ACADEMY OF SCIENCES)—*The Earth and Human Affairs*
The Alaska Oil Pipeline

MORGAN-GRAMPIAN (PROFESSIONAL PRESS) LTD—*Civil Engineering and Public Works Review*
Foulness: Some Geological Implications
Foundation Conditions and Reclamation at Maplin Airport

NEW YORK ACADEMY OF SCIENCES—*Transactions of the New York Academy of Sciences*
Geology and Our Cities

SATURDAY REVIEW CO.—*Saturday Review/World (Science)*
Tambora and Krakatau: Volcanoes and the Cooling of the World

VAN NOSTRAND REINHOLD, INC.—*The Encyclopedia of Geochemistry and Environmental Sciences*
Medical Geology

W. H. FREEMAN AND COMPANY (for SCIENTIFIC AMERICAN, INC.)—*Scientific American*
Quick Clay

SERIES EDITOR'S PREFACE

The philosophy behind the "Benchmark Papers in Geology" is one of collection, sifting, and rediffusion. Scientific literature today is so vast, so dispersed, and, in the case of old papers, so inaccessible for readers not in the immediate neighborhood of major libraries that much valuable information has been ignored by default. It has become just so difficult, or so time consuming, to search out the key papers in any basic area of research that one can hardly blame a busy man for skimping on some of his "homework."

This series of volumes has been devised, therefore, to make a practical contribution to this critical problem. The geologist, perhaps even more than any other scientist, often suffers from twin difficulties—isolation from central library resources and immensely diffused sources of material. New colleges and industrial libraries simply cannot afford to purchase complete runs of all the world's earth science literature. Specialists simply cannot locate reprints or copies of all their principal reference materials. So it is that we are now making a concerted effort to gather into single volumes the critical material needed to reconstruct the background of any and every major topic of our discipline.

We are interpreting "geology" in its broadest sense: the fundamental science of the planet Earth, its materials, its history, and its dynamics. Because of training and experience in "earthy" materials, we also take in astrogeology, the corresponding aspect of the planetary sciences. Besides the classical core disciplines such as mineralogy, petrology, structure, geomorphology, paleontology, and stratigraphy, we embrace the newer fields of geophysics and geochemistry, applied also to oceanography, geochronology, and paleoecology. We recognize the work of the mining geologists, the petroleum geologists, the hydrologists, the engineering and environmental geologists. Each specialist needs his working library. We are endeavoring to make his task a little easier.

Each volume in the series contains an Introduction prepared by a specialist (the volume editor)—a "state of the art" opening or a summary of the object and content of the volume. The articles, usually some thirty to fifty reproduced either in their entirety or in significant extracts, are selected in an attempt to cover the field, from the key papers of the last century to fairly recent work. Where the original works are in foreign languages, we have endeavored to locate or commission translations. Geologists, because of their global subject, are often acutely aware of the oneness of our world. The selections cannot, therefore, be restricted to any one country, and whenever possible an attempt is made to scan the world literature.

To each article, or group of kindred articles, some sort of "highlight commentary" is usually supplied by the volume editor. This commentary should serve to bring that article into historical perspective and to emphasize its particular role in the growth of the field. References, or citations, wherever possible, will be reproduced in their entirety—for by this means the observant reader can assess the background material available to that particular author, or, if he wishes, he, too, can double check the earlier sources.

A "benchmark," in surveyor's terminology, is an established point on the ground, recorded on our maps. It is usually anything that is a vantage point, from a modest hill to a mountain peak. From the historical viewpoint, these benchmarks are the bricks of our scientific edifice.

RHODES W. FAIRBRIDGE

PREFACE

The term *environmental geology* has been known only since the early 1960s, when problems created by use and abuse of the environment became a common concern and stimulated numerous disciplines to identify their particular roles in the study of these problems. However, as a subject, environmental geology can be traced in the literature from early in the nineteenth century.

The number of older writings cannot be determined, because the subject did not appear in bibliographies, and book and article titles usually gave no clue to it. The literature on environmental geology has grown rapidly in recent years, but there are still problems in identifying the subject from titles. The term is still used for a variety of purposes.

The readings in this volume provide a sample of environmental geology, necessarily limited in number and length by the available space, limited also to a number of topics that have to do with the purpose, scope, and methods of the subject. There are 30 readings—complete texts or representative portions—grouped in seven sections, which are not wholly independent. The emphasis in a reading determined the section to which it was assigned. Comments by the editor include cross-references to assist in linking sections. The sections conclude with lists of supplemental readings, totaling more than 300 titles.

Among the 30 readings, 21 originated in the United States. About 70 percent of the supplemental readings were published in the United States, although almost 85 percent of those readings that date from before the early 1960s came from other countries.

I am indebted to numerous geologists, as well as specialists in other fields that deal with the physical environment, for discussing with me their views on the concept of environmental science in general, and environmental geology in particular, and for calling to my attention literature that I might wish to include in the volume.

I express my gratitude to Marie Siegrist, bibliographer known to geoscientists around the world, for her invaluable assistance in making it possible for me to examine many publications that otherwise would have been inaccessible.

FREDERICK BETZ, JR.

CONTENTS

PART I: OVERVIEW OF ENVIRONMENTAL GEOLOGY

PART II: LAND-USE PLANNING

PART III: MILITARY GEOLOGY

PART IV: SOME GEOLOGIC ENVIRONMENTS AND RESOURCES

PART V: HAZARDOUS EFFECTS OF GEOLOGIC ENVIRONMENTS

PART VI: HUMAN INFLUENCES: BENEFICIAL AND DELETERIOUS

PART VII: DATA GATHERING AND PRESENTATION

CONTENTS BY AUTHOR

INTRODUCTION

The use of the term "environmental geology" is so widespread among geologists today that it seems hardly believable that the term is less than ten years old.

The first use is credited to James E. Hackett. In a personal communication received from him (1973), Hackett states: "The term Environmental Geology was initiated by me to identify a new orientation for the study and use of geology in a coordinated and integrated manner." It was applied to programs conducted by the Illinois Geological Survey, beginning in 1962. As a documented term, Hackett introduced it in a paper presented to a conference on Water Geology and the Future, held at Indiana University in April 1964; the published proceedings appeared in 1967. Hubert E. Risser (1969) and John B. Ivey (1969) confirm that Hackett originated the term. Peter Flawn et al. (1970) attribute the term to Hackett and John Frye.

In 1969, William C. Hayes and Jerry D. Vineyard remarked:

> Indeed, environmental geology is so new that the dictionary doesn't include it. It defines environment as "the complex of climatic, edaphic, and biotic factors that act upon an organism or an ecological community . . . ," and geology is termed "a science that deals with the history of the earth and its life especially as recorded in rocks . . . ," but no comprehensive definition is given. The geologist of today is writing his own definition. He is applying geologic principles to improve man's environment through the wise use of natural resources. He is making the new discipline—"environmental geology"—functional in metropolitan areas, the suburbs, rural areas, and the wilderness.

Looking somewhat further into expressed concepts of "environment," we may take note of remarks by Rhodes W. Fairbridge (1971):

"The concept of environment as commonly employed today is essentially one aspect of the integrated biologic/geologic global system, the physical ecologic habitat of *Man.* In this sense we exclude the social and cultural environment."

Robert F. Legget (1971) observes that "*Environment* has very recently become one of the words of the age. It is a splendid term for use in connection with the interdisciplinary approach to the protection and conservation of renewable natural resources that is at long last being recognized in the public domain after the long years of pleading by the few."

From the viewpoint of a soil scientist, Robert F. Reiske (1968) says that "Environment is the soil landscape, which results from the interaction of climate, relief, organic materials (vegetation), parent materials (geology), over a period of time. . . . Since there are many kinds of climate, relief, vegetation, and geologic materials many variations in the environment occur on the land." Reiske defines a soil landscape as "a three-dimensional segment of the land having similar parent material, soil profile, slope grade and position, drainage pattern, and aspect. [Here] vegetation is also included as part of the soil landscape."

These few samples suffice to indicate the diversity of opinion that exists on the definition of environment. On inspection, similarities can be seen, but the definition reflects the personal interest of the definer.

We could move on to statements from chemists, oceanographers, social scientists, and representatives of many other fields. In passing, I note that, as in the case of Reiske's definition, attention is generally focused on the land. I have mentioned oceanographers, who indeed have a concern with environment. We cannot overlook the marine environment by simply drawing a line between land and sea.

Since it is virtually impossible for any individual or group to deal with the whole of environment and arrive at a synthesis of all pertinent knowledge, we must fall back on consideration of single factors or groups of factors that can yield meaningful syntheses. It is to be hoped that different types of syntheses can be worked into broader understanding of the total environment.

We can readily concentrate on the physicial factors of environment in a geologic investigation and set aside the ecologic, social, and cultural factors. However, in studying the physical factors, it is unreal to think only of those that are regarded as natural factors, or of certain aspects of the biologic world. Thus, man-made features at and below the earth's surface frequently simulate natural features. In the preparation of an environmental map the complex of an urban center should not be indicated by the use of a distinctive color or pattern that would signify an area which is not being evaluated. There is a useful term, "surface geometry," in common use in military evaluations; it stands not only for natural landforms, but also for positive and negative relief created by man. In the present volume, surface geometry is not considered as a major topic, but human influence as a geologic agent is examined. Just as man-made structures can be integrated into a study of the physical environment, so, too, the physical aspects of biologic features can be included. Regardless of the interests of vegetation specialists in the life and habits of their objects of research, the profile and ground outline of a forest can be treated in the same manner as landforms composed of rock or soil. In a consideration of microrelief, for instance, the fairly uncommon work of animals, the beaver dam, is seen as a part of the areal physiography, and no special concern need be given to the nature of beavers, aside from recognizing them as the builders of the structure.

To return to the immediate subject of this volume, Paul H. Moser commented in 1969:

> Environmental geology uses the principles of geology, hydrology, engineering geology, geophysics and associated sciences and disciplines to determine how the resources of an area may be developed for the maximum benefit of man. It is a science that studies the environment in relation to man and its reports are helpful not just to the scientist but to all concerned with the growth of an area.

Peter T. Flawn et al. (1970) summarize more extensively and explicitly:

> Environmental geology deals with the entire spectrum of man's use of the earth, both in cities and in rural and primitive regions—it includes the location and exploitation of natural resources, the disposal of wastes, the effects of both mass movements and tectonic movements on structures, and the effects of subtle variations in the composition of earth materials on health. It involves the oceans and atmospheres as well as the solid earth—the effects on the earth of the great columns of heat and smoke produced by massive concentrations of people and industry fall within its domain. The key word in environmental geology is *application.*

It is obvious that man's use of the earth has become concentrated in urban centers and, now, conurbations in the developed countries. Therefore, it is not surprising to find that statement by William J. Wayne (1968): ". . . the geology of man's environment becomes most important in and near urban centers; the term *urban geology* is virtually synonymous with *environmental geology.*" Flawn et al. (1970) agreed: "Urban areas are currently the prime focus of environmental geology . . . ."

Nevertheless, it is quite clear that environmental geology is not synonymous with urban geology. In spite of their great importance, urban problems are only one aspect of land-use planning, which stretches from unoccupied areas to small settlements, to cities, to regions, to entire countries, and, possibly in a very general sense, to still larger units combining occupied and unoccupied tracts (e.g., major river basins, distinctive climatogeomorphic belts). It is also important to recognize that environmental geology has many facets, as cited in the remarks quoted from Flawn et al. but still others that they did not mention. In the present volume, not all these facets can be displayed, and indeed some have not yet emerged in the literature of environmental geology.

The term "environmental geology" has also been adopted by engineering geologists, who sometimes tend to create the impression that we are dealing again with synonyms to designate one field. The opinion of these engineering geologists is supported by the importance of their work both in connection with urban problems and in construction problems away from urban centers. However, as we have already learned from the authors cited, environmental geology stands for much more than engineering geology. The latter converts the implications of the geologic environment into forms usable by engineers. Legget has expressed the same view, somewhat differently, and finds it to be well recognized everywhere.

Furthermore, some geologists see a close linkage between economic geology and environmental geology. But, widely practiced, economic geology seems to be only distantly related to environmental geology. Such a relationship does exist when economic geologists are concerned with the pollution caused by the exploitation of mineral resources both in the immediate proximity of deposits or farther away from them, the destruction of the landscape as a result of strip mining, the results of underground mining, or the effects of the extractive industries on presently active environmental processes. A great range of problems may be involved, to be dealt with variously by mining geologists or mining engineers, sedimentologists, geomorphologists, health and sanitation specialists, and landscape architects.

There are objectors to the term "environmental geology." For instance, Gordon B. Oakeshott (1970) states: " 'Environmental geology' is a ridiculous term!" Continuing, he says, "*All* geology is environmental, and the most basic element in every man's environment is the geologic factor." Having spoken, Oakeshott concedes that the term is too widely accepted to be eradicated.

Legget (1971) dwells more emphatically on his dislike of the term:

> Geology *is* the study of man's physical environment so that to connect the two words is to be guilty of a terminological inexactitude that does real disservice to one of the greatest of the natural sciences. More than this, however, the very use of the term infers . . . that prior to the use

> of the term geologists were not concerned with the environment, and that those who do not use the term now are likewise not socially oriented. Both of these inferences are so patently absurd, even though they are now current, that the term itself should be very quietly droped . . . ."

Exceptions can be taken readily to the views of both objectors. Is *all* geology environmental? Is the geologic factor the most basic in man's environment? Does the term environmental geology really infer that prior to its adoption geologists were not concerned with environment? Is use of the term necessary to vouch for the social orientation of a geologist? Is social orientation the only orientation of environmental geology? In short, objectors appear to have misunderstood the meaning of environmental geology, and it seems strange that they wish the concept and term to be concealed in the preserves of fields with specific interests totally unrelated to environmental concerns.

The concept of environmental geology is, nevertheless, still in a nascent phase, and personal views play an important part in defining the subject. As Moser (1972) says: "Environmental geology has been called a 'state of mind'." Who said that first is not reported, but both proponents and opponents of the term should agree with the judgment.

A consensus seems to be developing among the supporters of the concept of environmental geology. They see it as earth science applied to the benefit of man and the living world, integrated application of the many branches of geology for this purpose; and coordination of geologic and other physical studies of the environment to deal with an expanding array of human and global problems.

This emerging consensus represents a return to the concept held by some geologists more than a century ago. They did not use the term "environment," but demonstrated a clear understanding of it in context with geology. Much, if not all, of the early discussion occurred in Europe. But until very recent time, when the term "environmental geology" came into use, it was difficult to identify relevant papers and books in bibliographic journals. This literature was distributed under existing index terms, for which bibliographers cannot be blamed. Today, environmental geology is an index term. That does not mean that the concept was previously ignored; it merely lacked a banner. Although some authors employ the term in the titles of papers that do not truly fall within the scope of environmental geology, as defined by the sources cited in the preceding discussion, the relevant literature is increasing rapidly. For the present, those interested in environmental geology should examine abstracts of oral presentations offered at professional meetings. They represent an overview of work in progress and a preview of papers and reports to come.

For the historical record, an outstanding book, *Die Geologie der*

Gegenwart (The Geology of the Present), by the German geologist Bernhard von Cotta, was published in 1866 (apparently a revised edition of an earlier work by the author that appeared in the 1850s). Cotta treated all aspects of environmental geology, except, of course, those dealing with problems generated by technology that had not yet evolved. A portion of Cotta's book is presented in Paper 1 in this volume.

In Europe, also, the term "anthropogenic sedimentation" has long been in use for those recent deposits related to human activity, for example, timber cutting, soil erosion, and animal domestication of the Bronze Age and later. The term "anthropogeology" is developed in a paper by Jäckli, most of which is reproduced in this volume (see Paper 21).

One of the earliest applications of geological factors to a practical purpose is an obscure paper titled "Verhältnis der Geognosie zur Kriegswissenschaft" ["Relationship of Geognosy (Geology) to Military Science"], by Johann Samuel von Grouner, published (posthumously) in 1826. Dealing mainly with the Alps, the discussion examines details of the mountainous environment affecting military operations. More is said about military geology later, and two representative examples are offered from the rich literature, unfortunately virtually unknown to most geologists.

There is much additional evidence of awareness of the geological environment in literature dating from the pre-historical episode of geologic science, that is, essentially before 1800. Georg Bauer (known better as Agricola) is remembered for his classical work, *De re metallica libri duodecim,* published in 1556, which summarized the knowledge of the time on mining and metallurgy. According to the great geological historian, Karl von Zittel, Bauer also "made a number of useful observations about springs, earthquakes, active and extinct volcanoes, volcanic rocks, the action of running water, and atmospheric movements." These observations appeared in other earlier works. Before geology had taken form, Wendelin Schildknecht published a work titled *Harmonia in fortalitiis construendis, defendandis et oppugnandis* (1652), in which he considered environmental influences on military defense. In the same vein, we have repetitive discussions in the *Kriegszbuch* by Leonhardt Fronsperger, published between 1565 and 1573. But this volume is not intended as a historical survey; interested readers are encouraged to probe the surprisingly extensive literature that demonstrates man's preoccupation with physical environment back to antiquity.

In considering environmental geology, we must think not only of the influence of the geologic environment on man, but also of the influence of man on the geologic environment. At one time, not too long ago, most geologists did not concede a role to man in shaping the geologic environment, although, if they had looked, the evidence was there to be seen. Reaching back into the literature, it seems almost obligatory to speak of the highly regarded work by George Perkins Marsh, which was published first as *Man and Nature* (1864) and later as *The Earth as Modified by Human Action* (1874). There were numerous printings and various revisions up to 1908. The original and a revised version were reprinted in 1965 and 1970, respectively. Much of the work has in fact little to do with the physical environment: such chapters as "The Waters" and "The Sands" are outnumbered by those on forests, vegetation, animals, and birds. Marsh, who was a lawyer, architect, state official, diplomat, and linguist, but never a practicing scientist, distilled vast quantities of literature and the observations made on his extensive travels into a simple, scholarly narrative. It merits attention, and its republication a century after its first appearance is more than corroboration.

One other book must be cited here: *Man as a Geological Agent,* by Robert L. Sherlock (1922). It may be preferred to Marsh's book because it focuses on the subject that geologists should examine most carefully and offers more specific details. For anyone interested in one aspect of English history, and especially of London, that is not included in most political or social historical studies, Sherlock has provided a very readable book. Whether he originated the expression used as the title of his book or not, it is heard widely today, but he appears to be little known.

The state of the art of a diffuse subject, such as environmental geology, is difficult to assess. Advances are being made in many aspects, but there is no general level of accomplishment. The emphasis on urban geology indicates widespread effort to provide useful information to planners. It is still reported all too often that the advice of geologists is either not sought or not used if requested and obtained. A common complaint is that the geologist often does not express his observations or explain his map legend in terms that the planner, architect, and engineer can understand. A goal of some land-use planning is the assessment of areas for recreational purposes. Geologists are being drawn into the study of this problem. Human influence on mass movement (soil or rock), earthquake occurrence, pollution, health hazards, and alteration of the surface geometry (whether through clearing of the land, mining, ill-advised tampering with the coastal terrain, or other causes) is occupying the time of numerous geologists. There is a great increase in activity in the study of energy problems. Earthquake prediction is coming under close scrutiny; subtle indicators are being examined, but the literature is sparse. The environmental effects of volcanic activity have been discussed more by laymen than by geologists. Military applications of environmental geology are being studied actively, but tend to become known in the open literature only after their usefulness to armed forces has ended.

The present volume can only be a sampling of the literature. Other collections of material published on environmental geology concentrate on certain topics and omit others entirely. In one case, urban geology and earthquakes receive major attention; in another, waste disposal is highlighted. In another there is no consideration of land-use planning, geothermal energy, health hazards, and the general concept of environmental geology. None seen so far deals with recreation, aesthetics, military geology, and data handling. Thus, the gaps have received attention here, and some topics in other collections are given less space.

In previously published collections, there are few, if any, examples taken into account of work performed in foreign countries. A false impression may be created that development of the field is essentially an enterprise of geologists in the United States. A fundamental reason for the lack of foreign literature is the continuing language barrier. Europeans have recognized this and are publishing increasingly in English to accommodate Americans. Foreign-language papers selected for the present volume have been translated.

It has not been possible to reproduce material that depends on colored illustrations. Readers should become aware of atlases (notably the national atlases of many countries) and other publications containing colored photographs, maps, and diagrams, which represent a high level of presentation of information for the benefit of nongeologist users.

Finally, the concept of "benchmark" papers must be modified in a volume on environmental geology to take into account the fact that the bulk of literature on the subject has been published in the last decade. It is certainly no simple task to obtain a perspective over that literature that will allow a sure identification of the papers of lasting influence. They will be seen in future reassessments, when the heritage of the developing state of the art can be attributed to the work of specific individuals and groups.

SUPPLEMENTAL READINGS

Fairbridge, R. W. 1971. Society and Geomorphology. In: *Environmental Geomorphology* (D. R. Coates, ed.), p. 215-220. Binghamton, N.Y., State University of New York. (Proc. 1st Ann. Geomorph. Symp.)

Flawn, P. T., W. L. Fisher, and L. F. Brown, Jr. 1970. Environmental Geology and the Coast—Rationale for Land-Use Planning. *J. Geol. Educ.,* **18**(2), p. 85-86.

Hackett, J. E. 1967. Geology and Physical Planning. In: *Water, Geology and the*

Future, p. 83-90. Indiana University, Water Resources Research Center. (Proc. Conf., Apr. 26-27, 1964.)

Hayes, W. C., and J. D. Vineyard. 1969. *Environmental Geology in Towne and Country.* Missouri Geol. Surv. and Water Resources Educ. Ser. 2.

Ivey, J. B. 1969. Definition of Environmental Geology and Purpose of the Conference. Colorado Geol. Surv., Spec. Publ. 1. (Governor's Conf. Environ. Geol., Denver, Apr. 30-May 2, 1969.)

Legget, R. F. 1971. Essay Review: Environmental Geology, Land-Use Planning, and Resource Management, by P. T. Flawn. *Amer. J. Sci.,* **271**(2), p. 187-190.

Marsh, G. P. 1864. *Man and Nature.* London and New York.

Moser, P. H. 1969. Alabama Investigates Resources to Determine Economic Benefits. *Appalachia,* **2**(9), p. 32-34.

——. 1972. *Environmental Geology Studies in Alabama.* 24th Intern. Geol. Congr. Proc., Sec. 13, p. 37-43.

Oakeshott, G. B. 1970. Controlling the Geologic Environment for Human Welfare. *J. Geol. Educ.,* **18**(5), p. 193.

Reiske, R. F. 1968. Forest Soils Workshop—Using Soils Information in Managing Forest Land for Recreation. Soc. Amer. Foresters, Div. Silviculture, Ann. Mt. (Philadelphia), 9 p.

Risser, H. E. 1969. Man's Physical Environment—A Growing Challenge to Science and Technology. *Missouri Acad. Sci. Trans.,* **3**, p. 5-14.

Sherlock, R. L. 1922. *Man as a Geological Agent.* London, Witherby.

Wayne, W. J. 1968. Urban Geology—A Need and a Challenge. *Indiana Acad. Sci. Proc.,* **68**, p. 49-64. (Comm. Sci. Soc., Publ. 3.)

Part I

OVERVIEW OF ENVIRONMENTAL GEOLOGY

It is evident from the discussion in the Introduction that we need to direct our attention to both the scope and content of environmental geology and the role of geologists in dealing effectively with the subject. We have seen that environmental geology has many ramifications and that a comprehensive summary is not obtained easily when such divergent views exist on the meaning of the subject. In established fields of geoscience, it is possible to present an overview that is both retrospective and relatively current. Textbooks are presumed to do this, although active workers in any field realize that such publications must always lag behind the actual state of knowledge and activity in any still vital field. The lag may be as much as five to ten years, if a truly comprehensive, worldwide appraisal is attempted by authors who are prepared to make extensive searches of the literature. It is quite apparent from a survey of the developing field of environmental geology that no textbook can yet offer a thorough overview.

Important syntheses of geologic knowledge are not apt to be textbooks but reference and reading material for practicing scientists and students. A classical work of this type is *Das Antlitz der Erde (The Face of the Earth)* by Eduard Suess, a genius in absorbing, organizing, and reconveying knowledge of the structural geology of the globe as far as it was known at his time. Other later examples dealing with segments of geoscience can be found, but their rarity attests to a lack of interest on the part of geologists in devoting time to reviewing the literature, as evidenced all too often by investigations that reflect a lack of awareness of past and current work.

A present-day trend is to publish selections of literature, which bring together a convenient overview of a subject. The trend seems to have been stimulated especially in Europe, where loss of libraries in World War II made much of the literature difficult to find. It has now spread to the United States, due mainly to the growth of many new institutions with limited library resources. Such collections afford a useful service for both investigators and students in developing fields.

I consider all the present volume to constitute an overview of salient aspects of environmental geology, some unique to the field, some shared with other fields. Most of the volume is concerned with selected parts of environmental geology, each of which generally combines a consideration of several features of the subject.

Readers will soon observe that the present part displays a diversity of views among authors, but this is precisely what is needed in a discussion of environmental studies. They will note that an older selection begins the group of readings, and that the rest are from the more recent time when environmental geology has become a recognized field.

The current literature of environmental geology includes a considerable amount of justification by authors to show the need from programs of investigation, the growth and extent of present activity, and the value of accomplishments so far. This literature is addressed primarily to present and potential sponsors; it is found especially in proceedings of symposia and conferences called by governmental bodies. Although undeniably very important to the success of an applied science, I have not thought it relevant to the purposes of the present volume.

SUPPLEMENTAL READINGS

Coates, D. R. (ed.). 1972-1974. *Environmental Geomorphology and Landscape Conservation.* Stroudsburg, Pa., Dowden, Hutchinson & Ross, Inc., 3 vols. (Benchmark Papers in Geology series.)

Cotta, Bernhard von. 1865. *Geology and History—A Popular Exposition of All That Is Known of the Earth and Its Inhabitants in Pre-historic Time* (translation by R. R. Noel). London, Trübner, 84 p.

Dunn, J. R. 1967. Environmental Geology in New York—Philosophy and Application. *Empire State Geogram* (Geol. Surv. N.Y.), 5(1), p. 1-4.

Ellison, S. P., Jr. 1970. *Some Environmental Geological Problems of Europe.* In: *Environmental Geology: Short Course Lecture Notes.* 8 p. Washington, D.C., American Geological Institute.

Flawn, P. T. 1968. The Environmental Geologist and the Body Politic. *Geotimes,* 13(6), p. 13-14.

Frye, J. C. 1971. *A Geologist Views the Environment.* Illinois Geol. Surv. Environ. Geol. Notes 42.

Hackett, J. E. 1967. Geology and Physical Planning. In: *Water Geology and the Future,* p. 83-89. Indiana University, Water Resources Center.

Hayes, W. C., and J. D. Vineyard. 1969. *Environmental Geology in Towne and Country.* Missouri Geol. Surv. and Water Resources, Educ. Ser. 2, 42 p.

Hilpman, P. L., and G. F. Stewart. 1969. *Geology and Concern for the Environment.* Nat. Res. Council, Highway Res. Record 271 (Planning: Conservation of the Physical Environment), p. 38-47.

Kraus, Ernst. 1923. Bemerkungen zur angewandten Geologie der obersten zehn Erdrindemeter. *Geol. Arch. (Königsberg),* **2**, p. 207-212.

LaMoreaux, P. E., P. H. Moser, et al. 1971. *Environmental Geology and Hydrology: Madison County, Alabama—Meridianville Quadrangle.* Geol. Surv. Alabama Atlas Ser. 1, 72 p.

Montgomery, H. B. 1974. *What Kind of Geologic Maps for What Purposes?* Geol. Soc. America Eng. Geol. Case Hist. 10, 8 p.

Philipp, Hans. 1921. *Die Bedeutung der Geologie für Handel, Industrie und Technik, Landwirtschaft und Hygiene.* Greifswald, L. Bamberg, 35 p.

Risser, H. E. 1970. Man's Physical Environment—A Growing Challenge to Science and Technology. *Missouri Acad. Sci. Trans.,* **3**, p. 5-14.

Rose, W. D. (ed.). 1971. *Environmental Aspects of Geology and Engineering in Oklahoma.* Oklahoma Acad. Sci. Ann., Publ. 2, 70 p. (Symp., Oklahoma Acad. Sci. and Oklahoma State Univ.)

Russell, R. J. 1956. Environmental Forces Independent of Man. In: *Man's Role in Changing the Face of the Earth* (W. L. Thomas, Jr., ed.), p. 453-470. Chicago, University of Chicago Press.

U.S. Soil Conservation Service. 1973. *Geology as Applied to Environmental Problems.* 25 p.

Wilser, Julius. 1921. *Grundriss der angewandten Geologie (unter Berücksichtigung der Kriegs-erfahrungen für Geolegen und Techniker).* Berlin, Gebrüder Borntraeger Verlagsbuchhandlung, 176 p.

Editor's Comments on Papers 1 Through 3

1 COTTA
Influence of Geologic Characteristics on Human Life

2 BROWN and FISHER
Excerpts from *An Approach to Environmental Geology*

3 PECORA
Resources and Environment—Quest for Balance

Bernhard von Cotta (1808-1879) was certainly one of the first geologists to recognize that environment has a profound influence on man. Cotta taught at Freiberg, participated with C. F. Naumann in producing the geologic (then called "geognostic") map of Saxony, produced a similar map of Thuringia, and wrote extensively on ore deposits in Europe and the Altai, based to a considerable degree on his personal observations. Important here is that Cotta began to write about environmental effects in the 1850s, first with a book titled *Deutschlands Boden: sein geologischer Bau und dessen Einwirkung auf das Leben der Menschen* (freely translated as *The Terrain of Germany, Its Geologic Structure and the Effect of the Latter on Human Life*). In 1866, part of this work was republished as *Die Geologie der Gegenwart (The Geology of the Present Day).* The entire book should be examined by anyone interested in environmental geology; but only one, particularly apposite chapter, translated into English and somewhat abridged, was included here as Paper 1.

Cotta's observations are a preview of growing modern interest in geologic effects on humans, animals, and plants. In Paper 1 he notes that geologic and geomorphic factors are involved in the location and form of settlements, the character of architecture, and the other types of aesthetic expression of regions. The characteristics that formerly prevailed because of local or nearby resources are today disappearing because of the long-distance importation of rock and other construction materials, together with new hydrologic and energy resources that introduce an alien facade to the environment.

From the text that follows, I draw attention to the following observation by Cotta: "Considering the manifold effects of external and internal [geologic] structure on human life, it is deplorable that the scientists who have concentrated on geology have neither taken note of, or been able to focus on, the influence of geology on human history." Is it a pious hope that modern environmental geology will correct this defect?

In November 1970, the American Geological Institute sponsored a short course of lectures on environmental geology. From among the lectures, we have included as Paper 2, in abridged form, the following consideration of an approach to environmental geology. L. Frank Brown, Jr., and William L. Fisher are both associated with the Bureau of Economic Geology and the Department of Geological Sciences in the University of Texas at Austin.

The authors discuss the attitude of geologists toward environmental geology and the adjustments needed when they bring backgrounds in traditional fields to the solution of its problems. They observe, as Moser suggested (see the Introduction), that "Perhaps the single most important factor to success is mental attitude or state of mind." A strong criticism merits quotation: "Geologists are fooling themselves if they believe that little more than a smattering of adequate environmental geologic mapping and analysis has been completed. The delusion that scientists understand the environment and need only to prescribe a solution is a critical misconception among many knowledgeable people."

Brown and Fisher are especially impressed with the necessity for improvement in the transfer of meaningful information to users of the geologic environment. They say, on this matter, "Geologists have talked to themselves too long. They must increase communication outside of the profession, and, in so doing, gain better insight into their role at this critical time." In the case of environmental geology, appropriate mapping is vital so that the basic data can be used in the preparation of derivative maps. More will be said on this subject in Part VII. Brown and Fisher demonstrated their approach to environmental geology with examples from the coastal zone of Texas. The examples are omitted here because of space limitations.

As we have seen, environmental geology has been and is concerned mainly with the influence of geologic factors on man and his activities. With increasing importance, however, man himself has been identified as a geologic agent. Urbanization has undeniably introduced intensification of some natural processes, and generally the effect has been detrimental. Unfortunately, the subject has become highly controversial, to the point that conservationists–preservationists tend to take exception to any alterations of the landscape and see destructive effects in a wide range of industrial practices.

William T. Pecora (1913–1972) was associated with the U.S. Geological Survey from 1939 to 1971 (Chief Geologist from 1964 and Director from 1965), and was finally Undersecretary of the Interior. In 1971, he delivered a significant commencement address, which is offered as the last paper in this part.

A main theme of the address is stated in the following:

> It is believed by many people that man alone is degrading and polluting his environment because of our modern industrial society. Some myths, however, need to be destroyed . . . . Those individuals who speak about our inherited environment of pure air, pure rain, pure water, pure lakes, and pure coastlines ignore the inevitability of nature.

Pecora concedes that in a specific or local context, man can be a major geologic agent. What he brings out is that natural forces ultimately have much wider and more severe effects on the environment than man can ever achieve. It is true that man has not yet introduced the same degree of hazards as in some naturally induced rearrangements of surface and hydrologic geometry. But, as Pecora says, the role of man as a geologic agent is one that imposes serious responsibilities.

1

This abridged translation was prepared expressly for this Benchmark volume by Frederick Betz, Jr., from "Einfluss des Erdbaues auf das Leben der Menschen," in Die Geologie der Gegenwart, *Leipzig, 1866, pp. 373–416.*

INFLUENCE OF GEOLOGIC CHARACTERISTICS ON HUMAN LIFE

Bernhard von Cotta

The influence of geologic characteristics on humans is highly varied, difficult to determine in all its details, and even more difficult to predict.

Man has spread like a fluid over a relatively unchangeable, rigid surface; however, where many generations have occupied the same segment of land, the natural setting has become visibly a part of their life and character. Man ultimately becomes a part of his environment—it becomes the native land in every sense, not merely in its climatic and external conditions, but also in the most specific characteristics. The ever-greater mobility of man in modern times may counteract the influence, but for the mass of the population it can never be eliminated.

The influence of the inner structure is more indirect than direct. The forms of the landscape are the result of inner structural and geologic processes, such as elevation, depression, and flooding.

The solid earth yields building stone and soil, metals, coal, and salts, a stable or unstable foundation, greater or smaller difficulties confronting transportation routes, many or few springs, good or bad drinking water, hot springs or mineral waters. It forces streams into very even or irregular flow, promotes or hinders navigation, increases or decreases the usefulness of water for power, acts as a good or bad heat conductor, and produces exhalations of various types of steams and gas. As a result, the inner structure of the earth is significant for the well-being of humans. I believe I can show that the inner structure of the earth has played an important part in human history, the separation of races and nations, their development, and their social, moral, and intellectual life. We may note that it has also had a relationship to politics and ultimately to military action.

This science does not deal with influences that are mystical and unrecognizable, but only with those that can be traced to specific material characteristics of rocks and their stratigraphic association and can be interpreted from these features. Surface form, soil fertility, formation of springs, and the technological usefulness of rocks or mineral deposits are the most important factors for making such influences visible. An investigation that claims to be scientific can be concerned only with such causal, provable effects.

A French geologist of the previous century claimed that the rock types even had an affect on the intellectual development of people in a region. He stood by his conclusion, based on some good observations, but did not use completely acceptable means for justifying it. He attributed a religious habit of mind to inhabitants of basaltic regions. It is possible that many attractive basaltic domes have become sites for the erection of picturesque chapels and religious monuments, and that in their pleasingly romantic atmosphere they have affected the religious feelings of observers. Nevertheless, the observation does not apply to basalt as a rock, but to the external form of the rock under certain circumstances. In no manner can one attribute a general influence to the rock.

It has been asserted that older formations have had a greater effect on humans than recent ones. In general, this is a totally vague interpretation, unreliable because the age relationship of rock formations cannot be correlated with specific human characteristics. It may be possible to find truth in this assertion, but it cannot be based on local observations. For instance, older rock formations usually form uplands. However, this is not the case in northwestern Russia. On the other hand, mountainous regions can be composed of younger rocks, as in the northwestern Alps. Also, the mineralogic nature of rocks is not completely dependent on age, as the Alps demonstrate, where the flötz formations are comparable to the oldest formations in northern Germany.

We may, therefore, regard the age of rocks or formations as having no general worth for our research unless it has a relationship to other specific characteristics, which is often the case.

There is a relationship between the age and the genesis of rocks. Eruptive rocks frequently form hills and mountains, but sedimentary rocks have also been elevated to mountains. And the very uneven surface of eruptive rocks has in places been leveled off subsequently. Nevertheless, there remains a significant difference of mass and chemical composition between eruptive and sedimentary rocks, which can be recognized by the fact that the first rarely form plains and that they are composed almost always of potassium- or sodium-bearing silicates, which in their decomposition yield on the average fertile soils.

In general, exact recognition of the natural resources of regions and their distribution is valuable. It shows how they interrelate and how geologic resources can be used most effectively.

It is important to consider natural geologic regions in fixing the boundaries of countries and their subdivisions. For example, in coal and ore regions, the reserves should be considered, since in a political unit any developing industry based on natural resources will be in a disastrous condition upon exhaustion of these resources.

The most effective size of areas has long been a problem of national economy. Undoubtedly, it is a factor of the nature of the ground. In flat areas, which promise little except natural vegetation, the individual owner of land needs a large area, whereas in areas that have other potentialities, such as mining and development of water supply, the owners can develop small properties as a side activity; conversely, farmers can consider working in industry as a side occupation. The formula to be determined is the effective size of an area for a given purpose.

Different rock types operate through different characteristics—chemical and mineralogic composition, compactness, thermal conductivity, reaction to water, etc. Some are useful as construction material, fuel, salt, or fertilizer. Some become good soil through weathering, others poor.

In some cases, the internal structure of the earth's crust works against these influences with great effect. We can differentiate the following influences:

1. Effect of geologic structure on quantity, distribution, and types of springs.
2. Influence of geologic structure on vegetation, greater or diminished fertility, and manner of use of the vegetative potential.
3. Influence of the ground on quantity and quality of human settlement: (a) size of population; (b) distribution; (c) configuration; (d) architectural style of dwellings.
4. Influence of geologic structure on activities and economic condition of the population: hunting, fishing, forestry, agriculture, cattle raising, mining, industry, trade.
5. Influence of land use on local and long-range transportation, both on land and water.
6. Influence of geologic structure on military action.
7. Influence of geologic structure on health and life span.
8. Influence of geologic structure on social conditions, general characteristics, and cultural development in arts, science, and recreation.

Even if the habitability of countries is dependent primarily on climatic conditions, both outer and inner geologic structure play an important role, since they control local fertility of the surface and accessibility, in various degrees, to the many elements suited for human and other use.

There can be no doubt that a plains region in the southern Russian steppe will not feed as many people as a comparable one in central Germany or England, even under completely similar climatic conditions. Not only the size of the population, but the type of settlement must be very different on account of different ground conditions.

A close relationship can be determined between geologic structure and distribution of settlements-Favorable factors are evenness of surface, solid ground, space for expansion, presence of springs or running water, protection against climatic hazards or biotic enemies, proximity of construction materials, fuels, and sources of specially needed mineral, animal, and vegetational materials. For most of these reasons, the nature of the rocks in the area is influential. This may not be imperative in the development of large cities, but attention should be given to location at terrain boundaries (rims of uplands and seacoasts), on navigable streams (especially where they have outlets to the sea), on protected harbors, in the middle of broad basins and, finally, a generally favorable location for communication with the outside world. Small- and medium-sized towns do not have to be dependent on geologic factors.

Unfavorable factors are marked relief, poor foundation or swampy ground, limited available space, lack of water, extreme altitude, lack of construction material, lack of fertile ground, and transportation difficulties. These conditions may be removed by technological skills and, thus, may not be absolute hindrances.

Industry is often stimulated by geologic structure and promoted by usable mineral deposits suitable for ceramics, glass, iron products, gas gems, paraffin, and fuels. Mining also promotes general industrial activity.

It would be fruitless to search for specific geologic causes to explain why lace and hosiery are important industries in the Erzgebirge, while damask is produced in the Oberlausitz, linen in the Teutoburg Forest, or why clocks are produced in the Black Forest and Jura. If we consider these and similar occupations as not being environmental industries, it is still easily understood why they have developed preferably in uplands, in which ground for living quarters and other related factors is inexpensive, and lightweight products can be transported out cheaply in relation to their value. Thus, local industries dependent on the geologic environment and other types of industries that are independent of the environment can be differentiated.

It is not only the elevation and external form of uplands that affect the stimulation of industry, but even more their internal structure, the variety of and monetary return from rocks in close proximity. These varied influences of geologic structure that favor or inhibit industry support or hinder each other reciprocally. In addition, there is the important factor of transportation and, perhaps also, the factor of national characteristics.

Traffic resembles a liquid that moves according to hydraulic laws. It moves from heights to depths, avoids highest elevations, crosses mountains in their deepest cuts, flows in channels, partly predetermined, and gathers in large basins.

The most important factors in the organization of the ground for the movement of humans and their material goods are water, flat ground, and uplands. Their interfaces are marked always by an increased friction of traffic, promoting or hindering it. For example, there is a decided movement problem on coasts, as well as on the periphery of small cities located on the margin of abruptly rising uplands.

The course of streams, the most natural transportation routes, and, even more, the nature of their channels—rapids, cataracts, widenings, narrowings, and islands—are largely the result of the inner structure of a region. The Rhine falls at Schaffhausen are caused by solid Jurassic limestone. The dangers of the Binger Loch (on the Rhine) were caused by the relatively solid Taunus quartzitic slate. The shallow, rocky rapids of the Elbe between Lobositz (Bohemia) and Pirna (Saxony) resulted from the presence of basalt, phonolite, and especially solid sandstone. The examples could be multiplied many times.

The inner structure and the variable solidity of rocks are of great importance not only for natural waterways, but for canals, railways, and roads. Softer rocks provide initially deeper cuts and easier mountain crossings; they are obviously excavated more easily.

Directly, geologic structure affects traffic only rarely with special problems, as, for instance, in road construction. Examples are found in some basins entirely lacking stone, as in the Pusztas of Hungary. Railways overcome the difficulties more readily than roads. Far greater and more general is the indirect effect caused by surface form, watercourses, and location of local industry.

Where eruptive rocks have come to the surface, mountains have been elevated or strata have been steeply tilted; here there are always movement problems. Inhabitants of some valleys in the Alps are extremely isolated, and it is obvious that the isolation has affected their entire development.

If we compare a railway map of Germany with a geologic map, it is soon evident that the most developed routes are located on nonvolcanic ground more often than on more readily available routes on other material.

Not only is movement on land dependent on past geologic processes and the composition of rocks, but their influence extends to water traffic. The general direction of streams, the curvature of their bends, the nature of their beds, the rapidity of currents, their cataracts, rapids, and sand banks are results of geologic

conditions. They often show very similar character in geologically similar areas—as in the tight bends in all streams in slate and greywacke plateaus, and in the cataracts crossing solid rock. Similarly, the form and composition of seacoasts, the depth of water, and the presence or absence of cliffs and reefs are clearly dependent on the nature and position of rocks forming the coast and sea floor.
the coast and sea floor.

That there are distinctly healthy and unhealthy regions is generally known. A main cause for these differences is climatic condition; but the nature of the ground exercises more or less indirect or direct effects through surface form, unequal heat capacity of the rocks, frequency and character of springs and stagnating waters, character of bedrock, availability of construction materials, and even through gaseous exhalations, materials absorbed by vegetation, and dust arising from mechanical destruction or chemical decomposition. The importance of the nature of the ground, especially in the distribution of groundwater, for health conditions is shown by Pettenkofer in his important investigation of the distribution of cholera.

It is most difficult, naturally, to demonstrate the effects of the natural environment on human temperament. Where so many influences are acting variably, it is hardly possible to single out individual factors. We may only generally claim that present-day man is the final product of physical and intellectual impressions resulting from effects on him and his forebears.

Spiritual development is affected more by the outer form of regions than by internal structure. However, insofar as we include social life and development of arts, very marked influences can be recognized at times.

In his work on his second trip to America, Lyell noted that whether fossil fuel is bituminous or anthracite is important, since the first produces smoke and often foul odor, whereas the second does not. To quote from this book [*Editor's Note:* Offered by Cotta in a German translation; reproduced here directly from Lyell's text:]

> Even in a moral point of view, I regard freedom from smoke as a positive national gain, for it causes the richer and more educated inhabitants to reside in cities by the side of their poorer neighbors during a larger part of the year, which they would not do if the air and the houses were as much soiled by smoke and soot as Manchester, Birmingham, Leeds, or Sheffield. Here the dress and furniture last longer and look less dingy, flowers and shrubs can be cultivated in town gardens, and all who can afford to move are not driven into the country or some distant suburb. The formation of libraries and scientific and literary institutions, museums, and lectures, and the daily intercourse between the different orders of society—in a word, all that can advance and refine the mind and taste of a great population, are facilitated by this contact of the rich and poor. In addition, therefore, to the importance given to the middle and lower classes by the political institutions of America, I cannot but think it was a fortunate geological arrangement for the civilization of the cities first founded on this continent, that the anthracitic coalfields were all placed on the eastern side of the Allgehany mountains, and all the bituminous coalfields on their western side.

That plastic art was initially dependent on stone being readily available I described repeatedly in my book *Deutschlands Boden.* Here I only offer Boué's remarks from his publication "On the Value of Geology":

> Had the Greeks not found in their country and on the coasts of Asia Minor such beautiful marbles and porphyries, their sculpture would not have taken the direction that led it to its peak. On the other hand, the idols in Mexico had to be coarser, because there trachyte had to be used, which is nowhere near as beautiful as the material used by the Greeks. Even the idols of Buddhists and other religious groups of India may have obtained their character partly from the plutonic or volcanic rocks used.
>
> How differently the architecture of Mesopotamia, Egypt, India, Greece, and Italy developed. In the Euphrates valley, clay, Tertiary limestone, alabaster, volcanic tuff, and pumice were available; in India and Egypt there were granites and other plutonic rocks, as well as sandstones and quartzites. These materials influence the indigenous architecture or even the excavation of rock for buildings. We find in certain regions, as in Asia Minor, European Turkey, and south-central France remains of troglodyte habitations because of fresh-water limestone or pumiceous conglomerate (S. Hamilton). The contrast between so-called cyclopean architecture and that of the Romans and Greeks is based partly on the type of available material. The first style uses rock that yields polyhedral stone, such as basalt, granite, porphyry, and some limestone. If there had been no Eocene nummulitic rock in Egypt, the famous pyramids would not have been erected, because only this soft rock type provided the possibility for their construction. This is proved by the absence of granite or syenite in these monuments, although the greater durability of the latter rocks was undoubtedly known to the Egyptians, and they made use of them for smaller monuments. They believed it possible to overcome the deficiency by making use of an artificial coating, but they miscalculated its effectiveness in places.

Hausmann dealt thoroughly with the influence of rock characteristics on architecture in "Transactions of the Royal Society of Goettingen." Of course, the development of transportation has reduced the need for using local materials more and more.

Considering the manifold effects of external and internal earth structure on human life, it is deplorable that the scientists who have concentrated on geology have neither taken note of, or been able to focus on, the influence of geology on human history. Consider the influence of iron and coal on the history of England, precious metals on that of the United States and Russia, and the presence of numerous geologic resources in Germany.

If we start only from specific differences of individual rock types and their distribution, their influence is very limited. Most meaningful are the abundance and special attributes of springs, fertility of soil, character of construction sites, construction materials, demand for heavy raw materials for locally based industries (iron ore, coal, salt, clay, etc.), and, finally, special traffic problems (obstacles or advantages).

In conclusion I believe that the present surface of the earth, with all its individual features, has been gradually developed in a continuing reciprocal relationship between man and nature.

REFERENCES

[*Editor's Note:* The bibliographic information that follows is all that is available.]

Boué, Ami. On the Value of Geology.
Hamilton, S. 1841. Researches in Asia Minor.
Hausmann. Schriften der Königlichen Societät zu Goettingen.
Lyell, Charles. 1850. A Second Visit to the United States of North America. New York, Harper & Bros.; London, John Murray.
Pettenkoffer.

2

Excerpts used with permission from *Environmental Geology: Short Course Lecture Notes,* American Geological Institute, Washington, D.C., 1970

AN APPROACH TO ENVIRONMENTAL GEOLOGY

L. F. Brown, Jr., and W. L. Fisher

At a time when scientific specialists are the rule in most disciplines, a major thrust toward clearly defining and delineating environmental problems is imperative, which should involve field-oriented geologists, biologists, and other scientists in preparing detailed maps of physical and biologic environments, as well as properly conceived and innovative geologic maps of the earth's surface and shallow subsurface. The nature and distribution of natural environments must be determined as quickly as possible if proper measures are to be devised for their protection; likewise, distribution of modern and ancient sedimentary facies or igneous and metamorphic genetic units is critical because of the common interrelationships between bedrock and environmental problems of many kinds.

Many specific solutions to environmental problems must come from specialists, but until the nature and extent of natural systems are better understood, mapped, and charted, a well-organized attack on these problems cannot be mounted. Geologists are fooling themselves if they believe that little more than a smattering of adequate environmental geologic mapping and analysis has been completed. The delusion that scientists understand the environment and need only to prescribe a solution is a critical misconception among many knowledgeable people.

WHAT MAKES GEOLOGY ENVIRONMENTAL?

Geology becomes "environmental" when geologists use their knowledge in outlining and solving environmental problems. If this sounds simple, it is. Environmental geology is, above all else, the practical or functional application of the science. Before more erudite colleagues abandon the environmental ship in anticipation (or apprehension) of a functional, practical application of their knowledge and energy, let us contemplate the alternative to safeguarding natural environments. For the colleague too interested in pH, *Heterostegina,* bulk modulus, or solidus–liquidus to be concerned with ecology, perhaps it would be sobering to consider that the problems related to the thin film of hospitable environments surrounding our sphere transcend all other problems in importance to the human race.

[*Editor's Note:* Original title: "An Approach to Environmental Geology with Examples from the Texas Coastal Zone"; publication authorized by the Director, Bureau of Economic Geology, The University of Texas at Austin. The excerpts presented here are slightly edited.]

L. F. Brown, Jr., and W. L. Fisher

PROFESSIONAL PROBLEMS

Most geologists working today on problems involving the environment were trained in traditional areas of the science. Pertinent curricula are only now being developed in some departments of geology, so that a steady flow of broadly trained environmental geologists is still in the future. In the meantime, geologists with other interests and specialties are being utilized in a growing number of projects in federal and state agencies. This utilization should increase if geologists prove their worth in interdisciplinary areas, [which they must prove by applying] those quantitative skills injected into curricula during the past decades along with more traditional geological skills. Perhaps the single most important factor to success is mental attitude or state of mind. Assuming average intelligence, a geologist above all must sense the relevance and excitement of an entirely new area to which he was probably not introduced during his formal education.

The fact that a growing number of geologists will align traditional geology to solution of contemporary problems of environment should not be an open invitation to incompetent or poorly motivated people. Functional or practical applications are incompatible with mediocre science. The environment in all its complexities and implications will require the best that geologists have to offer in research and application. Here is an opportunity for a mature science, to some perhaps a bit seedy, to apply its several hundred years' heritage to the most vital problem ever faced by the human race. Geology and geologists have some unique and potentially powerful tools to turn on pollution and environmental destruction. The tradition of four-dimensional thinking, process-related approaches, and appreciation of complex natural systems, coupled with a history of common sense and perspective of time and scale, should provide a head start in this very critical race against the pollution of our planet.

BASIC GEOLOGIC INPUTS

The application of geology toward solution of problems arising from the interaction of man and the earth does not necessarily require a new and unique body of knowledge but rather the selection of basic geologic skills, approaches, and know-how which can be brought to bear on a specific environmental problem. Traditional sources of information, as well as new areas of data, must be considered, and geologic mapping should possess a strong bias toward fundamental process-defined genetic rock units devoid of the many trappings of nomenclatural hierarchy, whether they be igneous, metamorphic, or sedimentary. The basic geologic map is the key to derivative mapping, which in turn translates data and ideas for other disciplines.

It has been stated that proper geologic mapping is the key to the initial phase of any environmental geologic application. It is imperative to know the distribution of the basic elements of the environments and their associated processes and modern and/or ancient facies. Important elements may include such diverse features as septic systems, oyster reefs, *Spartina* marsh, sand dunes, agricultural lands, zones of shoreline erosion, ancient fluvial channels, point-bar depositional sites, shallow aquifers, igneous intrusives, or potentially active faults. Perhaps an environmental problem may be a 10,000-square-mile general survey of possible sanitary fill sites, or it may be a detailed groundwater pollution study of very local extent; in either case appropriate mapping of the components involved is the first logical procedure.

Mapping that provides results that are functional and applicable is the ultimate goal, whether mapping modern environments and facies or ancient analogues. It is unfortunate that many geologic maps are not functional. Maps of the zone of *Cravenoceras* or *Schwagerina,* for example, offer little knowledge of the genetic constitution and distribution of rocks in an area of groundwater pollution or catastrophic engineering failures. No more helpful are highly subjective and philosophical time-stratigraphic maps which cover many walls. In fact, many traditional formations on geologic maps are nothing more than a name applied to undifferentiated rocks between two "contacts" or marker beds. There is a need for maps of lithofacies, metamorphic facies, igneous rock zones, landforms (yes, geomorphology *is* alive and well!), all types of man-modified features, floral and faunal distribution, and any other feature that is genetic and, hopefully, process-defined.

When mapping modern features such as active landslide areas, tidal flats, or floodplains, processes responsible for and operating in or on an environment can be reliably inferred. An understanding of these processes enables the geologists to understand better the genesis of the landslide area, tidal flat, or floodplain. Consideration of process and resulting genetic facies of floodplain muds and silts, for example, provides useful information and improves the predictability of composition, permeability, shrink–swell, uniformity or erratic nature of sediment distribution, and engineering properties.

The environmental geologic map is a record of the status of dynamic environments and processes, as well as a permanent record of exploitation, erosion, and human modification. Maps of physically dynamic areas such as rivers, coastal zones, slumping mountainsides, and active faults provide a base line or reference by which approximate rates and kinds of changes can be determined. The evolution of a dynamic geologic environment can be best monitored by periodic mapping which shows the average or significant direction and rate of change, rather than small-scale, commonly reversing changes recorded from day to day by monitoring devices.

Also important to quantitative studies of this nature are the processes and resulting changes in interrleated adjacent environments. A total system of interrelated modern sedimentary processes and environments can be monitored efficiently and inexpensively by sequential mapping, especially using aerial photographic and other remote-sensing techniques, such as infrared and radar photography. Precise quantitative measurements made at instrumental stations can be related properly and in perspective with the whole system of related modern environments. Mapping systems of ancient facies in terms of genetic systems also provide insight into the rate and kind of processes responsible for the rocks, giving the geologist a powerful tool to estimate properties of bedrock significant to environmental problems, such as porosity, permeability, geometry, strength, and probable hydraulic character, among many others.

The technical input from all kinds of scientists can best be integrated and applied through the display of the information on maps of modern and ancient facies and their inferred genetic processes. Significantly, economists, planners, public utilities specialists, power suppliers, sanitary engineers, lawyers, legislative councils, and untold numbers of other responsible persons in the governmental and private arena at every level, can plot, plan, refer, and digest available environmental input when it is projected and displayed in relation to the basic geologic units for the area or region. The environmental map is a common language and vehicle which can bring together the many divergent specialists and allow their collective contributions to be simultaneously focused on a problem.

DERIVATIVE MAPS

The environmental geologic map, because it displays the distribution of earth materials either areally or spatially, can occupy a unique and important role in environmental studies by serving as the prime source of easily derived interpretative maps. Because soils, engineering properties, vegetation, subaerial and subaqueous fauna, and many other similar features are so closely tied genetically to modern and ancient facies, the distribution of these facies is also the distribution of many properties critical to environmental workers. The environmental geologic map serves as the guide to accurate and speedy mapping of any significant properties that are primarily determined by the underlying sediment or bedrock. The importance of accurate genetic facies mapping for environmental studies should be obvious when it is understood that derived maps of environmental properties can be no better than the basic map.

It is conceded that maps which exhibit the distribution of vital properties of engineering significance, wildlife management, hydrologic nature, and other factors can be obtained eventually by analysis of data from closely spaced coring or observation stations. Tens of thousands of shallow cores or data localities in an area of 20,000 square miles, for example, would be necessary to prepare a map based entirely upon a network of sample and observational stations. If geologists map the various modern and bedrock facies using extensive aerial photographic interpretation and knowledge of the geometry and genetic nature of the facies, vital properties related to the composition and other characteristics of the geologic units determined at selected coring and sampling sites can be extrapolated to other similar facies in the area. These properties also can be reasonably interpolated for all areas within the boundary of the facies. Local variations in the properties of a single genetic facies are probable, but it is likely that the map would display critical characteristics, such as shrink-swell or high permeability. On-site drilling would be necessary before construction or other utilization of the area, but the map of shrink–swell or permeability provides an inexpensive, relatively detailed map for environmental and land-use planning and utilization.

COMMUNICATION WITH NONGEOLOGISTS

Geology possesses more powerful tools than most nongeologists realize, or, for that matter, than most geologists recognize. For engineering uses, nothing takes the place of detailed tests of a core, but geologists can help the engineer eliminate or grade vast areas, thereby eliminating test costs and time in reconnaissance and moderately detailed evaluation. Environmental geologists must talk to engineers and other specialists in their own language and from the nongeological point of view to learn what these disciplines need that can be supplied from the geologic side of the technical fence. A geologist certainly cannot convince an engineer that geology has something to offer if the geologist is ignorant of engineering problems and responsibilities. Geologists should attempt to appreciate engineering problems and then turn to sources of geologic data to see if the information can be derived or extracted.

Unless something is learned of engineering limits, wildlife tolerances, acceptable permeabilities, and similar properties vital in other areas, geologists cannot contribute to them. Engineers may with considerable reason be suspicious of a geoologist who thinks he can contribute to an engineering problem. Some geologists may hedge on specific interpretations, to be less than candid about their limitations, and appear to be overly hypothetical to the engineer, who must construct or design a suitable structure based on fact and not theory. This reticence on the part of engineers should be understood and lines of communication should be cultivated.

The derivative map is a means of translating basic geology into a form that can be utilized by a wide variety of persons interested in the environment and its proper exploitation and conservation. For example, public utilities are plagued in some areas by corrosion of pipes and cables. Corrosion is, in part, a second-derivative property of the soil and its ions, which in turn relates directly to bedrock. A map of the bedrock or underlying sediments on which highly corrosive soils develop becomes a derivative map of corrosibility. When selected testing shows that corrosion is primarily restricted to soils on one or more facies, these geologic units then define the corrosion derivative map unit.

Many other examples show that geologic information can be translated into very useful data supplied in the language of the public utilities people, the sanitary engineer, or others. Derivative maps can be as esoteric as a map of areas favorable for the growth of an ornamental plant species to maps as practical as load-strength limits for heavy construction. It is up to geologists to communicate and translate for the nongeologist. Any hesitancy on the part of other scientists and environmentalists to acknowledge that geology can provide hard, useful, reliable facts must be overcome with practical examples.

GENERAL OBSERVATIONS ON THE STATE OF ENVIRONMENTAL GEOLOGY

Geologists have talked to themselves too long. They must increase communication outside of the profession and, in so doing, gain better insight into their role at this critical time.

Geologists should exhibit flexibility and innovation by moving into many newly developed areas of application and advisory potential. Geologists can carve out entirely new areas of contribution if they display initiative and insight into contemporary environmental problems.

Geologists can and should relate to rapidly developing social and economic problems. Geologists can contribute in many nongeological areas by integrating data so that clear-cut recommendations can be made.

Geologists need to present a rational viewpoint of the environmental crisis to government leaders, legislators, and responsible agencies in a language that cuts through the jargon of the profession and comes to grip with basic problems.

Geologists can volunteer their expertise to planning agencies—local, regional, or national; they must not sit back and let others testify and recommend in areas where significant contributions can be made.

Finally, geologists should accept every opportunity to talk with engineers, soil scientists, economists, lawyers, and businessmen about the role geology can play in the environmental battle. Geologists should offer to help in any way possible just as long as it contributes to the ultimate goal of balanced natural systems and wise use of resources.

3

Reprinted from *Mining Congr. Jour.*, **56**(8), 65–70 (1970)

Resources and Environment —Quest For Balance

By W. T. PECORA,
Director,
U. S. Geological Survey

ANCIENT MAN LIVED IN HARMONY WITH NATURE. His existence was precarious but he accepted the good and the bad as qualities beyond his control and he stood in awe at natural phenomena he could not understand. From this humble beginning evolved our present society which now indicts man for all environmental ills and assumes that nature can be shaped to meet his every need. The ability to maintain an acceptable environment can be hindered by failure to recognize basic earth processes and quality patterns beyond our control. With the intellectual development now achieved by man, it is inexcusable that we should fail to predict responses of nature consequent to our own development. Environmental degradation is a natural process on earth. Man, however, is beginning to contribute to that degradation in large measure in certain areas. Man has begun to develop an awareness that better housekeeping of the earth must be practiced as he continues to take from the earth the things he needs and uses.

Planet Earth is man's abode

For some 5 billion years the planet Earth has revolved about the sun; and there is good reason to believe its journey will extend beyond another 5 billion years. Throughout this period the earth has undergone constant change—mountains have risen where oceans formerly existed; animal and plant species have flourished and become extinct; earthquakes and volcanoes have always been with us; rivers and plains have appeared and reappeared; and glaciers have covered large segments of the planet many times. Although on Earth but a few million years at the most, man has in the past 200 years unraveled a great deal of earth history and learned how to use the planet to meet his growing needs for survival.

As earthbound residents, we look constantly, nevertheless, to other planets. One, the Moon, satellite of the earth, has already been visited and found to be totally hostile to man. The surface of Venus is too hot for us, and Mars offers little, if any hope. The other planets are simply out of the question. Man, indeed, is earthbound and we must learn to accept this inescapable circumstance.

Of the billions of galaxies that exist in the universe, perhaps there is at least one other solar system like ours with a planet in the same solar position. Wherever that may be, it is beyond our reach, however great our expectation. We must learn to live on this planet throughout our full existence as one species.

Civilization always moves forward

Man has achieved phenomenal advance over 10,000 years in the face of a world population increase from a few million then to 3½ billion today. The complex development of society over this period was accomplished because man has an intellect that could innovate, plan, acquire information, store it, pass it along to succeeding generations, and increase the level of its systemic intellect through research and development. What has often been called intellectual curiosity, as directed toward our total environment, is really a necessity by society if it is to avert disaster.

Man now truly inhabits the entire planet. He has crossed mountains and oceans, explored the poles, and burrowed deeply underground. The simple but astute primitive observer of nature and natural processes has developed into the creative scientist who serves man's mind in seeking to feed the technologic engines of modern society. Those unique quinqueremes of ancient

[*Editor's Note:* All figures have been deleted, owing to limitations of space.]

times have developed into jet aircraft; simple mathematical devices that were developed separately in different civilizations have grown into complex modern computer systems; signal drums have blossomed into telecommunications systems that link hemispheres. Real time for man now has real meaning, and you are part of all of this. You cannot ignore nor escape your role. Your generation will do deeds only dreamed of by mine, just as my generation made a giant step from my father's. The status quo may have meaning for other species on earth; but for man there is no status quo because of his intellectual capability.

Projected resource needs of the United States

Let's take a look at the resource needs of the society that makes up the United States of America. We are at the apex of civilization, and yet within the life span of 200 million people now living in the United States, this nation will consume from the earth:

6½ *quadrillion gallons of water*
7½ *billion tons of iron ore*
1½ *billion tons of aluminum ore*
1 *billion tons of phosphate rock*
100 *million tons of copper*
and so forth . . .

In 40 years, our population will double. Just think of the added requirements of the next generation!

Water usage will triple by the year 2000
Energy requirements will triple by the year 2000
By the year 2000 *we will have to construct as many houses and other facilities as now exist in the United States.*

This staggering amount of natural mineral resources upon which the sustenance of the Nation depends imposes a tremendous task of new discovery, and new development. How can we do this without changing the character of our environment; for society must also provide against excessive noise, excessive pollution, excessive degradation of the landscape, water-scape, and seascape. We do want the best of all worlds!

If this be the situation for the United States, certainly resource needs for the rest of the world command even greater attention. Developing nations seek fulfillment in health and economic betterment. The crust of the earth is worldwide and knowledge gained in one country can be used to good advantage by scientists and engineers in others. The crust of the earth has full potential to provide for man's needs if we have the motivation to procure and develop. The problem for mankind is universal—planetary—not national. Certainly international competition cannot go on forever; wars must cease and man's society must be planetized if the species, Homo sapiens, shall persist on this earth.

If we must therefore take from the earth to provide for ourselves we must employ value judgment and trade-off concepts in deciding how much to take from our environment, where to take it, and how to leave it in the taking and using. Take and use we must or we cannot survive as a species on earth.

The need for research

If the earth shall provide the materials for the survival of man's society, then a prudent society must provide for an intimate understanding of the earth, inquiry into geologic processes that have operated over the span of earth history, and operate today, continuing inventory of current and potential resources, and continuing effort to develop new techniques for information-gathering systems. Research and technologic development are costly investments; but they pay off handsomely in long term benefits. Too often a society thinks only of "now." The cumulative benefits of early endeavors, on hindsight, are superb demonstration that today's long term is tomorrow's short term. Time, for man, is a long continuum.

The best example of this is the basic mapping systems that have been developed in the past. These include base topographic maps, geologic maps, hydrologic maps, geophysical maps, geochemical maps, and thematic, environmental, special subject maps. All of these are the products of intensive research effort. I am concerned that the pace of doing this kind of work has slowed down in recent years and I predict that our society will suffer for it. Our priorities will have to be reordered because of increasing needs of a World society that has a divine right under God to utilize our planet's total resources and to better the lot of man on earth.

Primary national resources goals

The United Staes, from its very inception, has been accused of placing too much emphasis upon the accumulation of wealth and too much effort in raising, through industry, its so-called standard of living. This view was first stated unequivocally in 1831 by Alexis de Tocqueville, a French nobleman who visited this country when it had 24 states and 13 million people. Similar views are being stated today, as we have grown to 50 states and 200 million people. The American democracy and its free enterprise system is a great and successful experiment, the first of its kind in the history of civilization. I, personally, see nothing wrong with dedication of individuals, or groups, toward amassment of wealth through honest industry. But in the process of achieving these goals our society unwittingly, or knowingly, has permitted deterioration of other values, not measurable in dollars or numbers, and which affect the quality of individual life in many ways. The growth of science and technology in discovery and utilization of our basic resources has failed in some ways and in many places to retain or fortify man's natural environment.

Some myths of our environment

It is believed by many people that man alone is degrading and polluting his environment because of our modern industrial society. Some myths, however, need to be destroyed. Let me cite a few natural earth processes to demonstrate that natural processes are by far the principal agents in modifying our environment. This is not to excuse or put aside what man does, but rather to put man's actions in proper natural perspective. Those individuals who speak about restoring our inherited environment of pure air, pure rain, pure water, pure lakes, and pure coastlines ignore the inevitability of nature.

It has been calculated that more than 100 million tons of fixed nitrogen in the form of ammonia and nitrates is annually transferred from the atmosphere to the surface of the earth as part of a natural precipitation process. In the United States alone there falls upon the face of our land annually more than 4 million tons of table salt, 2½ milion tons of sodium sulphate, and 36 million tons of calcium compounds—all in rain water.

Particulate matter and natural gases dispersed from the volcanoes is a continuing phenomenon. From three eruptions alone, the Krakatau eruption in Java (1883), the Mount Katmai eruption in Alaska (1912), and the Hekla eruption in Iceland (1947) more particulate matter in the form of dust and ash and more combined gases were ejected into the atmosphere than from all of mankind's activity. Add to current volcanic processes the normal action of winds, forest fires, and evaporation from the sea, and we can readily conclude that man is an insignificant agent in the total picture, although he is becoming an important agent in extremely local context.

We have long been led to believe that water issuing from natural springs is pure and beneficial to health because of its purity. The springs issuing into the Arkansas and Red Rivers carry 17 tons of salt per minute. In the Lower Colorado River salt springs carry 1,500 tons of salt per day. The Lemonade Springs in New Mexico carry 900 pounds of sulphuric acid per million pounds of water, which is ten times the acid concentration of most acid mine streams in the country. Hot Springs in Yellowstone Park is likewise many times more acidic than the typical acid stream in a coal mining district. The Azure Yampah spring in Colorado contains eight times the radium that the Public Health Service sets as a safe limit. These are but a few examples of the kind of pollution that goes on continually from natural springs.

The lakes and ponds throughout geologic history have gone through a life cycle of birth, maturity, old age, and disappearance. No lake is truly permanent. Some of our inland lakes during their mature stage become more salty than the ocean itself. The Great Salt Lake is nearing its dying stages. Once 20,000 sq. mi. in area, (Lake Bonneville), it is now only 950 sq. mi. in area. Many thousands of years ago it was essentially a fresh water lake, fed during the pluvial period of the Great Ice Age, and now it is about ten times as salty as sea water.

We frequently hear that Lake Erie is dead. This is pure rubbish. Lake Erie is the shallowest of the Great Lakes, was created about 20,000 years ago and, barring another Ice Age, has several thousands of years yet to go before senility. The western part of the lake is extremely shallow and receives a large amount of natural organic material transported from the surrounding terrain. Here is where the algae growth has always been present. Lake Erie has continually produced about 50% of the fish catch of the entire Great Lakes system, consistenly over the past 100 years. This is not a mark of a dead lake. Green Bay, of Lake Michigan, so named by the first settlers because of the green color of the algae so prevalent in the Bay is, like the western shallow part of Lake Erie, the source of a great amount of organic matter. The food supply for aquatic life is high in these environments. The oxygen supply, unfortunately, diminishes as algae growth increases, as this portion of the lake becomes more and more shallow and as organic material is swept into the water, whether from natural or human sources. Every lake or pond whether natural or man-made, faces a similar life history. Man can certainly better or worsen a natural situation like this.

The rivers of our nation are being called dirty because of the works of man. We must understand that the river systems of the land are the natural transport systems for sediment washed downhill by the rains that fall upon the land. It is estimated that the Mississippi River carries into the Gulf a load of more than 2 million tons of sediment per day. This is equivalent to the load of 40,000 freight cars. The Colorado River carries into Lake Mead about 40,000 tons per day. The Paria River in Arizona is probably the dirtiest river in the world. It carries 500 times as much sediment as the Mississippi River per unit volume of water. This is a continuing condition year after year. Chemicals are also transported by streams in phenomenal amounts. The Brazos River of Texas, for example, transports 25,000 tons of dissolved salt per day. Peace Creek in Florida carries twice the concentration of fluoride that is harmful to teeth. Many rivers and streams throughout the nation have natural qualities that do not meet the public health standards for drinking water.

The ocean has been the natural waste sink for the large-scale, natural pollution process of the earth. The character of the ocean itself has changed slowly throughout geologic time, while it has continually supported abundant life of all varieties.

Man as a geologic agent

It must be quite evident that, although natural earth processes dwarf the actions of man in a total context, man can become a major geologic agent in a specific or local context. This inter reaction of man with nature is without question a most important issue of future years. In a society that has reached maturity in the industrial sense, the issue of environmental alteration becomes more and more acute. It is within this framework that certain actions 100 or 200 years ago are now considered sinful.

The philosophy of engineering project costs is being modified to allow for certain actions which in the past were not factored into our cost analyses. A mineral resource, for example, should not be developed unless it is rich enough to support proper restoration or reutilization of the land. A major pipeline traversing Alaska some 800 miles should not be constructed without added safeguards to protect the natural environment. Offshore drilling for petroleum should not be endorsed without the added cost of providing maximum safeguards against pollution. Cities and industries should not use the water available from natural sources without factoring in the cost of returning the water to a usable state. The smokestacks of our refineries and energy plants must not treat materials that put unwanted matter into our air. Reservoirs are constructed and rivers are diverted. These problems and others like them are familiar to all of you. The science and technology which has made possible the great advance of mankind can surely pay attention to these matters and resolve them.

Who will bear the cost, however? Who will make the policy decisions? We know that government agencies have been urged by the people to regulate these matters, but who will pay the cost, I ask, for what I am convinced science and technology can do to ameliorate the situation. It must be the people. Whether through their tax participation or whether it be in the higher cost of a product, the people must pay the cost both in dollars and in landscape changes for taking from the environment what mankind needs. We who represent science and technology can show how resources can be utilized with minimal alteration or degradation of the environment. We can do this, however, only by making a complete inventory of cause and effect—in other words—spelling out our costs and trade-offs in more than just dollars. President Nixon and Secretary Hickel have spoken of this on many occasions.

A greater role for engineering

Throughout history engineering accomplishment has been a magnificent measure of human achievement, particularly where new challenges were posed. Daring projects, planned on basic natural laws, carefully monitored in the design and construction stages, became on hindsight well-conceived operations. For those times the ancient irrigation system of Mesopotamia, the Roman aqueduct, the pyramids were utterly fantastic operations. The hot water supply at Bath, England still functions perfectly almost 2,000 years after its construction. Today the Hoover Dam, the Open Pit at Bingham, Utah, the Golden Gate Bridge or the Hudson River Tunnel mark great ventures of our own era.

Although the engineering profession today is recognized as one which gets things done, it is also recognized as one that is essentially pragmatic in acquiring maximum integrity of structure and function at minimum cost. A new ethic for engineering is evolving. More and more concern is being directed to areas of social impact arising from waste products disposal and environmental degradation.

In the 1969 Christmas Pageant of Peace in Washington, D.C. pine trees marked a path for the States of the Union—each tree for a state. These trees had been supplied by a private company in the Midwest and grown on restored lands where ugly coal strip mines had disrupted the landscape. The mining venture here was but an incident in a time plan. How proud I was to see this; but so few people know of it and other cases like it. So many are angry over chemical wastes dumped into lakes and rivers and gaseous products from smelters and coal-burning smoke stacks emitted to the air. And their numbers are increasing!

The value of the extractive mineral resources in the United States is currently in the order of $30 billion annually. This includes both the energy and hard mineral resources. The impact of that part of the engineering profession responsible for this mineral production, and its subsequent use in the economy, is profound on the maintenance of our way of life. American industry is in trouble, however, because of past abuses to the environment which admittedly have been magnified or distorted by some critics. Profit has become an ugly word and the concept of service to mankind derived from engineering practice is being ridiculed in many places. We are in the midst of a conflict between the need to develop our resources and the need to preserve our environment—both for the benefit of mankind. In your chosen profession of engineering you will be very much involved—you will be subject to pressures from both sides and you must be responsive. Do not be silent under attack.

You young graduates will one day be leaders in your profession. The principal obligation of leadership is to be alert to the issues, to make tough value judgments, to be forthright in your expressions to management and to the community, to seek facts but, in the absence of total facts, to have the courage to make gut decisions with a constant edge toward people and people's problems. This is your quest—a quest for balance.

You may have read the words expressed by H. L. Keenleyside of Canada before a United Nations Scientific Conference. They came from a poem of James Russell Lowell in his tribute to Cromwell. More than anything else they represent the philosophic view I am presenting to you today. Let me give those words to you now.

"New times demand new measures and new men;
The World advances, and in time outgrows
The laws that in our father's day were best;
And doubtless, after us some better scheme
Will be shaped out by wiser men than we,
Made wiser by the steady growth of truth."

God be with you in your life's endeavor. Never, Never, shade the Truth, Thank you!

Part II
LAND-USE PLANNING

Land-use planning is an inclusive term. Within this category we find *regional planning* and *urban planning*, both indicating an areal extent. The term "land planning" is also used, but seems to be less explicit than *land-use planning*.

The boundary between regional planning and urban planning is indistinct in many cases. In the United States, a prominent example is the Northeast Corridor, stretching along the eastern coast from the Boston area to south of Washington, D.C., about 600 miles in length and 50 to 100 miles in width. This region is rapidly becoming a continuous built-up strip. Planning for the cities within the region may constitute urban planning, but in many instances different governmental units have regional planning activities, and in some measure problems in the entire corridor are being examined.

The clustering of population in areas of continuous settlement is a characteristic that has been developing for a longer time in smaller countries. The term "conurbation" was coined more than 50 years ago in England to designate a merging of cities, towns, and rural areas in social and economic units. The Netherlands is such a small country with such high population density that it is difficult to get out of sight of signs of habitation.

Geologists have adopted the term "urban geology" to identify their function in urban planning. It has been concerned largely with construction-foundation, water-supply, and waste-disposal problems. Urban geology is treated by some geologists as synonymous with environmental geology. To others the latter is synonymous with engineering geology. That these are not valid conclusions has been discussed in the Introduction. Nevertheless, engineering geologic determinations must be important in urban planning and can be significant in many regional planning efforts.

That geology is only one input in land-use planning is evident. A serious problem for all scientific and technical participants in the land-use planning process is the weight given their interpretations and recommendations by those who make ultimate decisions on land use. It is reported that one center for urban and regional studies found that, generally, the natural landscape, except for extreme conditions of slope, soil, and drainage conditions, has only limited effect on site selection by developers. Another finding is that sites are dependent more on social-class considerations than on environmental conditions. Thus, land-use planning may be a desperate exercise, in which opposition to findings may come from sources of prejudgment.

Four of the five papers that follow are concerned with city problems. They portray a variety of situations, ranging from those of established locations that are susceptible to expansion and modernization, to others that have been in need of reconstruction because of earthquake damage or extensive war damage and postwar expansion, and finally to the question of siting new centers.

As an indication, admittedly all too brief, of another concern of geologists in the problems of land-use planning, there is a paper on siting an important construction project. Regrettably, because of space limitations, no paper on the areal problem concerned with a subject of growing importance, recreational areas, is offered; references are given in the Supplemental Reading List.

SUPPLEMENTAL READINGS

Association of Engineering Geologists, Los Angeles Section. 1965. *Geology and Urban Development.* 19 p.

Association of Engineering Geologists, San Francisco Section. 1969. *Urban Environmental Geology in the San Francisco Bay Region* (E. A. Danehy, ed.).

Bakhtina, I., and E. Smirnova. 1969. Methods for Estimating the Suitability of Natural Conditions at Various Stages in the Drawing Up of Architectural Plans. In: *Land Evaluation* (G. A. Stewart, ed.), p. 179–186. Macmillan of Australia. (CSIRO Symp., Canberra, 1969.)

Bartelli, L. J., et al. (eds.). 1966. *Soil Surveys and Land Use Planning.* Soil Sci. Soc. America and Amer. Soc. Agronomy, 196 p.

Born, S. M., and D. A. Stephenson. 1973. *Environmental Geologic Aspects of Planning, Constructing, and Regulating Recreational Land Developments.* Upper Great Lakes Regional Commission, 97 p. (Inland Lake Demonstration Project.)

——, and D. A. Yanggen. 1972. *Understanding Lakes and Lake Problems.* Upper Great Lakes Regional Commission, 40 p. (Inland Lake Demonstration Project.)

Branch, M. C. 1962. Rome and Richmond—A Case Study in Topographic Determinism. *Amer. Inst. Planners J.*, **28**(1), p. 1-9. (See also *Planning—Aspects and Applications*, p. 109-120. New York, John Wiley & Sons, Inc.)

Clark, S. H. B. (chairman). 1973. *Carrying Capacity of Outdoor Recreation Resources.* Federal Work Group E—National Outdoor Recreation Plan. 46 p.

Cratchley, C. R., and B. Denness. 1972. *Engineering Geology in Urban Planning with an Example from the New City of Milton Keynes.* 24th Intern. Geol. Congr. Proc., Sec. 13, p. 13-22.

Espey, W. E., Jr., C. W. Morgan, and F. D. Masch. 1965. *Study of Some Effects of Urbanization on Storm Runoffs from a Small Watershed.* University of Texas, Center Res. Water Resources, Tech. Rept. HYD 07-6501, CRWR-2, 109 p. (Same title. Texas Water Develop. Board, Rept. R23, 110 p., 1966.)

Frye, J. C. 1974. *Land Resource—Its Use and Analysis.* Illinois Geol. Surv. Environ. Geol. Notes 70.

Gill, J. E. (ed.). 1972. *Engineering Geology.* 24th Intern. Geol. Congr. Proc., Sec. 13, 307 p.

Gross, D. L. 1970. *Geology for Planning in DeKalb County, Illinois.* Illinois Geol. Surv. Eng. Geol. Notes 33.

Hackett, J. E. 1966. *An Application of Geologic Information to Land Use in the Chicago Metropolitan Region.* Illinois Geol. Surv. Environ. Geol. Notes 8.

Hilpman, P. L., and G. F. Stewart. 1967. *Environmental Geology and Land-Use Planning.* Kansas State Geol. Surv., 29 p. (Paper delivered to 1967 Ann. Mtg., Geol. Soc. America.).

——. 1968. *A Pilot Study of Land-Use Planning and Environmental Geology.* Kansas State Geol. Surv. Planning for Development 701 Project, Kans. P-43, Rept. 15-D, 63 p.

Jäckli, Heinrich. 1961. Aktuelle Beziehungen der Quartärgeologie zum Bauwesen. *Naturf. Ges. Zürich, Vierteljahrsschr.*, **106**(2), p. 253-275.

Jolliffe, I. P. 1968. *Planning and Research Problems in the Exploitation of Coastal Areas.* 23rd Intern. Geol. Congr. Proc. Sec. 12, p. 96-103.

Knight, F. J. 1971. Geologic Problems of Urban Growth in Limestone Terrains of Pennsylvania. *Assoc. Eng. Geol. Bull.*, **8**(1), p. 91-101.

Knill, J. L. 1968. Geochemical Significance of Some Glacially Induced Rock Discontinuities. *Assoc. Eng. Geol. Bull.*, **5**(1), p. 49-62.

——, D. G. Price, and I. E. Higginbottom. 1968. *Aspects of the Engineering Geology of the City of Bristol.* 23rd Intern. Geol. Congr. Proc., Sec. 12, p. 77-88.

Kusler, J. A. 1972. *Carrying Capacity Controls for Recreation Water Uses.* Upper Great Lakes Regional Commission, 71 p. (An Inland Lake Renewal and Shoreland Management Project Report.)

Legget, Robert F. 1973. *Cities and Geology.* New York, McGraw-Hill Book Company, 624 p.

Leighton, F. B. 1971. The Role of Consulting Geologists in Urban Geology. In: *Environmental Planning and Geology* (D. R. Nichols and C. C. Campbell, eds.), p. 82-89. Washington, D.C., U.S. Geological Survey and U.S. Department of Housing and Urban Development. (Symp. Eng. Geol. in the Urban Environ., Assoc. Eng. Geol., Oct. 1969.)

McGill, J. T. 1964. *Growing Importance of Urban Geology.* U.S. Geol. Surv. Circ. 487, 4 p.

McHarg, I. L. 1971. *Design with Nature.* Garden City, N.Y., Doubleday & Company, Inc., 198 p.

Matula, Milan. 1968. *Fundamental Problems of Regional Engineering Geology in the Czechoslovak Carpathians.* 23rd Intern. Geol. Congr. Proc., Sec. 12, p. 89-103.

Miller, R. D., and Ernest Dobrovolny. 1959. *Surficial Geology of Anchorage and Vicinity, Alaska.* U.S. Geol. Surv., Bull. 1093, 128 p.

Moser, P. H. 1968. *Environmental Geology as an Aid to Urban and Industrial Growth in Northwest Alabama.* 4th Amer. Water Resources Conf., Proc., p. 392-398. (Amer. Water Resources Assoc. Proc., Ser. 6.)

Nichols, D. R., and C. C. Campbell (eds.). 1971. *Environmental Planning and Geology.* U.S. Geological Survey and U.S. Department of Housing and Urban Development. (Symp. Eng. Geol. in the Urban Environ., Assoc. Eng. Geol., Oct. 1969.)

Parizek, R. R. 1971. An Environmental Approach to Land Use in a Faulted and Folded Carbonate Terrane. In: *Environmental Planning and Geology* (D. R. Nichols and C. C. Campbell, eds.), p. 122-143. Washington, D.C., U.S. Geological Survey and U.S. Department of Housing and Urban Development.

Quinn, Ann. 1970. *Ecology, Conservation and the Landscape.* Planning for Amenity, Recreation and Tourism, vol. 2: Model Development Studies. Dublin, Ireland, An Foras Forbartha.

——. 1971. *Conservation and the Landscape.* Dublin, Ireland, An Foras Forbartha, 11 p.

Risser, H. E., and R. L. Major. 1967. *Urban Expansion—An Opportunity and a Challenge to Industrial Mineral Producers.* Illinois Geol. Surv. Eng. Geol. Notes 16, 19 p.

Rockaway, J. D. 1972. *Evaluation of Geologic Factors for Urban Development.* 24th Intern. Geol. Congr. Proc., Sec. 13, p. 64-69.

Schlocker, Julius. 1968. *The Geology of the San Francisco Bay Area and Its Significance in Land-Use Planning.* Bay Area Regional Planning Programs, Suppl. Rept. IS-3. (U.S. Geol. Surv. Regional Geol. Suppl. Rept. 15-3.)

Steinitz, Carl. 1970. Landscape Resource Analysis. *Landscape Architecture,* **60**(2), p. 101-105.

U.S. Soil Conservation Service. 1966. *Guide to Making Appraisals of Potential for Outdoor Recreation Developments.*

Wayne, W. J. 1960. *Geologic Contributions to Community Planning.* Am. Assoc. Advan. Sci., Sec. O (Symp. on Land Zoning in Relation to Agricultural, Industrial, and Recreational Needs of the Future), 17 p.

——. 1968. *Urban Geology—A Need and a Challenge.* Indiana Acad. Sci. Proc., p. 49-64.

Withington, C. F. 1967. Geology—Its Role in the Development and Planning of Metropolitan Washington. *Washington (D.C.) Acad. Sci. J.,* p. 189-199.

Záruba, Quido (ed.). 1968. *Engineering Geology in Country Planning.* 23rd Intern. Geol. Congr. Proc., Sec. 12, 227 p.

Zisman, S. B., and D. B. Ward. 1968. *Where Not to Build—A Guide for Open Space Planning.* U.S. Bur. of Land Management Tech. Bull. 1.

Zonneveld, J. I. S. 1970. The Physico-Geographical Aspects of the Randstad Holland Area. In: *Geological and Geographical Problems of Areas of High Population Density,* p. 89-105. Assoc. Eng. Geol. Proc. Symp., Washington, D.C.

Abstracts (1972–1974) from Geological Society of America, Abstracts with Programs

Brown, L. F. 1973. Environmental Geology and Land Management. 5(5), p. 381.

Emrich, G. H., Feodor Pitcairn, and David Witwer. 1974. Geologic Environmental Planning in a Major Metropolitan Area. 6(1), p. 22–23.

Everett, A. G. 1972. The Geologist in Environmental and Resources Decision-Making. 4(7), p. 501.

Matthews, R. A. 1973. Impact of Environmental Geologic Study on the Selection and Evaluation of Waste Disposal Sites, Lake Tahoe Area. 5(1), p. 78.

Montagne, John, and Clifford Montagne. 1974. Geologic Mitigation in the Planning Process. 6(5), p. 460.

Editor's Comments on Papers 4 Through 8

In Paper 4, Clifford A. Kaye offers observations on the historical aspects of siting cities on the eastern coast of the United States. Kaye has been associated with the U.S. Geological Survey since 1946, and for some time has studied the urban geologic problems of the Boston area. He points out that the early settlers, beginning in the seventeenth century, sought sites for settlement that had good harbors, land with good foundation for building, and sources of potable water. Until the beginning of the nineteenth century, most cities, however, "handled their environmental problems on a pragmatic day-to-day basis." At about that same time, geology emerged as a science, but does not appear to have been used to any extent in city planning until 1900.

Kaye observes the growing need for urban geology and asks the following question of geologists: "What can we do in the urban crisis? What is the most effective role that we can plan?" He stresses the need for mapping.

Ernest Dobrovolny and Henry R. Schmoll have published a paper, included here as Paper 5, that covers the range of geologic applications to urban planning in the Anchorage, Alaska area, which was badly damaged in an earthquake in 1964.

The authors open their discussion with a statement that bears noting:

> Geologists, of course, have understood that geology deals . . . with dynamic processes, and that these processes occur with as much vigor today as in the past. Except for application to engineering works of great magnitude, however, developers and the general public have often failed to recognize the relevance of these processes to many other facets of human endeavor involving use of the land.

They call attention to a report on the area, with useful information on development planning, prepared by the U.S. Geological Survey in 1959 (See Miller and Dobrovolny in Supplemental Readings), which "reached few, if any, individuals directly responsible for local planning."

Paper 5 discusses the different types of special maps being prepared. This material should be read in conjunction with Part VII.

Dobrovolny has been with the U.S. Geological Survey since 1944, working on oil and gas problems, engineering geologic projects, and land-use management. Schmoll, also of the U.S. Geological Survey, is especially interested in the application of geology to land-use planning.

Hamburg, Germany, was very severely damaged in World War II, but has been almost entirely restored. F. Grube states in Paper 6 that

> Because of the traffic situation, the building of new underground [subway] stations must take place in the centre of the city. When new projects such as underground railways, river tunnels for express highways, underground garages, etc., are planned in an area with difficult geological structures and soils . . . the subsurface situation has to be investigated very carefully to prevent or reduce damages to older buildings by soil compaction, landslides, etc.

The author makes pertinent comments on the role of the urban geologist, notably in his contribution "to the preservation of valuable landscapes and individual natural monuments."

Grube joined the Geological Survey of Hamburg following some years in academic work. He is concerned with engineering geology, Quaternary geology, and urban and environmental geology.

A different urban geologic problem is discussed in Paper 7 by Peter Hall, professor of geography in the University of Reading, England. Hall deals with the "new towns" in Great Britain, of which 28 existed by mid-1970. The concept of new towns, once called "garden cities," originated in the late nineteenth century, and, after early experiments, legislation in 1946 allowed many of them to be sited and built. The portions of the paper offered here display the different geologic and geographic considerations that planners took into account. So far, Hall says, the new towns have not experienced difficulties with their physical sites. The paper might give food for thought to developers of new towns in the United States. There is as yet no coordinated plan for new towns in the United States or for preparation of the geological basis for their site selection.

To meet the need for an additional airport to serve London, it was proposed to build one at some distance from the metropolis on Foulness Island (about 90 kilometers from mid-London), which lies near the Thames estuary. J. T. Greensmith and E. V. Tucker dealt with the geologic implications in a paper published in 1968 (Paper 11A), which indicated that at the time not enough had been done to consider the effect on erosion and deposition in the estuary of mass dredging and removal of offshore sand.

A second paper, published in 1973 (Paper 11B), shows the great value of additional exploration of the area by drilling. The original proposals have been modified as a result, and alternative sites have been considered.

4

Reprinted from *Trans. N.Y. Acad. Sci.*, Ser. II, **30**(8), 1045-1051 (1968)

GEOLOGY AND OUR CITIES*†

Clifford A. Kaye
U.S. Geological Survey, Boston, Mass.

For most of us, geology has always meant the great open spaces—intellectual pursuit in the vast out-of-doors. As any traveler knows, the species *field geologist* is found sparsely in the mountains, more sparsely on the plains, and almost never in the cities. This nation-wide distribution pattern is rational enough when viewed in the light of the social and intellectual priorities of the past century and a half—that is, since the entry of the geologist onto the national scene. Then the country was expanding; and exploring and finding were the order of the day. Now, however, our national concerns are fast changing. The problems of the cities rank with those of the farmers, the miners, and the manufacturers interested in raw materials. I think it will be agreed that it was largely the needs of these groups that motivated much of our past geological activity, for geology, like all human pursuits, responds to the social and economic concerns of the times. We are now well launched into the era of the cities. The fact is already having an effect on applied geology; it will, I think, continue to do so at an ever increasing rate. With this in mind I would like to trace briefly the history of city geology in this country and hazard a guess or two as to its future.

Most of our east coast cities came into being because of their topographic setting, which is another way of saying because of their underlying geology. The founders of these 17th-century cities came to the New World in boats, only to anchor, disembark, and dig in where harboring was satisfactory and the surrounding terrain not too forbidding. Thus, our chain of good harbor cities (Salem, Boston, Plymouth, Newport, New York, Philadelphia, etc.) and fall line cities (Trenton, Baltimore, Richmond, etc.). In some cities somewhat more sophisticated geologic considerations were involved in the selection of city sites. The site of Boston, for example, was selected not only because of the good harbor there, but also because of the absence of rock outcrops and, more important, the presence of an excellent supply of ground water. The story of the founding of Boston is worth telling in this connection.

The settlers of Boston first landed in Salem, about 15 miles to the north, in June 1630. Although Salem possesses a good harbor, the land is forbidding in aspect with many large, barren rock outcrops. When we consider that the settlers who came over in 1630 were mostly from southeastern England, a region of deep soil and rich farmland, we can readily imagine their dismay on arriving in the New World at having to build a new life in a hostile wilderness of barren rock. Within a week their leader, Governor John Winthrop, decided to look elsewhere for a site for his colony.

He traveled south across the narrow rocky belt in which Salem was located and dropped down onto the lowlands of the Boston basin, a structural and

*This paper was presented at a meeting of the Section on May 6, 1968.
†Publication authorized by the Director, U.S. Geological Survey.

topographic depression where deeply eroded argillaceous rocks are buried beneath thick drift. Here, on the lower slopes of a drumlin, in what is now Charlestown, a second start was made. No sooner had the colonists set up their tents and started to build when sickness spread through the group. Bad water was blamed, for a single spring served the entire settlement and this was along the river in the intertidal zone.

It was fortunate that across the narrow Charles River from the Charlestown settlement there was a lone Englishman, William Blackstone, who lived on the western end of the hill now called Beacon Hill. He met the newcomers and told them that he had an excellent spring on his place. In consequence, Winthrop ordered his group to pack up once more and to move across the river to share Blackstone's good fortune. It was this third site that became the city of Boston for, indeed, the colonists found water in abundance. Besides Blackstone's spring, several other springs were found and shallow dug wells produced water of good quality under artesian pressures. The reasons for this hydrologic situation have only lately been revealed by studies of the geology exposed in excavations for new large buildings. Briefly, it is because the area is underlain by a sandwich of thick highly pervious gravels between till and clay. This sequence, which underlies most of the area, has been deformed and folded up into Beacon Hill, a complex anticline.

Undoubtedly other cities were located because of special ground water conditions that the first settlers found there. The point is that the ground on which they were built was an important factor in these early cities, whether or not the founders were aware of this under the heading "geology."

The city's environment is, after all, dominated by the land underlying it. A city is interested in its stability and its foundations; it is interested in its source of construction materials and its sources of potable water. The cost of building a city includes the cost of foundations, and this is no mean portion. Every building, tower, pylon, tank, pier, bridge, plant, yard, street, and sidewalk has a foundation. The cost of all these foundations makes up between 8 and 20 percent of the whole.

It was not until the beginning of the 19th century that geology was recognized as a separate science. Until then our early cities handled their environmental problems on a pragmatic, day-to-day basis. In Boston, a search of the town records of the 17th and 18th centuries yielded many geologic fragments: deliberations on water wells, drainage, and supplies of gravel for streets and for land reclamation. The foundation of the city was taken for granted, and although permission had to be obtained from the Board of Selectmen for digging, filling, and the removal of gravel, this was not a matter of conservation so much as an expression of orderliness of administration, a feeling for the need to maintain equity among townsmen and, not the least, a respect for property and the public domain.

Geology and geologists began to make their appearance in the first few decades of the 19th century, and we find several attempts to study the geology of city areas. In Boston there was the curious work of two physicians, James F. and Samuel L. Dana,[1] which in the main consisted of an attempt to classify all the rocks and minerals in the Linnaean system. It was accompanied by a colored geologic map of sorts. Gerard Troost[2] reported on the geology of Philadelphia in a now rare, privately printed work, and Edward Hitchcock[3] described the geology around Portland, Maine. The first work to deal with the interaction of geology and engineering in a city was by Issachar Cozzens,[4] who aptly described the geology of Manhattan Island, New York City. His colored geologic map (approximate scale 1:90,000) had eight units and still makes good sense.

As the century developed, geology's interests broadened from its early preoccupation with mineralogy and paleontology to a concern for stratigraphy and structural geology. Moreover, the importance of geologists in the finding and developing of mineral deposits became well established. In consequence, by the second third of the century, geologists accompanied all exploring expeditions sent out by the Federal Government (viz, U.S. Exploring Expedition, Mexican Border, Colorado River, Railroad Route from the Mississippi to the Pacific, etc.), and most states had active geological surveys. There was, of course, a strong economic basis for this governmental support of geology, particularly in the post-Civil War era. This was the time of unprecedented national growth. The great increase in industrial productivity and wealth in the late 19th century had as its basis the exploitation of natural resources within the nation's boundaries. During these years mineral deposits were being discovered and highly productive mines were being opened at an accelerating rate. It was with these raw materials that the great metallurgical and petrochemical industries were built. Geology and governmental geological surveys, both state and federal, literally held the key to the wealth and well-being of the nation.

In all this 19th-century geologic activity, the application of geology to the building of the rapidly burgeoning cities was only manifest toward the end. At about the turn of the century, the U.S. Geological Survey had a fairly ambitious program of geologic mapping of city quadrangles under way. These maps were published in the folio series at a mile-to-the-inch and two miles-to-the-inch scales. The list of cities covered includes, among others: Sacramento; Chattanooga; Knoxville; Boise; Washington, D.C.; Chicago; New York City; Milwaukee; Passaic; Philadelphia; El Paso; San Francisco; Minneapolis-St. Paul; and Detroit. While the motivation for the mapping of a few of these quadrangles may have been dominantly minerals, most of the quadrangles were obviously done with a view to helping the cities directly. Interestingly enough, mapping of the highly urbanized strip along the eastern seaboard, now referred to as the Northeast Corridor, was nearly completed in the New York City to Washington D.C. area.

For the cities, the 20th century brought with it skyscrapers, subways, large bridges, tunnels and deep excavations for various purposes, superhighways, heavy industrial plants, airports, and many structures that stressed the ground beneath the city in a way that was entirely new. The foundation of the city was now critical to all this engineering activity. The demand for construction materials—sand, gravel, clay, and general fill—increased exponentially from decade to decade. This, then, was the century when geology would, of necessity, become a full-fledged partner with engineering in city building. Although to some of us this seems self-evident, regrettably there has been a lag in the achievement of this promise.

For example, in the first decade of this century, when the need for this partnership was already manifesting itself strongly, geologists were still working in the mountains and the plains. When they were not absorbed in field mapping and in what is now called basic research, they were fully occupied by mineral deposits and mining problems. In large measure this was justified for the great mineral wave of the 19th century was still rolling strong and, in fact, was reinforced by the emergence of the oil industry as a major employer of geologists. Nevertheless, civil engineering needed geology, whether or not geology needed civil engineering.

What happened was inevitable. Another discipline responded and filled, as best it could, the function that civil engineering was asking of geology. Soil mechanics, an offshoot of civil and materials engineering, was born. This momentous development we owe largely to one man. In the first decade of this century the late Karl Terzaghi, then a young Austrian engineer with a strong interest in geology, came to the United States to work for several years with the Reclamation Service (as the Bureau of Reclamation was then called). He had hoped to find here new ideas and techniques which would blend geology with engineering in the broad range of problems of large-dam construction. Disillusioned with the absence of quantification in geology, he single handedly developed in the succeeding decade many of the basic ideas on the mechanics of unconsolidated earth materials. The civil engineering profession was not long in appreciating his contribution, and today we see the second and third generation of soils engineers in command of a field that might have been geology's. The point is worth noting here, for on it hinges a problem to be mentioned below. Only a small percentage of geologists have been interested in studying the impact of geology on civil engineering, particularly within the cities. This minority was indeed miniscule up until World War II. For the New York-Boston area, I think of C. P. Berkey, T. Fluhr, W. O. Crosby, and Irving Crosby. There were others, but the fact remains that the geologists who made a career of working with engineers on urban problems were loners among their colleagues in the local universities. In this connection I should like to draw attention to a revealing fact or two. Both Crosbys taught in the geology department at MIT. Today the geology department of MIT does not employ an engineering geologist. In the Berkey volume published by the Geological Society of America in honor of Charles Berkey,[5] there is no article devoted to city geology as such. The undeveloped state of city engineering geology today is somehow reflected by these facts. This is particularly true in the eastern part of the country. This is sadly true in the light of the magnitude of the task that is before us.

The magnitude of the task is summarized by the following statistics used by the President of the United States in his message to Congress on "Cities," March 3, 1965. "Over 70 percent of our population—135 million Americans—live in urban areas. A half century from now 320 million of our 400 million Americans will live in such areas. And our largest cities will receive the greatest impact of growth. . . . Between today and the year 2000, more than 80 percent of our population increase will occur in urban areas. During the next 15 years, 30 million people will be added to our cities—equivalent to the combined population of New York, Chicago, Los Angeles, Philadelphia, Detroit and Baltimore. Each year, in the coming generation, we will add the equivalent of 15 cities of 200,000 each. . . . In the remainder of this century—in less than 40 years—urban population will double, city land will double and we will have to build in our cities as much as all we have built since the first colonist arrived on these shores. It is as if we had 40 years to rebuild the entire urban United States."

The dollar value of this work can be roughly estimated if we consider the cost of heavy urban construction at the present time. The amount of money spent on major urban construction in 1967 (from data in *Engineering News-*

Record, Jan. 25, 1968) is estimated to have been about 25 billion dollars. If we consider that 10 percent of this is the probable cost of foundations alone, we get some idea of the money value of good geological advice, for we know from experience that foundation costs can double and even treble if the geological conditions encountered are not those anticipated by the designer or the contractor. In tunneling and other forms of deep and essentially underground construction, the importance of ground conditions on construction costs is even greater. From my own experience in Boston and elsewhere, I estimate that at the present time at least 10 percent of current foundation costs, on a nationwide basis, is spent unnecessarily because of false starts and changes of plan resulting from unanticipated geologic conditions. From this we can get a rough monetary measure of the value of urban geology to our economy. At the present time, as a profession, we have the potential of effecting a national savings of something on the order of 300 million dollars annually. This will increase in the future.

Since the need for urban geology is great, it is time that geologists asked themselves, "What can we do in the urban crisis? What is the most effective role that we can play?" My own feeling is that first of all each of our cities should have large-scale geologic maps showing, in as much detail as graphic techniques will permit, the material factors of interest to the city. The reason that geologic maps are essential as the first step in a general campaign is that engineering geologists in private practice in cities (and there are some, although not nearly enough) are unable to find time for anything but the exigencies of their immediate assignments. Because of this, they welcome an overall picture of the geology that shows the full distribution, variations, and peculiarities of the earth materials underlying the city. What is generally needed is the broad picture—to help depict the forest as well as the trees. Such maps have been made for most large European cities and have proved their usefulness to planners and site engineers alike.

For these maps, how large is large-scale? Although that depends somewhat on the geological grain of the city and the density of data, as a general rule the maps should permit clear definition of individual building sites for large structures. This probably means that 1:12,000-scale would be a minimum and considerably larger scales would be preferable. The geological picture conveyed should be three-dimensional, and, depending on the geologic characteristics of the city, should show bedrock as well as surficial geology. Graphic means of depicting geology three-dimensionally are still in the experimental stage (recent Czechoslovakian maps are particularly interesting). As part of an expanded program in urban geology, the profession will have to give more thought to graphic techniques in the making of geologic maps.

Who will sponsor all of this geologic mapping? Means at our disposal are federal, state, county and city agencies, local professional societies and universities. Financial support from the Federal Government may be broadly required, just as it is for so much of our urban rebuilding at the present time. The U.S. Geological Survey is already embarked on an urban program. To date, work has been done, or is being done, in the following major cities: Washington, D.C.; Boston; Denver; Los Angeles; San Francisco; Portland, Ore.; Seattle; and Salt Lake City. In addition, research on specific problems connected with urban geology is being carried on in U.S. Geological Survey laboratories in Denver, and in Menlo Park, California.

Some geological mapping of cities is also being done by organizations other than the U.S. Geological Survey. In this category are the cities of: Los Angeles (mapped in part by the city); San Diego (State survey); Austin, Texas (State survey); and Kansas City, Mo. (University of Missouri). In addition, mapping (scale 1:24,000) has been completed for Chicago, St. Louis, and Minneapolis.

It has been estimated (C. G. Johnson, U.S. Geol. Survey, personal communication) that about 10 percent of the 7 1/2-minute quadrangles of heavily built-up city areas of the nation have been, or are being, mapped geologically. It should be noted that almost all of this mapping is on a scale of 1:24,000 or smaller. From this it can be seen that most cities are still without published or unpublished geologic information on a scale that is large enough to furnish planners, developers, and engineers with information in the detail that is needed. There is a good likelihood that existing programs will enlarge as realization increases of the importance of geologic knowledge to the most effective development of urban areas. It is not beyond the realm of possibility that within a decade means will be made available to have mapping parties in a hundred cities. This would mean that 500 to 1,000 geologists, most of them new to the profession, would be working in urban geology within ten years.

Where will these geologists come from? This brings up the problem of staffing and training. At the present time we would have difficulty staffing a sudden increase in urban geology mapping projects. Well-trained mappers are not presently available in sufficient numbers, and, even more important, many geologists are reluctant to exchange the open spaces for the confines of the cities. It appears to me that we are going to have to start from scratch and train a new generation of urban geologists. At the present time few universities have geological programs geared to an urban specialty. The additional demands of such a specialty are, over and above the usual concepts of geology, such technical specialties as soil mechanics, foundation engineering, and hydrology. The city geologist should be able to make himself understood by engineers, planners, and city administrators.

Moreover, the special requirements of this type of work may mean that we will have to reach out and bring in a new kind of student, one who until now has gone into engineering or some allied science. Such young people will, over and above learning to manipulate the tools and special mental aptitudes of field geology, find an especial fascination in being able to uncover nature deep under the artificial landscape of cities. The thought is promising because it may open a whole new reservoir of talent that up to now has left the pursuit of geology to others.

The conclusion to be drawn from these few remarks must now seem evident. Geologists must adjust to the era of cities, and cities must recognize the benefits to be derived from geology.

References

1. DANA, J. F. & S. L. DANA. 1818. Outlines of the mineralogy and geology of Boston and its vicinity, with a geological map. Amer. Acad. Arts, Memoir 4: 129–223.
2. TROOST, GERARD. 1826. Geological survey of the environs of Philadelphia. Philadelphia, Pa.
3. HITCHCOCK, EDWARD. 1836. Sketch of the geology of Portland and its vicinity. Boston Jour. Natural History 1: 306–347.
4. COZZENS, ISSACHAR. 1843. A geological history of Manhattan or New York Island. New York, N.Y.
5. PAIGE, SIDNEY, ed. 1950. Application of geology to engineering practice; Berkey volume. Geol. Soc. Am.

5

Reprinted from *23rd Intern. Geol. Congr. Proc.*, 12, 39-56 (1968)

Geology as Applied to Urban Planning: an Example from the Greater Anchorage Area Borough, Alaska*

ERNEST DOBROVOLNY and HENRY R. SCHMOLL
U.S.A.

Abstract: A geologic report on the Anchorage lowland, published in 1959 as U.S. Geological Survey Bulletin 1093 reached few, if any, individuals directly responsible for local planning. The report contains information useful in development planning and specifically identifies the Bootlegger Cove Clay, which underlies much of Anchorage, as an unstable material when wet, that can be dislodged along steep slopes by some triggering action. However, the bulletin cannot be considered as a warning to the local planning profession.

In consequence of the geologic effects of the Alaska earthquake and its impact on land development, local government officials requested the U.S. Geological Survey for additional geologic work deemed necessary for Borough-wide planning. To meet this request a general geologic map is being prepared of the Borough, which encompasses about 1,750 square miles. A general geologic map is basic to any rational understanding of the ground conditions, but does not, by itself, convey information to planners that is directly useful in their work. Special purpose interpretive maps are, therefore, being prepared to provide data in a form meaningful to planners. These include maps on mineral and construction resources, terrain, ground stability, surface water, and ground water.

Land development for human needs in Alaska is still in its infancy. Planning based on understanding of the geological environment can aid in anticipating and providing pre-development solutions to land-use problems now faced elsewhere by many large metropolitan communities.

Introduction

Geology commonly has been considered a study of rocks having as its main purpose the deciphering of past events, usually very remote ones. Although widely recognized as of great economic importance in the search for mineral and fossil fuel deposits, geology seemed to have little relevance to the general public and the everyday events of the present. Geologists, of course, have understood that geology deals as well with dynamic processes, and that these processes occur with as much vigor today as in the past. Except for application to engineering works of great magnitude, however, developers and the general public have often failed to recognize the relevance of these processes to many other facets of human endeavor involving use of the land.

---

* Publication authorized by the Director, U.S. Geological Survey.

It takes a dramatic and devastating event, like the Alaska earthquake of March 27, 1964, to forcefully draw the attention of the public to the dynamic processes of geology and the impact these processes can have on everyday life and on the economy and well being of a region. Although destruction from the earthquake was widespread throughout south-central Alaska, and some towns, notably Valdez and Seward, sustained much loss of life and property, the greatest volume of damage occurred in and around Anchorage, the largest city in the State and center of the greatest population concentration. Anchorage is located at considerable distance from the epicenter, but parts of the community were developed on areas particularly vulnerable to strong seismic shock of the duration encountered in the 1964 earthquake. Understandably, the first general public attention focused on geologic hazards associated with the seismic event, such as landslides, ground fracturing, generation of seismic sea waves, lurching, and shaking. This led in turn, however, to a wider if less immediate understanding of the role geology can and must play in the orderly development of our land and economic resources.

Existing geologic report

Anchorage was fortunate in having a geologic report covering the city and surrounding area (Miller and Dobrovolny 1959), which suddenly became a best-seller at least locally, and was widely cited as the chief source of background geologic information in the profusion of studies, geologic and otherwise, that rightfully accompanied the recovery from disaster. This new prominence for a geologic report, together with personal contacts with many geologists, served to make public officials aware of the importance of geology in fulfilling their responsibilities. The earthquake also caused a certain amount of soul searching on the part of the geologists, however, for it had become apparent that although the geologic report had correctly assessed potential dangers in the event of a strong seismic shock, little heed was paid to this warning. Obviously unless a different approach were adopted in the future there would be little need for preparing another geologic report only to have it go unused and unheeded. With this in mind it is instructive to review here the history of the old geologic report which emerged from a geologic mapping project started in 1949. A separate study of the occurrence, availability, and quality of ground water was started later the same year (Cederstrom, Trainer and Waller 1964; Waller 1964).

Anchorage was selected for an urban mapping project because the community was growing rapidly and because surficial geologic maps with engineering interpretations are of maximum usefulness in planning if available prior to a major period of development. Anchorage and vicinity was then the largest and by far the most rapidly growing community in Alaska. Much of the land surrounding the city limits was inaccessible to vehicular travel, but access roads were planned, and each year extended farther into areas suitable for homesteads and other development. And the more the surrounding land was opened up, the more Anchorage would grow as a center, with increasingly greater development of major support facilities. Thus data on sources of natural construction material, depth to bedrock, and foundation characteristics of unconsolidated surficial deposits were needed for the design of highways, railroads, bridges, airports, hydroelectric plants, water-distribution and sewage-disposal systems, and large buildings with special foundation requirements.

A map and brief description of the geology summarizing the fieldwork of 1949 was made available in 1950. The final report was published in 1959 as U.S. Geological Survey Bulletin 1093. In addition to the regular sales and exchange distribution, about 100 copies were distributed to individuals, post offices, and institutions in Alaska.

Bulletin 1093 contains a geological map at a scale of 1 inch to the mile (1:63,360), other illustrations, tables, and a text. The map shows 29 surficial and 2 bedrock units, the division made largely on the basis of age and origin. The text describes in some detail the fairly thick sequence of surficial materials upon which Anchorage is built and refers briefly to the bedrock exposed in the adjacent mountains. Of all the map units described, only one was given a formal name, the Bootlegger Cove Clay, a light-gray silty clay of Pleistocene age that is conspicuously exposed in the bluffs along Knik Arm. The physical properties and relationship of the Bootlegger Cove Clay to other units are presented in tables and diagrams. A text section on economic geology discusses known or potential mineral resources, construction materials, and engineering problems. Among the latter are: foundation conditions, excavation, slumps and flows, drainage, and frost heave. As an example of local conditions that should be considered in development, Figure 7 (p. 95) illustrates the relationship of groundwater movement to the upper surface of the Bootlegger Cove Clay where overlain by gravel, and the associated potential health hazard resulting from contamination of the water. Bluff recession caused by active shoreline erosion is documented in Figure 6. Conditions under which landslides can be activated in the Bootlegger Cove Clay are described under slumps and flows. These problems are closely related and in one way or another have a bearing on the geologic effects produced by the 1964 earthquake.

Awareness of the 1950 and 1959 reports was limited to parts of the engineering and geologic professions. The publications served as a guide for the Alaska Road Commission in selecting parcels of land underlain by sand and gravel to be set aside as a reserve for future highway construction needs. They were used as background information by foundation engineers in preliminary site studies for some of the larger buildings, and by geologists of the Alaska Department of Highways. The bulletin was quoted in a geological field guide with respect to geologic history, and has served as a teaching aid. For several reasons the bulletin was not used as background for planning. The planning department was relatively new and its early problems concerned more pressing matters. The report is a general treatment and did not zone or classify the ground except by geologic map units. The map is not a document the planner can use directly without interpretation by a geologist. There were no geologists on the planning staff.

After the earthquake of March 27, 1964, Bulletin 1093 was read by many and frequently cited for its warning about the susceptibility to landsliding in the Bootlegger Cove Clay. A report by the Engineering Geology Evaluation Group (1964, p. 8–10) established that all of the major landslides in the city of Anchorage were related to the physical properties of the Bootlegger Cove Clay and its distribution in relation to local topography. In a similar way Bulletin 1093 was used by the Scientific and Engineering Task Force of the Federal Reconstruction and Development Planning Commission for Alaska as a basis for preparing a preliminary map classifying "high risk" areas (Federal Reconstruction and Development Planning Commission for Alaska, 1964, maps, p. 59, 61). The principal criteria for the classification were the presence of the Bootlegger Cove Clay along steep slopes and fractures associated with landslides induced by the earthquake. The preliminary map was later modified as more definitive information derived from an emergency program of detailed geologic investigation became available.

It is evident that prior to the earthquake there was little or no communication between the authors of the geologic report and the planners and other public officials who could have used the information in it. The geologists produced their report, which was published in customary manner, and they went on to other assignments. The planners zoned the area under their jurisdiction for various and seemingly appropriate uses. The two groups went on their separate ways, essentially in ignorance of each other, largely because there was no common meeting ground for them. The geologic map was a classification of the various deposits largely in terms of origin and age; engineering characteristics that were given were largely buried in information that seemed nonrelevant to the unitiated (nongeological) reader. Furthermore, the report not only was not read by the nongeologist, but was probably not even known to him. Thus, the geologists were writing only for other geologists, as is their custom. The planners meanwhile were proceeding, as is their custom, without much consideration of the naturally occurring materials that underlie the land, without sufficient awareness of the importance of such a consideration, and without the understanding that they did need to know about the geologic map and its accompanying descriptions.

The events of the earthquake brought the two groups face to face, working in their own ways to repair the damage and look to the future. In this process the planners and other public officials recognized the need for geologic information, and the geologists saw that the conventional format of the standard geologic report was not adequate to meet this need. Consequently in any new geologic work to be undertaken, new products would have to be designed to present geologic information relevant to planning in a form that would be more readily intelligible to the nongeological professional worker. It was in this context that the Chairman of the Greater Anchorage Area Borough, a new political unit much more extensive than the city, asked the U.S. Geological Survey to begin geologic investigation that would encompass the entire Borough. Interpretive maps were especially requested, though the specific kinds of interpretations could not be defined at the time the request was made.

New geologic projects

The Greater Anchorage Area Borough is located in south-central Alaska (Fig. 1) and occupies most of a roughly triangular piece of land between the two estuaries at the head of Cook Inlet, Knik Arm to the northwest, and Turnagain Arm to the southwest. The western apex of this triangle and a narrow strip along Knik Arm lie within the Cook Inlet-Susitna Lowland physiographic province (Wahrhaftig 1965) and contain the city of Anchorage and its environs, including most of the land suitable for urban expansion. The remainder of the Borough lies within the Kenai-Chugach Mountains province, and is dominated by rugged, partly glacier-clad mountains with relief of as much as 6,000 feet; narrow valleys and some glacially planed shoulders provide only scattered and discontinuous areas suitable for extensive development. Thus, the greatest interest lies in a relatively small part of the Borough near the city of Anchorage, but as the Borough develops further, more and more of the marginally suitable land within the mountainous area will be brought into use, and it is vital to have basic information on this area at hand before unplanned opportunistic development is allowed to spread into unsuitable territory. The total area of the Borough is approximately 1,730 square miles (4,480 km^2), of which about 240 square miles (620 km^2) lies in the lowland and the remainder in the mountains; 330 square miles (855 km^2) of this part is underlain by glacier ice.

Of the two coordinated and in part interdependent U.S. Geological Survey projects now in progress in the Borough, one is directed toward the general geologic mapping of the area, with emphasis on the engineering aspects thereof, and the other toward the hydrologic aspects of the geology, especially the subsurface geology of the unconsolidated deposits in the lowland. The objective of the engineering geology project is the preparation of detailed and reconnaissance general-purpose geologic maps to provide data basic to understanding the physical features of the ground and to establish a framework within which rational land development plans can be formulated by borough officials. The hydrologic project aims to assemble and integrate the facts about the hydrologic system so that appropriate city and borough officials and their technical associates can design procedures for orderly development of the water resources.

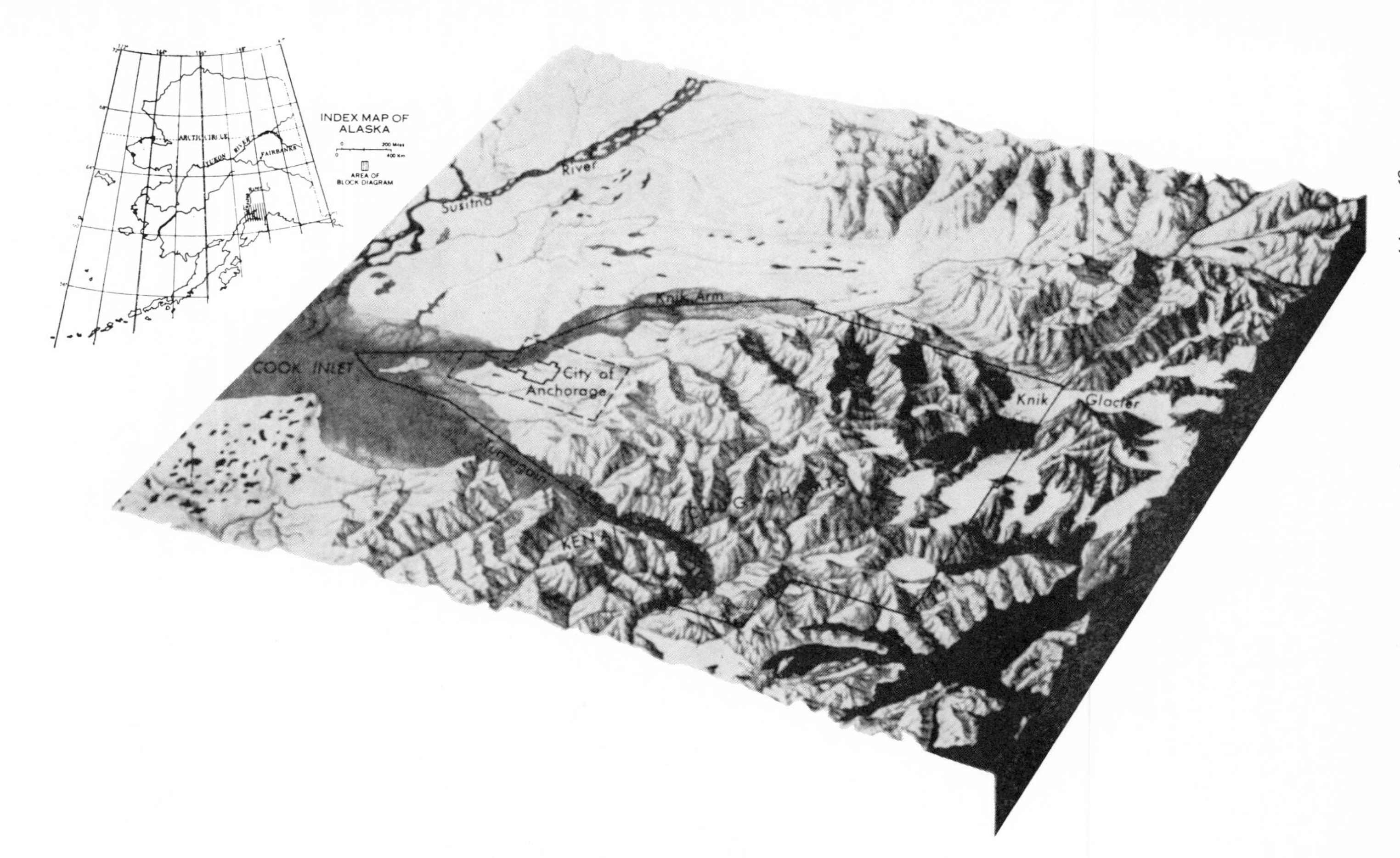

Fig. 1. Perspective diagram of upper Cook Inlet region showing physiographic setting of Anchorage
Boundaries of Greater Anchorage Area Borough and City of Anchorage shown by *solid line*; area of Figures 2, 3, 4, and 5 shown by *dashed line*

The use of geologic maps in the search for mineral and fossil fuel deposits is well known. Geologic maps and interpretations drawn from them are also widely used in the engineering profession, generally in connection with site selection and location of construction materials for major engineering projects. Geologists are usually employed on such projects to do the geological work. The use of geologic maps in regional planning, however, has not yet become standard practice, nor, with a few notable exceptions, are geologists normally included on planning staffs. This is perhaps because the field is rather new and still developing, and because the need for this service has not had wide enough demonstration. Geologic maps are recognized by the planning profession, but not significantly discussed in their publications (Chapin 1965, p. 254). There is further reason for neglect of geologic maps by planners. To the untrained user geologic maps are complicated; geologic information is usually plotted on a topographic base, requiring that the user see through a mass of three-dimensional data presented on a flat surface. Geologic units are shown by different colors and patterns to increase legibility, but on many such maps where contacts are intricate, the maps have the appearance of inscrutable clutter.

Nonetheless it is apparent that use of geologic maps would be of considerable benefit to the planner. Among the many considerations of regional land utilization planning are (1) development of the land for agricultural, residential, industrial, recreational, and other uses; (2) site selection for planning engineering works such as dams, bridges, highways, and other major structures; and (3) assessment and development of the hydrologic system for efficient use of water resources. Intelligent planning for land utilization therefore requires knowledge of features of the ground which may be categorized as topography, geology, and hydrology. Geologic maps on the topographic base, together with accompanying hydrologic information, provide just such knowledge, and are basic to any rational understanding of the ground conditions that the planner must have.

By itself, however, the geologic map does not convey to a planner information that is directly useful in his work. It is therefore necessary to interpret the geologic map for the planner. In parallel with the other uses of geologic maps cited above, it would be logical for the planning staff to include a geologist to interpret the geologic map at any time and for any purpose, and to participate fully in the planning process. This practice would indeed be desirable, and in the case of some large planning staffs it is followed; in the usual case of a smaller staff, however, this may not be practical. A useful alternative is the interpretive map, or a series of such maps, designed for the special needs of the planner, and based on the facts contained in the geologic map. It is this new and not at all standard approach that is being employed by the current U.S. Geological Survey projects in the Anchorage area, and interpretive maps are being produced to meet the needs of planners and other professional workers, and to some extent, the general public. Maps of this type have not had wide use in civilian application in the

United States, although similar maps have been used by the military for many years (Hunt 1950).

The primary product of the engineering geology project will be the general geologic map of the entire Borough at a scale of 1 inch to the mile (1:63,360). The hydrologic project ultimately plans to produce an electric analog or mathematical model of the hydrologic system. The model can be used to examine changes in the system that would result from various stresses applied, so that reasonable forecasts can be made of the effects of various methods of operating the system for years in advance.

Each interpretive map is restricted in coverage to single or closely related topics. All maps will have a similar format, and are designed to overlay each other for easy reference. The series will be designed for publication at a scale of about 1 inch to 2 miles (1:125,000) so that the entire Borough can appear on a single sheet of manageable size. Maps of selected topics of particular importance for the lowland area may have the larger scale of 1:24,000. Some of the topics to be covered in this series are:

Slopes
Construction materials
Foundation and excavation conditions
Stability of natural slopes
Recreation areas
Areas of potential development of ground and surface water
Availability of ground-water
Depth to unconfined aquifers
Depth to confined artesian aquifers
Water-table contours
Piezometric surface
Saturated thickness
Principal recharge areas
Chemical quality of water

Each project will also produce a final report in one of the Geological Survey's standard series of papers, in which the general geology and hydrology will be discussed, and the applications thereof summarized. This report will discuss only the topographic, geologic, slope, slope stability, and construction materials maps.

Topographic maps

The topographic map is an absolute necessity as a base on which to plot data about the land. Planimetric maps used in this report were constructed from a topographic map. Figure 1 illustrates the configuration of land surface in the area. The Greater Anchorage Area Borough is covered by topographic quadrangle maps

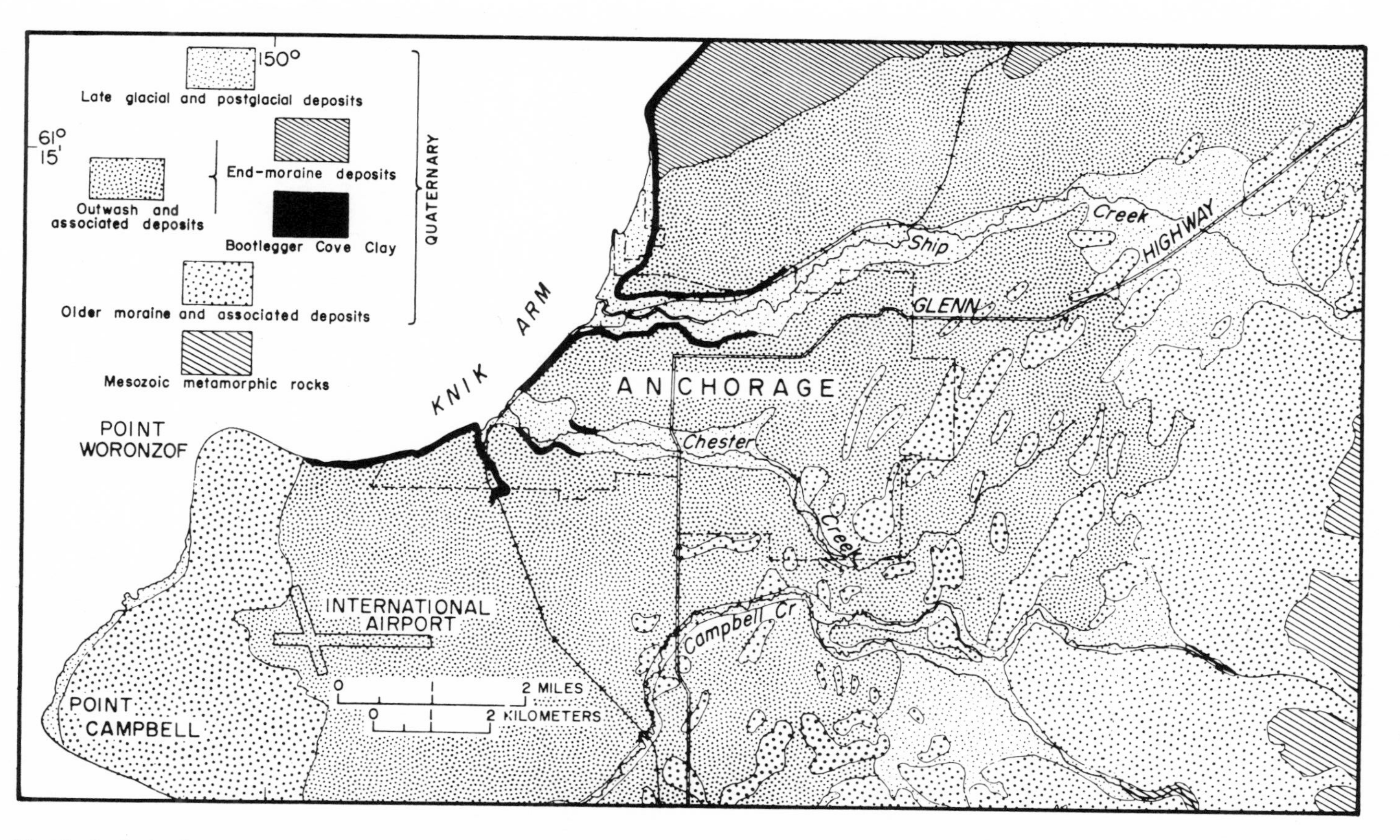

Fig. 2. Geologic sketch map of City of Anchorage and vicinity, Alaska

at a scale of 1:63,360 with a 100-foot contour interval in the mountains and 50-foot contour interval in most of the lowland. A special map at a scale of 1:24,000 with 20-foot contour interval covers the Anchorage metropolitan area in the lowland. Neither map shows changes caused by the 1964 earthquake. Separate topographic maps will not be included among the other maps in the series to be produced, as these maps are already widely distributed and readily available.

Geologic map

For most of the Greater Anchorage Area Borough a conventional geologic map at a scale of 1:63,360 is adequate to show the geology. Emphasis is placed on the surficial deposits, as these are quite widespread even in the mountains in this glaciated terrane, and comprise most of the materials most critical to development. The bedrock units, on the other hand, are relatively uniform over wide areas, and for the purpose of the current project are being mapped in reconnaissance fashion. The preliminary version of the geologic map currently under revision includes six bedrock units of Mesozoic and Tertiary age, chiefly eugeosynclinal sedimentary and volcanic rocks; 23 units of Pleistocene age, chiefly of glacial origin; and 28 units of Recent age, including glacial, alluvial, and colluvial deposits. Figure 2 is a simplified version of a part of this map, at reduced scale.

Slope map

The generalized slope map (Fig. 3) is a special kind of topographic map that summarizes the continuously variable slope information shown on the standard topographic map. Categories of slope can be chosen to meet various needs, and the map divided into units with the same slope characteristics. Such a map is necessarily generalized, because except at very large scales there usually are local slopes that do not fall within the assigned category. Inasmuch as the distribution of slopes is not random but systematic, however, it is practical to make a meaningful map. Though the slope map can be made directly from and entirely dependent upon the topographic map, its accuracy can be considerably enhanced by field observation and use of aerial photographs.

For planning purposes three primary categories of slope have been chosen: (1) slopes less than 15 percent; (2) slopes 15 to 45 percent; (3) slopes greater than 45 percent. In addition, the upper and lower categories have been subdivided, so that large areas with slopes less than 5 percent, and significant areas with many slopes greater than 100 percent are also shown. The most important division here is at 45 percent, which is the reasonable upper limit for maneuvering tracked vehicles. Figure 3 shows a part of the slope map that has been made for the entire borough.

The principal reason for making a slope map is to identify areas having the

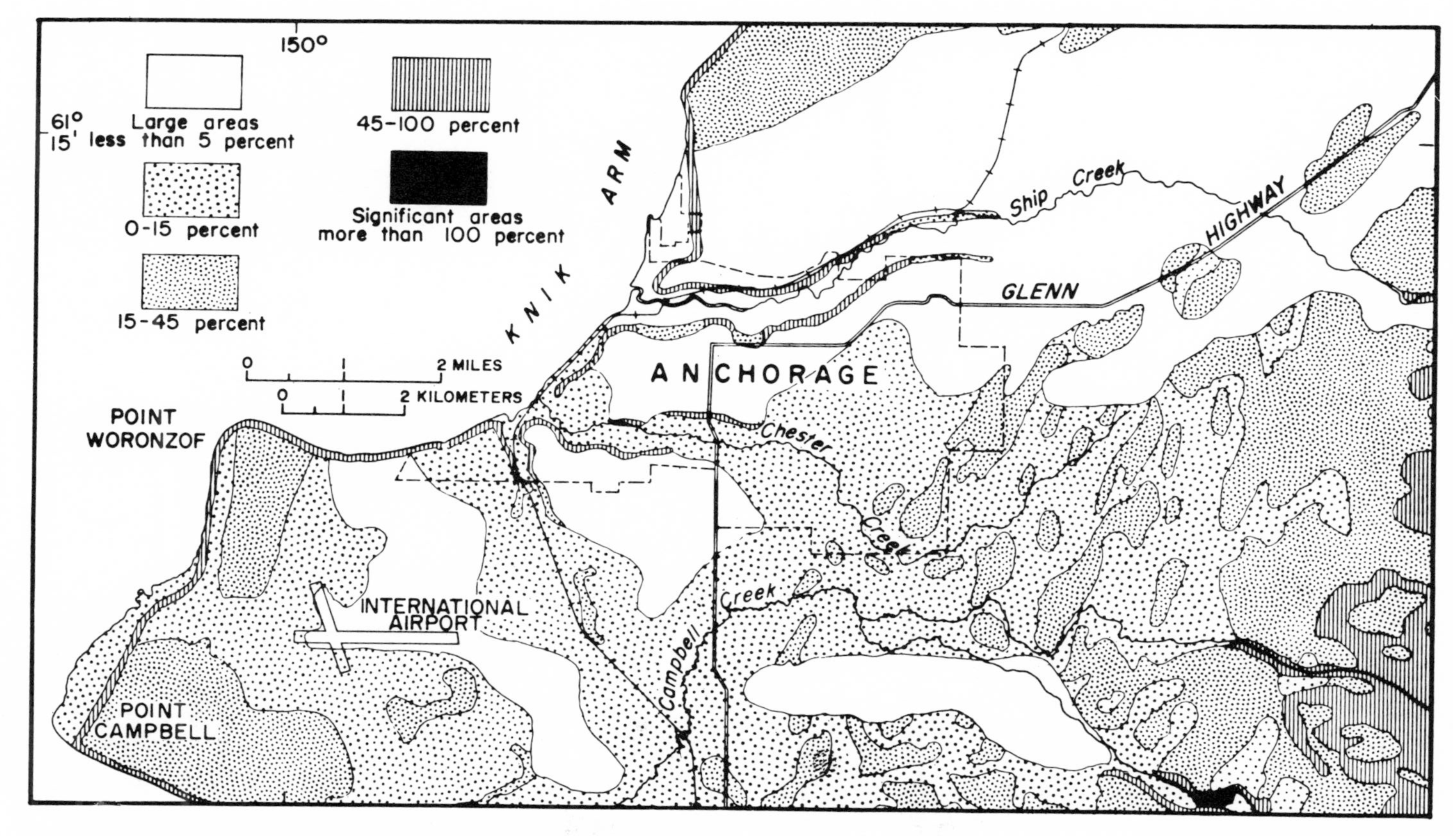

Fig. 3. Generalized slope map of City of Anchorage and vicinity, Alaska

same range of slope percent, because the angle of slope is a major factor in estimating slope stability. Landslides occur mainly on steep slopes and along sharp topographic discontinuities such as bluffs. They do not occur in areas of low relief that are far from breaks in topography. Identification of steep slopes is the first step in the process of isolating areas where landslides are more likely to occur. Used in conjunction with the geologic map, the slope map is a tool that leads to appraising the landslide potential of natural slopes.

Slope stability map

An interpretation of the stability of natural slopes in the Anchorage area is shown on Figure 4 and the map units are summarized in Table 1. The potential for landslides, especially in response to seismic shock, is of principal concern here, but other types of ground movement, such as solifluction and rockfall, are also considered. The map is derived essentially from the slope map and the geologic map. The primary criterion for determining instability is the degree of slope; areas of steep and very steep slope will be the chief sites of instability. However, the degree of instability depends considerably on the geologic material underlying the slope. Thus by combining elements of the two maps the slope stability map

Table 1—Description of map units for stability map of naturally occurring slopes

Map unit	Slope stability	Landslide potential	Solifluction potential	Slopes (percent)	Geology
1	High	Very low	Very low	<45	Undifferentiated
2	Moderately high	Generally low	Generally low	45–100	Generally stable bedrock
3	Moderate	Moderately low to moderately high	Moderately low	45–100	Medium- to coarse-grained unconsolidated deposits
4	Moderately low	Moderately low to moderately high	High	45–100	Soft, frost-susceptible bedrock, chiefly argillite
5	Low (This unit is not present in the area covered by Fig. 4)	Moderately high	High (chiefly rockfalls and snow avalanches)	Commonly >100	Bedrock, fractured and in part faulted
6	Very low	High	High (chiefly slumps and earthflows)	15–>100	Unconsolidated deposits underlain by sensitive clay and silt, or incorporating ice

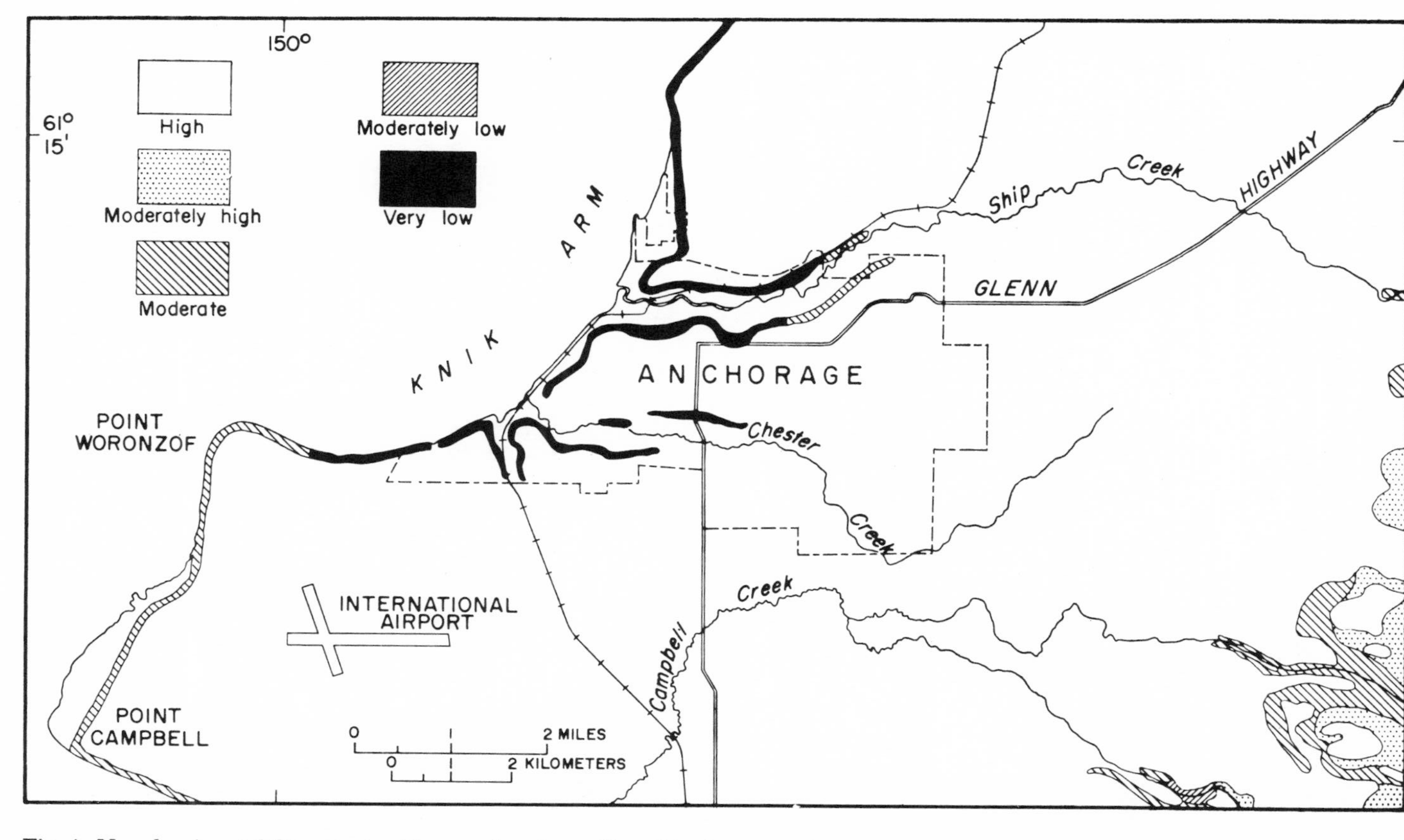

Fig. 4. Map showing stability of naturally occurring slopes, City of Anchorage and vicinity, Alaska Map units are described in Table 1

can be prepared, showing varying degrees of stability. Generally the areas of low and moderate slopes are lumped together as having high stability. Parts of these areas adjacent to zones of low stability, however, may have potentially low stability because, for example, of the possibility that a landslide can work back from an area of steep slope and "eat" into an adjacent area of low slope; this actually happened during the 1964 earthquake. Also, the low slopes extending outward from the base of a bluff may be overridden by landslide debris or disturbed by pressure ridges. Such features cannot be projected on the scale of a map used here, but the potential problem area is outlined as to general location.

Figure 4 clearly shows (a) those areas which are unsuited for most development because of slope stability problems that cannot feasibly be surmounted except in unusual circumstances (units 5 and 6); (b) those areas in which slope stability presents problems that must be considered in planning and design but for which a solution can be worked out (units 3 and 4); and (c) those areas which are relatively free of slope stability problems. It must be emphasized, however, that the map is generalized, that locally within each unit the described situation may not exist, and that for particular development detailed site investigations must be made. The map is designed primarily to delineate those areas in which particular problems can be expected.

Construction materials map

The construction materials map is one of the most direct interpretations that can be made from the geologic map. It is essentially a lithologic map, in which geologic units of various ages and origins that have similar lithologies, and therefore similar utility as construction materials, are grouped together in a single map unit. In some cases single units on the geologic map have been split into more than one lithologic unit, reflecting usually a gradational change of material within the geologic unit. Sometimes such a breakdown within a geologic unit requires additional fieldwork, and even so may have to be made arbitrarily. Thus, the transformation is not purely mechanical but requires refined geologic interpretation.

A construction materials map has been made for Anchorage based on the geologic map published in Bulletin 1093, and is confined largely to the lowland area; it has been produced at a scale of 1:24,000 and required additional field checking and laboratory analysis beyond what was done in the first survey. Six of the seven map units are for unconsolidated materials; the seventh includes all bedrock, which covers only a small part of the area and has no particular utility as a construction material. The six unconsolidated units range from gravel and commonly interbedded gravel and sand, which are of great economic importance in the area, to sand, silt, till, and swamp deposits which have relatively little utility. During the current project the entire Borough will be so mapped at smaller scale, and the

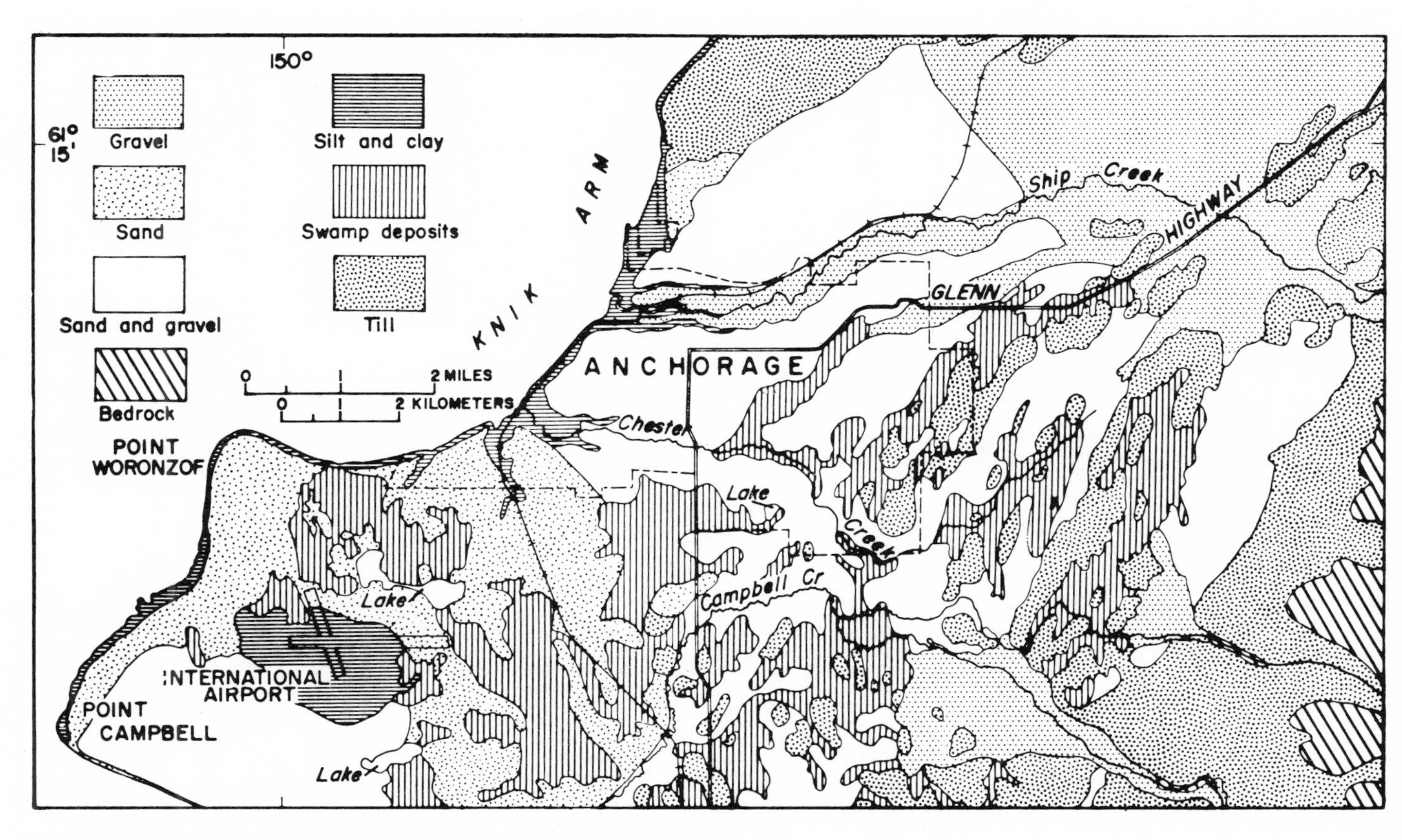

Fig. 5. Construction materials map, City of Anchorage and vicinity Alaska

large-scale map may be further refined. Figure 5 shows a part of the small-scale map.

On such a map, areas that are of interest for potential use as sources of granular construction materials can be readily seen. Though much of the land has already been preempted for other purposes, this map can still be used to select from the unoccupied areas those deposits which appear to warrant more detailed investigations. The more desirable deposits can be reserved for future use and, as part of a systematic plan, exploited prior to development of the land for other uses. It is not the intent of the map to serve as a resource map in the sense of estimating gravel reserves, or detailing the precise uses for which a particular deposit is most suited. More detailed site investigations are required for such determinations. The map does serve primarily as a guide so that the planner can know which areas are to be selected for further investigation, and which have no construction materials and can be considered for other types of development.

Recreation areas

It has been considered appropriate to present a map showing areas particularly suited for recreational purposes, particularly those areas that are of interest for their scenic beauty or natural history, but including other areas that are less spectacular but well suited for such activities as hiking and camping, and able to accommodate large numbers of people. Chiefly esthetic considerations are involved here, and these are necessarily subjective in nature. They can less readily be derived precisely from measurements of slope or behavior of geologic materials. Nonetheless, elements of topography, geology, and hydrology play a major role in determining which areas have high esthetic appeal, and we regard that in any study of these fields such considerations should not be overlooked.

The Greater Anchorage Area Borough is fortunate in having many areas of unusual scenic interest, some of which are widely known and frequently visited; others are presently inaccessible to most people and totally undeveloped. Of special interest is Lake George, a multistage glacial lake that forms when water from a tributary valley is impounded behind Knik Glacier in the spring. In early summer the water overtops the glacier, and the river cuts a narrow gorge between steep rock and ice walls, almost emptying the lake.

The recreation potential map is intended as a guide to the planning staff, to aid in assigning priorities to development and determining degree of development that should be permitted, so that the recreation potential of the area can be realized in an orderly manner.

Conclusions

There are many elements of an economic, social, and political nature that must be considered in the planning process; these have been deciding factors in the past, and no doubt will continue to be so. Nonetheless, the physical environment deserves increased consideration, especially in areas where that environment is not as hospitable as it is in other areas where more planning experience has accumulated. This is true of large areas in the western United States that are in a general way characterized by dynamic geological processes of greater magnitude and intensity than those in the more highly developed East. It is also true, however, that in the older established cities the less problematical areas have been developed first, and those areas into which new development will extend are more likely to present physical problems. Thus there is need for wide application of the principles of planning in this era of rapid urbanization, and need for these principles to include consideration of the natural physical environment. This is true for those areas that are already well developed such as the United States, and also for the less well developed nations that will grow at ever increasing rates in the near future. And the greater the intensity of development, the greater the need for this type of planning.

In order that considerations of the physical environment gain greater prominence, geologists and others who study the physical environment should promote understanding of their work. They should meet with planners and other responsible public officials on a personal working - level basis, they should learn the jargon and needs of those individuals, and they should design their own products to be of optimum use. This statement does not imply that there should be any lessening in their own scientific pursuits, without which their work would have less and less value, but the geologist does have to bear the added burden of carrying his work to the user so that its potential utility can be fully realized. It is with this goal in mind that we believe the work of the geologic and hydrologic projects in the Greater Anchorage Area Borough can make a contribution to the wider application of the geological sciences in the expanding area of orderly urban development.

References

Cederstrom, D. J., Trainer, F. W. and Waller, R. M. (1964): Geology and ground-water resources of the Anchorage area, Alaska. U.S. Geol. Survey Water-Supply Paper 1773, 108 p.

Chapin, F. S., Jr. (1965): Urban land use planning (2d ed.). Urbana, Univ. Illinois Press, 498 p.

Hunt, C. B. (1950): Military geology, *in* Paige, Sidney, chm., Application of geology to engineering practice. Geol. Soc. America, Berkey Volume, p. 295–327.

Engineering Geology Evaluation Group, 1964, Geologic report – 27 March 1964 earthquake in Greater Anchorage area. Prepared for and published by Alaska State Housing Authority and the City of Anchorage, Anchorage, Alaska, 34 p., 12 figs., 17 pls.

Federal Reconstruction and Development Planning Commission for Alaska, 1964, Response to

disaster, Alaskan earthquake, March 27, 1964. Washington, U.S. Govt. Printing Office, 84 p.
Miller, R. D. and Dobrovolny, E. (1959): Surficial geology of Anchorage and vicinity, Alaska. U.S. Geol. Survey Bull. 1093, 128 p.
Wahrhaftig, C. (1965): Physiographic divisions of Alaska: U.S. Geol. Survey Prof. Paper 482, 52 p. [1966].
Waller, R. M. (1964): Hydrology and the effects of increased ground-water pumping in the Anchorage area, Alaska. U.S. Geol. Survey Water-Supply Paper 1779-D, 36 p.

[*Manuscript received July 28, 1967*]

6

Reprinted from *24th Intern. Geol. Congr. Proc.*, 13, 30-36 (1972)

Urban and Environmental Geology of Hamburg (Germany)

F. GRUBE,
F. R. Germany

ABSTRACT

The ground available for building in Hamburg is underlain by more than 400 m of Quaternary and Tertiary unconsolidated materials. The Board of Works must use various methods for the construction of new underground tubes, river tunnels for express highways, etc., in the city, because of the multiplicity of weak unconsolidated layers. These layers must be carefully investigated in order to prevent damage to new and older buildings. Middle-aged town-moats are filled with anthropogene and other sediments, up to 20 m in thickness. Peat and mud from the Holocene and Eem-interglacial are deposited in depressions up to 20 m under the surface area. Marine and glacio-lacustrine clays may be about 100 m thick. A series of structure maps at scales of 1:250 or 1:1000 were made, showing all important formations in large areas of the city of Hamburg; about 20 km of quarry walls were mapped at 1:100.

To obtain the highest security against hydraulic subsidence, it is necessary to investigate all aquifers. Sudden intrusions of water into excavations in complicated moraine-systems occurred. If older buildings on wooden piles are within the cone of influence of the wells it is necessary to create new foundations. In open excavations, especially in shield tunnelling, large erratic blocks in the boulder clay are a great obstacle.

Areas planned to be waste dumps must be intensively mapped by geologists. In order to recognize the danger of contaminating groundwater, all regions where polluted water could reach lower groundwater zones are being explored.

GEOLOGICAL AND GEOMORPHOLOGICAL MAPPING, structural geology, hydrology and engineering-geology are of importance for urban geological work in the area of a modern city. In Hamburg, problems of engineering and environmental geology and hydrogeology are most commonly investigated.

GEOLOGICAL MAPPING

Because of the traffic situation, the building of new underground stations must take place in the centre of a city. When new projects such as underground railways, river tunnels for express highways, underground garages, etc., are planned in an area with difficult geological structures and soils (unconsolidated material, high groundwater, etc.) the subsurface situation has to be investigated very carefully in order to prevent or reduce damages to older buildings by soil compaction, landslides, etc.

In the centre of Hamburg, an underground station for three railway lines is being built in an area (Fig. 1) with upper Miocene solid marine clays, moraines of several glaciations of northern Europe with different physical properties, sand horizons with groundwater and compressible soils (peat, mud, etc.). An exact geological study (for example, thickness and expansion of peat and mud in the range of meters) is more complicated in an urban locality than in an open area, especially by drilling. In cellars down to a depth of 21 m, drilling done by hand has proved useful, whereas drilling machines can be set only in a few places to bore in streets, footpaths, parking places, etc., because of a dense system of supply pipes, etc. In waterproofed basements, it is not possible to explore the underground directly. Seismic and geoelectric methods cannot be used because of soil disturbance, stray currents, etc. On the other hand, static penetration tests and gamma measurements proved successful, whenever it was possible to compare the results with those of drilling in the neighbourhood.

All geological facts are recorded on structure maps, which have been especially developed for all geological horizons in Hamburg. The three-dimensional structures of sedimentary rocks cannot be understood correctly without intensive studies of the geological development of an area and of geological processes such as dissolution of gypsum, fluvial erosion, glacial disturbances, etc. These data must be presented at a scale of 1:1000 or 1:250 because of the abundance of data and the accuracy demanded by the building engineers. Operations with computers and drawing machines for the construction of isolines of base or thickness are projected. It was possible in smaller areas to obtain an accuracy in the range of meters with structural maps. This is important for shield tunnelling or beneath houses with unstable foundations.

HORIZONS WITH COMPRESSIBLE SOILS

An important task of urban geology is the mapping of compressible soils so as to prevent damage to older houses, etc., before beginning construction. From experience, damage by differential soil movement is not rare in areas of older houses. In the city of Hamburg, three horizons with compressible soils of organic origin are known.

(1) Anthropogene sediments in middle-aged town-moats, up to more than 20 m thick (mud, peat, boulder clay, etc.);
(2) Holocene and Weichsel-interstadial peats, muds, humic clays and sands in valleys, potholes, etc.;
(3) Eemian interglacial peats, muds, etc., in river-beds and other depressions from the Saalian glaciation (Riss).

Settlement can be calculated for the foundation of a new building if the nature and thickness of the compressible soils are uniform. If, however, thickness of a weak layer is variable in the area for a new foundation or if interfingering of various soils (peat, mud, clay, humic sand, for example) occurs, the soil must be replaced or a pile foundation must be used. A decision about the method of foundation is possible only after accurate geological investigations. In contrast to the situation in open land, replacement of soils in the urban situation is uneconomical because of the high costs of securing houses in the neighbourhood and for the construction of excavation-walls.

In several cases weak layers were only discovered after the beginning of earthwork. Time-lag and additional high costs (groundwater wells, securing of excavation-walls, etc.) must be allowed for.

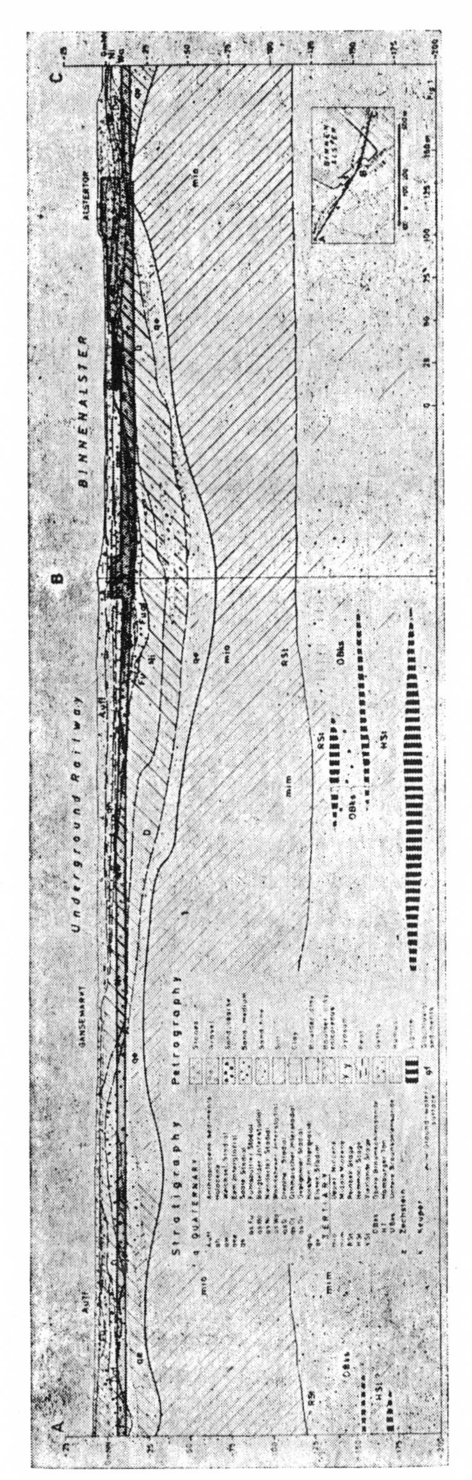
A
B
C
Underground Railway
BINNENALSTER
Stratigraphy
Petrography

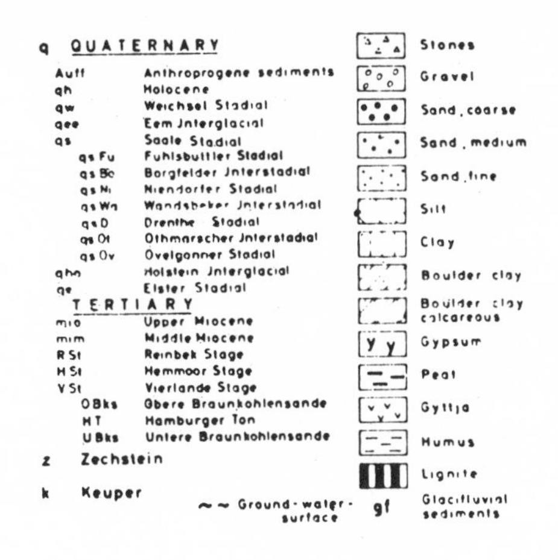
q QUATERNARY
Auff Anthropogene sediments
qh Holocene
qw Weichsel Stadial
qee Eem Interglacial
qs Saale Stadial
qs Fu Fuhlsbuttler Stadial
qs Bo Borgfelder Interstadial
qs Ni Niendorfer Stadial
qs Wa Wandsbeker Interstadial
qs D Drenthe Stadial
qs Ot Othmarscher Interstadial
qs Ov Ovelgonner Stadial
qho Holstein Interglacial
qe Elster Stadial
TERTIARY
mio Upper Miocene
mim Middle Miocene
R St Reinbek Stage
H St Hemmoor Stage
V St Vierlande Stage
OBks Obere Braunkohlensande
HT Hamburger Ton
UBks Untere Braunkohlensande
z Zechstein
k Keuper
Ground-water-surface
Stones
Gravel
Sand, coarse
Sand, medium
Sand, fine
Silt
Clay
Boulder clay
Boulder clay calcareous
Gypsum
Peat
Gyttja
Humus
Lignite
gf Glacifluvial sediments

FIGURE 1.

A zigzag, leveled middle-aged town-moat of 25 m width was first discovered during construction of a new railway tunnel between two frequently used railway tracks. The driven H-steels (Rammträger) of the Hamburg strutting were not in the natural boulder-clay. A special steel-support construction was used to restore the security of the excavation. During the earthwork for a skyscraper an Eemian interglacial peat, about 30 m wide with a thickness of 5 m, was discovered when a lorry sank in the soil. The peat was replaced by sand to stabilize the ground.

After the beginning of construction of the new underground main station of Hamburg-Nord, an Eemian interglacial peat was also discovered, in a round depression of Saalian dead ice, with a diameter of 50 m. After intensive mapping of the underground it could be proved that some of the buildings bordering the excavation are resting on boulder clay on one side and on peat 5 m thick on the other side. Extensive remedial work, such as placing of new foundation piles, may be necessary to avoid further sinking of the buildings (compaction up to 0.3 m occurred at the outset).

If a humic horizon occurs in the underground and is not discovered during the placing of the foundation and building, damages to buildings are possible. A railway bridge in Hamburg sank about 1 m, 40 years after completion. New drilling showed mud of Eemian interglacial material beneath 12 m of sands of the Weichsel-stadial. The primary drillings were only 10 m deep.

For a pile foundation it is necessary to know exactly the base of the horizon with compressible soils and the condition and thickness of the solid soil beneath the unfavourable layers, in order to determine the load-limit of the piles. The humic sediments of the Holocene and Eemian interglacial can be mapped relatively easily in the Hamburg area because of simple layering. The deposition of pushed peat- and mud-lenses of the Holstein-interglacial is however, very complicated, as they are scattered throughout the Saalian moraines. Glaciolacustrine sediments (Lauenburger Ton) have a thickness up to 80 m in the glacial canyons (Rinnen) of the Elster-glaciation; a pile foundation is thus not economical for skyscrapers, etc. If glaciolacustrine sediments are discovered in open excavations, the slopes can be stabilized by reducing slopes, driven steel H-beams, drainage, etc.

HARD LAYERS, BOULDERS

Local consolidation by calcium or iron is practically without importance in the area of Hamburg. On the contrary, petrified strips in the mud layers of the Eemian interglacial can simulate solid rock. A characteristic sign of all moraines is the abundance of various boulders of granite, gneiss, etc. up to 3 m in diameter. The presence of such boulders in open excavations creates difficulties for earthwork; demolishing of driven H-piles and driven H-beams, water-intrusions, blasting of stones in close quarters, problems of transport, etc. Large boulders are especially unwelcome surprises in shield tunnelling for underground railways and other tunnels (blasting of stones in the earth, partly outside the shield, stopping of the work, driving problems, etc.). Compaction of soils (sands, moraines, etc.) was measured up to 20 mm around full mechanical shields and up to 100 mm around hand-shields.

KARSTIFICATION

A salt dome must be investigated up to the base level of karst erosion if large tunnels, skyscrapers, etc. are proposed on the top of gypsum, salt, etc. In the first stage of the building of the new Hamburg-Othmarschen hospital several borings

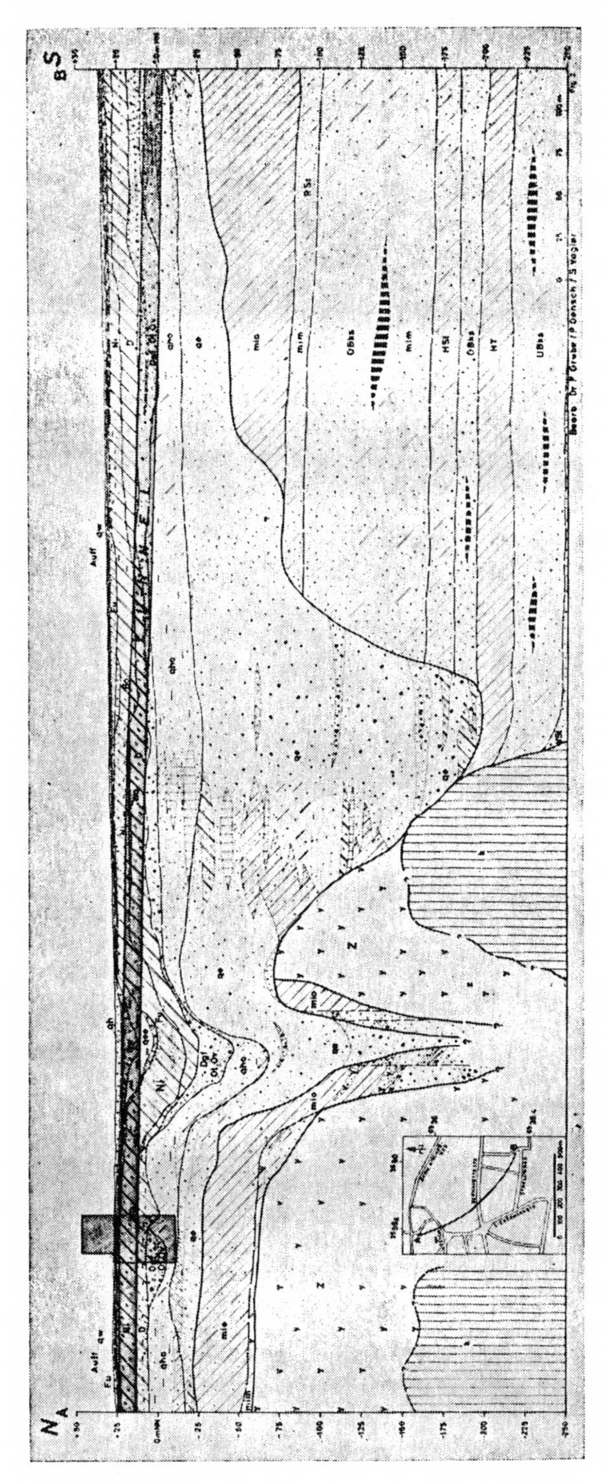

FIGURE 2.

up to 300 m were undertaken in order to study the condition of the Zechstein-gypsum series and the karstification. With seismic measurements, subterranean karst hollows were discovered. The southern border of the Othmarschen-Langenfelde salt dome was also studied (Fig. 2) with drillings up to 300 m in order to decide the building methods (tunnelling shield outside the salt dome and open excavation with ferro-concrete on the top of the gypsum) for a new Elbe River express highway tunnel (length 3209 m; three tubes of 10.8 m diameter each).

GROUNDWATER AND ENGINEERING GEOLOGY

For the floating method of buoyant tunnel elements (for example the southern part of the new Elbe tunnel) it is not necessary to drill groundwater wells, but it is indispensable to have a dry excavation for the traditional building methods. It is very practical for the foundation of a new building to collect the water outside the excavation by groundwater wells. An accurate study of all aquifers is necessary in order to have a rational system of wells. Especially the coarse base of an ancient river bed is suitable to collect the drainage water. The so-called residual water must be pumped out by vacuum wells or drained. A rational system of wells is dependent on the geologic structure, the permeability of soils, and the time between beginning of water pumping and beginning of earthwork. Hydraulic calculations are wrong if a thin strip of aquifuge is overlooked, for example silt layers of a few cm in glaciofluvial sands and gravels. Sand and gravel-lenses in moraines require a close net of drill holes with a spacing of 10-25 m (groundwater and sediment intrusions up to 700 m^3 have happened in shield tunnelling. A special problem of urban geology is the choice of the location of groundwater wells. Because of the close settlement, it may not be possible to place wells in the optimum geological location (set of wells inside an excavation!).

Before undertaking water-pumping, the investigation of the natural groundwater level is very important for deciding whether dewatering is necessary or not. The cone of influence of the wells is dependent on the complicated structure of the Quaternary sediments. Continuous measurements of water-levels are drawn on piezometric maps in order to follow the development of the cone of influence. Damages to agriculture, forestry and gardens are possible in the area of a cone of influence. Shallow groundwater wells can also be dried up in the area. In the regions of cities, the built-up areas must be observed, especially those within the first zones of a cone of influence, with compressible strata like peat and mud of various types and thicknesses, and with a direct aquifer contact in the subsurface. The presence of such critical zones in the neighbourhood of a large excavation is dangerous for many buildings with unstable foundations because of the movement of soils. For example in the centre of Hamburg-City (Fig. 1) such preventive measures as injections into the underground of the buildings, elevation of the streets, changing of endangered supply network and pipes, repair of macadamized roads, irrigation of weak humic layers, etc., were required during construction of an underground railway. In spite of all difficult preliminary investigations, we have experienced considerable damage by sudden inflows of groundwater, where artesian aquifers were not recognized. The boulder-clay of the Drenthe-stadial (Saalian-glaciation) is especially very dangerous for open excavations. Practical experiences have demonstrated that this firm boulder-clay can suddenly lose its structure under the influence of groundwater and flow into excavations as a mud stream! In one such case, the driven steel beams (Spundwände) of an excavation in the city of Hamburg were destroyed and the driven H-steels (Rammträger) of the Hamburg strutting were bent.

ENVIRONMENTAL GEOLOGY AND HYDROGEOLOGY

The urban geologist, besides his contribution to the preservation of valuable landscapes and individual natural monuments, like tunnel valleys, end moraines, eskers, dolines, etc. as recreation resorts in the near surroundings of cities, must also carry out exact geological mapping in areas, in which refuse and industrial waste depots are planned. If no danger of contamination of groundwater exists, which is indicated by clay and silt layers of large thickness in the underground, monotonous outwash plains or marsh plains can be transformed in close collaboration with landscape architects into attractive landscapes or peaks by heaping up waste and thus creating artificial hills.

An important task of environmental geology is the recognition of "windows" with large thicknesses of sands and gravels (absence of cover-moraines), in which polluted water can reach lower groundwater zones. Industry and inhabitants of Hamburg are supplied with fresh water only from the aquifers of Miocene (Braunkohlensande) and of the Quaternary Elster-glaciation (glacial and glacifluvial "Rinnen", channels of glacifluvial streams, up to 400 m under the surface). Water works of Hamburg pump about 15 millions m^3/a and industry about 55 millions m^3/a, none of which is obtained from the polluted water of the Elbe or other rivers, by aquaduct from the Harz of Scandinavia, or from fresh-water regeneration of salt or sea-water.

7

Reprinted from *Geological and Geographical Problems of Areas of High Population Density* (Proc. Symp. Assoc. Eng. Geol.), 1970, pp. 27-29, 31-33, 45-56, 59-62

GEOLOGICAL AND GEOGRAPHICAL FACTORS IN THE SITING OF NEW TOWNS IN GREAT BRITAIN

Peter Hall
University of Reading, England

ABSTRACT

Against a brief introductory background on their history and administration, this paper makes a systematic comparative analysis of geological and geographical data on the 28 British new towns existing in summer 1970. The analysis considers date of designation and purpose for which the new town was designated, location of the towns in relation to the major conurbations (urban agglomerations) of Great Britain and to national and regional communications systems, size of towns in terms of their base populations (existing populations at date of designation) and planned target populations, character of the sites, geological basis of the sites with special notes on some new towns where special geological problems were involved, and the agricultural value of the land. The general conclusion is that few British new towns have experienced any physical difficulties with their sites but, where these do occur, it is necessary for the community to evaluate these carefully against the value of agricultural land on alternative sites.

Introduction

The British new towns have excited worldwide interest, not least in North America, as one of the boldest experiments in social planning carried out during the twentieth century within the framework of a democratic society. Statistically, their achievement is impressive enough. By September 1969, only 23 years after the 1946 New Towns Act which allowed them to be brought into being, 15 new towns were almost completed and another 12 had started building (a thirteenth, Central Lancashire, was designated in March 1970); 191,000 new homes had been built in new towns, 180,000 of them by the public developmnet corporations and 11,000 by private enterprise; there were 48 million square feet of factory floorspace, 4.3 million square feet of office floorspace, and 6.5 million square feet of shopping space; and altogether 945,000 people lived in new towns, excluding those where the Development Corporation had not yet started building (1). Thus in less than a quarter of a century, the British new towns have already reached a point where 1 person in every 60 lives in a new town. By 1990, this proportion should be considerably greater.

The purpose of this paper is to make a systematic comparison of geographical and geological data about the 28 British new towns existing in the summer of

1970. First, however, it may be useful to give some preliminary background about the purpose and the organization of the new towns. Their origin lies in a remarkable book published by a social reformer, Ebenezer Howard, in 1898. Originally called *Tomorrow: A Peaceful Path to Real Reform,* it was republished four years later under the title by which it is generally known, *Garden Cities of Tomorrow* (2). Surveying the evils of the overcrowded Victorian cities and the stagnation of the countryside (at that time suffering from agricultural depressions due to the influx of cheap foodstuffs from the New World), Howard called for the creation of Garden Cities that would be self-contained communities outside the existing urban agglomerations. They would contain both homes and workplaces and would be planned within permanent green belts that would provide a sound basis both for agriculture and the recreational needs of the inhabitants of the Garden City. A practical reformer, Howard worked until his death in 1928 for the establishment of Garden Cities through private capital but succeeded in establishing only two: Letchworth (began 1901–1902) and Welwyn Garden City (1919–1920), both in the county of Hertfordshire north of London. Howard also founded the Garden Cities Association, later renamed the Town and Country Planning Association, to campaign for garden cities to be established by government action, but, up to World War II, the call was not heeded.

During the war, however, a radical change in official attitudes took place. The Royal Commission on the Distribution of the Industrial Population (the Barlow Commission), which was set up before the war in 1937, reported in 1940 recommending a policy of restricting further suburban growth of the major urban agglomerations and controls on the location of new industry, at least in certain areas, together with the establishment of an effective system of town and country planning (3). Further official reports, which appeared during the war, called for a radical reform of the issues of compensation and betterment as they affected town planning (4) and for more effective policies for protecting valuable agricultural land (5). A Ministry of Town and Country Planning was set up in 1943 in the middle of the war. The idea of new towns as an answer to the problem of urban growth was generally accepted, and, in 1945, a committee was set up under Lord Reith to consider:

> The general questions of the establishment, development, organization and administration that will arise in the promotion of New Towns in furtherance of a policy of planned decentralization from congested urban areas; and in accordance therewith to suggest guiding principles on which such Towns should be established and developed as self-contained and balanced communities for work and living (6).

The essential characteristics of garden cities (or new towns, as they then came to be known), as set out in these official terms of reference, are exactly as put by Ebenezer Howard in 1898: they were mainly to receive the *overspill of population* from congested urban areas where slum clearance would reduce population densities, and they were to be *self-contained and balanced communities* providing both homes and employment within their boundaries to reduce the burden of commuting, with a social balance of people from different social and economic groups of the population. These have been the ideals of the garden cities movement throughout its history to the present day.

The recommendations of the Reith Committee were embodied in the New Towns Act which Parliament passed in 1946. New towns were to be set up and built by autonomous Development Corporations appointed by the Government, one for each new town, and financed by the central Exchequer. Thus they were to exist side by side with the local government structure in each area. This was held to be necessary because the democratic local government system would still be necessary to provide ordinary local services but could not be expected to shoulder the burden of a building operation of this scale. The Development Corporation is a typically British administrative device very similar to the corporations which run nationalized industries, such as the railways, the Post Office, the coal mines, the electricity and gas industries, and BOAC. It is a public corporation that is free from detailed Parliamentary interference in its day-to-day management, though it must submit an annual report to Parliament for debate. The new structure for the United States Post Office, perhaps, provides a useful parallel. One detail left unsettled in the 1946 Act was the management of the new towns when the Development Corporations had completed them. It was generally imagined that the assets would revert to the local authorities concerned, but, in 1957, the Government resolved instead to transfer them to yet another public corporation, the Commission for the New Towns. Four towns have so far been transferred to the Commission on being officially completed.

[*Editor's Note:* Material has been omitted at this point.]

There are some other new communities in Britain that are new towns in fact if not in formal title. They are towns built since World War II by administrative agencies other than the Development Corporations. One group of these consists of the so-called expanding towns, which are existing towns expanded by agreement between their local authority and some other local authority in an area wishing to export population, usually in one of the larger urban agglomerations. The machinery for this, which provides for Government assistance, was created by the Town Development Act of 1952. Though many of the expanding towns are small and their scale of expansion modest, some are ambitious enough schemes to rank with new towns. They include the town of Winsford in Cheshire, which is being expanded by agreement with authorities on Merseyside, the towns of Basingstoke and Andover in Hampshire, and the town of Swindon in Wiltshire, all of which are being greatly expanded by agreement with the Greater London Council. Yet another group, a small one, consists of new towns developed solely by local government for the solution of local problems. These include the towns of Cramlington and North Killingworth in Northumberland, which were developed to accommodate overspill population from the Tyneside conurbation, and the ambitious new town of Thamesmead, which was developed by the Greater London Council on previously undeveloped marshland within the Greater London conurbation.

The analysis that follows, however, is restricted to the new towns in the strict sense: those built by Development Corporations and still either managed by them or transferred to the Commission for the New Towns. In England, Wales, and Scotland, there were 28 of them by summer 1970.

Date of Designation and Purpose

Designation of a new town is the making of an order by the relevant Minister, who is the Minister of Housing and Local Government in England, the Secretary of State for Wales, or the Secretary of State for Scotland, setting up the Development Corporation and conferring on it special powers over the so called designated area of the new town, including the right of compulsory purchase. It follows a long preliminary period involving public hearings against objections to the new town conducted by an independent inspector. With only one exception, the towns fall chronologically into two main groups.

The first group, often known in Britain as the "Mark One" new towns, consists of 14 towns—11 in England, 1 in Wales, and 2 in Scotland—all designated in the relatively short period of November 1946 to April 1950. With one exception, this group may be divided into two clearly defined subgroups according to the purpose for which the towns were designated. The larger of the subgroups consists of the new towns that were intended to achieve the specific objective of Ebenezer Howard and the Garden Cities movement, that is, the planned dispersal of population from overcrowded urban areas, which had been given official approval in the Barlow Commission report of 1940. Eight of the English towns were specifically set up to receive overspill population from London as specifically recommended in Sir Patrick Abercrombie's official Greater London plan that was prepared for the central government in 1944.

[*Editor's Note:* Material has been omitted at this point.]

The second group of towns includes all the 13 new towns designated between Skelmersdale in Lancashire in 1961 and the Central Lancashire new town in 1970.

[*Editor's Note:* Material has been omitted at this point.]

The new towns of the 1960s can be classified simply as to purpose. With a single exception, all of them were intended to solve specific overspill problems of the major conurbations. The exception was Newtown, a very small new town development designated in 1967 to help solve the problems of a thinly populated hill area of mid-Wales. Three new towns were designated to take London overspill in addition to the existing eight. They were Milton Keynes, Peterborough, and Northampton, all designated in 1967–1968. Farther north in England, new towns were designated for the first time to take overspill from the West Midlands, Southeast Lancashire-Northeast Cheshire (Greater Manchester), and Merseyside (Greater Liverpool) conurbations. These were Telford (1963) and Redditch (1964) for the West Midlands, Skelmersdale (1961) and Runcorn (1964) for Merseyside, and, somewhat belatedly, Warrington (1968) and Central Lancashire (1970) for Greater Manchester. In the northeast of England, which is a Development Area, local problems of overspill from the Tyneside conurbation led to the designation of the new town of Washington (1964). This new town was designated also to promote regional development as part of an axis of growth for the region stretching alongside twin modernized highways between which the new town stands. Lastly, two other new towns were designated in Scotland, Livingston (1962) and Irvine (1966), to cope with the continuing overspill from Glasgow and to promote growth in this Development Area.

[*Editor's Note:* Certain figures and some text material have been omitted at this point.]

The new towns of the 1960s show two striking tendencies, both of which mark a clean break with new town traditions and which have been criticized in consequence by some garden city champions. They are (1) a tendency toward a sharp increase in the target population of the town and (2) a tendency toward a sharp increase in the size of the base population. These are, of course, associated. Whereas the typical new town of the 1940s might have started with a population of 10,000 persons distributed in a small town and a nearby village and proceeded to a target population of 50,000, a typical new town of the late 1960s might start with an existing county town of 100,000 people and proceed to a target population of 200,000 to 300,000. However, the new towns of the latter group are much more diversified in size, and generalization is more difficult. Some of the new towns designated for overspill from the provincial conurbations—Skelmersdale and Runcorn for Merseyside, Redditch for the West Midlands, Washington for Tyneside, Livingston for Clydeside—followed the traditional formula but with modest increases in the planned target populations to levels of 80,000 to 100,000, reflecting a general belief among planners in early 1960s that a town of this size was the minimum necessary to achieve an adequate level of durable goods shopping and other specialized services. Several of the Mark One new towns had their planned populations targets raised to about 100,000 at this time. Irvine in Scotland, which was expanded from a population of 36,000 to a target population of 116,000, is really a slightly enlarged version of the formula earlier used at Hemel Hempstead. Three other towns, however, demonstrate a quite new formula. Peterborough and Northampton, designated for London overspill, and Warrington, for Manchester and Salford overspill, are using the new town machinery to expand large towns (with a population of 84,000 in the case of Peterborough, 131,000 in the case of Northampton, and 124,000 in the case of Warrington) to population targets of 190,000, 300,000, and 200,000, respectively. Most radical of all is the most recent new town designated in Britain, Central Lancashire, which will take a large region some 15 miles long and 5 miles broad, including one large and two smaller towns having a combined population of 240,000, and expand them to 430,000. During the 1960s, then, the concept of a new town has evolved into the concept of a new town added to an old town and then to the concept of a planned city region. This last, it must be noted, is not far removed from the social cities idea enunciated by Ebenezer Howard in 1898.

Geographical Basis of the Site: Character of the Land

Turning to the more local considerations that have influenced the choice of site for the new towns, one connection should be noted between regional and local considerations. The essential purposes of new towns, it has been seen, are either to provide for planned overspill of people from the major urban agglomerations or to provide for the restructuring of industry in a problem area where economic reinvigoration is desired. In either case, whether or not there was substantial local population in the area of the new town, there was substantial regional population. Only rarely, as in the case of valuable mineral deposits, have large regional populations occurred in areas of difficult topography. There are a few such examples in Britain. They include the South Wales coalfield, where linear urban growth follows valleys cut deep into a plateau in a manner reminiscent of the Pennsylvania and West Virginia coalfields; the West Yorkshire and North East Lancashire textile areas, where water power and then coal deposits have attracted industrial growth at the edges of an upland region and even in the valleys which dissect it; and the western part of the Durham coalfield and iron working area in northeast England. It so happens, however, that all these areas have been either stagnant or in decline industrially since World War II, and the problem of accommodating increased population has not been great. Insofar as new towns have been designated or planned to promote industrial reinvigoration in zones such as these, they have normally been planned in the more accessible lowland areas immediately adjacent to them where communications are easier and building costs lower. Thus, Newton Aycliffe and Peterlee are both in East Durham, Cwmbran is south of the scarp that marks the southern rim of the South Wales plateau, and Llantrisant, if built, will be in a similar location. The result is that all the British new towns have been located in lowland areas; as Figure 7 shows, in few of them does any part of the designated

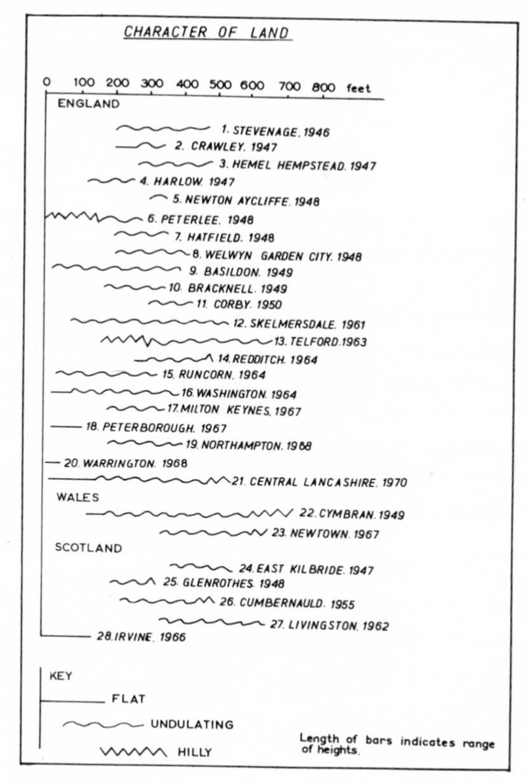

Figure 7

area exceed 500 feet above sea level. Normally, too, the towns have been built on gently undulating land giving good drainage conditions, which fortunately is the most usual type of topography found near the major industrial conurbations. The examples of hilly sites, with the exception of parts of the sites of Crawley in the south and Telford in the Midlands, are all in the north in Wales and in Scotland, and, even there, they usually occupy only a part of each site. Flat sites, probably because they tend to be alluvial river bottoms liable to flooding, have been avoided. Where flat areas do form parts of new town areas, as at Peterborough or Warrington, they will not normally be used for construction.

Problems of microclimate, in consequence, have been few, though they increase in number and intensity northward. In southern England, the biggest problems arise from inversion fogs and frost pockets in small depressions in the usually undulating surface. The largest areas liable to such problems, however, are the floodplains, which in any event would be avoided for building, such areas can be usefully used, as park zones for summer recreation, as in the plan for Peterborough. In northern England and in Scotland, there are problems of winter wind and snow drifting. Although westerly winds are more common throughout Britain, especially in the wintertime when they are associated with the passage of barometric lows, and although they also bring rain and high humidity (with associated lowering of the sensible temperature), the most difficult weather conditions in the northern half of Britain are undoubtedly associated with north and east winds caused by incursions of polar continental or arctic airstreams, which often lead to snow. Cumbernauld, where about 60,000 of the 70,000 inhabitants in the plan are located on an exposed whaleback hill with a steep north slope, is exceptionally exposed to such winds and has achieved a poor reputation among many of its inhabitants in consequence.

The Geological Basis

The same considerations that have given a rather uniform character to the geographical basis of the new town sites have also given a rather restricted range to their geological basis. As is well known, lowland England is composed largely of sedimentary strata of Permo-Triassic, Jurassic, Cretaceous, and Eocene age, mainly with a rather gentle dip but somewhat affected by the Hercynian orogeny in the north and west (Pennine and Welsh massif flanks) and by the Alpine orogeny in the extreme south. The Coal Measures, the basis for eighteenth- and nineteenth-century industrialization, are found at the borders of lowland and highland Britain in northern England and in south Wales and in the rift valley of central Scotland. These solid rocks are normally masked by a variable covering of glacial drift north of a line a few miles north of London. In lowland England, also, the gentle fall of the east-flowing rivers (and of a few west-flowing rivers, such as the Mersey and Ribble) has produced wide alluvial floodplains and bordering terrace gravel deposits. These gravels have been disproportionately important in the settlement history of lowland Britain because they afforded dry, well-drained sites with good water supplies. The most important settlements tended to develop at bridgepoint sites where gravel terraces approached close to each other on opposite banks of the river.

Given these facts, the pattern of geology by Figure 8 is simple enough to understand. The Mark One London new towns tend to be located either on the chalk backslope of the Chiltern hills, which is the dominant feature of the passage northwestward from London toward the Midlands and which is heavily masked by glacial and periglacial deposits as at Stevenage, Hemel Hempstead (Figure 9), Hatfield, and Welwyn Garden City, or on the Eocene rocks (especially the London clay) of the London Basin itself either with drift cover as at Harlow or without it as at Basildon and at Bracknell (Figure 10). The new town of Crawley south of London is on older Wealden Beds of early Cretaceous age. The three more recent London new towns are well north of London on Jurassic rocks that have a considerable amount of drift cover. The Jurassic rocks are either clays that form rather flat land or limestones that form low hills. In the case of two of these more recent new towns, which were expanded from large existing settlements (Peterborough and Northampton), the alluvial valley of the River Nene is also an important feature. The same pattern of Jurassic deposits masked by drift is also characteristic of the steelworks new town of Corby.

The new towns of the West Midlands and the Northwest are based on the Permo-Triassic rocks of the Midlands plain, which extends via the Cheshire Gate into West Lancashire and is generally covered by a heavy drift overlay. At Skelmersdale (Figure 11), Telford (Figure 12), and Central Lancashire, they also extend on to the neighboring Coal Measures, which have been worked and are still being worked

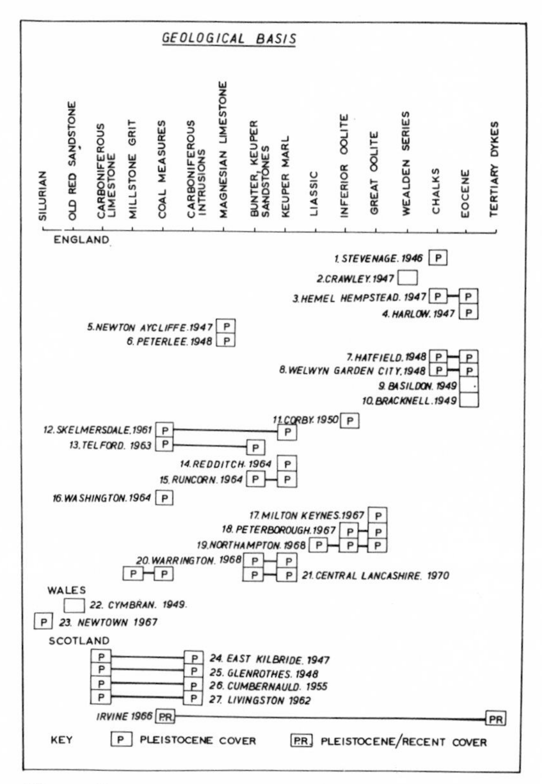

Figure 8

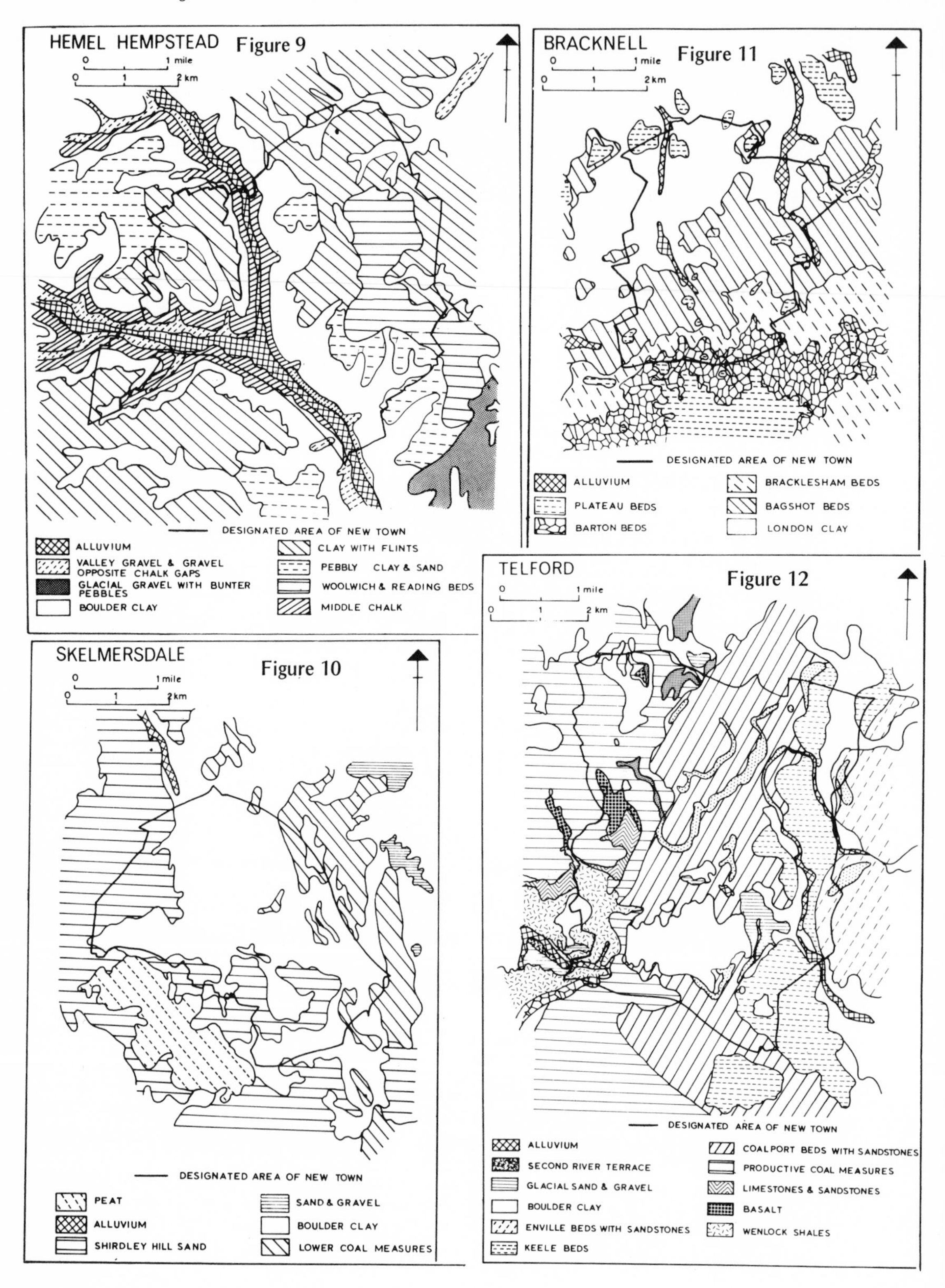
HEMEL HEMPSTEAD
Figure 9
0 1 mile
0 1 2 km
DESIGNATED AREA OF NEW TOWN
ALLUVIUM
VALLEY GRAVEL & GRAVEL OPPOSITE CHALK GAPS
GLACIAL GRAVEL WITH BUNTER PEBBLES
BOULDER CLAY
CLAY WITH FLINTS
PEBBLY CLAY & SAND
WOOLWICH & READING BEDS
MIDDLE CHALK
BRACKNELL
Figure 11
0 1 mile
0 1 2 km
DESIGNATED AREA OF NEW TOWN
ALLUVIUM
PLATEAU BEDS
BARTON BEDS
BRACKLESHAM BEDS
BAGSHOT BEDS
LONDON CLAY
SKELMERSDALE
Figure 10
0 1 mile
0 1 2 km
DESIGNATED AREA OF NEW TOWN
PEAT
ALLUVIUM
SHIRDLEY HILL SAND
SAND & GRAVEL
BOULDER CLAY
LOWER COAL MEASURES
TELFORD
Figure 12
0 1 mile
0 1 2 km
DESIGNATED AREA OF NEW TOWN
ALLUVIUM
SECOND RIVER TERRACE
GLACIAL SAND & GRAVEL
BOULDER CLAY
ENVILLE BEDS WITH SANDSTONES
KEELE BEDS
COALPORT BEDS WITH SANDSTONES
PRODUCTIVE COAL MEASURES
LIMESTONES & SANDSTONES
BASALT
WENLOCK SHALES

in the northern part of the designated area of Telford, bringing subsidence problems. The two new towns of Runcorn and Warrington in the Mersey Valley include a wide stretch of river alluvium in their designated areas. In the case of Warrington, this belt runs east-west through the middle of the designated area. Central Lancashire alone of the northwestern new towns on its east side borders the foothills of the Pennine hills, which are composed of Millstone Grit of Carboniferous age.

Perhaps the simplest pattern geologically is provided by the new towns of northeast England—Newton Aycliffe (Figure 13), Peterlee, and Washington. Here, heavy boulder clay cover overlies Magnesian Limestone of Permian age with the limestone strata dipping very gently eastward off the eastern Pennie flanks. Productive Coal Measures of Carboniferous age lie some distance beneath the limestone and have been worked actively beneath Peterlee and Washington, resulting in considerable problems of subsidence from old workings. In Peterlee, indeed, the phasing in of the construction of the town and the mining of the coal presented a considerable planning problem.

Cwmbran in south Wales and Newtown in mid-Wales have the most ancient geological foundations of any of the British new towns. Cwmbran is founded on the Old Red Sandstone, which occurs extensively on the southeast edge of the Welsh massif, and Newtown is being built on the ancient Silurian limestones and shales of the Welsh massif itself, here dissected by the upper Severn Valley. It is no accident that these two towns have some of the most accented topography and the steepest building slopes of any new town in Britain.

The Scottish new towns, with one exception, have absolutely similar geological settings. All are being built on the Carboniferous Limestone series that occurs so widely in the Central Valley of Scotland and which is invariably accompanied by intrusions of Carboniferous age in the form of basalt or dolerite. Needless to say, in the case of Glenrothes, the neighboring productive Coal Measures were the basis of the original economy of the new town. At East Kilbride, Glenrothes, Cumbernauld (Figure 14), and Livingston (Figure 15), the Carboniferous Limestone gives undulating land that is generally good for building, although the numerous intrusions naturally give a more accented topography than is usual in the English new towns. A thick mantle of drift that is characteristic of all these towns makes for rather heavy land, although the degree of slope gives reasonably good drainage. The exception among the Scottish new towns is the most recently designated. The seaside town of Irvine is founded mostly on glacial and recent deposits, including extensive areas of raised beach, overlying productive Coal Measures that are intruded by dikes of Tertiary age, which are so characteristic of this western littoral of Scotland.

Generalizing, it can be seen that all the new town sites in Britain have been carefully chosen from a geological point of view. Few of them present serious problems for building operations, and this is important because the cost yardsticks for public housing, which are applied nationally, put a close ceiling on building costs. Most of the land within the new town designated areas has few severe slopes and is well drained as well as providing good foundations. Where occasional tracts are hilly or liable to flood, these will often form attractive recreational areas in the plan for the town. The only really serious problems that have been encountered have been concerned with the effects of working for minerals, particularly coal

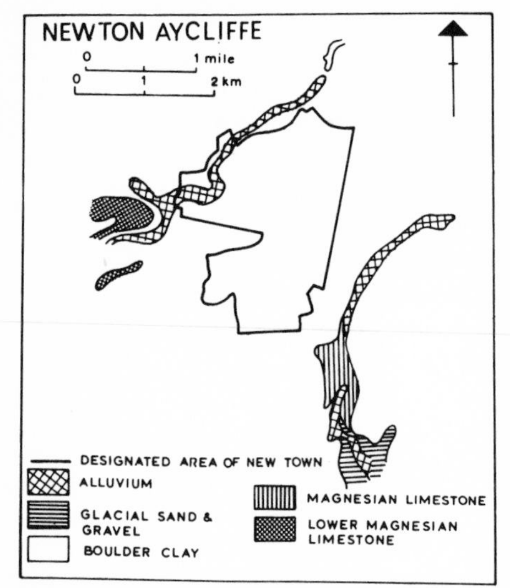

Figure 13

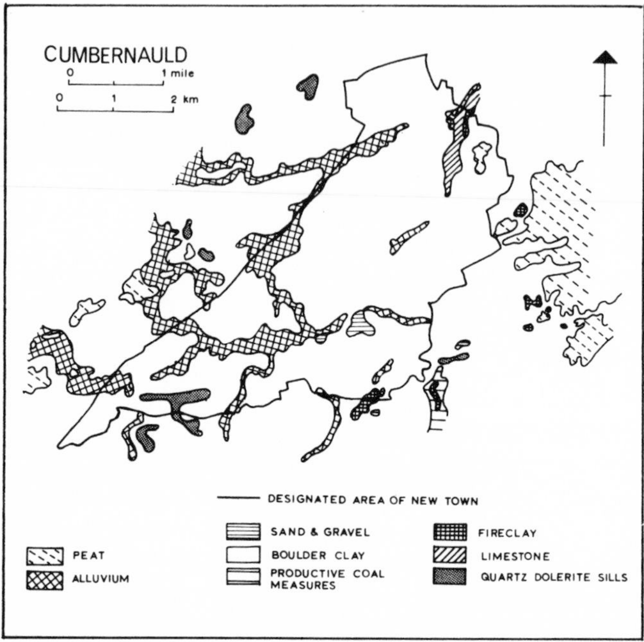

Figure 14

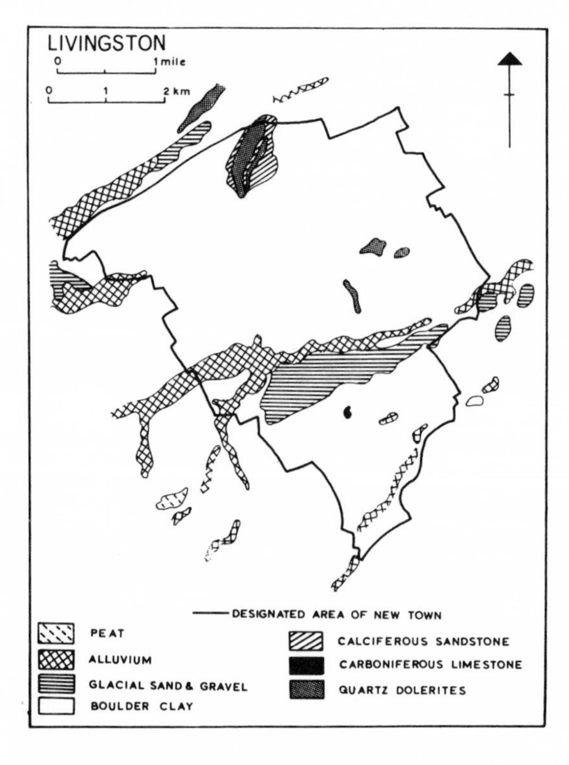

Figure 15

but also ironstone, at those few towns whose areas extend onto productive Coal Measures or onto ironstone. In order of their designation, the towns affected by these problems have been Peterlee, Corby, Skelmersdale, Telford, and Washington. Some of their particular difficulties are described below.

Some Particular Problems

Peterlee. Peterlee is sited on glacial cover, mainly boulder clay, over heavily faulted Magnesian Limestone that, in turn, overlies Coal Measures at a depth of 300-400 feet. The main productive Coal Measure strata, though, are at greater depths, 800-1,700 feet below the surface. The boulder clay cover is insufficient and the Magnesian Limestone too weak and powdery to give adequate support and protection against the subsidence, where material has been removed by mining in the underlying Coal Measures. Surveys at the time of the Draft Outline Plan showed that subsidence to the extent of 60 percent of the thickness of the extracted seams normally took place within 2 to 5 years of working. Between 1912 and 1946, an actual subsidence of 5 feet and more had taken place. The problems of relating surface development to mining operations proved very complex at the time the town was designated because it was thought that the coal might not be fully worked until 2000 A.D. or afterward, and the preparation of the Master Plan was delayed for this reason. The Plan finally provided that a residential density of 65 persons per acre with some apartment blocks could be provided in stable areas, but that the maximum density elsewhere would be only 35 persons per acre, with detached and semidetached houses. Industry was located in the northern part of the site on flat ground where light factories could be erected (7).

Corby. The problem here was to coordinate construction of the town with the open-pit mining of the iron ore in deep pits, which are filled afterward. As arrangements for future mining had been made for over a decade ahead of the time of preparation of the first Master Plan in 1952, the planning consultants had to adhere to them. They also had to ensure that houses were built at economic cost, however, without elaborate structures on land not liable to subsidence. Inasmuch as the pattern of ironstone mining progressed outward from the center of the town, the development of the town had to follow it (8).

Skelmersdale (Figure 11). A large part of this town site overlies productive Coal Measures that have been worked in the past both by shallow workings and by deep mining. There was one mine operating at the time of the original planning proposals—a small private one operated by agreement with the National Coal Board—and the consultants recommended that this be shut down. The deep mines were all flooded, and the consultants concluded that the risk of subsidence was not great provided appropriate structural precautions were taken. Shallow shafts, however, would need filling and capping. The greatest problem arose with old shallow workings within about 100 feet of the surface, but these areas were relatively restricted (9).

Telford (Figure 12). Telford, known previously as Dawley, has the most problematic site of any of the British new towns after Peterlee. Almost one-third of the Designated Area corresponding roughly to the outcrop of the productive Coal Measures is disturbed ground arising from the working of Carboniferous strata, not merely coal with its associated ironstone and clay but also limestone. At the time of designation of the extended new town area, there was one large National Coal Board mine still working in the northern part of the area and one small private one. Although extensive coal deposits remained under much of the site, they were unlikely to be worked. The main problem came from old workings (10). These were of three types. First, there were shallow underground workings over much of the exposed coal field, which formed 19 percent of the designated area of the town. These workings were generally within 60 feet of the surface and were liable to collapse under load. Filling would be necessary, at least in part. Second, there were former open-pit workings that had been restored only to agricultural standard and were liable to further subsidence if building operations took place over them. Here, rafted foundations would be necessary for most buildings together with either a height limitation or the use of lightweight structures. Third, there were open-pit workings yet to be restored. Although higher standards for filling were possible here, the consultants concluded that settlement problems could persist for a long time. Special precautions, therefore, would be necessary over most of the area, including thorough site investigation, careful consideration of local hydrologic patterns (because of the movement of water into and through the backfill), and special consideration of building foundations, height, flexibility, and other design features (11).

All this presented considerable problems because new town architects must work within close cost limits which are imposed nationally. The consultants, in a joint exercise with the Development Corporation for the town, therefore carried out a special cost study. This considered two alternative solutions for a representative site that suffered problems from both past and present mine workings and from previously existing development with different degrees of structural fitness. One was a clean sweep of the site with regrading; the other involved partial clearance but avoiding major spoil heaps and structures that might be saved. The first proved more expensive because of the cost of clearing property with a useful life and of regarding spoil heaps. But whatever the treatment, cost comparisons with unaffected land showed that the extra cost of rafted foundations, filling and capping old shafts, grading spoil heaps, and limited filling of old shallow galleries could be as much as $500 (US) per dwelling. In some areas, the extra costs were so high as to make alternative layouts or land uses necessary, but this appeared only after detailed site examination. When this was the case, plans could be prepared to use the most disturbed land for planned open space. Such detailed survey, though, could be undertaken only after site acquisition, and, even after that, the normal programming for undisturbed land (two and one-half to three years from planning to occupation) might be doubled because of the structural difficulties involved, particularly as agreement by the central government would be necessary on the extra costs involved (12).

A critical final question, because it examines the wisdom of developing a site like Telford in the first place, is whether the extra costs are justified by the saving of agricultural land elsewhere. The consultants concluded that they were not (13).

The justification for developing sites like Telford, therefore, must depend on the community's evaluation of the aesthetic and social benefits of rehabilitating such a scarred landscape of the early industrial revolution. The same dilemma must occur to regional development and planning agencies in similar areas all over the world.

Washington. Although Washington is only about 10 miles northwest of Peterlee and its geological setting is very similar, it does not have the same problem of large-scale current mining operations. There are considerable problems of subsidence from old workings, but the consultants concluded that the load-bearing properties of the subsoil would be sufficient for normal loadings over much of the area. Foundations for heavier structures would have to be specially designed. Because of a great variability of soil conditions, however, detailed soil surveys would be necessary in advance of development. The consultants worked closely with the National Coal Board in reviewing the likely pattern of subsidence during preparation of the Master Plan and took it into account in their recommendations on programming, but continuous review would be necessary throughout the construction period (14).

[*Editor's Note:* Material has been omitted at this point.]

Conclusions

It must be recognized that in the long run, the decision to develop a site for a new town is not an economic but a political one. Economic factors will and should enter, but the methods currently available for cost-benefit analysis are at present too imperfect, particularly in the controversial field of amenity, for total reliance to be put on them. The strength of public feeling as registered through organizations and pressure groups must be allowed to have its effect. In Britain, there are several well-documented examples of new town sites that were abandoned after local opposition. Lymm itself, which was twice rejected by the central government after public local inquiries, is one; Hook in Hampshire, which was abandoned after intense opposition from local landowners, is another; and proposals in the 1964 *South East Study* (15) for developments at Newbury, Ashford, and Ipswich were later abandoned by the government after local opposition had been aroused, although in each case there were good objective reasons for doubting the wisdom of the original proposal.

This does, however, involve one danger. Because in Britain there is a wide choice of new town sites that are physically feasible, the central government may follow a course of least resistance by permitting developments on sites that are least demonstrably unpopular, while abandoning the rest. The least unpopular sites may be not worth defending because they are poorly sited in relation to communications, or in need of economic assistance, or unattractive environmentally—perhaps a combination of all these.

Extra costs in rehabilitation of a derelict environment may be involved, as in several new town sites. This may well be a price which the community on due reflection thinks worth paying. But it is extremely important that the end result in a planned environment that is not only adequate for the demands of a free consumer society but is also demonstrably superior to the mass of development elsewhere. That, after all, is the justification for the type of planned community development represented by the British new towns.

REFERENCES

1. Frank Schaffer, *The New Town Story* (London, Macgibbon and Kee, 1970), Appendix. Figures are for September 1969; updated figures appear each year in the January issue of the journal *Town and Country Planning.*
2. Ebenezer Howard, *Garden Cities of Tomorrow.* Republished, with an introduction by F. J. Osborn (London, Faber & Faber, 1946).
3. *Report of the Royal Commission on the Distribution of the Industrial Population* (Cmd. 6153; London, H.M.S.O., 1940).
4. *Final Report of the Expert Committee on Compensation and Betterment* (Cmd. 6386; London, H.M.S.O., 1942).
5. *Report of the Committee on Land Utilization in Rural Areas* (Cmd. 6378; London, H.M.S.O., 1942).
6. *Final Report of the New Towns Committee* (Cmd. 6876; London, H.M.S.O., 1946), p. 3.
7. *Peterlee—Summary of the Draft Outline Plan,* 13 March, 1950 (mimeo), pp. 3-5.
8. *Corby New Town: A Report to the Development Corporation on the Master Plan,* by William Holford and H. Myles Wright, December 1952, pp. 9-16.
9. *Skelmersdale New Town Planning Proposals,* by L. Hugh Wilson for Skelmersdale Development Corporation, December 1964, p. 5.
10. *Telford, Development Proposals,* Vol. I. The John Madin Design Group. A Report to Telford Development Corporation (1969), pp. 221-222.
11. *Ibid.,* pp. 222-224.
12. *Ibid.,* pp. 224-225.
13. *Ibid.,* p. 225.
14. *Washington New Town, Master Plan and Report,* prepared for Washington Development Corporation by Llewelyn Davis, Weeks and Partners, December 1966, pp. 33-35.
15. *The South East Study, 1961-1981* (London, H.M.S.O., 1964).

8A

Reprinted from *Civil Eng. Public Works Rev.*, **63**, 525, 527–529 (May 1968)

FOULNESS

Some geological implications

J. T. Greensmith, B.Sc., Ph.D., F.G.S. and
E. V. Tucker, B.Sc., Ph.D., F.G.S.
Department of Geology
Queen Mary College
University of London

SELECTION of part of the intertidal zone flanking Foulness Island, Essex (Fig. 1) as a possible site for the third London airport has been done apparently with little regard to the geological conditions existing there or further offshore. Neither is there much evidence to indicate that the possible consequences of mass dredging and removal of offshore sand to the airport site have been thoroughly considered. That the geological problems inherent in the proposed site can be overcome is not in doubt, but the resultant effect on erosion and deposition in and around the outer Thames estuary may create subsidiary local problems not immediately apparent and certainly difficult to predict.

Geology of the Foulness area

Though the basic geology of the area is relatively simple, research work carried out there over the last six years has shown that the processes at work during the accumulation of the more recent surface and near-surface deposits have been very complex. The bedrock of London Clay, not exposed at the surface on Foulness Island, has an irregular surface at depth and is overlain almost completely by a blanket of variably consolidated younger sediments which have accumulated during the last half million years (Fig. 2). These sediments include sands and gravels of fluvial and glacial origin, some of which have been reworked from deposits originally

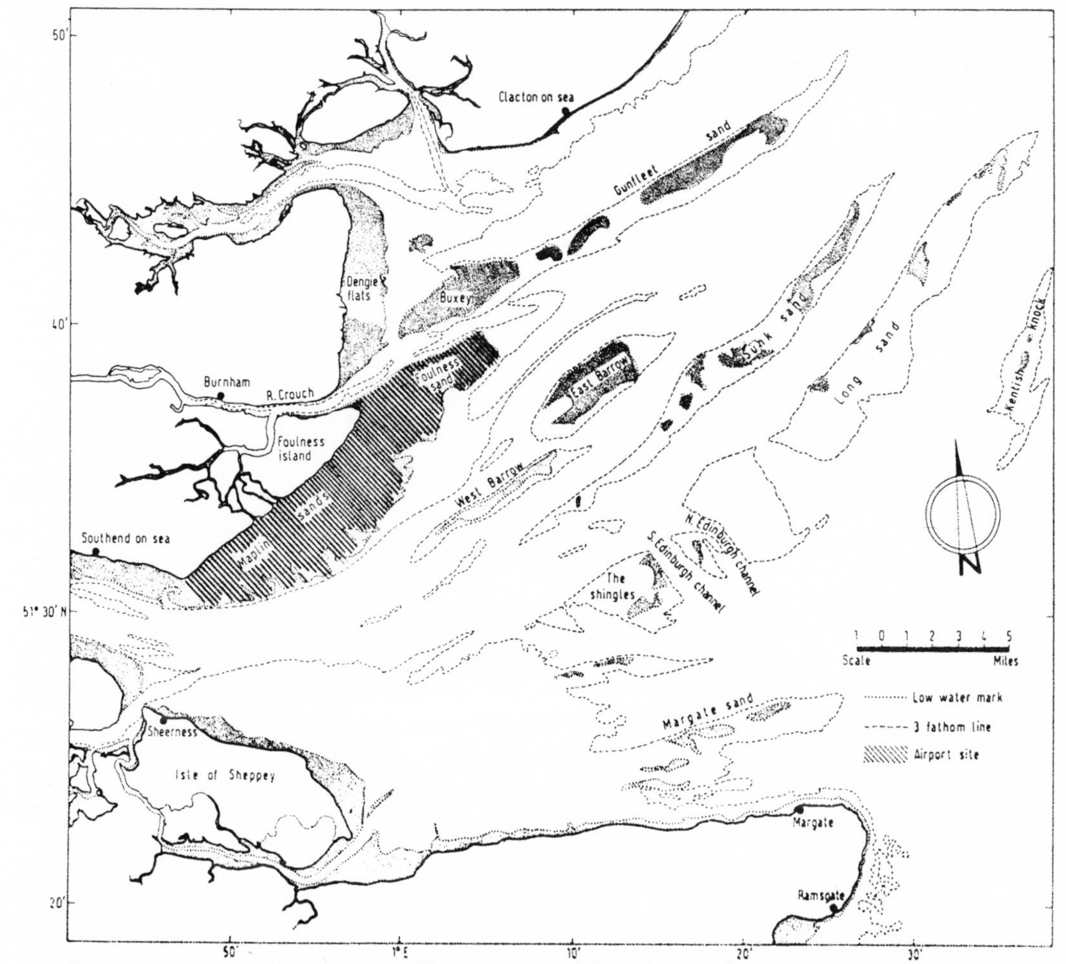

Fig. 1. Distribution of sand banks in the outer Thames estuary in relation to the proposed site.

[*Editor's Note:* Certain figures referred to in the text have been deleted, owing to limitations of space.]

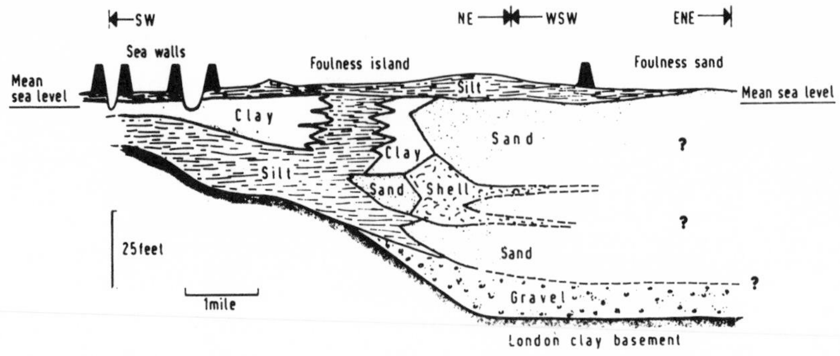

Fig. 2. Generalised cross-section of the foundation geology of Foulness Island.

laid down at the southern edge of an extensive ice sheet which covered most of East Anglia and adjacent North Sea areas in Pleistocene times. In addition, there are more recent sediments of fluvial and marine origin including salt marsh clays and silts, as well as sands, silts and shell beds of intertidal or subtidal origin (1)*. Characteristically this sedimentary pile changes rapidly from one type to another both vertically and laterally. That being so, it is clearly an unwarranted supposition, until deep borehole evidence proves otherwise, that the relatively firm sand layer which exists at the surface at Foulness and Maplin Sands is likely to persist downwards as far as the underlying London Clay surface. It is much more likely that deposits of other types, some more plastic or thixotropic and others more permeable in nature, are interbedded with these firmer sands at no great depth.

Although most of the material resting on the London Clay in the vicinity of Foulness is of shallow water origin deposited adjacent to the contemporary shorelines, intermittent subsidence has permitted accumulation of a considerable thickness of sediment. Over 100ft of recent sediment has been proved immediately north of the River Crouch and upwards of 70ft beneath the estuary and at Foulness Island. The thickness at Foulness and Maplin Sands is difficult to predict but is probably of the same order or may well be tens of feet more. Hence it is clear that the foundations for the airport are unlikely to rest directly onto the more consolidated, less compactible London Clay.

Subsidence of the outer Thames estuary region, accompanied by complementary rises in sea level, has proceeded in a pulsatory fashion over the last few thousand years at least. Phases of stationary sea level, or even minor lowering of level, have separated periods of rising sea level and active encroachment of the North Sea across low-lying land surfaces. Around the southern margins of the North Sea there is evidence indicating up to six phases of marine regression and transgression during the last 2,000 years. At present the net annual rise in sea level is of the order of 3mm, but this is probably a very conservative estimate. In arriving at such a figure it is necessary to take into account subsidence caused by regional downwarping of the southern North Sea, accentuated by natural compaction of soft fluvial and marine sediments, as well as the worldwide rise in sea-level resulting from melting of Polar ice caps. Local compaction tends to be more pronounced in reclaimed ground especially when extensive artificial drainage is introduced.

The fluctuating sea level has helped to determine the patterns of sedimentation in the past, as it does at present. At times the net input of sediment onto the coast has been considerable and wide belts of accretion some mile or so in width have formed. Seaward extension of salt marsh occurs at these times. The bulk of the inwards transported mud, silt and sand is derived from glacial and fluvioglacial deposits originally laid down on the shallow bed of the southern North Sea, only comparatively small quantities being introduced by rivers. Other immediate but probably minor sources are the London Clay and younger löess, boulder clay and fluvioglacial sands and pebble beds exposed in cliffs along the Essex coast. Periodically the salt marshes have been reclaimed, notably during the seventeenth and eighteenth centuries, and much of this ground has now subsided many feet below high water mark at spring tides.

At other times erosion of pre-existing salt marsh and tidal flat muds, silts and sands has been dominant with the partly compacted sediment being reworked back into the deeper subtidal parts of estuaries. Presently we are passing through such an erosional phase with the salt marshes being reduced in area comparatively rapidly (2, 3). Moreover, up to 1ft of sediment has been removed from the top surface of parts of the Maplin Sand in very recent years (Figs. 3 and 4). Consequently the increased rate of siltation induced in adjacent estuary channels is having serious effects on certain local shell-fish industries.

A further product of erosional phases, particularly evident along the coast at Foulness, are sheets and ridges of shell and sand. These are also present at the surface inland and at depth on Foulness Island; one sheet at a depth of 23ft has a thickness of 21ft. In the nearby Burnham Marsh a similar sheet some 22ft thick occurs at a depth of 45ft intercalated between loosely consolidated sediments of other kinds.

Shoreline problems

Reclamation of Foulness and Maplin Sands poses problems which are probably more important in the long term than those connected with the engineering geology of the airport site. Some of these problems are related to the effects upon adjacent coastlines and estuaries. Most reclamation schemes, as in The Wash and Low Countries, involve areas where sediments are accumulating inshore at a sufficient rate to extend the coastline outwards over the sea bed. In these areas the mud, silt and sand is being supplied from the sea in such quantities that the load of sediment in transport consistently exceeds the carrying capacity of the sea water and hence is deposited freely. In contrast, at Foulness, the supply of offshore sediment is relatively low at present and the current velocities appear to be more than adequate to keep the load in suspension. Because of this, sediment is being removed from the shoreline.

Inevitably the reclamation of some 8,000 acres of the intertidal zone will divert large volumes of sea water into adjacent estuaries, but what the effects of this diversion will be can only be conjectured. Two interrelated aspects need consideration, the depositional and the erosional.

It is known that sediment, albeit in reduced quantities at present, is being transported into the estuaries adjacent to Foulness mainly from the North Sea. Much of this sediment is funnelled through the numerous deep channels which exist between the offshore sand banks. In general the finest particles in suspension are taken furthest inshore, penetrating far up the estuaries. This is a function of the diminution of current velocities and carrying power of the sea water as it traverses the estuaries from mouth to head. Dense, turbid suspensions of clay and silt are known to be present near to the bottom of estuary channels and it is from these that soft mobile black mud is deposited. The average surface velocity of the flood tide is 2 knots. By analogy with water movement studies in major salt marsh creeks, the flood pushes the bed load up the

* *Figures in parentheses indicate References following the article.*

estuaries as well as carrying particles in suspension, whereas the flushing action of the ebb flow is inadequate to move comparable loads seawards.

If the Foulness reclamation diverts more water into the estuaries then it is probable that the carrying capacity of the flood tides will be increased sufficiently to introduce fine sediment at a faster rate than present. Additionally, it is probable that the faster currents will cause enhanced scour of unprotected salt marshes in the middle and lower reaches of the estuaries. The combined effect of these two features is likely to be increased deposition within the deeper channels.

Alternatively, it is possible that changes in tidal flow patterns adjacent to Foulness could accentuate the flushing effect of ebb tides, especially along the pre-existing channels. Hence, fine grained sediment comprising the sides and floor of the channels might be scoured, the material being returned into suspension and in part re-deposited on the flanks of the estuaries.

Of these alternatives the first is considered more likely but it is clear that theoretical considerations are no substitute for facts. And the facts can only be obtained relatively quickly by the use of laboratory models simulating the conditions likely to be produced in the estuaries when adjacent coastlines are speedily extended. The models may demonstrate that the new conditions will affect adjacent estuaries in different ways. It may be anticipated that siltation will occur much more rapidly in the River Crouch, whose catchment area supplies little water, than in the Thames with its considerable input of river waters.

Offshore problems

One of the factors influencing the Foulness scheme is the close proximity of large quantities of sand in the form of offshore banks. The project, as presented, envisages the use of this sand for fill purposes. Several million tons will be dredged to raise the level of Foulness and Maplin Sands by approximately 22ft. The sand banks are finger-like structures partly exposed at low water, extending northeastwards from the Essex coast for 30 to 40 miles (Fig. 1). Between them are channels of varying degrees of asymmetry and depth, some of which provide the main navigation routes into the Thames. The topography of the seabed determines the flood and ebb tide flow patterns although the full extent of this controlling influence is not yet clear. A complicating factor is the introduction of flood tide waters into the southern part of the outer Thames estuary from the English Channel which divert certain ebb currents and assist in keeping open transverse channels, or swatchways, across the sand banks. There is also a return of North Sea waters on the ebb tide. The Edinburgh Channels are prime examples of swatchways and much tidal and sedimentological information has been collected on them over recent years by the Hydrographic Department of the Ministry of Defence.

The surface form of many of the banks has been a subject of very close study and changes in topography have been plotted for the last 100 years. The surface and near-surface sands have been sampled extensively and prove to be very mobile, especially along the flanks (4). Within the Edinburgh Channels sand movement has been plotted in detail yet it is still difficult to predict the rate or direction of movement. Even the reasons for the movement are not too well understood.

Knowledge of the internal structure and composition of the banks is very limited at present and until deep cores are taken down to the London Clay surface it is difficult to predict their nature. Geophysical surveys over certain offshore banks suggest that sand may form a large proportion of the bank sediment. Some of the sands appear to be cross-laminated, a function of current direction during the period of sand accumulation, and it can be inferred that the current directions have fluctuated during growth of the bank.

It is possible that some of the banks to the north of the Edinburgh Channels, including Foulness Sand, have a core of gravel. The proximity of the banks to pebble and sand deposits (laid down from ice sheets) on the mainland together with their orientation parallel to the former margin of the ice sheets, suggests that their position may have been predetermined by ice-dropped material. A firm core of this glacial material might account for the apparent stability in position of the infrastructure of the banks.

Alternatively, the banks may be sited on and around a nucleus of old river terrace gravels flanking a series of early to middle Pleistocene channels cut into London Clay when sea level was much lower than at present.

If either of these suppositions is correct then the removal of the top layers of sand and silt and exposure of the gravel-rich cores may well induce undesirable changes in local depositional conditions. For instance, the sand, silt and clay 'fines" would be washed out of the gravel layers, taken up by the currents circulating locally and redeposited either in adjacent channels or on nearby beaches. A similar situation could arise if the coarse material were dredged and the 'fines' returned to the bank (5). Hence, in an environment comprising surface sediments which are mobile and subsurface sediments which would become mobile on exposure by dredging the chances of creating unwelcome side-effects are potentially high.

Analogies have been drawn between the proposed reclamation scheme and similar projects undertaken in the Netherlands. Although the projects have technical points in common a strict comparison is not feasible. In general the Netherlands reclamation schemes have involved tracts of ground ready, in a sedimentological sense, to be retrieved. They were, and are, areas of accretion where sediments accumulate at a relatively rapid rate. In many cases this sediment, forming mud flats and salt marshes, received protection from wave attack by long lines or strings of offshore sand banks and barriers growing parallel to the coastline. These barriers partially closed pre-existing embayments of the coastline and man subsequently has assisted nature by producing an even more sheltered environment in which sediment could be preserved and into which sediment can be pumped. Amsterdam airport is situated on such reclaimed ground but is some 20 miles or so from the open North Sea coast.

The outer Thames region does not conform to this picture as evidenced by the present marked erosional effects around the whole estuary. If wholesale removal of banks in the outer estuary is envisaged before their constitution and history are fully investigated then the resulting effects on navigation channels and the surrounding coastline may well be profound.

Conclusions

The reclamation of Foulness and Maplin Sands for the site of a major airport has many attractive aspects by no means the least being sociological, with aircraft noise being dissipated over water and very little good agricultural land being lost. The civil engineering problems of the site can probably be fairly easily overcome with modern expertise gained from vast reclamation works in the Netherlands such as the 'Delta Project'. But the usefulness and amenity value of the adjacent seaways may well be impaired or destroyed if insufficient consideration is given to the consequent effects on local sedimentation patterns. Certain of these effects are mentioned in this article. The significant issue made apparent in the course of this discussion is the present lack of fundamental geological information which has an important bearing on these problems. The information can be acquired in various ways but, in view of the urgency of the matter, long term studies are inappropriate. The quickest methods would be first, to institute a systematic deep boring programme on the site and adjacent sea area and second, to construct scale models of the outer Thames estuary and attempt to simulate the new sedimentological conditions created by the reclamation scheme. Unless these preliminary exploratory procedures are put in hand there can be no doubt that environmental consequences may follow affecting large tracts of the coastline and having a major impact on communities fringing the Thames estuary.

References

1. Greensmith, J. T. and Tucker, E. V. Morphology and evolution of inshore shell ridges and mud-mounds on modern intertidal flats, near Bradwell, Essex. *Proc. Geol. Assoc.*, 1966, 77, 329.
2. Greensmith, J. T. and Tucker, E. V. Salt marsh erosion in Essex. *Nature*, 1965, 206, 606.
3. Davis, D. S. The physical and biological features of the Mersea Flats. C. E. R. Laboratory Note, 1964, RD/L/N 131/64.
4. Stride, A. H. Current-swept sea floors near the southern half of Great Britain. *Quart. J. Geol. Soc. Lond.*, 1963, 119, 175.
5. Cloet, R. L. Determining the dimensions of marine sediment circulations and the effect of spoil dumping and dredging upon them and on navigation. Brit. Nat. Conf. Tech. Sea and Sea-bed. Proc., 1967, S.B. 27.

8B

Reprinted from *Civil Eng. Public Works Rev.*, **68**, 349–350, 352 (Apr. 1973)

Foundation conditions and reclamation at Maplin Airport

J T Greensmith PhD FGS
E V Tucker PhD FGS
Department of Geology, Queen Mary College, University of London

It is now some three years since the geology and soils conditions of the Foulness-Maplin site were reviewed in *Civil Engineering and Public Works Review*. At that time it was not clear if Maplin would be considered as a site for the third London airport, but events moved quickly thereafter, and by January 1969 the first series of boreholes requested by Roskill had been completed. Little geological information on Foulness and Maplin had been published prior to that date in the public field, though some exploratory holes had been drilled by private contractors in 1968.

The first group of Roskill deep boreholes, eight in number, were sited on Maplin Sands and widely distributed over the intertidal zone from some 2km offshore to near high-water mark. These holes proved 13 to 29m of superficial beds overlying London Clay and indicated prospective foundation problems, more especially with regard to the variable depth at which the firm London Clay occurred. A second series of deep holes were drilled in May 1969. These were located on the eastern side of Foulness Island and within a zone 1km wide on the adjacent sands. The reasons for the more restricted spread of holes were twofold; firstly, to establish the form of the London Clay surface more satisfactorily and, secondly, because even at this early stage it was thought likely that the airport would be sited on or adjacent to the island.

As consequence of the borehole programme a considerable quantity of new information became immediately available, the geological data being supplemented by a significant body of soils mechanics information.

Geology of the site

Shell and augar equipment was used for drilling and samples were taken at a minimum interval of 1·5m, more frequently when rapid sediment changes occurred, as was often the case. U4's were taken at regular intervals where the sediments allowed retrieval of the core. Excluding the difficulties of working in an intertidal zone, one of the main problems facing the drilling crews involved the identification of London Clay. As all major structures on the airport site are likely to be piled down to the clay surface it was essential to establish its position at depth with some accuracy. Generally, recognition was based on a combination of visual inspection (the fine grained clay being pale brown at the top surface and grey beneath) and resistance to penetration. However, within the superficial deposits, at depths of 19m and below, there occur over-consolidated clayey silts similar in appearance to the London Clay, even to the point of having comparable soils mechanics characteristics. At certain localities the thickness of these firm superficial beds proved to be at least 4m and as a result drilling sometimes stopped prematurely before reaching London Clay.

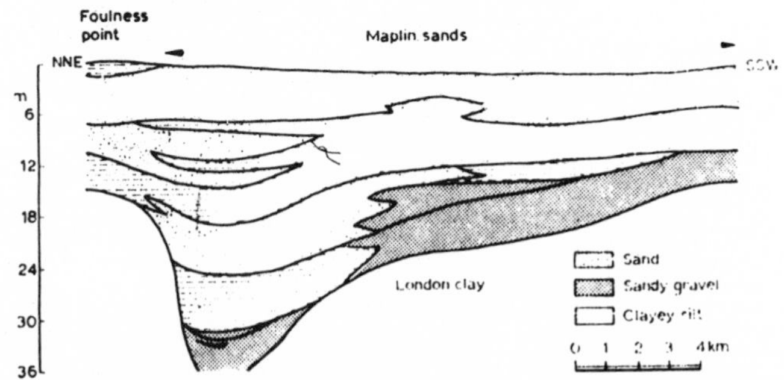

Fig 1. Surface contour map of London Clay.

The distinction between firm superficial beds and London Clay is unquestionably difficult on site. A common geological field technique is to identify any contained fossil organisms; the London Clay forms are different from those in the overlying sediments. But this method fails beneath Foulness and Maplin because the top layers of the Clay are so strongly weathered that all fossil remains have been leached away. The only satisfactory, but laborious, method of distinction is by chemical and mineralogical analysis. The quantities and ratios of minerals, such as the clay minerals illite and montmorillonite, and trace elements, such as copper, differ markedly between the two groups of rocks.

A further hazard in pin-pointing the Clay surface successfully is the presence of blocks and sheets of Clay incorporated within the lowest superficial deposits. These blocks and sheets were incorporated during deposition of the superficials by erosion and slipping along the flanks of ancient river channels carved into the Clay. The thickest discovered in the exploratory programme was a mere 0·5m, but the possibility of sheets attaining a thickness of several metres can not be ruled out. If of slip origin identification should be straightforward, as it would be anticipated that they would contain internal slip-planes, fracture surfaces and distorted laminae.

Despite these drilling uncertainties, exacerbated at certain sites by gravels blowing and sands boiling, in-place London Clay was proved in 15 out of the 21 deep holes requested by Roskill. The geological structure of the Clay beds could not be determined, though evidence from further offshore suggests that they are folded and faulted. However, what could be determined using the new and older information was the form of the London Clay surface at depth beneath the whole of Foulness Island, Maplin Sands and adjacent areas. The relevant part of this surface, contoured in metres below Ordnance Datum, is illustrated in Fig 1. This map gives a somewhat generalised picture and will be modified in detail as more information is acquired.

An important point emerging from the map is that the Clay surface consists of a network of ancient buried channels separated by ridges, all trending in a west-east direction, similar to the modern rivers. Some of the deeper channels vary in width between 2 to 4km and are infilled with 30m and more of superficial deposits. Somewhat unexpectedly, one of the channels beneath Foulness Island and adjacent Maplin Sands proved to be at least 35m deep. Two boreholes sited on the axis of this channel, one near to the sea-wall and another 3km offshore, failed to penetrate through the infill into London Clay. As the northern part of the airport site is located over the channel it is clear that careful consideration will have to be given to the position of major structures within the airport complex.

Because the London Clay surface varies in depth between 15m and 35m (·) the superficial sediment cover fluctuates in thickness. In general, the superficial deposits are thickest beneath Maplin Sands, thinning westwards towards the higher land bounding Foulness Island. The thickness also reduces eastwards into the outer parts of the Thames estuary.

At and towards the base of the superficial deposits under most parts of the airport site are sandy gravels, usually medium to coarse grade with a high water content (under pressure). They vary between 1 to 10m in thickness. The thickest beds proved occur beneath the south-east side of Foulness Island and adjacent sands reaching seawards as far as a position below mid-tide level (Fig 2). Certain of these gravels were very difficult to penetrate. Indeed, on some occasions the difficulties were such as to warrant cessation of boring and removal of the rig to another site. The problems were caused in two ways. Firstly, a weak lithification of the gravels by iron and calcareous cements, in effect making the rock harder. Secondly, the infiltration of clay and fine silt particles into the permeable fabric of the gravels, producing a more coherent, denser rock. Both the cementation and infiltration processes are secondary and reflect geological events some considerable time after the gravels were first deposited.

As the gravels are indurated in places it is probable that they would meet the minimum requirements for the foundations of some surface structures. Thus, it is clearly important that the depth of their top surface and their areal distribution should be established. In fact, the top surface of the gravels over practically the whole site is at a depth of 12 to 16·5m. The main exceptions are in the axial areas of the buried channels (surface at 24 to 32m), beneath the northern extremity where gravels are only intermittently present, and at the southern extremity where gravel is proved at only 10m depth.

The pebbles constituting the gravels are predominantly reworked flints with minor amounts of chert, vein quartz, quartzose sandstone and ironstone. They have a complex history but ultimately, some 14 000 to 30 000 years ago, were laid down by rivers flowing eastwards across Essex into the North Sea. Sea-level was much lower at the time. At certain periods after deposition the pebbles were subject to intensive sub-aerial frost action so that many of them are now shattered and very angular. In the last 10 000 years there is evidence of reworking by the sea which flooded westwards progressively submerging the Essex coast. The infiltration of clay and silt into the gravels probably dates from this period.

Interstratified clayey silts and sands overlie the gravels. In general, sand predominates in the top part of the succession and clayey silt predominates towards the base (Fig 2). The sands compare with those exposed at low water over the whole width of Maplin Sands excluding a narrow salt marsh zone. They vary in thickness between 3 to 28m. They also vary lithologically, some having a higher silt and clay content than others. Shells of marine organisms are common and at certain depths beds of shells 3m thick exist within or at the base of sand units. These shelly levels are very permeable locally influencing groundwater circulation.

The clayey silt beds, between 2 to 15m thick, present some of the more difficult foundation problems at the site. They vary considerably in degree of consolidation, some being soft and plastic with apparent cohesion values of 0·14 to 0·30kg/cm² and others being firm with values of 0·44 to 1·27kg/cm² (the London Clay locally varies between 0·88 to 2·55kg/cm²). The firmest levels occur at depths below 20m this, in practice, being the deeper parts of the buried channels. It is at these deep levels that the thickness of the firm beds is at least 4m. At higher levels in the succession, that is between depths of 0·6 to 17m, the firm beds are rarely thicker than 2m. The distribution of the firm layers has not yet been fully established.

Fig 2. Geological cross-section of airport site illustrating the geometry of the superficial deposits. The vertical scale is considerably exaggerated.

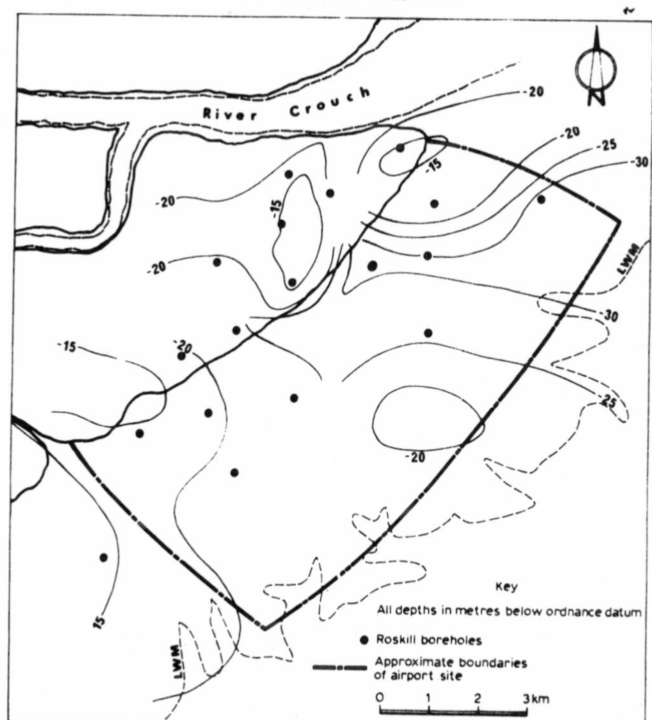

These firm levels are a consequence of overconsolidation by subaerial exposure and desiccation. They represent phases of eastward coastline extension which punctuate the inundation of Essex by the North Sea during the last 10 000 years. Because they relate to specific geological events their presence at depth is to some degree predictable. It is anticipated that virtually all future deep boreholes will penetrate into or through them. Again, within limits it should be possible to anticipate how many levels are likely to be penetrated and at approximately what depth.

A greater problem arises with the more plastic and dominant type of clayey silt where the effects of differential loading as a consequence of surcharging the airport site with fill, 2 to 5m thick, and ultimately concrete pavements are likely to be significant. The variable geometry of the clayey silts, interstratified between gravels and sands, creates difficulties in predicting the rate and amount of settlement. Excessive differential subsidence is a problem facing any constructional project located on low-lying coastal sites around the British Isles. In such locations peat seams commonly aggravate the problem. Along the margins of the Thames and Blackwater estuaries peat seams up to 2m thick occur within the superficial deposits. However, at Maplin such seams appear to be very thin, posing no problems in themselves. But, until a more detailed borehole programme is completed, it would be unwise to assume total absence of thicker seams, although the geological history of the site suggests that their presence is unlikely.

Reclamation

Much of the material to be used for reclamation and surcharging will be silt, sand and gravel pumped and dredged from the seabed in the outer Thames estuary. The bulk of this material is likely to be derived from the offshore deeps, but it will also include sediment dredged from new shipways leading to the proposed dock complex at the southern end of Maplin Sands. It is of acute concern to all interested parties that the bottom configuration of the estuary should be disturbed as little as possible by these operations. The 'balance of nature' is not an idealistic concept in this context and it is probable that there would be undesirable repercussions within existing shipping channels if excessive dredging was allowed. On the other hand, it is estimated that the total quantity of mobile sand in the outer estuary is 30 000 million cubic metres. So, if careful extraction of fill is assumed to be only of the order of 200 million cubic metres, then it is improbable that side-effects will be very significant.

Concluded on page 352

Although much land reclamation has been undertaken in the past on this coastline of Essex it must be appreciated that, even within historical times, this reclamation was preceded by a natural phase of accretion which produced extensive salt marshes. For three decades at least this same coastline has been retreating at a measurable rate due to a local imbalance between sediment supply and wave activity; simply, this can be regarded as a marine transgression.

The airport would have been best sited entirely on Foulness Island, because this would not have affected the present tidal regime and sedimentary budget. Conversely, the least desirable would have been several kilometres to the east. As it is, a compromise site has been selected, economic and social reasons being the important controlling factors. Even this site could lead to tidal complications, particularly with respect to the enhanced funnelling of water which could occur along the rivers Crouch and Roach. Under normal tidal conditions, the effects upstream are likely to be insignificant, but equinoctial spring tides and storm surges could create additional flood control problems and necessitate raising sea-walls along the full tidal extent of the rivers. The shape of the northern reclaimed section of the site is critical in this respect.

The maintenance of adequate sea-walls falls at present within the purview of County River Authorities and periodic heightening and strengthening is commonplace. This is a consequence of many factors, some local such as subsidence due to the draining of reclaimed ground or changes in rates of sediment deposition on the seaward side of walls, others regional due to long-term subsidence of the southern part of the North Sea basin. It is estimated that the minimum rate of regional subsidence in the outer Thames estuary is 2mm per year. But it is not always realised that superimposed onto this subsidence are long-term rises and falls in absolute sea-level. At the present time there is evidence that the region is passing through a phase which might be ascribed in part to a rise in absolute sea-level. This marine transgressive phase is expressed by marked erosion of salt marshes along the coast and the development of extensive shell ridges at high-water mark. The latter are well developed at the northern end of the airport site adjacent to Foulness Point, a locality designated as a site of scientific importance.

In summary, the reclamation will create certain difficulties most of which can be counteracted by adequate planning, based on experimentation and calculation, but there always remains a slight doubt that even the most sophisticated of plans can not take fully into account the vagaries of nature.

Conclusions

Of all the sites considered by the Roskill Commission the Maplin site is undoubtedly the most difficult in terms of actual and potential problems. Most of the foundation difficulties were broadly predictable prior to the preliminary exploration programme. But the drilling programme produced invaluable factual detail hitherto not available. Despite the undesirable characteristics proved for some of the sediments it is anticipated that there will be no insuperable foundation problems. The main difficulties foreseen relate to the shape and position of the reclaimed area and the inevitable disruption of the delicate balance of natural processes at work within the outer Thames estuary.

The authors are grateful to British Airports Authority and the Ministry of Public Works and Buildings (Department of the Environment) for making information readily available during the period of the Roskill Inquiry. The deductions made and opinions expressed are entirely their own.

Part III

MILITARY GEOLOGY

Military geology is a field not known to many geologists. It has been practiced mainly in time of war, when geologists were drawn into national service, either as civilians or uniformed members of armed forces. Since they come from unrelated peacetime activities, rapid indoctrination into the problems they will face is required.

A chief reason for the absence of military geology in university curricula is that military requirements for geologic advice are constantly undergoing change. It is obviously impossible to disclose these requirements for use in the open classroom.

Some geological surveys and other civilian organizations conduct research projects supporting military needs, and the results, at least in part, do appear in accessible literature. Also, some military establishments have developed their own staffs of geoscientists, who now have an opportunity for careers in the field of military geology.

What, then, is the importance of military geology in a consideration of environmental geology? A brief historical review will examine the question.

The origins of military geology date from the time when geologic science came into being. Among the first geologists, most of whom were eagerly pursuing what became the traditional activities of their science, a few were concerned with practical applications.

In the Introduction, mention was made of Johann Samuel von Grouner, who can be credited with having written the first examination of the relationship between geology and military science (1826). Grouner was trained as a geologist at Freiberg. During his career he engaged in a variety of technical and military tasks in Bavaria and Switzerland, which explains his principal interest in mountainous regions. This career was typical of the early military geologists in western Europe. In the mid-nineteenth century, the most prominent was Joseph E. Portlock, who also became a major-general and headed the Ordnance Survey of Ireland.

Military geology received little actual use until World War I, when it was used to great effect by the British and German armies. As a result, after the war, former military geologists in numerous publications stressed the applicability of the wartime experience to civil projects. We can attribute the emergence of the field of engineering geology to the demonstrated effectiveness of wartime work in studying construction problems (see Paper 27; there are many other examples). Hydrogeology gained stature from intensive study of problems of water supply, drainage, and sanitation.

The mobile warfare of World War II changed the scope of military geology, a change for which the newly gathered military geologists were not prepared. Their hasty orientation at the beginning of the war did not include consideration of problems to be found in the blitzkrieg, and they encountered numerous other unfamiliar problems as the war progressed. Primary concern was evaluation of possibilities for movement, both on roads and cross country, amphibious landings, and aerial landings. Engineering geology and hydrology were always important, especially in the more remote and unfamiliar environments, for example, the deserts of North Africa, the coral reefs of the Pacific, and the jungles of Burma and New Guinea.

Military geology in World War II grew out of the confines of purely geologic application. The term "military geology" was no longer truly indicative of the scope that developed.

Evaluation of terrain became the blanket term for the function of military geology. We should note here that "terrain" acquired a new meaning, differing from the one long used by military men (essentially equivalent to topography) and from the term as used by geologists (also in the spelling, "terrane," which has a vague stratigraphic-lithologic meaning, once discredited but now again in use).* To military geologists, terrain stands for the composite of features at and near the earth's surface, as far as the limits of man's activity extend, as stated by Betz and Elias, and Van Lopik (see the Supplemental Readings). Thus, terrain is essentially the same as *physical environment,* and it is a broader concept than that embodied in environmental geology.

To satisfy the requirements of terrain evaluation, military geologists recognized the need for multidisciplinary studies. They took the lead in bringing soil scientists, hydrologists, plant ecologists, foresters,

*"Terrain" in French is traditionally used in this sense, e.g., "Paleozoic terrain"; an approximate synonym would be "Paleozoic country."

meteorologists, climatologists, geographers, and other specialists into team efforts. They were in the forefront in developing air-photo interpretation (as contrasted with air-photo reading practiced by military personnel) so as to present armed forces with evaluated data on inaccessible areas. They opened the study of analogous areas, a concept that had not been thoroughly examined before.

It took more than two decades for geologists as a community to become aware of environmental sciences, in spite of the pioneering work done by military geologists. Geologists have now responded to the broad public demand for attention to environmental development and conservation, which all scientists are heeding.

Unlike the burst of literature on concepts of military geology, displays of performance, and applications for other purposes that followed World War I, there has been a paucity of publications on these topics since World War II. Thus, lack of recognition may be partly a fault of military geologists; but the opportunity has been there for others to observe what military geology produced. Copies of reports and maps have been deposited in numerous libraries, and this material should serve as evidence of the relationship between military geology and environmental geology. Both Parts I and VII include papers that show the concern of geologists with the problem of preparing derivative maps and other forms of communication intended for nongeologists, with which military geology has coped successfully.

Although the open literature on military geology must now exceed 1,500 titles, much of it consists of lengthy, once "classified" studies and "technical reports" that are not easily obtained. Furthermore, recent military geologic literature shows preoccupation with techniques rather than concepts.

SUPPLEMENTAL READINGS

Anonymous. 1948. Caratteristiche Militari dei Terreni Tipici dal Punta di Vista Litologico. *L'Universo,* **28**(2), p. 153-164.

Betz, Frederick, Jr., and M. M. Elias. 1957. Relationship of Geology to Terrain. *Geol. Soc. America Bull.,* **68,** p. 1700-1701. (Abs.)

Brooks, A. H. 1920. *The Use of Geology on the Western Front.* U.S. Geol. Surv. Prof. Paper 128-D.

Bülow, Kurd von, Walter Kranz, Erich Sonne, et al. 1938. *Wehrgeologie.* Heidelberg, Quelle & Meyer Verlag, 170 p. (An English translation by K. E. Lowe was issued by the U.S. Army Engineer Research Office, ERO 30, 1943. It is available in various libraries.)

David, T. W. E., and W. B. R. King. 1922. *Geological Work on the Western Front.* Chatham, Institution of Royal Engineers, 71 p. (*The Work of the Royal Engineers in the European War, 1914-1919,* vol. 7.)

Davies, Arthur. 1946. Geographical Factors in the Invasion and Battle of Normandy. *Geograph. Rev.*, **36**(4), p. 613–631.

Friedrich, Alexander. 1925. Das Gelände in seiner militärischen Bedeutung. *Verein Erdkunde (Dresden), Mitt.*, **3**(5/6), p. 54–93.

Genez, A. F. J. 1914. *Historique de la guerre souterraine.* Paris, Berger-Levrault, 297 p.

Grouner, J. S. von. 1826. Verhältnis der Geognosie zur Kriegswissenschaft. *Neue Jahrbücher Berg-Hüttenkunde (Nuremberg)*, **6**, 187–233.

Günther, Siegmund. 1918. Beziehungen zwischen Krieg und Erdkunde. *Deut. Rev.*, **43**(2), p. 134–142.

Johnson, D. W. 1921. *Battlefields of the World War—Western and Southern Fronts: A Study in Military Geography.* New York, Oxford University Press, 648 p. (Amer. Geograph. Soc. Res. Ser. 3.)

Keller, Gerhard. 1936. Die geologischen Voraussetzungen für den Minenkrieg im Wytschaetebogen. *Zentr. Geol., Paleontol., Mineral.*, Abt. B(6), p. 235–242.

Kranz, Walter, 1916. *Geologie und Hygiene im Stellungskrieg.* Stuttgart, E. Schweizerbart, 66 p. (Part in *Centralblatt Mineral., Geol., Paleontol.*, 1916, p. 270–276, 291–300.)

——. 1927. *Die Geologie im Ingenieur-Baufach.* Stuttgart, Ferdinand Enke Verlag, 425 p.

Kraus, Ernst, et al. 1941. *6. Wehrgeologischer Lehrgang in Heidelberg 14, bis 20, XII, 1940.* Berlin, Reichsdruckerei, 156 p., 96 plts.

Launay, Louis de. 1915. Les Champs de bataille predestines—histoire et geologie. *Rev. Deux Mondes*, 6 ser., **25**, p. 35–63.

Philipp, Hans. 1921. *Die Bedeutung der Geologie für Handel. Industrie und Technik, Landwirtschaft und Hygiene.* Greifswald, Germany, L. Bamberg, 35 p.

Schuring, H. A. 1958. *De invloed van het landschap op de militaire operaties tijdens de tweeden wereldoorlog.* Aalten, The Netherlands, De Graffschap, 127 p.

Shotton, F. W. 1944. The Fuka Basin. *Roy. Engr. J.*, **58**, p. 107–109.

U.S. Geological Survey, Military Geology Branch. 1952. *Geology and Its Military Applications.* U.S. Dept. Army Tech. Manual 5-545, 356 p. (*Editor's Note:* A later publication using the same serial number should not be examined as equivalent to this authoritative work, since the scope and contents have been extensively altered.)

Van Lopik, J. R. 1962. Optimum Utilization of Airborne Sensors in Military Geography. *Photogram. Eng.*, **28**, p. 773–778.

Wasmund, Erich. 1933. Seeablagerungen als Rohstoffe, produktive, technische und medizinische Faktoren. *Arch. Hydrobiol.*, **25**, p. 423–532. (See p. 523–529, dealing with military geology.)

Werveke, Leopold van. 1916. Die Ergebnisse der geologischen Forschungen von Elsass-Lothringen in ihrer Verwendung zu Kriegszwecken. *Wiss. Ges. Strassburg Schriften*, 28, 73 p.

Wilser, Julius. 1921. *Grundriss der angewandten Geologie (unter Berücksichtigung der Kriegserfahrungen für Geologen und Techniker).* Berlin, Gebrüder Borntraeger Verlagsbuchhandlung, 176 p.

Editor's Comments on Papers 9 and 10

9 KING
The Influence of Geology on Military Operations in North-West Europe

10 HUNT
Excerpt from *Military Geology*

Much has been written on the influences of geology in northwestern Europe on military operations. An excellent sample is Paper 9 by W. B. R. King, which was his presidential address to Section C of the British Association for the Advancement of Science. He states that "The object was to point out the ways in which the goelogical history of an area has controlled the military strategy and produced a pattern which tends to repeat itself on successive occasions. With changing military arms and equipment the details change but the basic pattern remains the same." He presents numerous examples from past wars, but then goes into illuminating detail on the effects in World War II. The two great overriding considerations, King observes, are "the availability of communications and the configuration and state of the ground," which control movement and actual deployment.

King (1889–1964) had the dual career typical of early military geologists. In both World Wars I and II, he was actively involved as a geologist working for the British forces. He was associated with the Geological Survey of Great Britain, and from 1920 with Cambridge University (Woodwardian Professor, 1943–1955).

A comprehensive review of military geologic activities during World War II has not been published, but one of the major organizations, the Military Geology Unit of the U.S. Geological Survey, has been discussed in some detail by Charles B. Hunt. The portion of this review reproduced here as Paper 10 deals with methods of applying geology and other sciences to military problems. Since field observations were generally impossible, numerous sources of data were sought and obtained; two of the most important were geologic and other maps and aerial photographs. The quality of the source materials was uneven and the coverage often incomplete.

Hunt was associated with the U.S. Geological Survey from 1930 to 1960 and headed the Military Geology Unit during part of the war. Since 1961 he has been a professor at Johns Hopkins University.

9

Reprinted from *Advan. Sci. (London)*, 8, 131–137 (1951)

THE INFLUENCE OF GEOLOGY ON MILITARY OPERATIONS IN NORTH-WEST EUROPE

Address by

PROF. W. B. R. KING, O.B.E., M.C., F.R.S.
PRESIDENT OF SECTION C

WHEN we look back into history, we see that the type of military operations in north-west Europe changes from time to time for various reasons, among which are primarily the location of the major centres of military strength and the kind of weapons used. Nevertheless, there appear to be two great overriding considerations which have influenced military operations ever since large armies were employed. Firstly the availability of communications, which allowed the movement of troops and equipment and made it possible to feed and clothe the army, and secondly the configuration and state of the ground, which controlled the actual deployment of the opposing armies in battle. Before railways were built, road and barge traffic shared in the problem of transportation, and fortifications were placed so as to control and command the few main roads and water transport systems. The reduction and capture of fortresses was not so much a military objective in itself as a means of freeing lines of communication, thus allowing the army to manoeuvre for position for the battle which followed as the main opposing forces came to grips. We need not go back many centuries to study the way the theme I am taking works out, but from the time of Marlborough onwards the pattern of operations and the control exercised by the geological features becomes clear.

The campaigns of Marlborough, apart from that which culminated in Blenheim, were mainly fought in Flanders and the adjacent areas of Brabant and northern France. Although the reduction of fortresses constituted an important part of each year's campaigning, Marlborough's great victories were won by skilful manoeuvring leading up to pitched battles in the open. In these, ability to move large bodies of troops was clearly influenced to a very marked degree by the position of naturally strong defensive river lines and the gaps between them, or possible crossing points not covered by a fortress.

Perhaps the most striking example of topographical controls affecting strategy is that of Ramillies, while Oudenarde and Malplaquet are further examples. The great defensive lines of French fortifications, first at La Bassée and then the 'Ne plus ultra' lines, illustrate cases at this period where natural features were strengthened by inundations and artificial earthworks to block movement from one area of open country to another. That they did not stop Marlborough is a testimony to his military skill rather than any indication that these were not naturally strong features based on geological considerations and were not the correct places for siting defensive lines. Rather do they illustrate that in 1711 as in the wars of 1914–18 and 1939–45 strong defensive positions, though obviously better than weak ones, cannot stand against first-class generalship combined with the will to break the position and backed by adequate equipment and men.

With the development of artillery the close-walled fortified city became of less and less importance and fortifications developed into a series of forts which were in the main protection for gun emplacements. Napoleon was a master in mobile warfare and it is significant that the great Napoleonic battles were in open country. The same may be said of Wellington. In the area we are considering these two did not meet face to face till Waterloo and the site of the battle is in the centre of this area.

In 1870 there was a change in the place of attack, the German columns descended on Sedan, avoiding Belgium, but in 1914 there was a return to form and the first battles followed a break through along the 'corridor of invasion,' across the 'cockpit of Europe,' spreading out over the open chalk lands of northern France with thrusts directed at Paris.

We come now to 1939 with Belgium and Holland neutral and with the French and British separated from the enemy over much of the front by great distances. In May 1940 Germany invaded the Low Countries and Belgium, whereupon the French and British, at the request of the Belgians, entered Belgium. During the previous winter this eventuality had been expected and plans had been worked out. Three possible lines of defence were considered : the most easterly line was the frontier of Belgium and Germany, and it was unlikely that the allied forces could reach this line in time ; the next, the 'D' line, ran from Antwerp using the line of the River Dyle (hence the 'D') across the 'corridor of invasion,' to Namur thence following the line of the Meuse till it reached the Maginot Line ; the more westerly line was that based on the River Escaut (called the 'E' line). I hope to show from a study of the geological controls that the 'D' line was the natural and obvious line to try to hold and this was the one which the allies took up in May 1940. Things, however, did not go as predicted and the flank of the position on the Dyle was turned in the south, and while a stand might have been made on the Escaut (the 'E' line) this again was turned by the rapid advance towards Arras and the coast. Withdrawal from Belgium was perforce necessary and the evacuation from Dunkerque took place.

A new line, the Weygand line, was for a time stabilised, based on the Somme, following the edge of the wooded plateau of Laon (formed by the Tertiary strata of the Paris basin) and running thence east to join the Maginot Line—again a line of strong natural defence which we shall see is controlled by the geology of the area. However, with the cracking of the Weygand line at Abbeville, again at a point where geology exerts a controlling influence, the whole front became fluid and the enemy overran the rest of the French Army.

The story of the 1944 invasion from this country across the Normandy beaches illustrates again the importance of a number of geological controls.

Let us now examine the general geological pattern of north-west Europe and see where the oft-repeated pattern of military operations can be co-ordinated with the physical geography of the area and the way in which this is clearly controlled by the geological history. The area falls into five natural regions :

(1) The low-lying area of north and north-west Belgium and the Netherlands formed of late Tertiary beds generally covered by alluvium or peat.

(2) The chalk and high-standing Tertiaries with inliers of Palaeozoic rocks forming an east-west belt north of the main mid-Meuse in Belgium and stretching away through northern France past Lille to Boulogne.

(3) The Palaeozoic and igneous rocks of the Ardennes and the Vosges with the intervening area of the Sarre and the hill country of Bunter sandstone.

(4) The main Paris basin in its larger sense, which can be divided into (a) the outer belt of the Trias, Jurassic and Lower Cretaceous strata, (b) the middle zone of the Chalk with the open Tertiary area of the Beauce, and (c) the central area of the Tertiary mass round Paris itself.

(5) The Palaeozoic area of Brittany and west Normandy.

Besides this lay-out of the 'solid' geology, the fact that the area was in the periglacial zone, lying immediately beyond the ice-front during the various ice advances, had far-reaching results on the final geography of the area.

It is not my intention to go into the details of the tectonics and stratigraphy of the various divisions but it is necessary to point out a few of the special characters of each subdivision.

(1) The low-lying areas in the north. These are formed of soft alluvial deposits

Map of the natural regions of North-West Europe

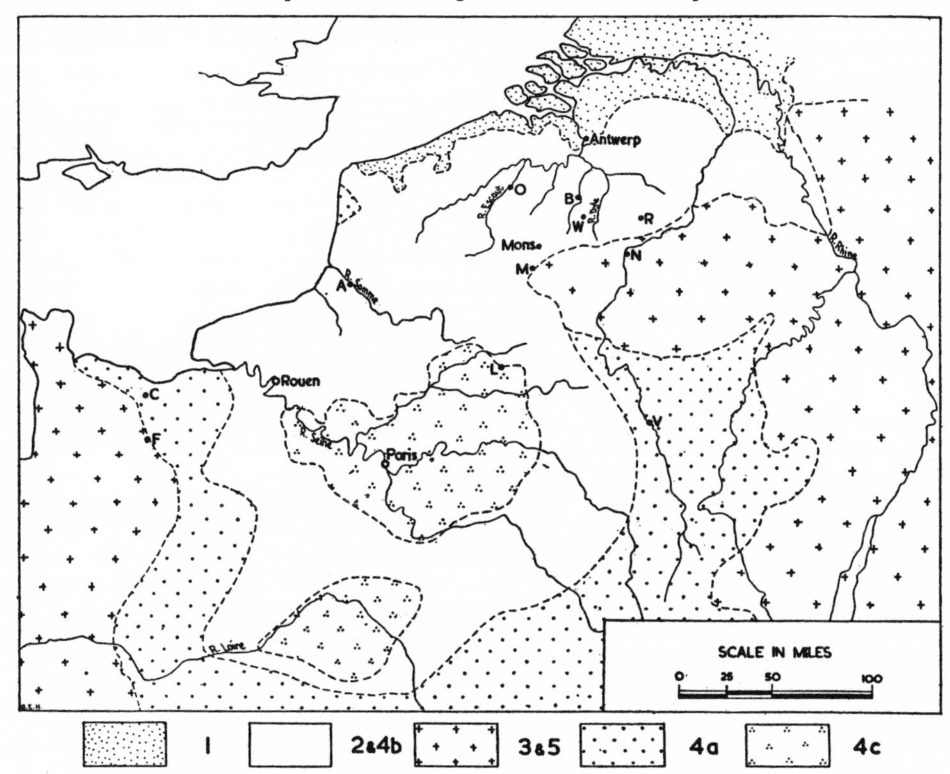

Numbers refer to the divisions noted on the opposite page.

Towns (West to East)

C = Caen; F = Falaise; A = Abbeville; L = Laon; O = Oudenarde; M = Malplaquet; B = Brussels; W = Waterloo; N = Namur; R = Ramillies; V = Verdun.

with wide, deep sluggish rivers and numerous canals, clearly making cross country movement difficult, while much of the area, being artificially drained, is therefore capable of being flooded. These conditions clearly aid defence and inundations if conveniently located, form parts of naturally strong defence lines. As an example of this, the flooded areas of the Yser in 1914–18 formed a strong defensive flank to the Ypres salient.

(2) The Chalk and high-lying Tertiary tract of Belgium and northern France. The fact that the area is high-standing and possesses a fair relief means that in summer it is well drained, but the clayey nature of the soil, due mainly to thick superficial deposits, results in wet slippery surfaces during winter and spring. However, particularly in the southern part of the area, we are near the watershed of the rivers converging on Antwerp on the one hand and those draining towards the Meuse and its tributaries on the other. This results in there being an east-west line where there are few major physical barriers, forming a natural east-west corridor for free movement.

(3) The Ardennes, Vosges and intervening area. These are either plateau lands with deeply incised valleys or steep mountainous country often heavily wooded. They form a belt of country on the east of France and southern Belgium which by its nature is extremely difficult country for the movement of large bodies of troops with heavy equipment, and when fortified in the modern manner of the Maginot line was considered impregnable. The weakness obviously lay in the fact that the boundary between France and Belgium divided this natural unit between two

countries, which precluded a joint plan for the proper layout of a continuous defensive system. Nevertheless, the area as a whole favoured defence, particularly along such lines as that where the Meuse cuts across the strike of the beds south of Namur in a deep, steep-sided gorge.

(4) The main Paris basin. The general geological arrangement of the Mesozoic and Tertiary strata is extremely simple, being that of a number of saucers of progressively smaller diameter ' nested ' one inside the other. The main complication is the failure of the Jurassic and Lower Cretaceous outcrop between the western end of the Ardennes and the Boulonnais. Thus from the Normandy coast round the southern side of the basin to the Ardennes we find the strata dipping gently inwards towards Paris, and since they consist of alternating limestones and clays in groups of strata each a hundred feet or so in thickness, there develops a pronounced dip and scarp topography the like of which, for regularity and persistence, is seldom found save in our own English Mesozoic rocks. The military significance of this is considerable, for each scarp overlooks the long gentle dip slope on its outer edge and, whereas the scarp edge of limestone is well-drained, the foot of the scarp and lower parts of the dip slope are liable to be wet and swampy clay lands. The configuration therefore tends to favour defence against anyone moving towards the centre of the basin. In our summary we divided the Paris basin into three sub-zones. The dip and scarp feature is best seen in the Jurassic and Lower Cretaceous belt, where many of the clay belts are heavily wooded and only the limestone tops are open country. In the middle zone, however, we find the Chalk forms large tracts in part covered by loams but in part bare Chalk downland, while into this zone may also be put the Tertiary lands of the Beauce where open well-drained country is as characteristic as in the Chalk areas. These open areas are the next best thing to the ' desert ' for open tank warfare, particularly in summer and early autumn when the superficial loam covering, where it exists, is dry and firm. This country is also ideal for the rapid construction of fighter airstrips. The innermost group is formed of practically flat-lying Tertiary deposits of massive limestone alternating with soft sands and clays. The result in a general way is a well-defined steep scarp edge surrounding an area of diversified woodland with well-entrenched valleys.

(5) The remaining division is the Palaeozoic massif of Armorica. The marked east-west strike implants a strong ' grain ' on the country which is seen in the elongated ridges of high ground where massive beds of resistant rock, such as quartzites, stand up as steep-sided ridges, as near Falaise, restricting north-south movement. The effect of this was seen in the way the United States forces, once to the south of the ridge of high ground at its western end, were able to sweep eastward with such speed along the strike of the softer beds.

Although the solid geology clearly has a great influence on military operations, the area we have been considering was subjected to all the effects of periglacial erosion and deposition during the Pleistocene Ice-age. In analysing the effects of this periglacial position, we note that in the first place there is abundant evidence that during the maxima of the ice advances there must have been a marked eustatic lowering of sea-level, and in general it is agreed that in the area of the English Channel and northern France and Belgium isostatic movements would not be expected to be significant. This area was possibly even in a zone of compensatory up-bulge around the main Scandinavian depressed area. We may with safety visualise a time when all the rivers, at any rate in their lower reaches, were undergoing marked rejuvenation. Whether the upper reaches of the longer rivers may not have been suffering from aggradation by excessive solifluxion need not concern us here. The point that is of importance is that the lower reaches of the Seine, Somme and all the rivers draining to the estuary of the Scheldt must have been markedly rejuvenated and must have incised their beds to a marked degree into the relatively soft Mesozoic and Tertiary rocks. This may have happened several times, for during each low sea-level period the rivers easily and quickly would clear out any deposits which had accumulated in the drowned valleys when water-level rose in the interglacial periods. Clearly therefore the effect of the latest lowering and refilling of the oceans, resulting from formation and melting of large ice-caps, left the most obvious results in the lower river valleys and coastal reaches. The 'Flandrian Transgression' has been responsible for the swampy, peaty infillings of so many of the lower river valleys and some coastal areas.

Whether the lower parts of a river valley became a marine estuary or merely a peat-filled fresh-water swamp appears to depend on the quantity of material brought down by the river at the time of rising sea-level. In some of the rivers such as the Seine there appears to have been sufficient gravel and sand in transport to prevent any peat swamps forming and so, although large alluvial and gravel flats occur, particularly where refilling the excavated deep-sided valley, the only real military obstacle over great lengths of the valley of the River Seine is the river itself.

Very different however is the case of the Somme valley and to nearly the same extent the Lys, Escaut, Dyle and many others. In these valleys the rivers flow through low-lying fen-like swamps from which peat has frequently been dug. The river itself is only a very small part of the actual obstacle from the military point of view. In the case of the Somme in particular the crossing points are few in number and are generally situated at points where river gravels have been brought down from tributary streams or where calcareous tufa has built up a deposit firm enough to carry a roadway. We thus see why the northern river lines are natural defence positions apart from the actual river itself.

The strength of the middle Meuse line above Namur depends on the fact that the river there is incised into an old uplifted plateau of hard rocks resulting in a gorge with almost precipitous sides, a line of great natural strength which should be able to be held by very few troops.

We must now consider certain features which have resulted from the very special deposits which formed in the periglacial area surrounding an ice-cap whose outer margin is on a relatively flat-lying land area. The meteorological conditions round the margin of the Scandinavian ice-cap have been considered by several workers and, whilst some claim that a glacial anticyclonic type of air circulation must develop with strong north-east winds in the periglacial area, others claim that the normal westerly circulation must have been maintained to nourish the ice-cap with moisture-laden winds. When speaking of the winds under the periglacial conditions pertaining at the time, it would perhaps be more accurate to call them 'moisture-bearing' rather than 'moisture-laden.' What is demanded is that they should have enough moisture to give precipitation when forced to rise and become chilled by the high surface of the ice-cap.

It must also be remembered that at the glacial maximum large tracts now covered by the sea had been laid bare and would probably be in a state ready to supply considerable quantities of dust to the winds whichever way they were blowing—a point made by several French authorities. However, whatever the meteorological conditions were, it is clear that the loess was deposited over much of the area we are considering. The characteristics of this deposit are of the greatest importance from the military point of view. Since it was wind-carried, it was spread as a great blanket over hill and vale alike but during the subsequent periods of greater rainfall it has been washed from the slopes and is now found only on the plateaux or partially filling the smaller side valleys. In its original state it is a fine calcareous loam with grains mainly of the silt grade. In common with nearly all loess deposits it has a tendency to a rough vertical jointing structure which allows it to stand almost vertically in open cuttings and, during dry weather to drain well, forming a firm even surface. When weathered it develops a 'loamy weathering' largely by decalcification and then the surface behaves rather more like a clay. But even so, on drying it does not shrink and crack in the same way as a true clay but when wet the weathered loess 'puddles up' to a fluid mud which becomes almost impassable to concentrated or heavy traffic. The loess-covered plateau has another feature—the soil produced on these surfaces is an excellent arable soil, and in France and Belgium forms the main plough lands, being cultivated in huge open fields without hedges or ditches. When dry, therefore, we find the vast almost horizontal surfaces of the plateau regions formed of firm, dry loam land, ideal for the siting of temporary airstrips and for rapid and uninterrupted movements of armoured vehicles. However, when the soil of these areas becomes soaked with water in the winter and spring months, they cannot be used for these purposes, certainly not for temporary airstrips, without considerable preparation and surfacing. In fact for most cross-country movement the area must be downgraded to the same category as pure clay lands. In most of the areas of chalk, the covering of loess overlies thick clay-with-flints which produces a much more intractable soil and so, if there had been no loess formation at some time in the past in this area, we would have had a very different agriculture. This would probably have been of the ' bocage ' type with its numerous small fields, high hedges and orchards, rather than these great open stretches of country.

As far as the land areas are concerned the main effects of the glacial period in the area we are considering from a military point of view were therefore, first, the formation of overdeepened valleys later filled with peaty alluvium forming well-defined transverse barriers, and secondly, the even covering of loess giving wide open spaces ideal for free movement during seasons when the soil was dry. This is particularly important in area 2 and to some extent in area 4, for where the solid geology gave a flat plateau topography, the loam covering gave an agriculture of large fields without hedges. In the inner zône of the Paris basin this advantage is masked by the deep dissection of the area, while in the two areas of old rocks the loess cover is usually absent or very sporadic.

On the coasts however we find another complication resulting from the movements of sea-level in glacial and post-glacial times. The effects of the Flandrian Transgression are well seen in the type-area of the Flanders coast lands but it is the latest oscillations of sea-level as they affected the coasts of the Channel and North Sea that produced certain outstanding conditions and deposits. At first sight a study of this movement of sea-level might be thought to be of purely academic interest and of no military significance. Let us look at it from the scientific standpoint first and see what can be deduced from the known sequence of events.

There is the generally accepted figure of 100 metres for the lowering of sea-level during the maximum of the last glacial episode but we need not argue about the exact figure. All that is relevant for the present argument is that there was an eustatic lowering of sea of about this order of magnitude. This must have resulted in a complete draining of the North Sea basin as far as the Dogger Bank and the English Channel to a point about the longitude of the Lizard, with perhaps an estuary running up the line of the Hurd Deep (Fosse Central). But apart from this time of extreme lowering, the sea-level must have been considerably lower than at the present time in late glacial and immediate post-glacial times. During this time the older beach deposits would be eroded and deep channels cut through the hard solid rocks of the former off-shore wave-cut platform where this consisted of limestone or other resistant rock ; the valleys themselves would be deepened ; new sea-beaches or coastal dunes would form and behind these, swamps and fen peats would develop, resting either on the remains of the older beach deposits or on the now exposed wave-cut platform of solid rock. As the sea-level reached its present position, the old peats and forests were submerged and the present beach developed, beach sands overlying the peats in those places where the invading sea did not remove them by erosion. A new pebble beach would develop at and above the new high-tide mark formed of stones, either driven landward by the sea or derived from denudation of the cliffs. Thus we find many of our own coasts and those of the French side of the Channel characterised by off-shore stretches of a wave-cut platform of solid rock with here and there deep passages cut through it, then, landward of this, patches (often extensive) of plastic clay and peat overlain by a veneer of modern beach sand of variable thickness and, finally, a storm beach of pebbles behind which the modern fen-like deposits are now forming.

It was these conditions on the Normandy beaches which gave our planning staffs such concern, since experiments on counterpart British beaches had shown that most vehicles could not travel over outcrops of peat or plastic clay and were even liable to cut through a thin cover of sand and become bogged in the underlying soft deposits. It was known from French records that peats existed on the Normandy beaches but in the early days of the planning, the beaches appeared from air photographs to be evenly covered by sand. After the storms of 1943, however, suspicious dark patches in the bottoms of the runnels suggested that peat might be exposed. Much careful work over the next six months culminated in the preparation of large-scale maps of the beach indicating those parts which were unsuitable for the passage of assault vehicles, and a pre-invasion reconnaissance by commando volunteers confirmed the existence of outcropping peat. The success of the landings was sufficient justification for the time spent on this problem. By far the greater part of the geological research was carried out by Professor Shotton who was then geologist at H.Q. 21st Army Group. The deep-water entrance to the 'Mulberry' harbour was the result of the erosion of a gap through the old wave-cut platform during the time of low sea-level, so that again we see the control exercised by the geological history of the area.

The presence of flat-lying Jurassic limestones covered by loess-like loam made it possible to construct, with remarkable rapidity, the necessary airstrips, for here we had the nearly ideal conditions of a permeable rock at depth stripped by weathering to a flat topography and covered by a loess-like loam which during summer months dried quickly to a firm soil. Owing to the lack of irregularities such as ridge-and-furrow or even hedges scarcely any work was necessary on levelling the sites. These airstrips allowed the fighter aircraft to be serviced near the frontline thus helping materially to maintain the air superiority with all that that entailed.

Had the return journey across France and Belgium been contested stage by stage, no doubt the geological controls mentioned previously would have operated but after the Battle of Falaise things moved so quickly that geology did not enter into the problems until the front was temporarily stabilised in Holland and at the German frontier.

The object of this address has been to point out the ways in which the geological history of an area has controlled the military strategy and produced a pattern which tends to repeat itself on successive occasions. With changing military arms and equipment the details change but the basic pattern remains remarkably constant.

10

Reprinted from *Application of Geology to Engineering Practice* [Berkey Vol.], Geological Society of America, 1950, pp. 295–306

MILITARY GEOLOGY[1]

By Chas. B. Hunt
U. S. Geological Survey, Denver, Colo.

INTRODUCTION

During World War II geology won its spurs as an important scientific tool in both planning and operations by the United States Army. This growth of geology was due to increased appreciation on the part of our military leaders of the importance of scientific techniques and information, and to the increased appreciation on the part of our scientists of the usefulness of their abilities in the solution of a large variety of very practical problems. It can be fairly said that at the beginning of the war neither the military leaders nor the geologists fully appreciated the manifold

[1] Published by permission of the Director, Geological Survey, U. S. Department of the Interior.

applications of geology to military problems. Basically this was because the geologist, prior to the war, had signally failed to give sufficient thought to the many ways in which geology can and should contribute to solving everyday social and economic problems. In engineering geology, for example, there were too few Berkeys.

An important factor contributing to this situation, and in my opinion the most important factor is the inadequacy of geologic mapping in the United States. In 1946 less than 10 per cent of the continental United States was geologically mapped on a scale as large as 1 mile to the inch. Even that scale is far too small to be useful in most practical matters other than general planning. The result has been that we have not had immediately available the basic scientific information to assist the solution of commonplace problems whenever and wherever they arose. We have not had a complete and ready file of geologic maps that could be consulted and used to furnish the scientific information needed for a specific interpretation. And most engineering projects could not be delayed while the tedious assembling of the scientific information was undertaken; so those in charge largely dispensed with geology.

It is a curious fact that our military geologists were more successful in overcoming problems abroad than in overcoming problems at home, because geologic mapping in most of the countries that became operational areas was far more advanced than in our own country. The application of geology to military problems was successful in proportion to the degree to which the basic scientific information, that is, the geologic mapping, had been completed in advance.

I strongly suspect that the applications of geology to civil-engineering problems would be found similarly dependent on the availability of geologic maps; that the engineering geologist is successful to the degree that he already possesses the basic information and can devote his full time and energies to the interpretations applicable to the problem at hand. This, in my opinion, is the most important lesson to be learned from the experiences of the military geologists in World War II.

APPLYING GEOLOGY TO MILITARY PROBLEMS

Applying geology to the solution of military problems involves consideration (1) of the techniques and uses of geology, and (2) of the administrative procedures by which the necessary information can be furnished to the right office at the right time without imposing a burden on an already complex military organization.

Three important ways in which geology can be used by the Army are: (1) map interpretation, especially of geologic and topographic maps; (2) photo interpretation; and (3) consulting service on the ground.

INTERPRETING FROM GEOLOGIC MAPS

Geologic maps used in conjunction with topographic maps can provide much information about the ground conditions of a given area. Whether the problem concerns the planning for a military operation in an area not yet occupied or the planning of an enterprise within our own country, much waste can be avoided if the geologic information is available during the early planning stage. A few examples will illustrate the kind of information that can be quickly gleaned from the two basic maps.

Figure 1 is a topographic map of an area. It is possible to project the contours, mathematically accurately, for any desired angle of view, and for orthographic, isometric, or perspective views. It is then possible for an artist to sketch the slopes

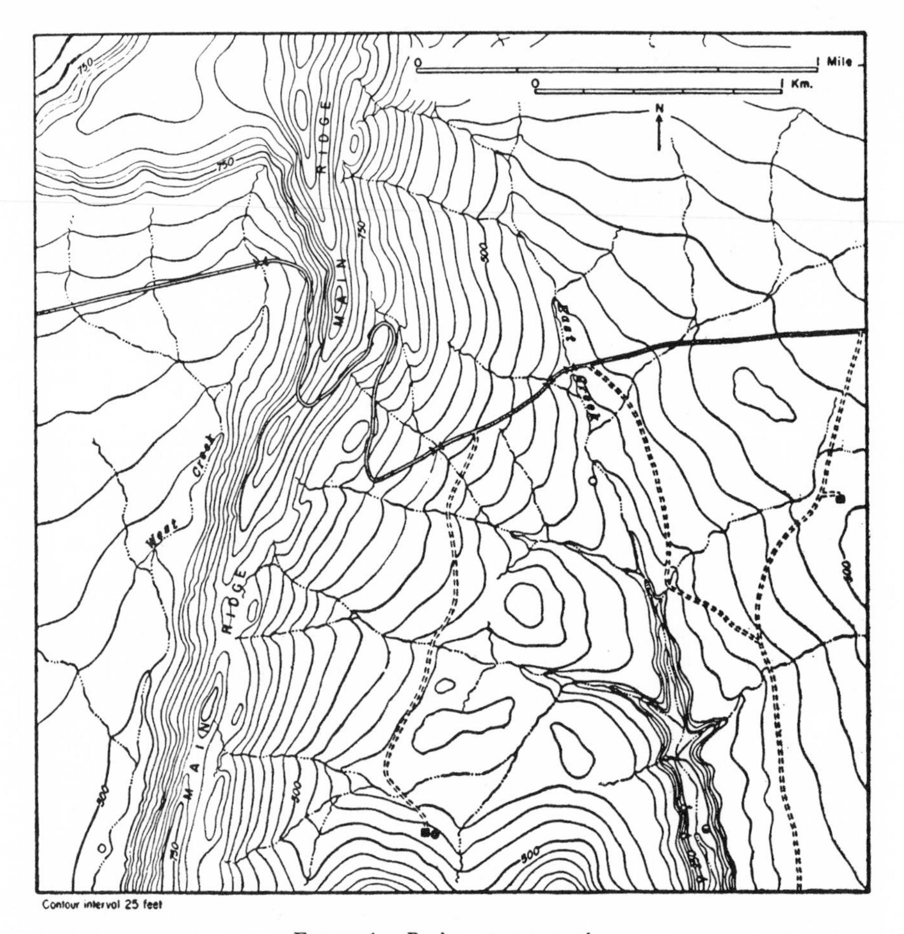

FIGURE 1.—*Basic map, topography*

between the projected contours and so produce a relief diagram. Figure 2 is an example.

To a large degree, of course, the accuracy of such a diagram depends upon the adequacy of the topographic base, but the limitation of a topographic map is that one must assume an even slope between contours. Knowledge of the geology, however, can add much detail that cannot be determined from a topographic map alone.

Figure 3 is a geologic map of the same area shown in Figure 1. By combining this geologic information with the information provided by the topographic map, it is possible to predict many terrain details that have military or other engineering significance. Some of these details are shown in Figure 4. Let us examine this figure closely to see what details are shown that are important in the problem of terrain appreciation—that is, the estimate of the terrain situation as it affects the movement, cover, and concealment of troops and supplies.

Just east of the crest of the main ridge and parallel to it is a continuous depression. The importance of this depression lies in its suitability for crestline movement of foot troops and light equipment. Moreover, troops moving along this depression would have good cover—they would be exposed only at the gully crossings. This

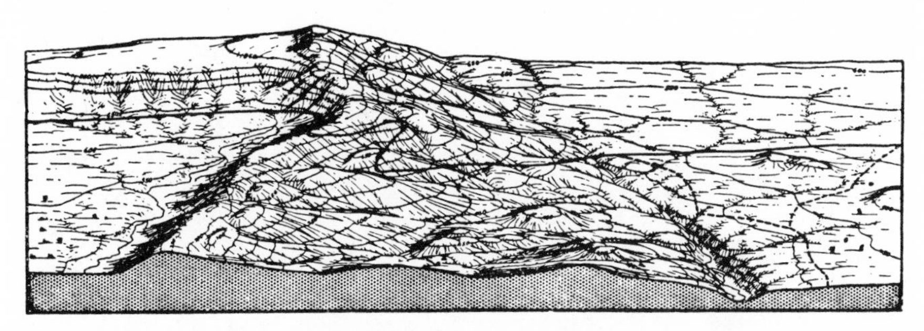

FIGURE 2.—*Terrain diagram*

Diagrams such as this are artist's sketches controlled by projected topographic contours. Contours can be projected for any angle of view and for orthographic, isometric, or perspective drawings. This diagram was sketched from orthographic projections of the basic topographic map (Fig. 1) viewed at a vertical angle of 13°. The vertical scale is not exaggerated.

depression fails to show on the topographic map (Fig. 1) because its depth is less than the contour interval (25 feet) except at the few localities where closed contours appear, but its presence is suggested by the geologic map because of the position of the contact between the hard sandstone D and the softer sandy shale of formation C. In Tunisia, Italy, and on Attu, combat troop movements commonly were along ridge crests in order to secure commanding ground or to drive an enemy from it.

Similarly, the geologic map reveals the presence of a low, asymmetrical ridge along the west side of West Creek, a feature entirely missed by the topographic map because the contour lines do not "catch" it. Though less than 25 feet high, the ridge is probably high enough to provide excellent cover against fire from the west.

At the south edge of the map two rounded hills are shown between East Creek and Main Ridge. These hills look alike on the topographic map, but the geologic map shows that the eastern one is capped by hard limestone. The Army of course is not interested in this fact as such but is interested in its significance—namely, that on this hill an observation post may not be able to dig in and may therefore be dangerously exposed. Effects of shell fire on the two hilltops would be quite different, with shallower penetration and greater fragmentation occurring on the eastern hill.

The topographic map shows that the walls of East Creek are steep; the geologic map shows that these walls are formed by the outcrop of hard, well-bedded sandstone which almost certainly forms low cliffs and ledges—the engineers will need to blast to construct a road across the valley.

Vehicles moving cross country along the east side of East Creek would have to cross the gullies that are tributary to East Creek. The gullies probably are narrow and steep-walled; they may be serious obstacles to the movement of even tracked

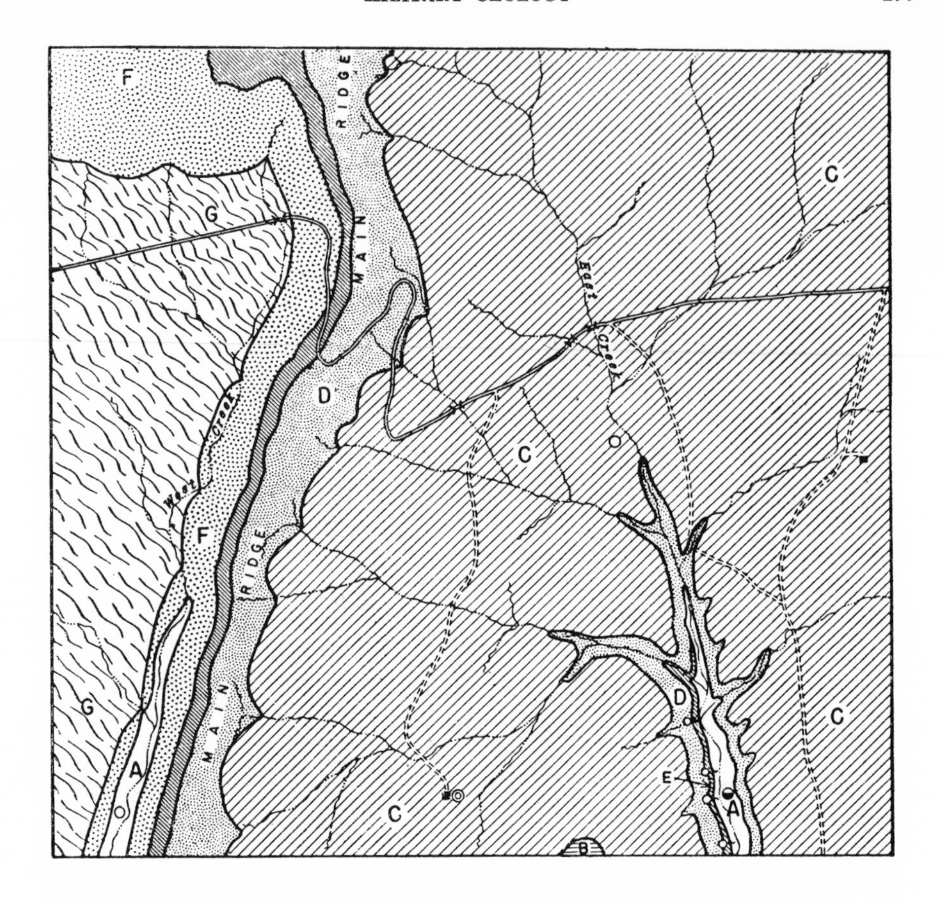

EXPLANATION OF GEOLOGIC MAP

SYMBOL	KIND OF MATERIAL	THICKNESS IN FEET	TOPOGRAPHIC EXPRESSION	DISTINCTIVE FEATURES	WATER-BEARING PROPERTIES
A	Alluvium; gravel, sand, silt, some clay.	< 50	Flood plain and valley bottoms.	Unconsolidated	Permeable, small underflow; perennial yield only where underlain by impervious rocks.
B	Limestone.	max. 45	Caps flat-topped buttes.	Thick massive beds; hard; resistant.	Impervious.
C	Shale; kaolinitic clay shale; lower 50' is sandy shale interbedded with sandstone in beds up to 20'	200	Low hills; lower sandy beds form minor ridges on east slope of Main Ridge.	Clay shale is finely laminated, easily eroded; sandstone is hard. Sharp contact with limestone above.	Shale impervious; basal sandy beds slightly permeable, very small yield.
D	Sandstone: quartz-sandstone, with lime cement.	125	Forms crest of Main Ridge.	Fine-grained; thick massive beds; resistant.	Coefficient of permeability 45; small yield; water is high in carbonates.
E	Shale; gypsiferous shale containing shaly sandstone lenses and pure clay layers.	200	Forms broken slopes along East Creek and west side of Main Ridge.	Sandstone lenses are fine to coarse-grained; poorly cemented.	Shale impervious; sandy beds slightly permeable; water is brackish.
F	Sandstone.	250 to 300	Moderately resistant; hard beds form ridge along West Creek valley and cap northwest upland.	Coarse grained (in part containing pebbles up to 1" diameter); poorly cemented; some hard, resistant beds.	Coefficient of permeability 350; large yields; water is low in dissolved minerals.
G	Schist: micaceous rock cut by irregular quartz veins.	900 + M	Forms western lowland.	Finely laminated in general; a few massive granitic layers.	Impervious; no water except in occasional open fissures within 200 ft. of surface.

FIGURE 3.—*Basic data, geology*

vehicles. The best place to seek a crossing is just above the mouths of the gullies, where they meander broadly and have emerged from the shale C and flow on the sandstone D.

Well-drained ground is not always available at the place where a foxhole must be dug, or at a site where a battery, a dump, a roadway, or an airfield must be located.

Poorly drained ground would not be selected, however, if well-drained ground is known to be available and is tactically equally suitable.

In the example before us the shale of formation C (Fig. 3) is increasingly sandy westward, where the ground is hilly, so that the clay soil formed from the shale

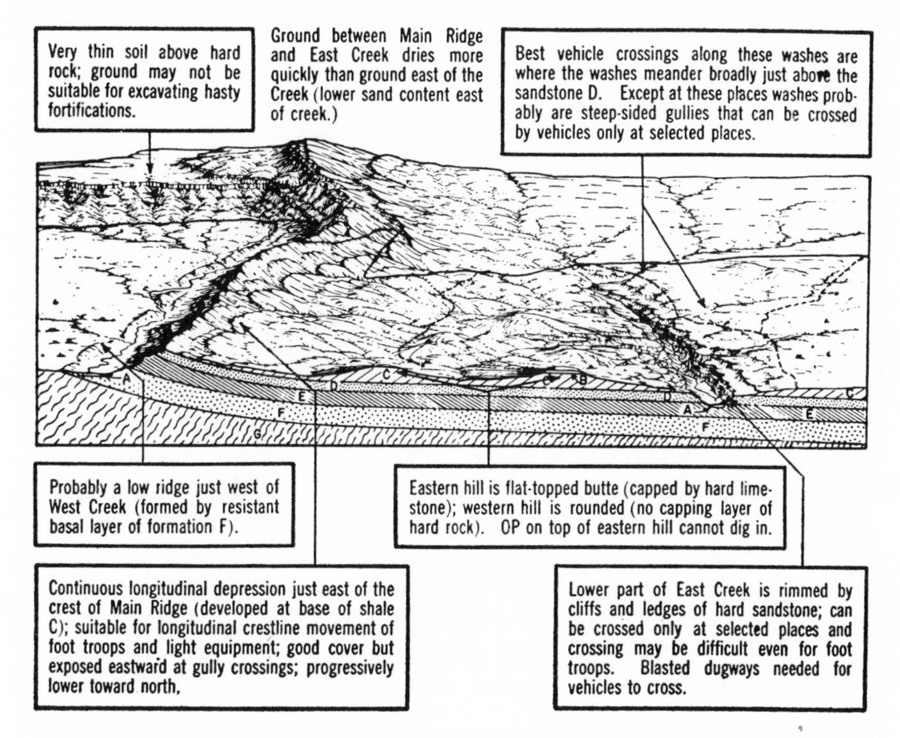

FIGURE 4.—*Terrain diagram*

Prepared by combining the geological information with the information provided by the topographic map. *Compare* with Figure 2.

likewise is increasingly sandy westward. Accordingly there are two reasons why the ground between East Creek and the Main Ridge is better drained and dries more quickly than the ground east of the Creek—the greater hilliness means increasing runoff and a lower water table, and the greater sand content means improved ground drainage. This has a major bearing not only on excavation and other construction problems but also on the trafficability of the ground. Following wet weather, mud would impede military operations or construction anywhere in shale C, but for shorter periods and less severely west of the creek than east of it.

The best-drained ground in the area is on the plateau formed by the sandstone F in the northwest corner of the area, but this ground, although an excellent natural foundation, is not suited for digging shelters, for the soil above the bedrock is thin. Moreover, because the bedrock is hard to excavate it will require blasting; but the rock is highly permeable so an excavation in it would drain well.

Foxholes or other shallow excavations in the clay ground in the eastern part of the area would, in wet weather, become cisterns. The more sandy clay ground on the hills east of Main Ridge and the micaceous silt ground on the flat to the west of the Ridge are somewhat better-drained, and excavations there are not so likely to collect water. For general trafficability, the eastern part of the area is less suitable in wet weather and more suitable in dry weather than the hilly belt and the flat ground farther west.

Terrain appreciation is not the only field in which such forecasting can be made successfully. In the problems of water supply, for example, available maps commonly show wells or springs but fail to record their discharge or quality. But by knowing depth, geologic location, and climatic environment of such water sources it is possible to estimate the limit of expectable yield. Figure 5 shows some examples of this type of prediction. Moreover, if these already developed water supplies are not adequate for the number of troops to be involved in the proposed operation, additional development of the water resources may be necessary. The geology indicates the yield and quality of water that can be obtained from the springs or from dug or drilled wells in various parts of the area. Figure 6 provides examples. This figure also shows a cross section of the geological formations that underlie the terrain and shows how they control the occurrence of underground water. Conversely, the geologic map also reveals in which parts of the area one can expect the greatest seepage losses of surface water, a factor to be considered in planning storage reservoirs or canals.

The success of many an operation is dependent upon the speed with which certain construction can be completed. In turn, the speed and also the economy of construction is dependent upon understanding and planning to meet the problems of the particular area. Here again knowledge of the geology is useful. Because the geologic map shows the outcrops of the different formations, it indicates the distribution of different kinds of materials needed for construction. Descriptions of the formations reveal their suitability for various construction uses. Also, because the kind of parent material is a major factor controlling soil development, the geologic map—where soil surveys have not been made—can be used to predict the natural foundation conditions. Figure 7 gives some examples of these uses for a geologic map.

In selecting a site for an airbase, the topographic map (Fig. 1) indicates no basis for choosing between a site west of West Creek and one in the northeast part of the quadrangle. But the geologic map shows two reasons for preferring the site in the northeast part of the quadrangle. (1) As pointed out earlier, water can be developed in the northeast part of the quadrangle, but formation G, in the west, is devoid of supplies suitable for an airbase. If a base is constructed on formation G, water will have to be imported. (2) The soil developed from formation C is easily compacted and will make good foundation for a compacted base of selected aggregate borrowed from the main ridge. The soil over formation G is elastic and would make a poor foundation for a compacted base because of the great thickness of base required.

The examples cited have been selected primarily to review the usefulness of geologic maps in planning military operations. But the same examples illustrate equally well the usefulness of geologic maps in planning civil enterprises that are influenced by ground conditions. The only difference is that in a military operation

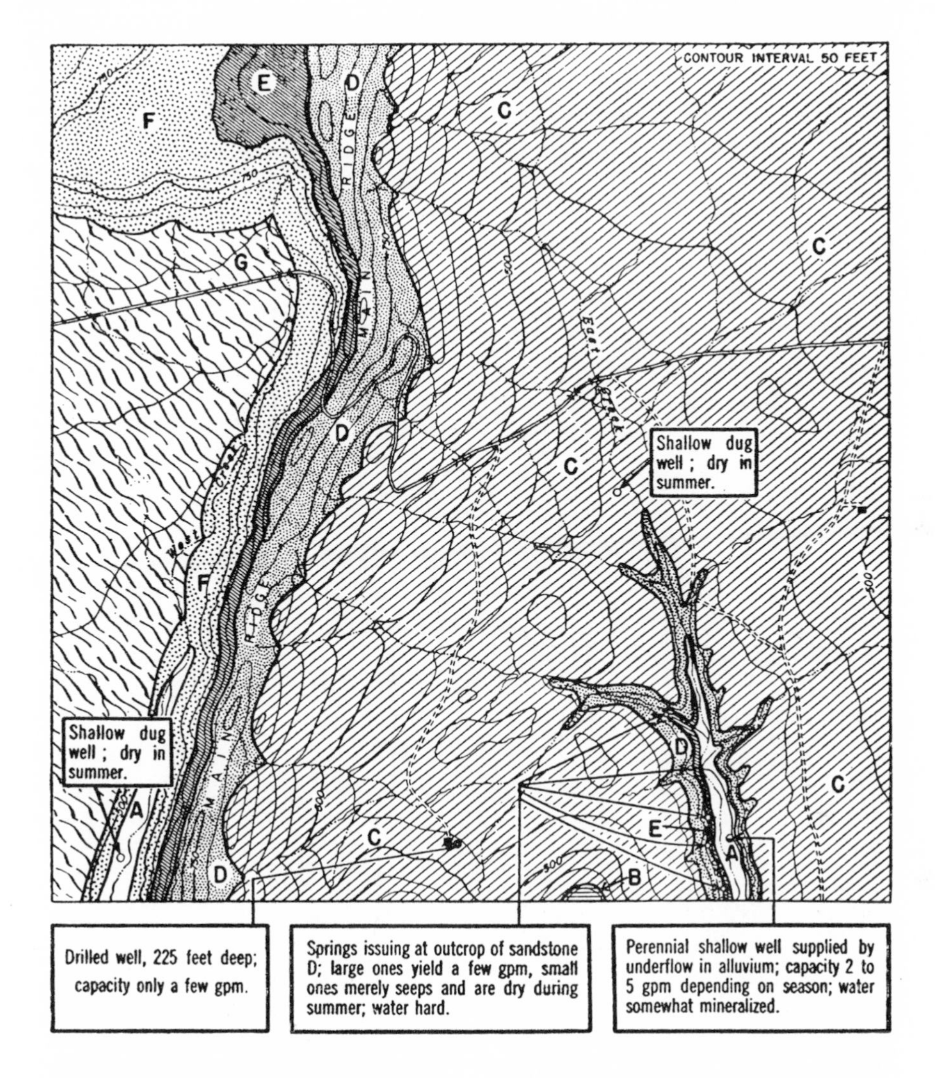

West Creek and East Creek are intermittent streams; have small flow for several months during rainy season. Other streams flow only after heavy rains. Sufficient water for combat units is obtainable from springs throughout year and from streams during rainy season, but existing water is not adequate for permanent installations. Diagram on the next page shows how adequate supplies can be developed by utilizing groundwater.

FIGURE 5.—*Predictable limits of existing water supplies*

the geology is useful primarily for achieving speed, whereas in civil enterprise the geology is useful primarily for achieving economy.

INTERPRETING FROM PHOTOGRAPHS

Many of the principles that make a geologic map useful for estimating the terrain situation apply with equal force to the usefulness of aerial photographs. The accuracy and completeness of the interpretation depend always on the skill of the

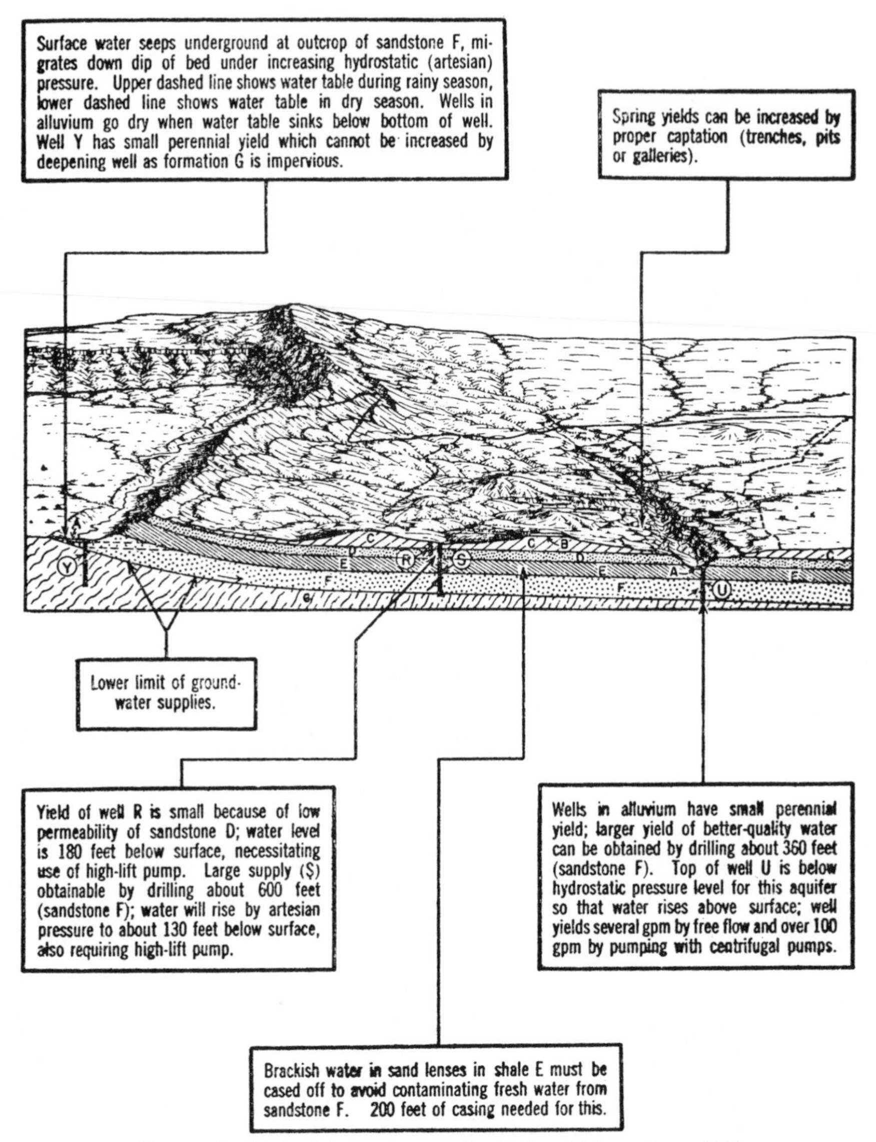

FIGURE 6.—*Undeveloped water supplies that could be made available*

interpreter—his knowledge of, and experience with, the natural processes and forces that determine the condition of a given piece of ground. Without professional knowledge of these geologic processes and forces many clues to ground situations that are militarily important will be overlooked.

For the preparation of strategic terrain intelligence, aerial photographs do not add a great deal of information, provided geologic maps are available on a scale commensurate with the strategic report. On the other hand, aerial photographs are indispensable for preparing tactical terrain intelligence. In strategic studies the

Subgrade problems along roads: West of West Creek subgrade lacks stability, is highly subject to frost heave; excessive maintenance needed. On Main Ridge soil is SP(A-3) or possibly SF(A-2) type; binder should be borrowed from shale east of the ridge; soil west of ridge unsuitable for binder. East of ridge soil can be easily compacted, is good foundation for a compacted base of selected aggregate, and possibly can be mechanically stabilized by adding aggregate borrowed from ridge; increasing amount of aggregate needed as subgrade is improved eastward.

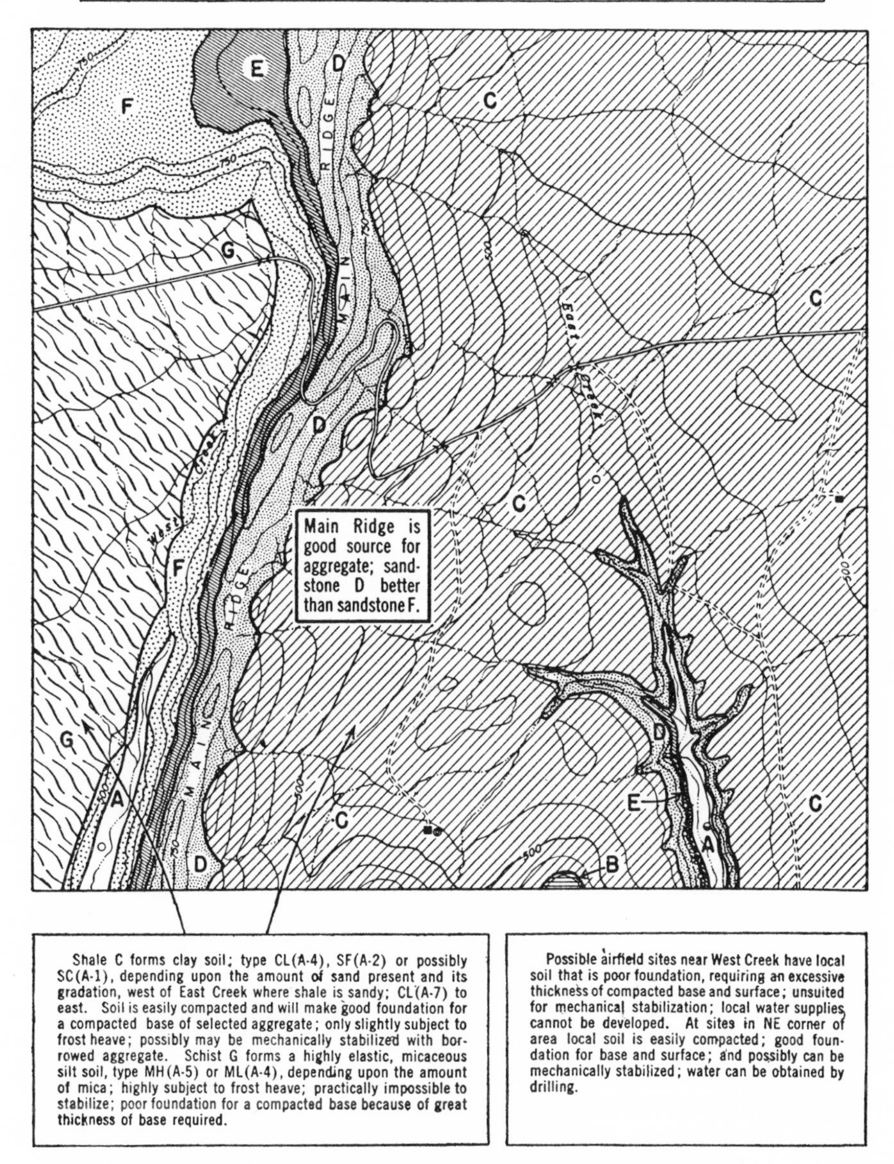

FIGURE 7.—*Predictable ground conditions affecting construction*

regional picture is more important than details of localities; in tactical studies the reverse is true.

The usefulness of aerial photographs varies therefore with the scale of report being prepared. The kind of uses made of photographs depends too on the adequacy of the existing map information. If geologic maps are available on a scale as large as the aerial photographs, the two can be used together to reveal many specific details that neither one alone can furnish.

But this ideal combination is rare. More commonly the geologic maps are on a scale that is half or a quarter that of the aerial photographs. In this situation primary dependence is on the photographs, but the geologic maps are on a scale sufficiently large to be of tremendous aid in determining the details of the geology as they appear in the photographs. With this combination of data, it is frequently possible to lean confidently on very minor variations in tone in the photographs or on variations in vegetation to predict major differences in ground conditions, because the geologic map forewarns the interpreter to look for the difference, and he has only to determine where it is located.

Even where geologic maps are lacking or are on such a small scale as to be practically useless for tactical intelligence, geologic principles can be applied with advantage to interpreting the terrain from aerial photographs. Little training in reading vertical photographs is required to recognize mountains, hills, lakes, rivers, woods, plains, or some kinds of swamps. But much more than that can and should be interpreted from the pictures for the purposes of acquiring complete terrain intelligence. It is essential to know the kind of hill, the kind of plain, the kind of river or lake, and so on, because by knowing this it frequently is possible to reconstruct the geology. The interpreter, with some confidence, can then make predictions as to water supply, the kind and depth of the soil, trafficability, ground drainage and other construction problems, construction materials, movement and cover, and many of the other elements that are essential to an adequate estimate of the terrain situation. In brief, therefore, aerial photographs are useful to the preparation of terrain intelligence insofar as they provide information on the geology of the area. Identification of a hill or other terrain feature is but a small part of the story that can be read from a photograph; all important is the recognition of the significance of the particular land form, in terms of kind of ground and slope.

CONSULTING SERVICE IN THE FIELD

Reading the face of the earth is geology; reading the face of the earth as it affects military operations is military geology. The surface of the land can be interpreted for military purposes by study of maps; it can be interpreted even more effectively by interpretation of aerial photographs and maps together. Both these methods have proved highly effective in practice, but they cannot compare in accuracy or completeness with observations made in the field.

A common military situation is that in which an attacking force confronts a strongly defended ridge or hill. Knowledge of the terrain beyond the hill is vital to the attacking commander in planning his next move. The terrain may be very different on the side that cannot be seen, yet very commonly the difference is clearly expressed in the geology of the slope that is under observation. Ground observation by a geologist combined with interpretation of aerial photographs and maps is the most efficient approach to the problem. This principle is applicable to many combat situations.

For obvious reasons, the services of geologists can be utilized to an even greater extent in the development of territory that has been secured. Results obtained by field geologists in the recent war have conclusively proved that, especially in site selection, the presence of a geological consultant can expedite military construction and render it more efficient. This is true whether the problem is siting a bivouac, a depot, a quarry or gravel pit, a source for water, an airfield, or a road. The unskilled observer must rely on the obvious features that are immediately observable, but very commonly the whole fabric of the terrain can be read quickly by ground observation of significant geologic features.

In addition to solution of specific problems, geologic analyses can be made, in advance, of areas where military construction or operations are anticipated. Such studies, as made by Geological Survey and War Department geologists since the end of the war, consist of preparing, on the ground, geologic and soils maps. These are used as basic data for preparing other maps showing the distribution of water resources, construction materials, and potential airfield sites, and the suitability of the ground for foundations and for cross-country movement by vehicles and foot troops.

Part IV

SOME GEOLOGIC ENVIRONMENTS AND RESOURCES

As its title states, this part is not a comprehensive review of settings and resources, but a representative selection. Aside from the limited space that can be allowed, there is the criterion of relevancy of literature to environmental quality and impact determination. Many geologic publications on resources show little awareness of the environment in which the resources exist, and this comment is even more applicable to settings.

The selection presented is a sampling of several situations. The emphasis is not equally distributed, which may be regrettable, but there is no rating system for settings. Resources are susceptible to rating in terms of supply and demand, but this consideration has not been uppermost in making the selections.

ENERGY SOURCES

Their importance is so great today that no defense is required for making this the initial topic. Among the various energy sources, geothermal energy is highlighted because of its potential in the face of critical conditions affecting other sources. For more than a decade, international interest has been growing. As evidence we can cite the United Nations Conference on New Sources of Energy (Rome, 1961), which dealt at length which geothermal sources, and the United Nations Symposium on the Development and Utilization of Geothermal Resources (Pisa, 1970), which yielded many papers. There is now a continual flow of literature on the topic.

Also in the area of energy sources is the hotly debated subject of oil pipeline construction, an interesting and, to many people, disturbing aspect of energy development. Here we arrive at a mixing of the subjects of this part and that of Part VI, human influences. The environmental impact is explored, and, as happens so often in controversial cases, the evidence receives different interpretations. There is a natural setting, a resource, and consideration of the effect that the proposed means of exploitation will have on the surrounding environment.

MINERAL RESOURCES

A narrowing of the vast topic of mineral resources had to be undertaken. The choice made was to examine submarine deposits. Minerals in the sea are a relatively untapped resource, but technology exists and is being developed further to make economic exploitation possible. Exhaustion of some resources on land makes it all the more important to look at the potential in the sea. The outlook may not be too bright according to some marine scientists, who entertain serious doubts about the size of reserves and the feasibility of making the materials attractive to venture capital. This does not apply to fuels, since, in a strict sense, they are not *mineral* resources. The future for fuels appears to be distinctly greater, perhaps in part because development is being conducted mainly by organizations that also control sizable land-based supplies, that are being exploited.

The topic of marine mineral resources appears on the scene at a time when the general environmental impact of exploitation of mineral resources is a live matter. As far as land-situated mining and quarrying operations are concerned, the problem is mostly one of attempting to restore damaged landscape, under governmental and public pressure, and not one of planning in advance to avoid damage to the environment, which must be seriously considered in reference to the untapped submarine resources.

LIMESTONE TERRANES

Compared to subsea mineral resources potential, the topic of limestone terranes does not appear to be of like urgency or future interest. However, these terranes present a number of intriguing problems in respect to geomorphic features, hydrologic peculiarities, and the presence of metallic mineral resources. As is well known, limestone is also a very important construction material as shaped stone and as an ingredient of manufactured products. Its many uses cannot be enumerated in this space.

Students of terminology will note once again that the use of the term "terrane" (also subsequently spelled terrain and even terraine) by geologists was as a vague stratigraphic–lithologic term; it was little used after the late nineteenth century, and today "terrane" usually has a distinctly other significance (a utilitarian, environmental sense; see also comments in Part III.) The term "terrane" is used here to accommodate to the language of present-day carbonate-rock specialists.

UNDERGROUND OPEN SPACES

Underground open spaces represent a topic of growing interest. They are formed both by natural processes and by human action. The geologic setting must be suitable for their creation, but they differ in usefulness and/or hazard after having been developed. Man-made spaces are deliberately produced for a number of reasons, the most familiar being to gain access to mineral and rock resources. From ancient time there has also been mining for military purposes; some of the most impressive tunneling occurred on the Western Front during World War I.

An interesting historical note is provided by the famous British author, Alec Waugh, in his volume *In Praise of Wine* (London, Cassell & Company Ltd., 1959), who notes that some champagne vintners of Reims "have use of the superb chalk cellars that lie under the city." The cellars were excavated by the Romans as mines for extracting chalk. Even earlier, Stone Age man mined this same chalk to obtain chert nodules for making his flint tools. In southern France at Roquefort-sur-Soulzon, natural caves in limestone cliffs have long been used to store and cure the famous Roquefort cheese, and some provided shelter for Cro-Magnon man (see Ellison in Supplemental Readings, Part I).

There are many examples of storage of art works and valuable records in underground spaces, such as salt mines, in several countries. Today, a salt mine in the area of Hutchinson, Kansas, located at a depth of 650 feet, has 30 miles of dry corridors, said by the operators of the storage facility to provide 50,000,000 cubic feet of space. Current mining is conducted so as to provide further storage area. A wide range of materials is being kept in the mine for 3,000 depositors from every state of the United States and 36 foreign countries.

AESTHETIC QUALITY OF GEOLOGIC ENVIRONMENTS

This topic is not represented by a paper, owing to space limitations, but comments are offered. It is rarely considered by geologists as a topic requiring their professional attention. Today, however, aesthetics has become a factor in land-use planning, especially in respect to siting and developing recreational areas (see brief comment and Supplemental Readings, Part II). For such areas, the aesthetics of the physical environment, with its geologic background, must be weighed against other resources that play a part in recreation, such as capacity and amenity.

The National Institute for Physical Planning and Construction Research (An Foras Forbartha) in Dublin, Ireland, has conducted continuing studies in which these factors are examined. Concerned with the development of tourism, this agency has published guides to determining the suitability of environmental settings. Factual observations are the basis for much of the mapping undertaken by the institute, but qualitative ratings appear in the aesthetic evaluations. As an example, scenic evaluation along a coastal road is given by the following color-coded indicators: "sublime, highly scenic, scenic, rural, eyesores" (see the Supplemental Readings).

In a paper on the aesthetic factors in rivers in Idaho, Luna B. Leopold (1969) (see the Supplemental Readings) comments that "It is difficult to evaluate the factors contributing to the aesthetic or nonmonetary aspects of a landscape." His study

> is a preliminary attempt to quantify some elements of aesthetic appeal while eliminating, insofar as possible, value judgments or personal preference . . . . Assignment of quantitative elements to aesthetic factors leads not so much to ratios of value as to relative rank positions. In fact, value itself tends to carry a connotation of preference, whereas ranking can more easily be used for categorization without attribution of preference . . . ."

Leopold notes that any ranking scheme must be based on a philosophical framework. He stresses uniqueness and human interest, and allows landscapes that are uniquely bad to be ranked as such. To one reader it is impossible to escape the conclusion that personal preference is being cloaked by numbers.

Aesthetic quality of the environment refers also, largely retrospectively, to the role of geology in developing the character of man's monuments and habitations. The topic was touched on in Paper 1.

From antiquity, the architectural style reflected building materials near at hand. Thus, the geologic environment and resources were factors in determining a concept of aesthetic quality in various societies and contributing to the cultural environment.

However, even in early times stone was brought from some distance to satisfy certain needs and tastes. The link between styles and local resources has been weakening progressively, especially during recent centuries. This trend has been hastened by the widespread introduction of manufactured materials. Today, typical regional styles of architecture are being replaced by styles that are virtually worldwide in use and acceptance.

In another paper that could not be reproduced here (see Supplemental Readings), Alois Kieslinger, a geologist in the Technical University at Vienna, tells of the restoration of the severely damaged St. Stephen's Cathedral. A great deal of sleuthing had to be undertaken to match the stone, which differed in different parts of the building owing to the prolonged period of construction during which different sources were used. Kieslinger has written at greater length on the subject in another publication. The engineering geologic problem was also of major importance, since the structure had been weakened and required emergency action to prevent further damage.

SUPPLEMENTAL READINGS

Agnew, A. F. 1967. *Water, Geology and the Future.* Indiana University, Water Resources Research Center, 117 p.

Armstead, H. C. H. 1973. *Geothermal Energy—Review of Research and Development.* Paris, UNESCO, Earth Sciences Series 12, 186 p.

Banwell, C. W. 1963. Thermal Energy from the Earth's Crust. *New Zealand J. Geol. and Geophys.*, **6**, p. 52-69; **7**, p. 585-593, 1967.

Barnea, Joseph. 1972. Geothermal Power. *Sci. Amer.* **226**(1), p. 70-77.

Bowen, R. C. 1973. Environmental Impact of Geothermal Development. In: *Geothermal Energy, Resources, Production, Stimulation* (P. Kruger and C. C. Otte, eds.), p. 197-215. Stanford, Calif., Stanford University Press.

Bullard, Edward. 1973. Basic Theories. In: *Geothermal Energy—Review of Research and Development* (H. C. H. Armstead, ed.), p. 19-29. Paris, UNESCO.

Bullard, F. M. 1962. *Volcanoes—In History, in Theory, in Eruption.* Austin, Texas, University of Texas Press, 441 p.

Hammond, A. L. 1972. Geothermal Energy: An Emerging Resources. *Science,* **117**(4053), p. 978-980.

Healy, James. 1970. Pre-investigation Appraisal of Geothermal Fields. *Geothermics,* Special Issue 2, p. 571-577. (United Nations Symposium on the Development and Utilization of Geothermal Resources, Pisa, Italy, 1970.)

Ireland, National Institute for Physical Planning and Construction Research (An Foras Forbartha). 1966. *Planning for Amenity and Tourism—Special Development Plan Manual 2-3.* Dublin.

——. 1970. *Planning for Amenity, Recreation, and Tourism,* vol. 2, pts. 1, 2, 3. Dublin.

Jaggar, T. A. 1945. *Volcanoes Declare War: Logistics and Strategy of Pacific Volcano Science.* Honolulu, Paradise of the Pacific, Ltd., 166 p. (See p. 115-119.)

Kieslinger, Alois. 1957. Gesteinskunde im Dienste der Baugeschichtsforschung. *Oester. Akad. Wiss. Anz. Phil.-Hist. Klasse,* 25, p. 399-404.

Kummer, R. E. 1964. Geological Considerations in Architecture. *Kansas Acad. Sci. Trans.,* **67**(2), p. 307-310.

Larsen, David. 1973. Back to the Salt Mines: The World's Biggest Chamber of Commerce. *Los Angeles Times,* Mar. 18, Pt. 2, p. 1, 2; also *The Washington* (D.C.) *Post,* Apr. 8, p. K-17.

LeGrand, H. E. 1969. Potential of the Ground Environment for Water Supply and Pollution Abatement—With Special Reference to the South Atlantic States. *18th Southern Water Resources and Pollution Control Conf. Proc.,* p. 137-145.

Leopold, L. B. 1969. *Quantitative Comparison of Some Aesthetic Factors Among Rivers.* U.S. Geol. Surv. Circ. 620, 16 p.

McKelvey, V. E. 1969. *Potential Ill Effects of Subsea Mineral Exploitation and Measures to Prevent Them.* U.S. Geol. Surv. Circ. 619, p. 14-16.

McNitt, J. R. 1973. The Role of Geology and Hydrology in Geothermal Exploration. In: *Geothermal Energy—Review of Research and Development* (H. C. H. Armstead, ed.), p. 33-40. Paris, UNESCO.

Marinelli, Giorgio. 1973. Deep Down Power: *Develop. Forum* (United Nations), **1**(3), p. 5, 9.

Pettyjohn, W. A. (ed.). 1972. *Water Quality in a Stressed Environment—Readings in Environmental Hydrology.* Minneapolis, Burgess Publishing Company, 309 p.

Revelle, Roger, Eugen Seibold, et al. 1970. Submarine Geology in Relation to the Use of the Earth's Mineral Resources. *Oceanology* (Amer. Geophysical Union), **10**, p. 749-760.

Stauffer, Truman, Sr. 1973. *Kansas City: A Center for Secondary Use of Mined Out Space.* Soc. Mining Engr. (AIME), Preprint 73-I-63, 26 p.

United Nations. 1964. *Proceedings of the United Nations Conference on New Sources of Energy, Rome, Italy, 21-31 August 1961.* Vols. 2, 3, New York, United Nations, E/CONF. 35/4.

United Nations. 1971. *Symposium on the Development and Utilization of Geothermal Resources. Proceedings, Pisa, Italy, 22 September-1 October, 1970.* New York, United Nations, ST/TAO/SER. c/126, 26 p. (Also see *Geothermics,* Special Issue 2, 1971.)

U.S. Senate Committee on Commerce. 1973. *The Oceans and National Economic Development.* 93rd Congress, 1st Session, 263 p. Washington, D.C., Government Printing Office.

Van Tuyl, D. W. 1951. *Ground Water for Air Conditioning in Pittsburgh.* Penna. Geol. Surv., 4th Ser., B. W-10, 34 p.

Waring, G.A. 1965. *Thermal Springs of the United States and Other Countries of the World.* U.S. Geol. Surv. Prof. Paper 492, 383 p.

White, D. E. 1965. *Resources of Thermal Energy and Their Utilization.* U.S. Geol. Surv. Circ. 519, 17 p.

Editor's Comments on Papers 11 Through 15

Several summaries on geothermal energy sources have been published. Our choice is an especially comprehensive and well-organized paper in two parts (Papers 11A and 11B) by L. Trowbridge Grose, who is a professor of geology in the Colorado School of Mines and a consulting geologist.

Dealing first with *geologic settings* favorable for exploration and development of geothermal resources, Grose notes in Paper 11A that "Exploration for geothermal steam in commercial quantities involves an integrated and coordinated program of application of many geological, geophysical, and geochemical techniques." Once again, this exemplifies environmental geology as characterized by Hackett, to add to those views presented in the papers in Parts II and III.

The worldwide review of information on areas that have current or potential production of geothermal energy is especially valuable as a resource survey. The benefits of geothermal energy are seen to be the relatively low cost production of electric power, relatively minor pollution and environmental effects of geothermal power plants, and potentially multipurpose use of developed geothermal resources. A valuable bibliography of almost 200 titles accompanies this paper.

The implications of the Alaska oil pipeline are brought out in Paper 12, taken from *The Earth and Human Affairs,* which was prepared by the Committee on Geological Sciences (National Research Council-National Academy of Sciences, Washington, D.C.). The construction in this environment may be expected to "disrupt the natural equilibrium between the insulating vegetation mat, the active layer, and the permafrost below." Effects of slope modification and subsequent runoff of surface water may be multiple. Earthquake activity in the region can endanger the pipeline directly and indirectly, and oil spills must be considered as a serious possibility.

Frank T. Manheim, a member of the U.S. Geological Survey, is an authority on the chemical composition and distribution of sea-floor deposits. A portion of his publication on mineral resources off the northeastern coast of the United States is reproduced as Paper 13; it deals with sand and gravel, which "represent not only a very large, but probably the most immediate resource likely to be exploited" in this region. In a statement on environmental factors, Manheim explores the various effects that the removal of sand and gravel will have on the environment. Once more we may note that setting, resource, and the effect of human influence will be inseparable when exploitation commences.

Paper 14, by Harry E. LeGrand, presents a discussion on karst regions, which not only portrays the geologic and topographic characteristics of these regions, but takes into consideration the interrelated factors that make the regions distinctive in terms of habitability and their overwhelming influence on hydrological sufficiency and ecological balance.

LeGrand was a research hydrologist with the U.S. Geological Survey at the time of his retirement, and is the author of various perceptive, well-written papers on the scientific and applied aspects of water-related problems.

Reproduced as Paper 15 is a portion of a comprehensive study of the secondary use of underground space in the area of Kansas City, Missouri. It refers to the largest development of mined-out limestone quarries in the United States. According to the investigator, Truman Stauffer, the mined-out and available underground sites have an area of 102.5 million square feet. The statistics are so impressive that it is difficult to appreciate the significance of this underground world. As an example, an underground refrigerated warehouse in the area can store over a pound of food for each person in the United States. Underground rail spurs can accommodate eighty freight cars, each with a capacity of 100,000 pounds of food. Manufacturing plants, storage of files and other records, and offices are located here.

Stauffer, of the Department of Geosciences, University of Missouri-Kansas City, has prepared a unified environmental study, combining consideration of the relevant geologic and geographic elements, the social and economic impact, and the mining technology.

11A

Reprinted from *Min. Ind. Bull.*, 14(6), 1-14 (1971)

Geothermal energy: geology, exploration, and developments
Part 1

by L. Trowbridge Grose
Colorado School of Mines

ABSTRACT

Geothermal energy is emerging as a potential major source of low-cost, relatively pollution-free electric power in the Western United States and in many other areas of the world. Geology, exploration, and initial developments of significant geothermal areas of the world are summarized in this report which is divided into two parts. Part 1 is a review of the geological and explorational aspects of geothermal energy development: the author also discusses areas of potential development in the Western United States. Part 2 is a review of significant developments of producing and potentially productive areas of the world, and includes an extensive and select bibliography covering Parts 1 and 2.

The most favorable geological environment for exploration and development of geothermal steam is characterized by Recent normal faulting, volcanism, and high heat flow. Successful exploration for steam consists of coordinated multidisciplinary application of geological, geophysical, and geochemical knowledge and techniques. These are reviewed. California leads in known geothermal reserves and is followed by Nevada, Oregon, and New Mexico. Specific prospective areas in these 11 Western States are described.

INTRODUCTION

The impending energy crisis in America is focusing increasing attention on possible new and potential sources of electrical energy: solar, tidal, wind, fission, fusion, and geothermal. In the last few years, the use of nuclear fission and of geothermal heat for generating electricity has proven economic in competition with energy produced from conventional sources: oil, gas, coal, and hydro facilities.

Geothermal steam and associated geothermal resources means "(1) all products of geothermal processes, embracing indigenous steam, hot water, and hot brines; (2) steam and other gases, hot water, and hot brines resulting from water, gas, or other fluids artificially introduced into geothermal formations; (3) heat or other associated energy found in geothermal formations; and (4) any byproduct derived from them" ("Geothermal Steam Act of 1970," Public Law 91-581, December 24, 1970).

The first electric power-generating facility in the world using natural steam as an energy source was established at Larderello, Italy, in 1904 (fig. 1). In 1958 electrical power from geothermal steam became commercial at Wairakei, New Zealand. In June 1960 the first commercial steam generating plant at The Geysers, California, began to produce electricity. At the present time, seven countries are commercially producing electric power from geothermal energy, and several more are planning to go on-stream within a year or two (fig. 1). Worldwide installed generating capacity has now reached nearly 900,000 kilowatts (kw), and probably can be increased at least tenfold under current economic conditions. The U. S. Geological Survey has recently outlined nearly 1,800,000 acres within the Western United States and Alaska that are potentially valuable for geothermal energy development, and that may contain an energy source estimated in the range of 15- to 30-thousand megawatts (MW). The "Geothermal Steam Act" signed into law by President Nixon on December 24, 1970, provides legislation for leasing, development, and use of geothermal resources: this long-awaited act has stimulated geothermal interests and investigations in the Western States.

This report is a summary of the geological, explorational, and initial developmental aspects of geothermal energy technology. Emphasis is placed on abundant reference to the most up-to-date literature on this rapidly expanding subject. The largest assemblage of data on geothermal technology is contained in the 190 individual reports of the United Nations Symposium on the Development and Utilization of Geothermal Resources, held in Pisa, Italy, in September 1970. The first significant and comprehensive series of reports on the subject emerged from the United Nations Conference on New Sources of Energy, held in Rome, in August 1961. Other papers that broadly cover the geology and exploration of geothermal energy include Kiersch (1964), McNitt (1965), White (1965), and Austin (1966). A comprehensive bibliography of geothermal phenomena, compiled by Summers (1971), was published in October 1971. A most recent paper by Rex (1971) forcefully presents the case for full development of the enormous geothermal resource base available in the Western United States.

This report is divided into two parts. Part 1 deals with

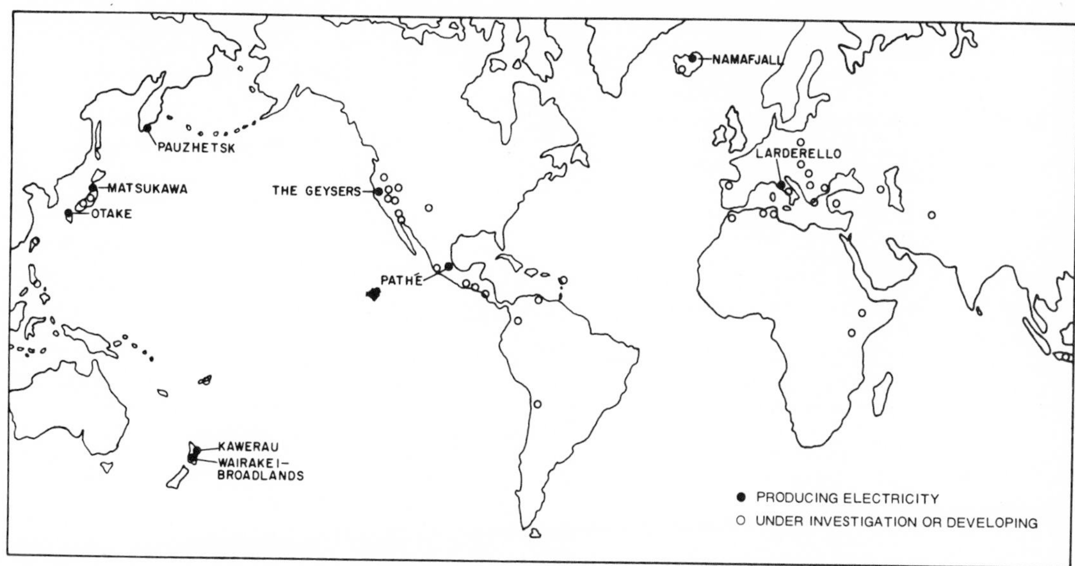

FIGURE 1.—Geothermal areas of the world.

basic geology, exploration methods, and geothermal resources in the Western United States. Part 2 includes a review of developments on a worldwide scale, and a selected reference list.

During the preparation of this report, Dr. George W. Berry of Cordero Mining Company, and Dr. George V. Keller of the Geophysics Department of the Colorado School of Mines provided encouragement and constructive criticism for which the author is grateful.

GEOLOGICAL CHARACTERISTICS OF GEOTHERMAL SYSTEMS

The distribution of high-temperature thermal springs in the world (Waring, 1965) indicates a close association with Late Tertiary and Quaternary volcanism and tectonism in areas of greater-than-normal heat flow. These regional geological conditions—volcanism, tectonism, and high heat flow—are very favorable for the existence of most geothermal systems with potential energy production. Volcanism, a surface expression of deep-seated magmatism and intrusion, is a major local direct cause of high heat flow. Tectonism is related by cause and result to the intrusion and extrusion of magma, and it facilitates movement of convecting fluids that carry heat upward and to the earth's surface.

On a world-wide basis, the tensional volcano-tectonic environments of the oceanic rises and continental rift zones (Lee and Uyeda, 1965), and the inner sides (concave sides) of compressional orogenic belts and island arcs (Oxburgh and Turcotte, 1970) seem to harbor areas of greatest positive geothermal anomalies. Various classifications of thermal water occurrences based on tectonic setting, type of heat source, and age of volcanics are presented by Frolov and Vartanyan (1970), Makarenko and others (1970), and McNitt (1970).

VOLCANIC ASSOCIATIONS

Most hot springs occur in volcanic districts and are related to demonstrable volcanic activity in the immediate area. Hot springs commonly emerge from faults along caldera margins and from within volcano-tectonic grabens. The grabens are partially filled with volcanic flows, pyroclastics, and continental and lacustrine sediments. Examples include Wairakei, New Zealand; Pauzhetsk, Kamchatka; Valles, New Mexico; and parts of the Basin and Range Province. Thermal springs and fumaroles are commonly located on the slopes of, and are related to, Late Quaternary volcanoes, such as those of Ahuachapan, El Salvador; Mt. Lassen, California; and Mt. Katmai, Alaska.

Some energy-productive hot spring areas are indirectly associated with surface or shallow subsurface, time-related volcanism evident not in the immedate hot-spring area, but several kilometers away in the greater region. However, the ultimate source of heat is believed in many, but not all, cases to come from deeper magmatic intrusions. The Larderello Field, Italy (Marinelli, 1964), where steam comes from capped, porous Mesozoic anhydritic sedimentary layers, is believed to be deeply underlain by cooling intrusions. The Geysers, California, produces steam from a long, hydrothermally altered fault zone and nearby areas in Jurassic-Cretaceous sediments, basalts, and serpentine into which magma is believed to be intruding at depth (McNitt, 1964; Koenig, 1970).

TECTONIC ENVIRONMENTS

The greatest hot spring areas of the world occur where tectonic tension is manifested by Late Tertiary and Quaternary normal faulting and accompanied by volcanic activity. On a worldwide scale, the great rift zones of the

East Pacific Rise, the Mid-Atlantic Ridge, and the Indian Ocean Ridge are characterized by high heat flow and geothermal phenomena. Where the East Pacific Rise system or associated spreading phenomena seems to underlie parts of the Western United States and northwestern Mexico, geothermal areas of the Basin and Range Province and the Salton-Imperial-Mexicali Valley occur. Where the Mid-Atlantic Ridge system comes to the surface in Iceland, numerous thermal areas occur. And where the Indian Ocean Ridge passes through the Gulf of Aden and the Red Sea, high heat flow and submarine hot brine areas occur. High-temperature regions of the USSR and the general relationships between continental drift and geothermal anomalies are discussed by Tamrazyan (1970).

In other areas tectonically contrasted with the worldwide rift system, such as orogenic belts and island arcs, geothermal phenomena occur in areas of normal fault collapse of volcano-tectonic grabens and related caldera, and in areas of normal faulting. The Late Cenozoic rifting of the Taupo volcanic zone of North Island, New Zealand, and the Trans-Mexican volcanic belt are examples.

Petrologic Associations

Specific petrologic association with modern geothermal phenomena has not been established systematically to date; however, in many areas acidic volcanics (rhyolite, ignimbrite, obsidian, dacite, etc.) seem to be more closely associated genetically with geothermal prospects than are the more basic volcanic rocks (basalt) (Healy, 1970). Rocks comprising geothermal reservoirs *per se* can be of practically any type or age if they are relatively porous and permeable, and preferably sufficiently brittle to sustain open fractures at elevated temperatures. A sealing or cap rock of low permeability and low thermal conductivity is usually essential to maintain a large and deep geothermal energy source.

Heat Flow and Hot Water Systems

More than 6000 heat-flow measurements have been taken on the continents and ocean floors, and they are increasing at the rate of over 600 per year (Blackwell, 1971). Although most of these are not at hot spring sites, they form the basis for better interpretations of the thermal regime of the earth. The worldwide average or normal heat flow is 1.5×10^{-6} calories per square centimeter per second, or 1.5 heat flow units (HFU). Hot spring areas in the upper 3000 meters (m) of the earth's crust discharge heat at 10 to 1000 times the normal rate. For example, the Firehole Geyser Basin of Yellowstone includes an area of 700 square kilometers (sq km) with an average heat flow of 67 HFU, or 45 times normal conductive heat flow (White, 1970). The Taupo graben of New Zealand includes 4000 sq km with average heat flow of 15 times normal (Studt and Thompson, 1969). Steamboat Springs, Nevada, has heat flow of 150 times normal over 5 sq km (White, 1968a).

Four types of thermal systems, as defined by White (1965), are summarized as follows:

(1) Regions of normal geothermal gradient and heat flow (1.2 to 1.5 HFU). These include the continental cratonic areas and most of the ocean basins.

(2) Areas of higher-than-normal gradient and conductive heat flow (about 1.5 to 3.0 HFU). These include the Gulf of California, parts of the Basin and Range Province, Hungarian Basin, Red Sea, and most areas of the world rift systems.

(3) Hot spring areas with convective transfer of most of the total heat flow in shallow depths by circulating water and steam. Most of the largest and hottest spring areas are of this type. Areas of intense heat transfer appear to demand magmatic temperatures and magma reservoirs nearby. Convective heat transfer upward by water and steam, which are heated by contact with hot rocks near a magmatic intrusive, produces a heat reservoir nearer the earth's surface than is possible by simple rock conduction without circulating fluids.

(4) Composite hydrothermal systems with convective and conductive heat transfer of aforementioned types (2) and (3). Little or no hot spring activity is evident at the surface. Heat is largely transferred by conduction through shallow rocks of low mass permeability that prevents appreciable discharge of water or steam. Temperatures immediately below the near-surface insulating layers are too high for conductive heat flow alone, and thus indicate presence of deep circulating fluids transferring heat convectively. The Salton Sea-Imperial Valley and the Larderello geothermal systems are of this type. Heat flow to the surface from composite systems is appreciably lower than that from hot spring systems with permeable upper parts. The insulating effect of relatively impermeable "cap rock" helps to maintain a large high-temperature geothermal reservoir at several hundreds or thousands of meters in depth.

Summary of Characteristics of Potentially Commercial Geothermal Reservoirs

A geothermal reservoir should have the following characteristics to have potential for large-scale power generation from direct use of steam under present conditions:

(1) Reservoir or base temperature of at least 200°C, which is necessary to sustain power generation of 100 MW or more in conventional steam plants (Bodvarsson, 1970a). Thermal waters at temperatures lower than 100°C may be used at low efficiency in low-pressure turbines, but technology is not established for major commercial production. Temperatures over 300°C may be undesirable because of high chemical content of fluids which causes scaling and corrosion.

(2) Reservoir volume of several to tens of cubic kilometers.

(3) Rock permeability sufficient to permit water and/or steam to flow into wells at high rate and large volume, and sufficient to permit recharge to the system during exploitation. Cumulative pay permeabilities should amount to 100 darcys x meters or more.

(4) Reservoir at economically drillable depth. Most reservoirs now undergoing exploitation produce from less than 1000 m. Some reservoirs produce from depths of 2000 or nearly 3000 m.

(5) Reservoir cap rock of low permeability and low thermal conductivity that prevents loss of fluids and heat, and that consequently maintains high temperature in the reservoir below.

(6) Sufficient fluid recharge into the reservoir at adequate rate and volume from surrounding areas with favorable geohydrologic and meteorologic conditions.

(7) Reservoir fluids that do not contain undesirable quantities of dissolved solids, mainly silica and calcite, and environmentally harmful chemicals, especially arsenic and boron. Reduction of temperature and pressure of reservoir fluids on production may cause excessive deposition and/or corrosion of equipment and create disposal problems.

(8) A large and potent heat source near enough to the reservoir to maintain high temperatures for at least a 20- to 30-year plant life.

EXPLORATION FOR GEOTHERMAL ENERGY

Exploration for geothermal steam in commercial quantities involves an integrated and coordinated program of application of many geological, geophysical, and geochemical techniques. These techniques should generally follow a plan of successive and partly concurrent stages, each dependent upon the analyses of results of knowledge acquired from previous stages. The modern exploration approach to geothermal energy is basically similar to that of metalliferous mineral deposits and oil and gas. However, the science and technology of development of geothermal resources is young and untested by long experience of success and failure.

Case histories are few and incomplete. Exploration is directed primarily to areas of surface heat leakage—such as thermal springs—which is analogous to the early days of oil exploration concentrated in areas of oil seeps. One of the greatest challenges in geothermal exploration is discovery of large, hidden, exploitable heat reservoirs where thermal manifestations at the surface are not obvious.

The science of exploration for geothermal energy sources is presented in this section for organizational convenience according to conventional disciplines—geology, geophysics, and geochemistry. A recipe for an optimal exploration program should not be organized necessarily along these classical lines. Rather, the program should be divided into three major phases—reconnaissance, delineation, and evaluation. Each phase should draw upon knowledge and methods from all three disciplines according to the local and unique situation of any given area (Banwell, 1970; Healy, 1970; White, 1970).

GEOLOGICAL METHODS

Various elements of geological exploration for economic geothermal energy are presented in many articles, most prominent of which are the following: Grindley (1964), White (1964, 1968a), Lovering (1965), McNitt (1965, 1970), and Healy (1970).

Exploration now and probably for some time in the future will be directed to evaluation of known thermal areas, rather than to search for undiscovered, completely hidden thermal anomalies. Of prime initial importance is surface geological mapping by use of all conventional tools, including photogeologic and remote sensing techniques, for the purpose of synthesizing basic detailed information on the following:

(1) types and geometry of structural features, especially intersections of fracture zones;

(2) lithology with emphasis on delineation of porous and permeable beds and impermeable interbeds in the stratigraphic section, and occurrence and age of all volcanic rocks;

(3) nature and extent of hydrothermal alteration and mineral deposition, and

(4) modern thermal springs, including location, size of area, elevation, flow volume, temperature, associated deposits and rocks, chemical analyses of the hottest springs, possible changes in behavior through time, etc.

Mapping in a thermal area and the immediate surroundings should be integrated with reconnaissance mapping of the tectonic, volcanic, and hydrologic features of the region. Especially important is the understanding of the relationships of most recent faulting and volcanic phenomena of the region to the local geology of the thermal areas (Grindley, 1964).

Geological mapping reveals the thermal history and evolution of the area and gives clues to how long the system has been active, how it may have changed location, where and of what type and magnitude the heat source may be, and how thermal activity relates to the geomorphic and hydrologic changes in the area (White, 1968a). The construction of geologic cross sections based on surface geologic data and well data (if any, in preliminary investigative stages) gives insight to the geometry, depth, and location of the hottest reservoir areas. Blind geothermal reservoirs located several miles eccentric to the thermal seep may be of greater economic interest than the immediate surface thermal area (Lovering, 1965). Subsequent data derived from various geophysical and geochemical surveys and test holes permit greater refinement of the subsurface thermal geology, which in turn guides further exploration and evaluation.

GEOPHYSICAL METHODS

Geophysical measurements applied in conjunction with geological and geochemical guidance provide a powerful means of exploring for geothermal energy sources. Numerous accounts of geophysical investigations of geothermal systems are available from many areas of the world; the best comprehensive summary is provided by Banwell (1970). Geophysical methods for detection, mapping, and evaluation of geothermal reservoirs include: (1) surface and shallow subsurface temperature and heat flow measurements; (2) electrical resistivity measurements; (3) gravity measurements; (4) magnetic measurements, and (5) seismic methods.

Surface Temperature and Heat Flow Measurements

Measurement and monitoring total heat flow of thermal

areas provide a direct way of estimating minimum continuous power potential, the enthalpy and size of the system, and the efficiency of extracting energy from the system (Dawson and Dickinson, 1970), as well as aiding in the selection of drill hole sites. Basic data on heat flow, problems and techniques of measurement, and heat-energy extraction are covered in several comprehensive papers in Lee (1965), and by Banwell (1963, 1964), and Dawson (1964).

Rapid and inexpensive mapping of the form and distribution of surface hot areas at Wairakei, by use of a 1 m thermocouple probe, is described by Thompson and others (1964). Measurements of geothermal gradients in holes less than 18 m deep at East Tintic, Utah, are described by Lovering and Goode (1963). Regional thermal surveys have been undertaken in the Larderello and Monte Amiata areas, Italy, by measuring gradients in 30-m holes at several hundred meters spacing (Burgassi and others, 1964, 1970). Local differences in total heat flow within thermal areas may be detected by different rates of melting of snow cover (White, 1969a).

Thermal radiation from first-order manifestations of thermal activity—hot springs, fumaroles, and steaming ground—are mappable from aircraft by infrared scanners sensitive to the 3 to 5 micron (μ) and especially the 8 to 14 μ spectral zones. Hot spring areas and some alteration areas at The Geysers, California, as well as some subtle radiation features in the area, were depicted on 8 to 14 μ band imagery, although no regional geothermal anomaly was detected (Moxham, 1969). The boundaries of some discharge areas of hydrothermal systems in New Zealand have been mapped by infrared imagery in the 4.5 to 5.5 μ band (Hochstein and Dickinson, 1970). At Reykjanes, Iceland, infrared thermal sensing agrees well with ground-temperature surveys, and provides a means of detecting changes in surface temperature when resurveyed in areas under exploitation (Palmason and others, 1970). Results of extensive research on multiband (visible and near infrared spectral regions) and intermediate infrared reconnaissance surveys in the Los Negritos-Ixtlan De Los Hervores geothermal area, Michoacan, Mexico, are provided by Gomez Valle and others (1970).

The ability of the infrared line scanner to detect and delineate shoreline and sublacustrine springs in the Mono Lake geothermal region of California is described by Lee (1969). Aerial infrared surveying has thus far been applied principally to regional thermal exploration in little-known geothermal areas, but rapid progress in quantitative radiometry will enhance the use of this tool for more specific and refined thermal differentiation. Other methods of remote sensing in addition to infrared—photographic, radar, and microwave—are discussed with reference to geothermal exploration by Hodder (1970), with particular application in the Casa Diablo and Salton Sea thermal areas of California and with reference to general exploration problems by Lyon and Lee (1970).

An intensive study of total heat flow of a thermal area from fluid flow and rock conduction involves measurement of surface temperature and discharge from springs and fumaroles, and measurement of thermal conductivity of the rocks, as well as information from magnetometer, gravity, resistivity, chemical, and geohydrologic surveys. As many of these data as possible should be acquired after geologic mapping and prior to deep exploratory drilling.

Electrical Resistivity Measurements

The application of several electrical methods in geothermal prospecting has proved extremely useful because of the direct relationship between fluid content, temperature, and electrical conductivity. Current conduction in rocks is carried out through the contained waters and is essentially directly proportional to the conductivity of the water, except for rocks of very high temperature near the melting point, and for compact shales, meta-shales, and metalliferous rocks. Conductivity of water increases with increasing dissolved salt content. Therefore, a geothermal anomaly of hotter, more saline water is characterized by a highly conductive (low resistivity) area in contrast to surrounding area of low conductivity (high resistivity) and cooler, less saline, or fresh water.

Keller (1970) reviews various electrical survey methods, theory, and results over a wide range of depths from several meters to as deep as the lower crust and upper mantle. He emphasizes the electromagnetic induction method as an effective method of searching for deep conductive zones. A brief review of electrical properties of various kinds of cover or overburden and problems of deep exploration by several electrical methods are presented by Strangway (1970). Mapping of heat-flow anomalies, faults, and lithologic-hydrologic changes to depths of 760 m in the center of Imperial Valley, California, where no hot springs occur, is described by Meidav (1970), who used the direct-current vertical sounding method in conjunction with gravimetry.

Also working in the Imperial Valley, McEuen (1970) reports on the relationship of apparent resistivity and airborne magnetic trends to the temperature variations at depth in thermal and nonthermal areas. Risk and others (1970) define the geometry and volume of the Broadlands geothermal field, New Zealand, to nearly 3 km in depth by application of direct-current resistivity surveys. Results of other electrical resistivity measurements in various geothermal regions can be found in reports by Battini and Menut (1964), Banwell and MacDonald (1965), Hatherton and others (1966), Hayakawa and others (1967), Banwell and Gómez Valle (1970), and Lumb and MacDonald (1970).

Gravity Measurements

Density contrasts between rocks produce gravity anomalies that are useful in delineating major structural depressions as well as local structural highs, buried volcanic or intrusive rocks, and areas of densification attributable to hydrothermal metamorphism—all of which may point up the presence of a local heat source within a depression. The value of gravity interpretations is greatly enhanced when considered together with magnetic and seismic information.

A residual gravity anomaly up to 10 milligal (mgal) in the Broadlands field, New Zealand, reflects the volume and degree of hydrothermal alteration (Hochstein and Hunt, 1970). Studt (1964) comments that gravity surveys in several areas of New Zealand show a correlation of minor positive anomalies with known hot areas, a genetic association that may be due to buried magmatic bodies and/or local metamorphism. In the Salton-Imperial Valley, Meidav (1970) reports that most, if not all, known residual gravity highs reflect metamorphism of loosely consolidated sediments associated with local hot water circulation systems. Another aspect of gravity studies of thermal areas, though it is not directly related to primary exploration, is the change in gravity over a reservoir during exploitation (Hunt, 1970). Over the Wairakei field, New Zealand, water removal versus replenishment gives measurable gravity change patterns during a period of several years.

Magnetic Surveys

Magnetic susceptibility of most rocks is attributable almost entirely to the magnetite content of the rocks. Magnetic surveys of geothermal areas, when combined with gravity, seismic, and electrical surveys, may provide information on variations of depth to basement and basement faulting, location and size of buried igneous masses, and areas of thermal metamorphism involving alteration of magnetite to nonmagnetic minerals.

In the New Zealand thermal fields, low magnetic field intensity is usually observed and is believed to be caused by conversion of magnetite to pyrite (Studt, 1964), although a direct relationship between magnetic anomalies and occurrence of geothermal steam has not been established (Hochstein and Hunt, 1970). At the Broadlands field, a first-order positive magnetic anomaly is believed to reflect a buried rhyolite mass which retained its magnetization. Surveys in the Tatun geothermal region of Taiwan reveal high magnetic values over andesitic terrain and low magnetic values in hydrothermally altered zones (Cheng, 1970).

McEuen (1970) describes his work in the Imperial Valley involving the use of airborne magnetic data to locate a deep fault zone along which resistivity surveys were run to outline highly conductive hot areas. A strong linear positive magnetic anomaly, which is believed indicative of a shallow (at least 2000 m in depth) intrusive, occurs in the Salton Sea geothermal area (Griscom and Muffler, 1971). The interpretation of magnetic data is especially problematical as pointed out by Banwell (1970), and these data must be evaluated in light of all available geophysical and geological data.

Seismic Methods

Normal reflection and refraction seismic methods (active) widely used in petroleum exploration for determining subsurface structure have had limited application and success in geothermal areas. Lithologic and velocity changes are complex in geothermal basins, and exceptionally high noise levels, damping and dispersive effects accompany hydrothermal activity. Seismic reflection soundings of high frequency and small amplitude in areas of three Japanese geothermal fields reveal structure and lithology to 4 km in depth that are believed related to a magmatic intrusion; low-frequency, large wave amplitude areas are believed related to fractured hydrothermal reservoir areas (Hayakawa, 1970).

Work with seismic refraction methods (Hochstein and Hunt, 1970) in the Broadlands field shows the configuration at shallow depth (900 m) of rhyolite domes and ignimbrite sheets, and changes in compressional wave velocities of rhyolite which indicate changes in rank of hydrothermal alteration. Hochstein and others (1967) and Grindley (1970) report on limited mapping of five velocity layers at Broadlands. Delineation of geometry of buried fault troughs in the Dixie Valley geothermal area, Nevada, using seismic refraction together with other geophysical methods, is described by Thompson and others (1967).

Passive seismic monitoring of geothermal areas includes microearthquake and ground-movement detection and geothermal ground-noise studies, both of which may prove to be practicable techniques for locating areas of high thermal activity. Microearthquakes believed to arise from fault movements concentrated in major geothermal areas have been described by Lange and Westphal (1969) at The Geysers, California; by Ward and others (1969) in Iceland; by Brune and Allen (1967) in the Salton-Imperial Valley, California; and by Ward and Jacob (1971) at Ahuachapan, El Salvador. Measurements of absolute ground movement at Waiotapu, New Zealand, show a threefold increase of amplitude in the thermal area (Whiteford, 1970). Geothermal ground noise, or hot water and steam-generated seismic noise, has been monitored at Wairakei-Broadlands (Clacy, 1968), in Iceland (Rinehart, 1968a), at Beowawe, Nevada (Rinehart, 1968b), and at Yellowstone National Park (Nicholls and Rinehart, 1967).

These initial studies suggest that individual hot water and steam systems have characteristic seismic signatures in some cases, and that these data may be useful in locating a heat source and in outlining the extent of a steam reservoir.

Geochemical Methods

The geochemistry of geothermal systems is extremely complex. Chemical data and interpretations contribute significantly to most aspects of exploration, evaluation, and exploitation of geothermal resources. Geochemistry should be used to the fullest extent in concert with geological and geophysical contributions. The most recent and comprehensive report on the application of geochemistry to geothermal development is by White (1970).

Geochemical data that are particularly useful reveal information on the type of geothermal system and on the subsurface and base temperatures. High-temperature geothermal systems have been divided into two types by White and others (1971): hot water and vapor-dominated (steam). Hot water systems appear to outnumber the vapor-dominated systems by about 20 to 1, but the vapor-dominated systems are considerably more attractive economically for power. The many characteristics of, and contrasts between, these two systems are discussed by White (1970) and White and others (1971). Perhaps the best preliminary indicator of the type of system is the chloride content of the spring waters. Usually high-temperature hot water systems have chloride contents greater than 50 ppm, whereas vapor-dominated systems have less than 20 ppm. Also, hot water systems display a range of total discharge from very small to thousands of gallons per minute, whereas vapor-dominated systems usually have low discharge of near-boiling water.

The subsurface temperatures of hot water systems can commonly be predicted from geochemical data derived from surface hot springs. Important general papers on chemical indicators and problems of geothermometry of geothermal systems include Ellis (1970), Fournier and Truesdell (1970), Mahon (1970), and White (1970). The best geothermometer is the silica content, which is largely controlled by variation in solubility of quartz in water as a function of temperature and pressure. Recent investigations of subsurface temperatures deduced from silica content are described by Fournier and Rowe (1966) and Fournier and Truesdell (1970) in Yellowstone, and by Arnorsson (1970a) in Iceland.

Next to the silica content, the sodium/potassium (Na/K) ratio of hot spring water is believed to provide the most reliable indication of underground temperatures. The Na/K ratio was indicative of the very high temperature (about 380°C) at Cerro Prieto, Mexico; and it was also useful in locating productive steam zones and in determining water distribution and migration (Mercado, 1970). Other chemical indices of subsurface temperature (White, 1970), which have been applied with variable success, include Ca and HCO_3 contents, Mg and Mg/Ca, Cl dilution, Na/Ca, Cl/HCO_3+CO_3, Cl/F, and H_2/other gases. Another promising approach to derivation of temperatures at depth is from isotopic compositions of spring water or steam flows, and of rock samples from wells. Studies have been made on isotope exchange reactions involving $^{13}C/^{12}C$ and $^{34}S/^{32}S$ (Ellis, 1970) and ^{18}O (Clayton and others, 1968).

Subsurface temperatures greater than 180°C are implied by deposits of sinter around hot springs and the presence of natural geysers, whereas travertine deposits usually indicate lower temperatures. Dilute spring waters that are neutral or alkaline and even of moderately high temperature in some areas do not have significantly higher temperatures in moderate depth (White, 1964).

As yet, no reliable geothermometer is known for vapor-dominated systems. However, most of these systems (excluding highly gassy ones) are believed, generally on the basis of the thermodynamic properties of steam, to have initial reservoir temperatures from 236°C to 240°C (James, 1968; White and others, 1971).

In addition to determination of type of geothermal system and of subsurface temperatures, numerous data on chemical compositions, reactions, and equilibria, chemistry of individual elements and isotopes, mineralogical transformations, etc., provide valuable information on the source, age, distribution, and interaction of fluids and rocks in a geothermal system. These data are useful in predicting size, extent, volume, and variations in permeability of the reservoir, and compositional changes, flow patterns, and recharge potential of the reservoir fluids. The multiplicity of applications of chemical studies to geothermal exploration and development is treated in many papers; some of the more inclusive and recent papers are authored by White (1969b, 1970), Ellis (1970), Mahon (1970), Tonani (1970), and Wilson (1970). Furthermore, geochemical studies during exploitation are very valuable with respect to depletion and energy changes in the reservoir, corrosion and precipitation problems in the well and power-plant facilities, effluent disposal at the surface or underground, and economic recovery of chemicals and/or fresh water from the geothermal brines. Finally, the increasing knowledge of geochemistry of modern geothermal systems can be effectively applied to the search for epithermal and mesothermal metalliferous ore deposits, such as mercury and porphyry copper (White and others, 1971).

GEOTHERMAL ENERGY RESOURCES IN THE WESTERN UNITED STATES

The Western States from the eastern front of the Rocky Mountains to the Pacific Coast are characterized almost everywhere (the Colorado Plateau excepted) by higher-than-normal heat flow (Sass and others, 1971) and seismicity, prevalence of Late Cenozoic tectonism, which is mostly tensional faulting, widespread Late Cenozoic volcanism, and an abundance of thermal springs — all augur well for occurrence of commercial geothermal resources. Waring (1965) lists about 1200 thermal springs in the United States; more than 1000 are in the Western States (fig. 2), including Alaska and Hawaii, and at least 100 of these are considered hyperthermal or having surface temperatures high enough to imply boiling water at depth. Koenig (1970) points out that the hyperthermal phenomena are closely associated with Quaternary volcanism allied with faulting as well as with faulting alone. This fact suggests two related heat sources: one derived from shallow intrusives, and the other from mass transfer and convective rifting processes beneath a thinned crust with only incidental youthful volcanism. In either case, heat transfer to the surface is by fluid flow largely along normal faults.

In the following subsections, geothermal geology and developments are succinctly presented for the Western States in alphabetical order.

ARIZONA

Several hot springs are localized in southeastern Arizona and very sparsely distributed elsewhere in the southwestern part of the state. The hottest surface water issues from Gillard Hot Springs in Greenlee County at a temperature of about 84°C (Haigler, 1969). The most recent volcanism in Arizona occurred about 900 years ago at Sunset Crater, 30 km northeast of Flagstaff. Mean heat flow in the southern half of Arizona is 2.0 HFU and some values approach 3.0 HFU (Roy and others, 1968). No geothermal wells have been drilled in Arizona.

CALIFORNIA

The only commercial geothermal power plant in the United States, The Geysers steam facility (fig. 3), is in California. California is also the scene of the most intense geothermal exploration at the present time. Three general areas have the most potential for geothermal development: (1)

FIGURE 2.—Thermal springs in the Western United States (Modified from Waring, 1965, p. 12.)

The Geysers — Clear Lake area, 120 km north of San Francisco in Franciscan and Lake Cenozoic volcanic terrain; (2) the Salton-Imperial Valley of southern California, a deep graben or rift containing very thick sedimentary fill; and (3) the eastern and northeastern parts of California, characterized by Late Cenozoic normal faulting and volcanism associated with the Basin and Range Province, and by the volcanic Cascade Range. (See fig. 4 and table 1.)

FIGURE 3.—The Geysers geothermal area, California. View to northwest. Units 1 and 2 of PG and E power complex in foreground began commercial operation in 1960 and 1963, respectively. On knoll in upper left center are Units 3 and 4 which went into operation in 1967 and 1968. Pacific Gas and Electric Co. photo.

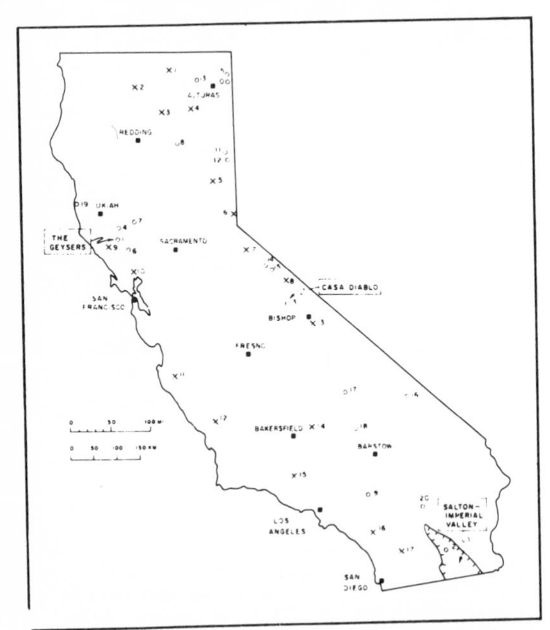

FIGURE 4.—Geothermal areas in California. (See table 1.)

TABLE 1.—*Geothermal areas in California (See fig. 4.)*

O Areas with exploratory drilling to October, 1969 (from Koenig, 1970)

Area No.	Locality	No. of Wells
1	The Geysers	78 (101 as of Sept. 1971)
2	Salton Sea	15
3	Casa Diablo	11
4	Clear Lake	4
5	Surprise Valley	4
6	Calistoga	3
7	Wilbur Springs	2
8	Mount Lassen	1
9	Arrowhead Springs	2
10	Cedarville	1
11	Wendel	1
12	Amedee	3
13	Kelley's Hot Springs	1
14	Fales Hot Springs	1
15	Bridgeport	1
16	Tecopa	1
17	Coso Hot Springs	1
18	Randsburg	1
19	Fort Bragg	2
20	Desert Hot Springs	~80

X Other potentially significant areas (from "Economic Potential of Geothermal Resources in California," 1971).

Area No.	Locality
1	Glass Mountain
2	Mt. Shasta
3	Big Bend Hot Springs
4	Bieber
5	Marble Hot Wells
6	Brockway Hot Springs
7	Grover's Hot Springs
8	Mono Lake
9	Skagg's Springs
10	Sonoma Valley
11	Tassajara Hot Springs
12	Paso Robles
13	Keough Hot Springs
14	Kernville
15	Sespe Hot Springs
16	Murrieta Hot Springs
17	Warner Hot Springs

The Geysers-Clear Lake Area

Geology and developments at The Geysers are summarized later in this report. The Geysers is in the western part of a much larger geothermal area of about 325 sq km. This area includes the Clear Lake volcanic field, where several wells have been drilled at Sulphur Bank in southwestern Lake County for a combined estimated energy potential of 54,000 kw (Geothermal Resources, 1967). However, a high percentage of boron (500 ppm) in the waters has discouraged further development (Geothermal Resources, 1967). The geochemistry and mercury mineralization of the Sulphur Bank hot springs are discussed by White (1955) and

White and Roberson (1962). Wilbur Springs, about 24 km northeast of Clear Lake, was drilled to 1133 m where a large flow of hot water with minor steam was encountered (Koenig, 1970). At Calistoga in northwestern Napa County, space heating is accomplished by directly circulating hot waters.

Salton-Imperial Valley Region

The Salton-Imperial-Mexicali Valley geothermal province (fig. 5) is the landward topographic continuation of the Gulf of California rift. It is roughly aligned along the East Pacific Rise system and the San Andreas Fault system. The depression narrows and topographically rises in the Coachella Valley north of the Salton Sea and is terminated by the junction of the San Jacinto Range with the Transverse Ranges. To the southeast, it gradually widens into the Gulf of California. The geologic framework of the region is presented by Dibblee (1954), Koenig (1967), and Biehler and Rex (1971), and portrayed on the Santa Ana, Salton Sea, and San Diego-El Centro geologic map sheets of California.

The Salton-Imperial rift is generally bounded by faults and diagonally transected by normal and strike-slip faults that are actively associated with modern seismicity (Brune and Allen, 1967). The trough is filled with water-saturated, weakly consolidated sand and silt of the Colorado River delta and coarse detritus of local marginal provenance which have been accumulating on granitic basement since earliest Pliocene time. Seismic refraction, gravity, and deep drilling (to 4097 m) data indicate that the sedimentary fill increases in thickness from about 4500 m at the Salton Sea to over 6100 m in the Mexicali Valley south of the international border (Biehler and others, 1964; Rex, 1970). This abnormally thick and young sedimentary prism rests on a crust 10 to 20 km thinner than crust in the bordering mountain ranges on the east and west.

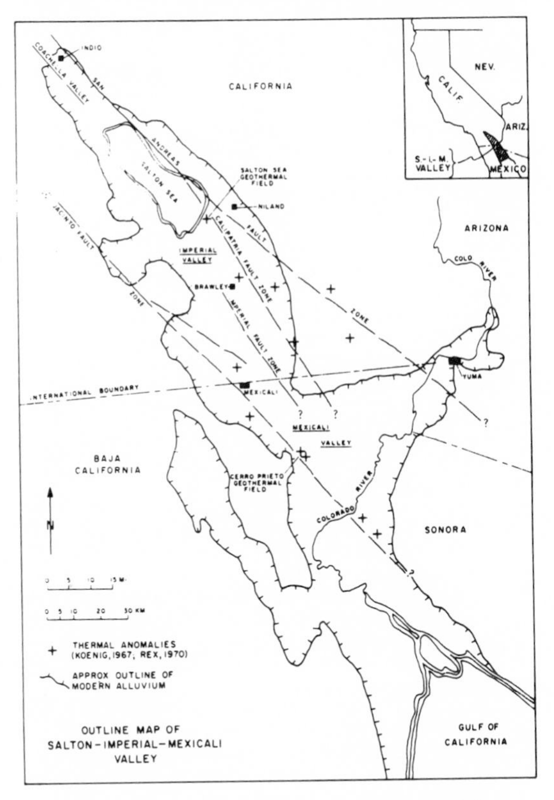

FIGURE 5.—Outline map of Salton-Imperial-Mexicali Valley.

The positive heat flow anomaly of the East Pacific Rise extends regionally into the Salton-Imperial-Mexicali Valley rift. It is reflected locally by at least 11 known areas, each of several square kilometers, where heat flows range up to 10 times normal (Muffler and White, 1969), and temperature gradients substantially exceed 6°F per 100 feet (Rex, 1970). Seven "hot spots" are outlined by Rex (1970) in the Salton-Imperial Valley. The Cerro Prieto anomaly (discussed later) in the Mexicali Valley includes at least four areas of groundwater isotherm highs (Koenig, 1967). More surface anomalies will probably be discovered as work progresses in the region.

The Salton Sea geothermal field (fig. 5) covers about 40 sq km, a distance of 8 km southwest of Niland at the southeast edge of the Salton Sea. Natural heat flow of 7.0 HFU (Rex, 1966) has surface expression of feeble hot springs (about 20 gpm), carbon dioxide springs, and five small rhyolitic domes which formed about 16,000 years ago (Muffler and White, 1969). Much information has been derived from 12 geothermal wells drilled to average depth of 1525 m and maximum depth of 2469 m (Koenig, 1970). Helgeson (1968), who studied the thermodynamics of the Salton Sea geothermal system, reports that the reservoir consists of over 610 m of arkosic sand overlain by 600 to 900 m of shale. Subsurface temperatures exceed 300°C at 900 m and 360°C at 2135 m. The thick insulating shale section overlying the sandstone reservoir has a conductive heat flow of about 17 HFU, or about eight times normal. Hydrothermal metamorphism produced highly indurated sedimentary rocks of the epidote-amphibolite (greenschist) facies by reaction of the hot brines with the sediments at depths as shallow as 1220 m (Muffler and White, 1969).

The Salton Sea geothermal brines contain 220,000 to 260,000 ppm total weight of dissolved solids, which are rich in chlorides of sodium, calcium, and potassium and have significant amounts of lithium, cesium, manganese, silver, iron, strontium, barium, lead, zinc, and copper (White, 1968b). This is one of the highest natural concentrations of salts known in the world. About 20 percent of the brine flashes to steam at the wellhead, leaving a residual bittern of 330,000 ppm concentration, which is nearly 10 times higher than seawater. The origin of the brines is believed to be essentially meteoric (Berry, 1966; White, 1965), with the dissolved salts being derived essentially from the containing rocks. Not all of the fluids in the Salton-Imperial-Mexicali geothermal system are as hypersaline as those found in the Salton Sea field. Cerro Prieto, 90 km southeast, yields fluids of only about 17,000 ppm solids.

Major problems of brine deposition, corrosion, and disposal have retarded development at the Salton Sea field. It is estimated that enough brine exists in this pool to yield over 500 MW for more than 50 years (Economic Potential of Geothermal Resources, 1971). Exploration activities are intensifying in the region. It is probable that more wells will be drilled on other thermal anomalies between the Salton Sea and Cerro Prieto. Data relating to a comprehensive investigation of potential power generation, chemical extraction, and desalination of geothermal fluids in the Salton-Imperial-Mexicali Valley are provided by Rex (1970, 1971) and in a Compendium of Papers (1970). Rex (1971) estimates a power potential in the Imperial Valley of 20,000 to 30,000 MW.

Eastern and Northeastern California

This third region of California contains several thermal areas (fig. 4) which have attracted exploration effort and which merit more. The Casa Diablo hot springs are in southwestern Mono County about 64 km northwest of Bishop on Highway 395. Geology and drilling in the area are described by McNitt (1963) and Rinehart and Ross (1964); gravity and related geology are described by Pakiser (1961). Casa Diablo lies in the western part of the 500-sq-km Long Valley structural depression marked by a gravity low and filled with Late Pleistocene acidic and intermediate volcanic rocks that may exceed 2500 m in thickness. About 11 wells have been drilled to depths less than 324 m (Koenig, 1970). Residual brine after minor (5 percent) steam flashover contains unacceptable quantities of arsenic, boron, and fluorine. The Long Valley depression lies 30 km south of the huge Mono Lake depression (Gilbert and others, 1968). Together these two areas have features that are attractive for further exploration for geothermal power. Two geothermal wildcat drilling sites are currently on the south shore of Mono Lake.

Some other potentially significant thermal areas in the Basin and Range portion of California, which have been drilled and explored inconclusively during the last 10 years (McNitt, 1963; Koenig, 1970), include: Surprise Valley (Lake City) in eastern Modoc County, Mt. Lassen area in Tehama and Plumas Counties, Coso Hot Springs in southwestern Inyo County, and other areas (fig. 4).

COLORADO

Many warm and hot springs occur scattered through different geologic environments mainly in the central and southwestern mountainous portions of Colorado. Waring (1965) lists 44 thermal springs, and George and others (1920) provide basic data on the geologic occurrence, chemical analyses, temperatures, etc., of "Colorado mineral waters." Highest temperatures are reported from Poncha Springs (75°C), 8 km west of Salida; Waunita (Tomichi) Hot Springs (72°C), 37 km east of Gunnison; and from Pagosa Hot Springs (72°C) at Pagosa Springs. The largest natural thermal-water flows by far are reported from Glenwood Springs: 3000 gpm at 41°C to 65°C; and from Steamboat Springs: 2000 gpm at 39°C to 65°C.

Tertiary volcanic centers and extensive volcanic fields occur widely in central and southwestern Colorado (Epis, 1968), but Quaternary volcanism is rare in Colorado. The Late Cenozoic Rio Grande Rift Valley extends northward from New Mexico through the Rockies of central Colorado, and together with the San Juan Volcanic Field of southwestern Colorado, represents a region of above-normal heat flow. The six reported heat-flow values in the region range from 1.52 to 2.4 HFU (Birch, 1950; Roy and others, 1968; Decker, 1969). Although discovery of commercial geothermal steam in Colorado seems relatively unlikely in the near future, these areas should not be overlooked, especially if new technology using heat exchangers for power production from hot waters becomes economic. Thermal areas in Colorado have been used in many places for bathing and recreational purposes.

IDAHO

About 200 thermal springs occur in Idaho (Waring, 1965; Ross, 1970). In a few springs the water is boiling, but most are low temperature; no steam has been observed. Idaho is underlain by large areas of granitic rocks of the Cretaceous Idaho batholith, where most of the springs occur, and by Tertiary volcanic rocks of the Snake River Plain. Preliminary heat-flow measurements (Blackwell, 1969) suggest that in most of Idaho the flow is above normal, about 2.0 HFU, similar to the Basin and Range Province as a whole. Although geothermal waters have not been seriously explored for electric power generation in Idaho, more than 200 homes in Boise are heated by hot water (77°C) from two 122-m wells (Koenig, 1970).

MONTANA

The mountainous western one-third of Montana is block-faulted and composed in part of Tertiary granitic and volcanic rocks. Forty thermal springs are reported by Waring (1965) mainly in southwestern Montana with temperatures less than about 80°C. The waters have been used locally only for bathing, resorts, and limited space heating.

NEVADA

Nevada ranks second to California in exploration for geothermal steam. Several discrete areas have been drilled as listed by Koenig (1970), three of which are especially noteworthy: Steamboat Springs, Brady's Hot Springs, and Beowawe. (See fig. 6 and table 2.)

The detailed investigations at Steamboat Springs, carried out over the last 20 years (White and others, 1964, 1967, 1968; Thompson and White, 1964), have contributed much to basic understanding of geology, geochemistry, geophysics, and fluid dynamics of geothermal systems in many parts of the world. Steamboat Springs is 12 km south of Reno (fig. 7). The area consists of numerous boiling springs issuing from a sinter deposit built up within at least 300 m of Plio-Pleistocene alluvial graben fill. The alluvium is normal faulted against Sierran granitic basement rocks. Middle and Late Tertiary trachytic volcanics are abundant in the region, but are rare in the thermal area itself. Small Quaternary cinder cones occur within a few kilometers of the spring area. About 36 exploratory and test wells (including hot water wells for resorts) have been drilled at Steamboat Springs, the deepest being 558 m. Well waters have a 5- to 10-percent steam flashover. Maximum or base temperature

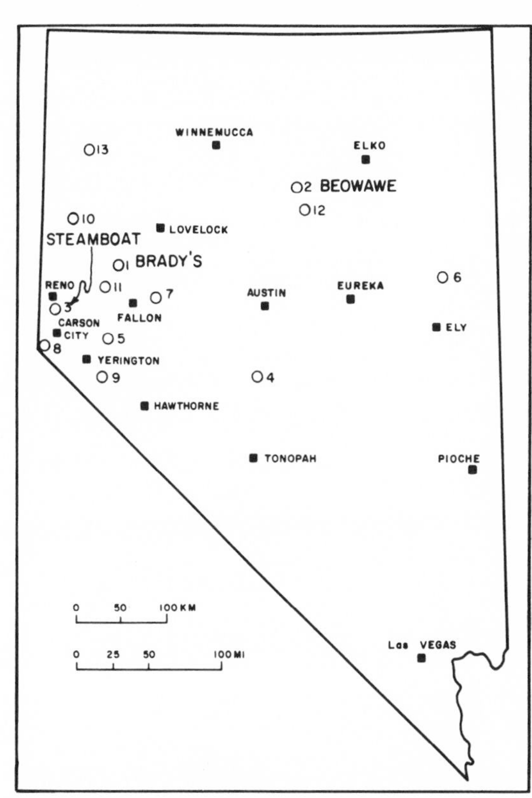

FIGURE 6.—Areas of geothermal wells in Nevada. (See table 2.)

TABLE 2.—*Areas of geothermal wells in Nevada (See fig. 6.)*

Areas of exploratory geothermal wells in Nevada (from Koenig, 1970)

Area No.	Locality	No. of wells
1	Brady's Hot Springs	9
2	Beowawe	11
3	Steamboat Springs	~36
4	Darrough	1
5	Wabuska	3
6	Monte Neva	1
7	Stillwater	1
8	Genoa	2
9	Smith Valley	1 ?
10	Pyramid Lake	3
11	Fernley-Hazen	1
12	Crescent Valley	1
13	Ward Ranch	1

approaches 172°C at 107 m in depth, below which the temperature remains essentially constant. Several dozen shallow warm and hot water wells have been drilled along a 15-km zone north into Reno; the hot waters are used for swimming pools, space heating, and industrial purposes.

The Brady's Hot Spring area is about 90 km northeast of Reno along the east side of Highway 40. The springs occur along a fault in a large region of high-temperature phenomena on the western margin of the Carson-Humboldt structural depression. Wells have penetrated Late Tertiary and Quaternary fanglomerate, lake beds, volcaniclastic sediments, and volcanic flows overlying an unknown basement. Nine wells, one drilled to 1542 m, have produced insignificant to appreciable volumes of hot water with 5 percent steam flashover (fig. 8) from reservoirs at temperatures of about 215°C (Koenig, 1970). According to Decius (1964), early drilling stimulated new fumarolic activity along 4 km of the main fault. Magma Energy Company of Los Angeles is re-reported to be considering construction of a pilot plant at Brady's to test the economic feasibility of a 10 MW power plant using hot water to flash isobutane in a heat exchanger for propulsion of a turbine (called the Magmamax Power Process) (Facca, 1970). If successful, this process could significantly encourage exploration and development of low-enthalpy thermal water systems for power generation.

FIGURE 7.—Steamboat Springs thermal area, Nevada. View to west. Light-colored spring deposits in middle-ground. Carson Range in background. L. T. Grose photo.

FIGURE 8.—Brady's Hot Springs, Nevada. One of early wells discharging to atmosphere. L. T. Grose photo.

The Beowawe geyser area is in north central Nevada, about 65 km west-southwest of Elko. The geology of the area and the thermal phenomena were described by Nolan and Anderson (1934). Studies of short-time temperature histories and microseismic signatures of the geysers have been reported by Rinehart (1968b). The geysers and hot springs are on a large sinter terrace along a fault zone in a volcanic and sedimentary series typical of graben fill in the Basin and Range Province. Eleven wells have been drilled at Beowawe since 1959, the deepest to 625 m. One well is reported capable of yielding over 50,000 lb of steam and 1,400,000 lb of hot water at 172°C per hour from a depth less than 213 m (Koenig, 1970). Although no wells are in production, Beowawe is regarded as an important geothermal prospect.

In addition to the above three areas in Nevada, Koenig (1970) presents summary data on nine other significant thermal spring areas that have been drilled and have yielded water at temperatures near and above the boiling point (fig. 6). Most of the 152 hot springs listed by Waring (1965) and the 185 hot springs located by Horton (1964) occur in the central and northern portions of Nevada.

New Mexico

Fifty-seven thermal areas are concentrated in the block-faulted volcanic terrain of the southwest and along the faulted western margin of the Rio Grande Rift Valley. According to Summers (1965a, 1965b, 1970a, 1970b), who has compiled extensive basic data on geochemistry and geology of hot springs in New Mexico, at least seven areas show some promise for steam exploitation. Most of these areas are in the Gila River Basin of southwestern New Mexico. One prominent and potentially productive area, the Valles (Jemez) caldera, 80 km northwest of Santa Fe in Sandoval County, has been explored recently by deep drilling. Five wells have been drilled there, one in 1970 to a depth of 2286 m, in which large volumes of dry steam and very hot water (up to 290°C) are reported to have been encountered in several horizons. The Valles caldera is an Early Pleistocene circular collapse following the explosive extrusion of a large volume of rhyolitic pyroclastics (Smith and others, 1961, 1970). Subsequent minor volcanic activity within the caldera, the youngest being 400,000 years before present (BP), has supplied heat to many hot springs issuing from faults along the western margin of the caldera rim. Tertiary and Quaternary volcanism is extensive in southwestern New Mexico and in the Rio Grande Rift Valley region (Kottlowski and others, 1969), and heat flow in these regions is high—up to 2.77 HFU (Roy and others, 1968; Decker, 1969).

Oregon

More than 40 percent of the land area of Oregon, mainly the southeastern lava plains and the Cascade Range, is underlain by Late Cenozoic volcanic rocks that locally are normal faulted and involved in volcano-tectonic and caldera collapse structures. Bowen and Peterson (1970) have located and provided precise temperature and flow data on 169 thermal springs and wells in Oregon. Seven prospective areas (fig. 9 and table 3) are described by Groh (1966) and four

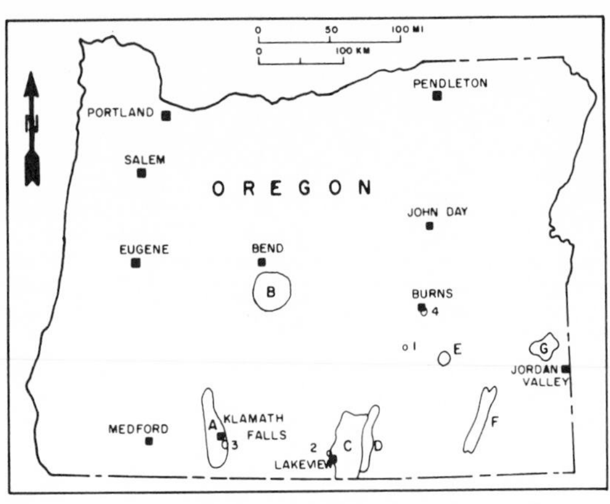

Figure 9—Prospective geothermal areas in Oregon. (See table 3.)

Table 3.—***Prospective geothermal areas in Oregon (See fig. 9.) (From Groh, 1966.)***

A Klamath Falls
B Newberry Volcano
C Warner Range
D Warner Valley
E Diamond Craters
F Alvord Valley
G Jordan Craters

O Areas with exploratory drilling to October 1969 (from Koenig, 1970)

Area No.	Locality	No. of Wells
1	Crump (Adel)	1
2	Lakeview	2
3	Klamath Falls	>350
4	Burns	1

have been drilled (Koenig, 1970) encountering temperatures over 100°C at shallow depths. Geothermal investigations thus far have been directed mostly to the Klamath Falls graben and the Goose Lake-Summer Lake graben complex (Lakeview region) of south-central Oregon (Peterson and Groh, 1967; Peterson and McIntyre, 1970). Fractured, scoriaceous, and permeable basalt layers interbedded with impermeable lake beds and cut by normal faults are typical of the region. At Klamath Falls, hot water slightly over 100°C is derived from about 350 wells, about 150 m in average depth, to heat over 400 buildings. As more basic data are accumulated on thermal geology of Oregon, several additional areas should emerge as highly prospective for geothermal power.

Utah

Utah has about 50 thermal springs. Most of them occur within a north-south zone 110 to 130 km in width, which is bordered on the east by the Wasatch-Hurricane normal fault zone and which lies entirely within the Basin and Range

Province. Geologic, chemical, and thermal data on these springs are presented systematically by Mundorff (1970), and the geothermal power potential of Utah is discussed by Heylmun (1966). Most of the springs are closely associated with Late Cenozoic faults, but only several are close to volcanic rocks. Only three springs have temperatures near the boiling point of water, the maximum recorded temperature of these being 86°C. Most of the higher temperature springs (over 38°C) occur along the Wasatch fault zone from Provo northward to the Utah-Idaho border. Two springs are regarded as potentially valuable geothermal areas (Mundorff, 1970); Roosevelt (McKeans) hot springs in north central Beaver County, and Abraham (Crater) hot springs in south central Juab County. Also, the geothermal anomaly in the East Tintic mining district (Lovering, 1965) may be another worthwhile target.

Washington

Campbell and others (1970) list 18 mineral spring areas in Washington; most are located in, and marginal to, the High Cascade volcanic range. Only one spring known to have a temperature approaching the boiling point of water is on the north slope of Mount St. Helens, a Quaternary volcano (Phillips, 1941); the maximum temperature recorded there is 88°C. Washington is mostly underlain by Tertiary volcanic rocks, and in the northern areas by Mesozoic granitic and metamorphic complexes. Several major Quaternary volcanoes cap the Cascade Range. Although young tectonic activity (normal faulting) and surface thermal phenomena are relatively minor in these regions, several heat-flow measurements (Blackwell, 1960) suggest that the region is basically thermally continuous with the Basin and Range Province. A review of the volcanic geology of Washington by Thorsen (1971) and of geothermal exploration by Crosby (1971) suggests that the basic conditions for occurrence of steam reservoirs are present in various places in Washington, but that basic data are thus far nonexistent.

Wyoming

Within the borders of Wyoming, the occurrence of geothermal phenomena can be divided into two distinct geological provinces: (1) Yellowstone National Park, a high volcanic plateau with nearly 100 separate localities of various types of intense, high-temperature hot spring and geyser activity; and (2) the rest of the state, which is mostly nonvolcanic Laramide Rocky Mountain ranges and basins with only about 20 scattered low-temperature springs.

The Yellowstone thermal region includes about 4100 sq km of voluminous, locally faulted, Late Tertiary and Quaternary volcanic rocks that are interpreted to overlie a batholith (Hamilton, 1960). A review of the igneous and geothermal geology of Yellowstone is incorporated in the Geological Society of America Field Trip Guide (1968). Basic data on the hot springs and geysers are presented by Allen and Day (1935). Since 1967, the U. S. Geological Survey has drilled 13 test holes at six localities in Yellowstone (White and others, 1968; Koenig, 1970). The deepest hole, 332 m, encountered the highest temperature, greater than 238°C. In many holes isothermal conditions were obtained which suggests a higher reservoir base temperature. One well produced dry steam; all others produced large flows of hot water with minor steam flashover. A recent article by White and others (1971) draws upon critical physical data from results of research drilling in Yellowstone that provides a better understanding of relationship between vapor-dominated and hot-water systems. In spite of the natural potential for dry-steam production in Yellowstone, thermal areas fall entirely within the National Park boundaries and are unavailable for commercial exploitation.

The few thermal springs known elsewhere in Wyoming (Waring, 1965) have temperatures less than 60°C. Heat flow in the region is about normal, 1.5 HFU (Blackwell, 1969).

[*Editor's Note:* The references for this article follow Paper 11B.]

11B

Reprinted from *Min. Ind. Bull.,* 15(1), 1-16 (1972)

Geothermal energy: geology, exploration, and developments

PART 2

BY L. TROWBRIDGE GROSE
COLORADO SCHOOL OF MINES

ABSTRACT

Part 2 is a review of the geology and geothermal energy development in the following countries, which are currently commercially producing electric power from geothermal steam: United States, Iceland, Italy, Japan, Mexico, New Zealand, and Russia. Six other countries should have geothermal energy production in a few years, and at least 19 others are in the investigative stages. Current economics and governmental legislation in the United States are summarized. A select bibliography on the geology and exploration for geothermal energy in significant areas of the world is included.

AREAS OF GEOTHERMAL ENERGY PRODUCTION AND POTENTIAL DEVELOPMENTS

At the present time, seven countries derive electric power from geothermal steam, and several more are developing their geothermal resources for commercial exploitation (fig. 1, Part 1). Geothermal energy developments in the producing countries are summarized by country in the following order: United States, Iceland, Italy, Japan, Mexico, New Zealand, and Russia.

The United States

Geothermal energy is economically converted into electricity at one facility in the United States at the present time — The Geysers, California (fig. 3, Part 1). The Geysers geothermal steam field is significant for the following reasons:

(1) it was the first geothermal field in the Western Hemisphere to produce electric power successfully;

(2) to the present time, it is the only geothermal power facility on-stream in the United States;

(3) it is the only dry-steam field in North America;

(4) it is the site of the deepest steam well in the world;

(5) it is the only geothermal facility in the world completely financed by private capital; and

(6) it is projected to become the largest geothermal power complex in the world.

The productive Geysers area includes over 2 sq km along Big Sulphur Creek at about 600 to 1000 m elevation in the rugged Mayacmas Mountains (fig. 3, Part 1 and fig. 10, Part 2). The area is about 120 km north of San Francisco. The geology of the area is described in the following reports: Bailey (1946), Koenig (1963, 1969), McNitt (1963, 1968).

The area is underlain by a mixture of graywacke, shale, basalt, and serpentine of the Jurassic-Cretaceous eugeosynclinal Franciscan Group. These rocks were severely compressed and overthrust from the east by the Great Valley sequence along the Coast Range thrust during Late Mesozoic and Early Cenozoic time. Late Cenozoic normal faulting

FIGURE 10.—Sulphur Bank area, The Geysers, California. Two wells, center and left, testing to atmosphere. Steam at right is from flow line of hole being drilled. G. W. Berry photo.

widely affected the area and produced a major northwest-trending, 40-km-long fault zone, along which occur The Geysers hydrothermal area and nine major mercury mines. A probable genetic relationship between vapor-dominated geothermal systems and mercury deposition with reference to The Geysers area is discussed by White and others (1971). Intersecting northwest- and northeast-trending normal faults localize the surface hot springs and areas of hydrothermal alteration in a northwest-trending graben.

Volcanic activity, dated from 3 million to about 50,000 years before present (Koenig, 1969), is evident on Cobb Mountain 5 km northeast of The Geysers and in the Clear Lake volcano-tectonic depression about 16 km northeast, where numerous hot springs occur. The volcanic rocks of the region include rhyodacite, dacite, andesite, and basalt in domes, flows, and pyroclastic accumulations. A close spatial and temporal relationship between volcanic activity, uplift, normal faulting, and geothermal activity suggests that the source of heat at The Geysers is a buried igneous mass of Pleistocene age, possibly associated at depth with the Cobb Mountain volcanic rocks.

The immediate thermal area of The Geysers is a hydrothermally altered zone 400 m long and 200 m wide. At least a dozen hot springs had a total estimated flow of 90 gpm at temperatures ranging from 50 to 100°C. Most of the springs are acidic (pH 2-3) and have chloride content similar to local rain water (2 ppm). The dry, slightly superheated steam is more than 99.9 percent pure water. Small, but possibly harmful amounts of boron in the steam condensate are returned to the subsurface via injection wells with no adverse effect on production from the steam wells.

Drilling, development, and general economics of The Geysers project are discussed by McNitt (1963), Koenig (1969, 1970), Barton (1970), McMillan (1970) and in a recent report to the California State Senate entitled "Economic Potential of Geothermal Resources in California" (1971). Modern drilling began in 1955, and to October 1969, 78 wells had been drilled of which only four were noncommercial. The status of drilling as of the end of September 1971 in The Geysers general area, which includes Sulphur Bank, Little Geysers, and Castle Rock Springs, is as follows (G. W. Berry, personal communication):

Wells completed	80
Producing wells abandoned	4
Holes drilled waiting	5
Dry holes (incl. 2 mechanical failures)	6
Holes drilling	2
Locations	2
Disposal wells	1
Wild well	1
Total	101

Recent wells have been drilled about 1200 to 2150 m deep and have averaged 100,000 lb of steam per hour (figs. 11, 12) which is derived from fractured and solution-fissured sandstones. The deepest steam well in the world, 2752 m deep, was drilled in August 1971 at The Geysers. This well is capable of producing 190,000 lb of steam per hour, or about 10,000 kw. Reservoir temperatures range from 236 to 288°C, and shut-in pressures of the deeper wells range from 450 to 480 psig. More recent drilling is approximately on 40-acre spacing. A steam phase for much of the field is indicated by shut-in pressures independent of well depth. Apparently the reservoir character and discharging steam from the deeper wells have remained essentially the same during the 16-year life of the field. Boundaries of The Geysers steam field have not, as yet, been determined by drilling. The steam reservoir is known to include an area of at least 26 sq km, and it may be greater than 50 sq km.

FIGURE 11.—Early producing well in foreground, with separator in flow line to remove rock particles, The Geysers, California. Main lines with expansion loops go to Pacific Gas and Electric Co. Plant No. 1. Cordero Mining Co. photo.

FIGURE 12.—Well head and separator assembly, The Geysers, California. D. A. McMillan photo.

Commercial geothermal power generation began in June 1960 with a 12,500-kw plant fed by 250,000 lb of steam per hour from four wells. Present production is 82,000 kw generated from four units. By the end of 1971, units 5 and 6 will be operational (fig. 13), bringing the capacity up to 192,000 kw. Annual increments added to the facility are expected to bring the on-line capacity up to 630,000 kw by 1975. Conservative estimates of total steam capacity based on yield of 150,000 lb of steam per hour per well are calculated to be about 1200 MW; liberal estimates range up to 4800 MW. A million people could be served by 1200 MW.

FIGURE 13.—Power Unit No. 5, 55 MW, Aug. 19, 1971, The Geysers, California. D. A. McMillan photo.

Three companies — Magma Power, Thermal Power, and Union Oil Company of California — are currently developing The Geysers field and selling steam to the Pacific Gas and Electric Company. Several other companies have leases in the region. Geothermal Resources International and Signal Oil Company have each drilled several wells with significant tested initial capacity. The total cost of geothermal energy at The Geysers, calculated on the actual price of the steam at $0.0026 per kwhr, is about $0.005 per kwhr produced (McMillan, 1970). This is more than hydro and less than fossil fuel power costs in California.

ICELAND

In Iceland the foremost use of low-temperature (80°C) thermal springs has been in space heating, rather than in power production. Over 250 thermal areas are known in Iceland. The geological, geophysical, and geochemical characteristics of many of these areas are presented by Bodvarsson (1964).

Geologic and seismic refraction studies indicate that Iceland is built up of typically inhomogeneous layers of Tertiary basalt reaching thicknesses of 3 km or more. These layers are overlain by Quaternary basalts mainly in the central and southern parts of the island. The mid-Atlantic Ridge passes through Iceland and effects measurable modern rifting, Recent volcanism, and markedly high heat flow. The Tertiary volcanic districts have minor Recent faulting and hot-water areas with temperatures below 150°C. The Quaternary districts occur along a major throughgoing northeast-southwest rift zone that bifurcates to the southwest. This rift contains spring areas with subsurface temperatures from 150 to 200°C and large areas of hot ground, natural steam vents, and thermal metamorphism. The highest temperature areas show a close association with centers of very recent silicic eruptions (Bodvarsson, 1970b). Individual surface thermal areas are controlled by intersections of contacts between permeable volcanic layers, dikes, and faults. Several thermo-artesian circulation systems are believed to come close to contact with magmatic intrusions, the ultimate heat sources for the high-temperature areas (Arnason and Tomasson, 1970; Bodvarsson, 1970).

Experience in drilling in many areas of Iceland shows that large amounts of high-temperature water can be produced from relatively shallow depth (less than 450 m). In some areas, deposition of silica and calcite contribute to a near-surface layer several hundred meters thick that acts as an impediment to hot water and steam circulation. High-production rates of individual bore holes are largely unrelated to natural surface heat discharge which implies large amounts of stored heat over large areas. Rate of use of this stored heat depends on availability of water.

Various geophysical and geochemical techniques involved in the assessment of geothermal areas have emerged from research in Iceland. Recent contributions include reservoir model studies by Bodvarsson (1970), and Thorsteinsson and Eliasson (1970), subsurface temperature measurements by Bodvarsson and Palmason (1964), infrared aerial surveys by Palmason and others (1970), microearthquake surveys by Ward and others (1969), trace elements in thermal waters by Arnórsson (1970b), and water-system differentiation by deuterium and chloride content by Arnason and Tomasson (1970).

Drilling for steam has been carried out mainly in two high-temperature areas: Namafjall and Hengill areas. A small 3000-kw plant has been operating at Namafjall since 1969, and plans call for a considerably larger natural steam plant at Hengill.

ITALY

Geothermal energy was used for the first time to generate electricity in 1904 at Larderello, Italy. By the late 1930s, capacity had increased to 100 MW, and at present the installed capacity of all Italian plants is 390,600 kw (fig. 14). A condensed review of the geothermal energy industry in Italy is provided in a booklet entitled "Larderello and Monte Amiata . . ." (1970).

FIGURE 14.—Power plant, three cooling towers, and substation in Larderello, Italy. Total capacity 69,000 kw in four 14,500-kw units and one 11,000-kw unit. Cordero Mining Co. photo.

There are nine thermal fields along a 500-km zone on the western side of the Apennine Range in central Italy (fig. 15). Larderello and Monte Amiata are at the northwestern end, 200 km and 130 km northwest of Rome, respectively. Relatively few papers in English have appeared on the geology of the thermal areas of Italy. The two most important areas, Larderello and Monte Amiata, are described by Burgassi (1964) and Marinelli (1964). Most recently Calamai and others (1970) described the geology, geophysics, and geohydrology of the Monte Amiata fields which were discovered in an area of very little surface thermal activity.

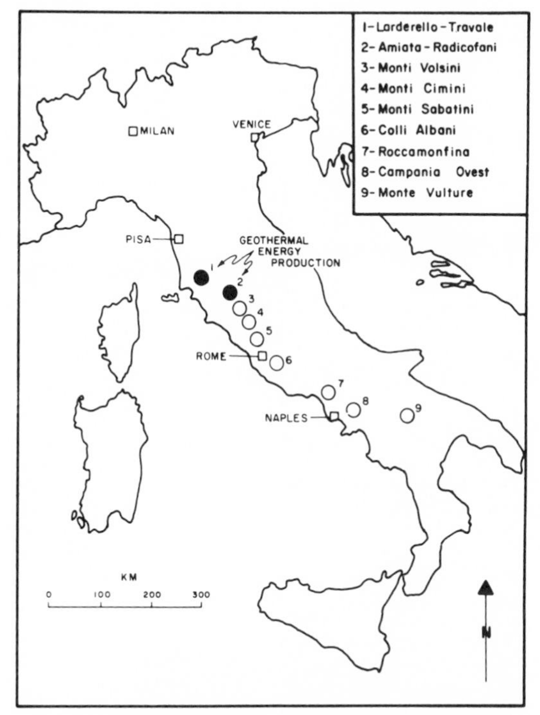

FIGURE 15.—Geothermal energy producing and developing areas in Italy.

At Larderello the rock sequence is sedimentary; igneous rocks are not directly evident in the area. Dry steam is produced from the Tuscan formation of Upper Triassic to Upper Jurassic age composed of highly pervious carbonate and anhydrite. The formation is underlain by a crystalline basement of phyllite and quartzite, and overlain by a chaotic thrust complex of clay, limestone, and ophiolite. Faulting in the basement allowed ascent of hydrothermal fluids into the porous anhydritic series, and the impermeable clays of the overlying thrust complex act as a cap rock by sealing in the fluids and heat of the geothermal system. The deeper faulting is pre-Quaternary and followed the Oligocene Apennine orogeny. The heat source is presumed to be a deep magmatic body of Miocene or younger age. Latest faulting of post-Pliocene age is tensional and associated with the post-orogenic collapse of the western margin of the Apennine Range. These latest faults form an intersecting network that localizes the recent hydrothermal activity. Natural outlets emitted largely steam and gas rather than hot water, although most have disappeared since exploitation of the Larderello steam. In the central producing area of Larderello, the geothermal gradient is 30°C per 100 m, or about 15 times normal.

The Monte Amiata area, 70 km southeast of Larderello, has a stratigraphic and structural regime generally similar to Larderello, except for the presence of ignimbrites and rhyolites of the post-Pliocene volcano, Monte Amiata. Volcano-tectonic collapse gave rise to a new fault system which controls feeble hot spring activity (including cold-gas outlets) on the flanks of 1500-m Monte Amiata. Within a 400-sq-km area around the Monte Amiata field, the geothermal gradient exceeds 10°C per 100 m, about five times normal.

In 1969 the greater Larderello area (Boraciferous Region) had a capacity of 365 MW. Dry steam was derived from 181 wells, with an average production per well of about 50,000 lb per hour at 150°C (fig. 16), (2500 kw), with shut-in pressure 5 atm and power station pressure 4.2 atm. Temperatures have increased during the time of exploitation, except in some local areas where recharging meteoric waters have effected cooling. The Travale and Monte Amiata regions together in 1969 had a capacity of 25 MW produced from 11 wells averaging 78,000 lb per hour. Planned expansion at Larderello should increase production there to about 415 MW. A comprehensive report on geothermal developments in Italy by Cataldi and others (1970) indicates that several thermal areas are being intensively explored and are in various stages of development. Substantial increase in geothermal power production in Italy must come from areas other than Larderello and Monte Amiata, since the limits of these fields apparently have been reached.

FIGURE 16.—View to north in Larderello geothermal area, Italy. Typical producing steam well and flow line in foreground. Cooling towers of Larderello No. 2 station at right. Town of Larderello in distance at right. Cordero Mining Co. photo.

Japan

The Japanese have been engaged in intensive research and development of their geothermal resources because fuels — coal, oil, and uranium — must be imported from long distances overseas, and hydropower sites are limited. At present two fields, Matsukawa and Otake, are producing electric power, and four other areas are being developed for commercial production. About 200 Quaternary volcanoes and more than 90 areas of hot springs with water temperatures exceeding 90°C are known in Japan. Results of investigations and attending problems related to geothermal energy production are presented by Sato (1970).

The general geology of geothermal fields in Japan is covered by Saito (1964) and Ishikawa (1970). The Matsukawa field is described by Nakamura and others (1970), and the Otake field by Yamasaki and others (1970). Japan comprises part of the Circum-Pacific volcanic-tectonic belt, and hence has geothermal environments basically similar to those of Kamchatka and New Zealand. Most of the hot springs in Japan are closely associated with Quaternary volcanoes and domes of rhyolitic, dacitic, and andesitic composition, rather than basaltic composition. Many of the thermal fields are also found in highly faulted Tertiary volcanic and granitic areas that are seemingly unrelated to Quaternary volcanism. The heat sources of the geothermal fields are related to young volcanic centers as well as to deep intrusives of Tertiary and possibly Late Mesozoic age.

The Matsukawa area displays relatively little surface thermal manifestation, but has a wide area of hydrothermally altered rocks. Hot water and steam are trapped in Late Tertiary porous pyroclastics, brittle, fractured welded tuff, and marine black shale and sandstone. This sequence, up to 1700 m thick, was normal step-faulted during Late Tertiary and Quaternary time. Andesitic lavas from the Pleistocene volcano, Marumori, which is 1 to 2 km from the Matsukawa field, form a 160-m-thick cap over the older volcanic reservoir rocks. The geothermal system at Otake appears basically similar to Matsukawa, except apparently for more urface flow of hot water and steam, and greater influence of intersecting faults on fluid migration. Thorough studies of hydrothermal alteration mineralogy and geochemistry of Japanese thermal areas appear in articles by Yamasaki and others (1970), Fujii and Akeno (1970), Koga (1970), Noguchi and others (1970), and Oki and Hirano (1970). Calculations on heat flow and energy reserves are presented by Hayakawa (1970) and Noguchi (1970).

The Matsukawa geothermal station in northeastern Honshu has a capacity of 20,000 kw, with a planned increase to 27,000 kw. The total potential of the greater Matsukawa area (including Takinokami 7 km distant) is estimated at 200 MW (Mori, 1970). The Otake plant in Kyushu has a capacity of 13,000 kw. According to calculations by Noguchi (1970), the total power potential of 124 magma chambers in Japan is 8400 MW for a few thousand years duration. The average generating capacity of all fields is about 70 MW. Many other uses of geothermal resources in Japan are reviewed by Komagata and others (1970). They include recreation and bathing resorts (150 million people annually visit hot spring sites), and agricultural, chemical, and secondary industries.

Mexico

Mexico is well endowed with potentially productive geothermal areas. More than 100 thermal areas and nine active volcanoes are known. Seven thermal areas in particular have been described (fig. 17). Pathé, the only currently producing area, generates 500 kw, although the turbine is rated at 3500 kw. Cerro Prieto is scheduled for a 75 MW operation in 1972.

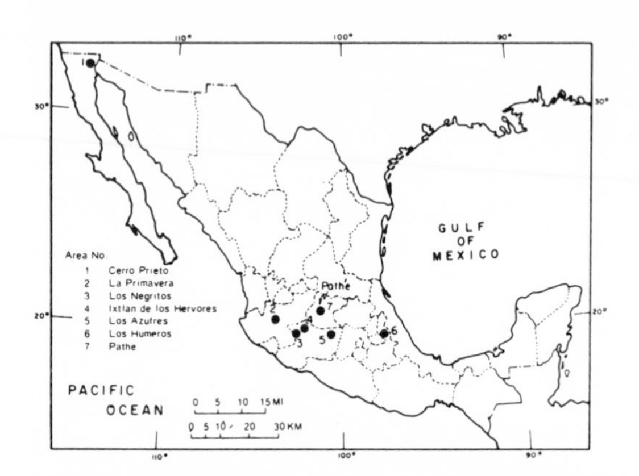

FIGURE 17.—Geothermal areas in Mexico (from Banwell and Gomez Valle, 1970).

The relatively recent developments in geothermal exploration in Mexico are presented by De Anda and others (1964), Alonso and others (1967), Alonso (1968), and Banwell and Gomez Valle (1970). There are six thermal areas in basically similar geologic environments within the trans-Mexican volcanic belt of Tertiary and Quaternary age. They are intimately associated with rhyolitic to andesitic flows, pyroclastics, and their sedimentary derivatives which accumulated in volcano-tectonic graben and caldera fault-block depressions. Surface phenomena, which are usually associated with Recent tensional faulting, consist of hot springs, steam vents, alteration zones, and hot ground. Surface temperatures range from 36 to 100°C (Banwell and Gomez Valle, 1970).

The Pathé field reservoir, which is a graben of intensely fractured and altered Tertiary andesites and basalts up to 1000 to 1500 m thick, overlies Cretaceous limestone formations and underlies Late Tertiary and Early Quaternary tuffs. Many wells have been drilled there, but only four wells have produced water and steam sufficient for power generation. The reason for the low productivity is believed to be lack of lateral permeability in the volcanic series. New drilling has been proposed to reach Cretaceous limestone where larger production may be expected (Mooser, 1965). More recently geological, geophysical, and geochemical investigations by the Federal Electricity Commission have been underway at the four other areas in southern Mexico. Preliminary data suggest that the Los Negritos field may have a potential comparable to the largest known geothermal fields of the world (Banwell and Gomez Valle, 1970).

The Cerro Prieto geothermal field (fig. 5, Part 1), in the Mexicali Valley about 30 km south of the USA-Mexico border, is part of the largest of all known geothermal systems in the world. Geologically it is an integral part of the enormous Salton-Imperial-Mexicali Valley geothermal system (Alonso Espinoza, 1966; Rex, 1970), which occurs in the sediment-filled graben associated with the northern part of the obliquely spreading Gulf of California and the East Pacific Rise system. The Cerro Prieto area is on the western side of the rift and on the general southern prolongation of the San Jacinto fault zone, where numerous normal and strike-slip faults step successive blocks down to the east. The basement of Cretaceous granitic rocks is overlain by about 2500 m of imperfectly compacted conglomerate, sandstone, and shale derived from erosion of block uplifts to the west and from deltaic sedimentation of the Colorado River to the east. These sedimentary rocks are locally intruded by Quaternary andesite and basalt manifest at the surface by the small Cerro Prieto crater 2 to 5 km west of the geothermal field. Boiling springs, mud volcanoes, and phreatic explosion vents are aligned in the area on a general north-south trend and extend into areas not fully explored.

Many wells have been drilled in the area to various depths up to nearly 2700 m. Usually, two steam and hot water producing zones of permeable sandstone interbedded with impermeable shale occur at depths of 600 to 900 m and 2400 to 2635 m. One well (M-20) drilled to 1385 m discharges a steam-water mixture at the rate of 1,500,000 lb per hour at a bottom hole temperature of 388°C. This is the greatest known natural steam-well discharge in the world (Mercado, 1969). Average well yields are about 500,000 lb per hour of moderately briney water (17,000 ppm total dissolved solids), with about 120,000 lb per hour flashing to steam. A geothermal plant with initial capacity of 75,000 kw is scheduled for completion in 1972. The power facility will tap steam from 15 wells drilled to average depth of 1375 m in an area of about 1 sq km. Residual brines will be drained initially into the Gulf of California, although the feasibility of an extractive chemical industry to process the brines rich in sodium, potassium, lithium, and calcium is being investigated. On the basis of geophysical data, the power potential of the Cerro Prieto field is well in excess of 1000 MW (Rex, 1971).

New Zealand

During the last 20 years, extensive development of geothermal resources has taken place in the Taupo volcano-tectonic depression in the North Island of New Zealand (fig. 18). A great deal of literature is available on many aspects of geothermal science and technology as developed by the New Zealanders in their classic areas. Basic geology, geophysics, and problems of energy extraction from heated rocks -- data which are applicable to many thermal areas of the world -- are provided by Banwell (1963, 1964) and Dawson (1964). Grindley (1965) in his paper on Wairakei presents one of the most extensive geologic studies of any to date on a productive geothermal field.

Many thermal areas are in the central 80-km-long portion of the 160-km-long Taupo volcanic zone (Healy, 1964).

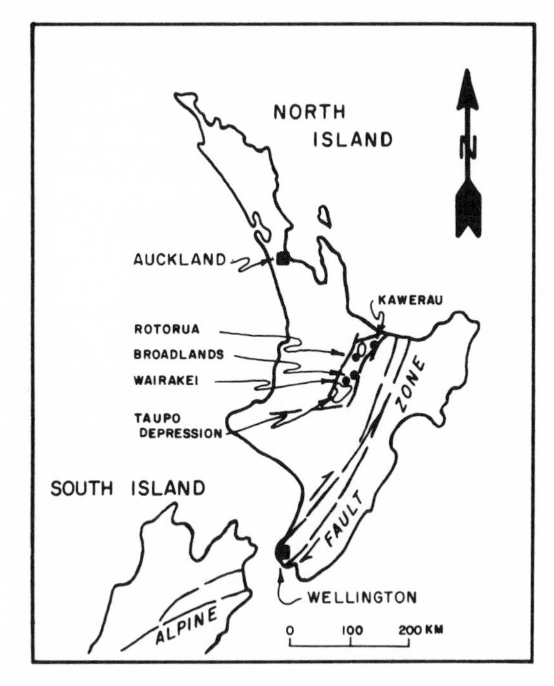

Figure 18 —Developed geothermal areas of New Zealand.

The Taupo volcano-tectonic depression occurs on strike with the Kermadec-Tonga submarine ridge and marginal to the regional Alpine fault. Numerous hot springs, fumaroles, and geysers occur in areas of abnormally high-temperature gradients. Features common to the hydrothermal fields on North Island (Grindley, 1965, 1966, 1970) include the following:

(1) reservoir rocks of nearly flat-lying Quaternary acidic volcanics up to a few thousand meters thick;

(2) a stepped aquifer system of several permeable pumiceous breccias and fractured flows interbedded with low-permeability mudstones, welded tuffs, and rhyolites;

(3) proximity to a rhyolitic eruptive center of Late Pleistocene age interpreted to overlie a granitic magma chamber;

(4) a northwest- and northeast-trending, conjugate, diagonal-slip fault system with fault intersections producing open fractures favorable for ascent of hydrothermal fluids and igneous intrusions; and

(5) direct association of areas of maximum heat flow with horsts and structural domes.

Many detailed geophysical investigations have been carried out in various thermal areas of New Zealand. Recent papers include direct and indirect methods of measuring heat flow (Dawson and Dickinson, 1970); seismic, gravity, and magnetic studies in the Broadlands field (Hochstein and Hunt, 1970); geothermal ground noise spectra (Clacy,

1968); electromagnetic induction methods for measuring resistivity (Keller, 1970); net mass loss from gravimetric measurements (Hunt, 1970); ground subsidence during exploitation (Hatton, 1970); absolute ground movement (Whiteford, 1970); d-c resistivity methods to 3 km deep (Risk and others, 1970); and near-surface (30 m deep) resistivity surveys (Lumb and MacDonald, 1970).

In the Wairakei area (figs. 19, 20), wells tap near neutral, weakly chloride waters at temperatures generally between 250 and 275°C. Steam discharge at wellhead is variable in different areas, and may range up to 50 percent. Only rarely is dry steam produced. The typical production well is 600 to 900 m deep. Average steam output of high-pressure wells (210 to 230 psi) is 52,000 lb per hour (4000 kw), and that of medium-pressure wells (75 to 85 psi) is 45,000 lb per hour (3000 kw). Average steam content at wellhead is 15 percent. Results of exploration drilling in recent years are described by McKenzie and Smith (1968).

FIGURE 19.—Standard wellhead equipment in Wairakei, New Zealand. Average discharge is 15 percent steam which is recovered from cyclone separator (vertical cylinder at left). High-pressure dry steam goes directly to plant. Waste water goes to silencers (steaming center and right).

FIGURE 20.—Steam sampler measuring discharge of a high-pressure well in Wairekei, New Zealand. A sampling nozzle is drawn manually across the jet discharging vertically to atmosphere, with time in each position proportional to area of annular ring at radius being sampled. Sample is measured in small calorimeter in portable hut at left. G. W. Berry photo.

Geothermal developments in New Zealand to 1970 are summarized by Smith (1970). The electric power industry of New Zealand is entirely state owned and operated. About 20 percent of the electric power consumed in North Island is derived from geothermal sources. Greatest production comes from Wairakei, where present installed capacity is 192,000 kw, and expansion to 250,000 kw is being considered (fig. 21).

At Kawerau, steam is used for processing in a pulp and

FIGURE 21.—View in producing field, Wairakei, New Zealand. Steam is mostly from cylindrical silencers where hot water is flashing to atmosphere. G. W. Berry photo.

paper mill, and about 10,000 kw of power is also produced for the mill. Recent deeper drilling has enlarged the potential at Kawerau, but increasing deposition of calcite in the plumbing system poses a problem. Rotorua continues to use geothermal steam mainly for heating purposes (Burrows, 1970). Broadlands is a new and significant field discovered by regional resistivity surveys in an area of minimal natural thermal activity at the surface (Grindley, 1970; Facca, 1970). A generating capacity of 120 MW is tentatively scheduled for 1976.

Data on the behavior of geothermal systems in New Zealand during exploitation (Bolton, 1970) are revealing some pressure and temperature decline locally. However, the power output of the geothermal plants at the present level is expected to endure for many years, and significant increases in power generation are probable.

Russia

Steady and impressive progress in the science and engineering of geothermal resources utilization in many forms has been taking place in the Soviet Union along rather original lines compared to other countries during the last 30 years. Many papers given at the Pisa Symposium in 1970 cover Soviet developments to which the reader is referred for details far beyond the scope of this summary article.

At present, 11 geothermal projects are operating in the USSR. One geothermal power plant with capacity of 5000 kw is in operation at Pauzhetsk, in the southeastern part of Kamchatka. The geology of the hydrothermal systems in Kamchatka is described by Piip and others (1961), and Vakin and others (1970). The Pauzhetsk geothermal field is within the Circum-Pacific mobile belt of active volcanism, tectonism, and high heat flow, and is similar to the geothermal environments of New Zealand. Nearly flat-lying Quaternary pyroclastic formations with varying primary and fracture permeability are interbedded with impermeable layers that fill elongate and circular, normal faulted, volcano-tectonic depressions to thicknesses averaging 1500 m. Recent hydrothermal activity is localized at fault intersections of caldera collapse margins, which are superimposed on older graben-synclines, and is localized along fault margins of horst-anticlines within the large regional depressions. The thermal activity is closely associated with Middle to Late Pleistocene acid volcanism, rather than Pleistocene and Recent andesitic and basaltic volcanism.

Surface thermal phenomena consist of hot and boiling springs, geysers, steam jets, and patches of hot ground. Extensive studies of the hydrodynamics of geothermal systems in Kamchatka include relationships among temperature, viscosity, pressure, thermo-artesian head, and other factors. These studies are presented in several articles by Averiev (1967). Maximum temperature recorded to about 1220 m in depth is 200°C, and the temperature gradient may be as high as 70°C per 100 m.

The present facility at Pauzhetsk is being expanded to a planned total output of 29,000 kw. The geothermal resources of the Kamchatka region are believed to have a potential capacity of 500 MW from operation of several power stations (Vakin and others, 1970). Another area, Kunashiry in the Kurile Islands, is being considered for construction of a 6000-kw station (Tikhonov and Dvorov, 1970). The potential elsewhere in Russia for electric power generation from geothermal energy appears to be enormous (Makarenko and others, 1970; Tikhonov and Dvorov, 1970).

Geothermal resources in Russia are being used for many other purposes, such as space heating, agriculture, miscellaneous industries, and extraction of chemicals. Multiple uses are described for the Georgia region by Buachidse (1970), and the use of thermal waters obtained with oil and gas production in the Caucasus region is discussed by Sukharev and others (1970). General, lengthy summaries of scientific investigations and engineering of geothermal resources in Russia are provided by Makarenko and others (1970), and Tikhonov and Dvorov (1970).

Developing Areas in Other Countries

The aforementioned seven countries, which currently produce electricity from geothermal sources, will be followed within a very few years by several other countries (fig. 1, Part 1) that have geothermal power plants under construction or in the planning stages (Facca, 1970). *Chile* plans for a 20,000 kw plant at El Tatio (Koenig, 1971). The Ahuachapan area of western *El Salvador* has been explored and drilled (at least six wells) and estimates indicate a 25 to 30 MW power potential (Tonani, 1967; Sigvaldson and Cuellar, 1970; Summers and Ross, 1971). A recent article on microseismicity along a fault zone at Ahuachapan describes a relatively new method of locating thermal fluid-bearing zones in the subsurface (Ward and Jacob, 1971). The French have discovered a steam field with estimated capacity of 30,000 kw on *Guadaloupe* in the West Indies volcanic island chain (Cormy and others, 1970). In the Legaspi area of extreme southeastern Luzon, *Philippines,* a 10,000-kw power plant is planned (Koenig, 1971). Extensive geothermal investigations in the Tatun volcanic area at the northern tip of *Taiwan,* described fully by Chen (1970) and by Feng and Huang (1970), reveal a potential of 80 to 200 MW; a 10,000-kw plant is planned initially. The Menderes Massif geothermal province of Western Antatolia, *Turkey,* has been geologically and geophysically investigated along several Miocene-Pliocene grabens associated with hot springs and economic mercury mineralization; but no appreciable recent volcanism has occurred (Ten Dam and Khrebtov, 1970; Ten Dam and Erentoz, 1970). At Kizildere, the Turkish government is planning a 30,000-kw power station, and other areas as well hold promise for commercial exploitation.

Many other countries (fig. 1, Part 1) are in various stages of investigation of potential geothermal resources for development of electric power (Facca, 1970). These countries are as follows: Algeria (Cormy and d'Archimbaud, 1970); Bulgaria, Colombia (Arango and others, 1970); Czechoslovakia (Kremar and Milanovic, 1970); Ecuador (De Grys and others, 1970); Ethiopia, Fiji, Greece, Guatemala, Hungary (Boldizar, 1970); Indonesia (Muffler, 1971; Zen and Radja, 1970); Kenya, Morocco, Nicaragua, Poland (Dowgiallo, 1970); and Spain, Tunisia, Venezuela, and Yugoslavia (Kremar and Milanovic, 1970).

CURRENT STATUS AND OUTLOOK

Increase in interest in geothermal energy development in the Western United States is accelerating as a consequence of the following factors:

(1) demonstration of low-cost electric power production from geothermal steam relative to conventional and atomic sources,

(2) realization of the relatively minor pollution and environmental effects of geothermal power plants, and

(3) realization of the potential of the multipurpose or multiple-use approach to geothermal resource development involving the production of electricity, chemicals, and fresh water.

Energy exploration companies are investing many millions of dollars in research, exploration, and development. The government has passed legislation on leasing and operating regulations, taxation policies, total environment impact, etc., that should affect development on federal lands.

Assessments of costs involved in the various operations and facilities connected with exploration and development of geothermal energy are described in general by Facca and Ten Dam (1964), Bradbury (1970), and Kaufman (1970); and The Geysers geothermal field is described by McMillan (1970). Total cost per kwhr at The Geysers is 4.91 mills, at Larderello over 3.2 mills, and at Matsukawa 4.6 mills; all three are dry steam-producing fields. Costs at fields yielding a steam and hot water mixture are higher, however, but still competitive with conventional power sources. Cost analyses reported by Kaufman (1970) reveal that a base-load plant of 1500 MW capacity (assuming a privately financed plant and a 90 percent load factor during the life of the plant) will have a total cost of 2.96 mills per kwhr using geothermal energy, 3.45 mills per kwhr using hydropower, 4.82 mills per kwhr using natural gas, 4.87 mills per kwhr using oil, 5.22 mills per kwhr using coal, and 5.49 mills per kwhr using nuclear energy. Geothermal costs could be increased as much as 1.5 mills per kwhr to accommodate transportation costs, engineering and disposal problems, etc., and still remain relatively inexpensive.

In response to the Geothermal Steam Act of 1970, the U. S. Geological Survey has delineated lands classified as "known geothermal resource areas" (KGRA) where "the prospects for extraction of geothermal steam or associated geothermal resources are good enough to warrant expenditures of money for that purpose" (Godwin and others, 1971, p. 8). These areas are listed in the "Federal Register" between the dates of March 26 and April 23, 1971, and are on figures 22, 23, and 24. Leases on federal lands within a KGRA can be acquired only by competitive bidding. Classification factors, as determined by geological, geophysical, and geochemical data that are considered in the definition of KGRA's, are described by Godwin and others (1971). In further response to the Geothermal Steam Act, a notice of proposed rule-making on geothermal resources leasing and operations on public, acquired, and withdrawn lands is presented in the Federal Register, v. 36, no. 142, July 23, 1971. A general discussion of the implementation of the Geothermal Steam Act is presented by Stone (1971).

In an effort to keep interested people abreast of significant developments in the geothermal resources field, such as meetings, hearings, legislation, publications, prospect developments, activities and contributions of the Geothermal Resources Council, etc., a newsletter entitled *Geothermal Hot Line* was started and is published by the Geothermal Resources Board of California. This newsletter can be obtained from Geothermal Hot Line, Division of Oil and Gas, 1416 9th Street, Room 1316, Sacramento, California 95814.

CITED REFERENCES

Allen, E. T. and Day, A. L., 1935, Hot springs of the Yellowstone National Park: Carnegie Inst. Washington Pub. 466, 525 p.

Alonso, H., 1968, Geothermal energy in Mexico: Geol. Soc. America 1968 Ann. Mtg., Mexico City, Field Trip No. 6, Geology and utilization of geothermic energy at Pathe, State of Hidalgo, 15 p.

Alonso, H., Fernandez, G., and Guiza, J., 1967, Power generation in Mexico from geothermal energy: World Power Conf., 7th, Moscow.

Alonso Espinoza, H., 1966, La zona geotermica de Cerro Prieto, Baja, California: Soc. Geol. Mexico Bol., v. 29, p. 17-47.

Arango, E. E., Buitrago, J. A., Cataldi, R., Ferrara, G. C., Panachi, C., and Villegas, V. J., 1970, Preliminary study on the Ruiz geothermal project (Colombia): U. N. Symp. on Dev. and Util. of Geothermal Resources, Pisa, Italy.

Arnason, B. and Tomasson, J., 1970, Deuterium and chloride in geothermal studies in Iceland: U. N. Symp. on Dev. and Util. of Geothermal Resources, Pisa, Italy.

Arnorsson, S., 1970a, Underground temperatures in hydrothermal areas in Iceland as deduced from the silica content of the thermal water: U. N. Symp. on Dev. and Util. of Geothermal Resources, Pisa, Italy.

——— 1970b, The distribution of some trace elements in thermal waters in Iceland: U. N. Symp. on Dev. and Util. of Geothermal Resources, Pisa, Italy.

Austin, C. F., 1966, Selection criteria for geothermal prospects: Nevada Bur. Mines Rept. 13, pt. C., p. 93-125.

Averiev, V. V., 1967, Hydrothermal processes in volcanic areas and their relation to magmatic activity: Bull. volcanol., v. 30.

Bailey, E. H., 1946, Quicksilver deposits of the Western Mayacmas district, Sonoma County, California: California Jour. Mines and Geol., v. 42, p. 199-230.

Banwell, C. J., 1963, Thermal energy from the earth's crust. Introduction and Part I: New Zealand Jour. Geol. and Geophys, v. 6, p. 52-69.

——— 1964, Thermal energy from the earth's crust. Part II. The efficient extraction of energy from heated rocks: New Zealand Jour. Geol. and Geophys., v. 7, p. 585-593.

——— 1970, Geophysical techniques in geothermal exploration (Rapporteur's Report): U. N. Symp. on Dev. and Util. of Geothermal Resources, Pisa, Italy.

Banwell, C. J. and Gomez Valle, R., 1970, Geothermal exploration in Mexico, 1968-1969: U. N. Symp. on Dev. and Util. of Geothermal Resources, Pisa, Italy.

Banwell, C. J. and MacDonald, W. J. P., 1965, Resistivity surveying in New Zealand thermal areas: Commonwealth Mining and Metallurgical Cong., 8th, Australia-New Zealand, New Zealand, Sec., 213, 1.

Barton, D. B., 1970, Current status of geothermal power plants at The Geysers, Sonoma County, California: U. N. Symp. on Dev. and Util. of Geothermal Resources, Pisa, Italy.

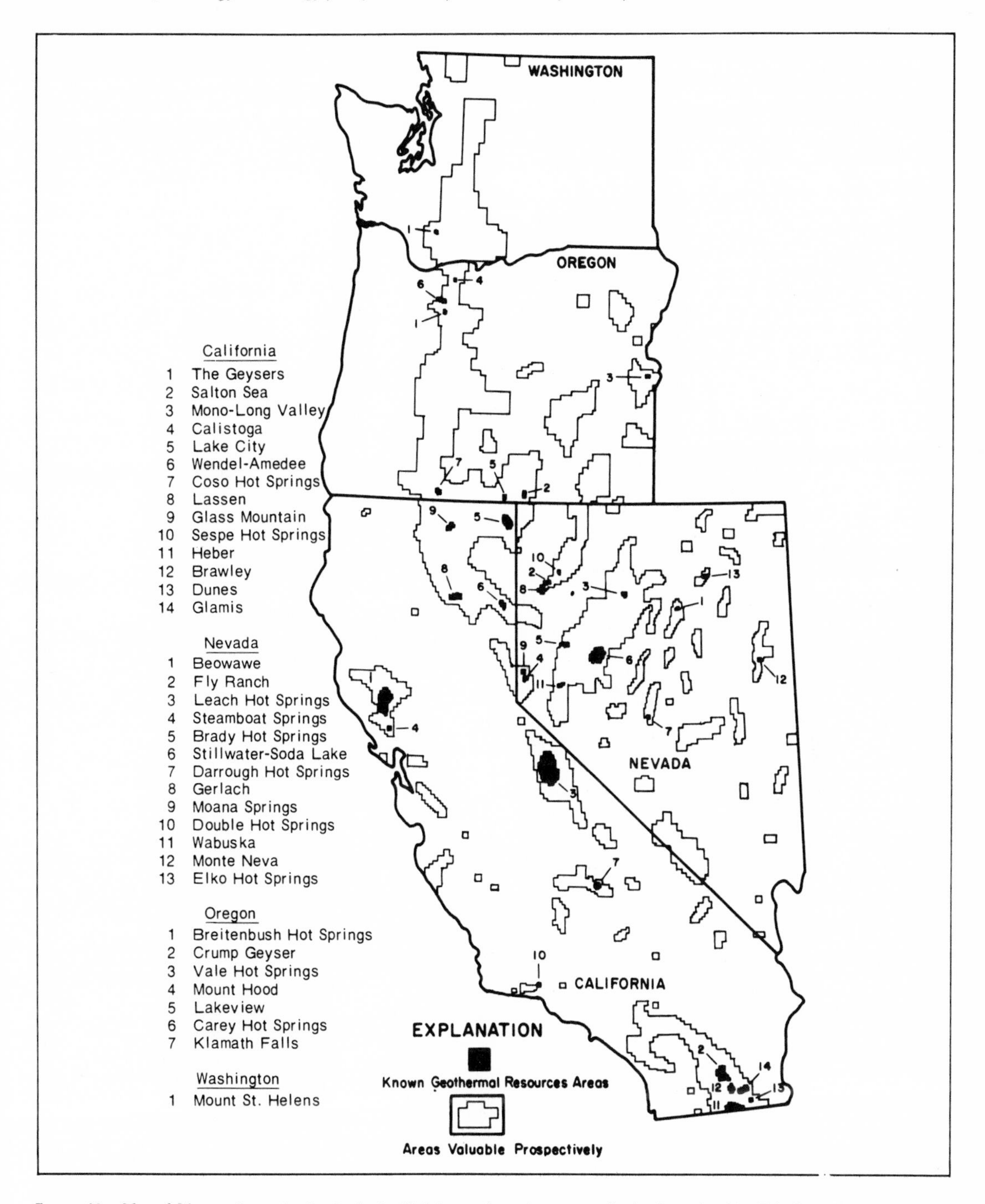

FIGURE 22.—Map of Western States showing lands classified for geothermal resources effective December 24, 1970 (from Godwin and others, 1971).

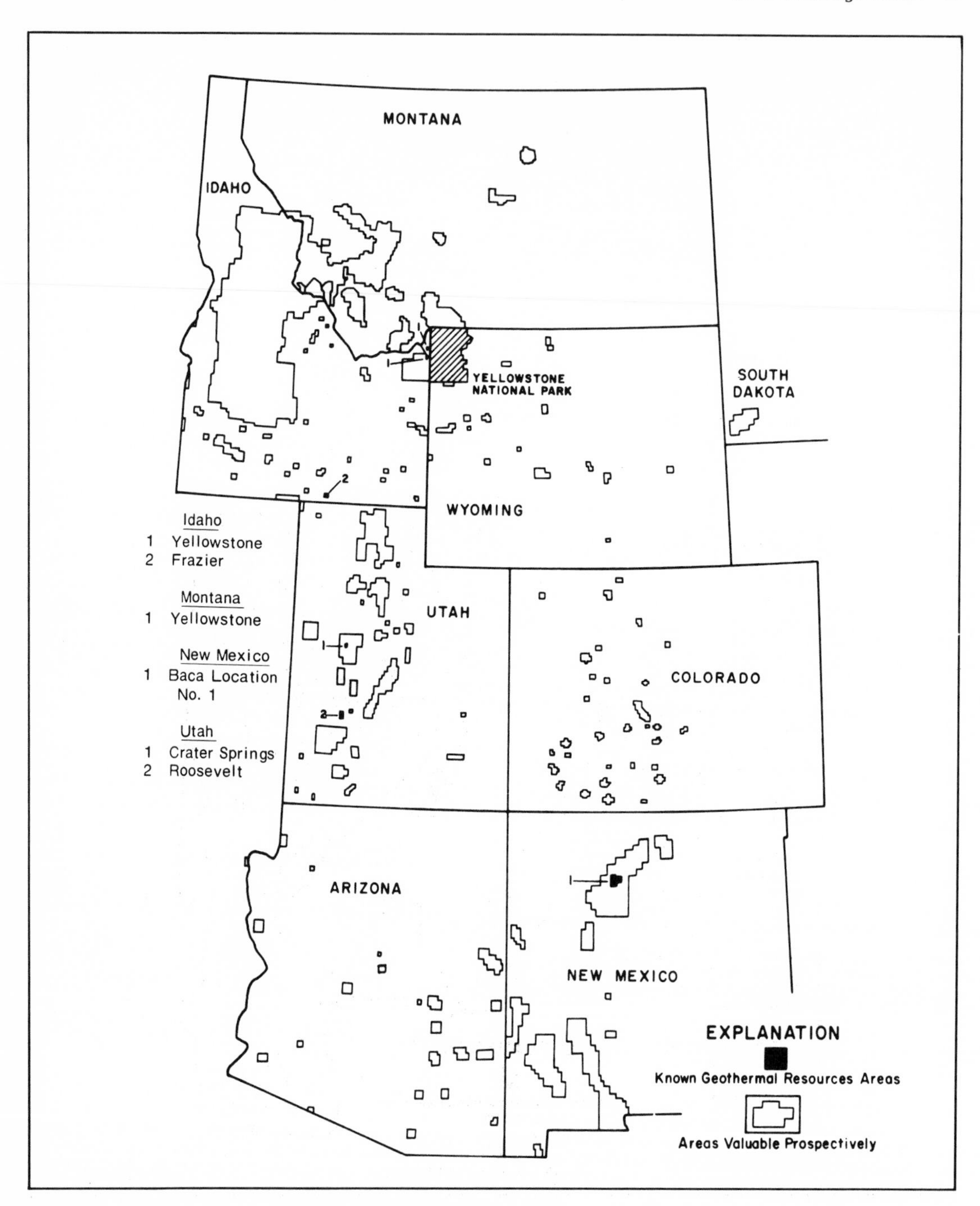

FIGURE 23.—Map of Western States showing lands classified for geothermal resources effective December 24, 1970 (from Godwin and others, 1971).

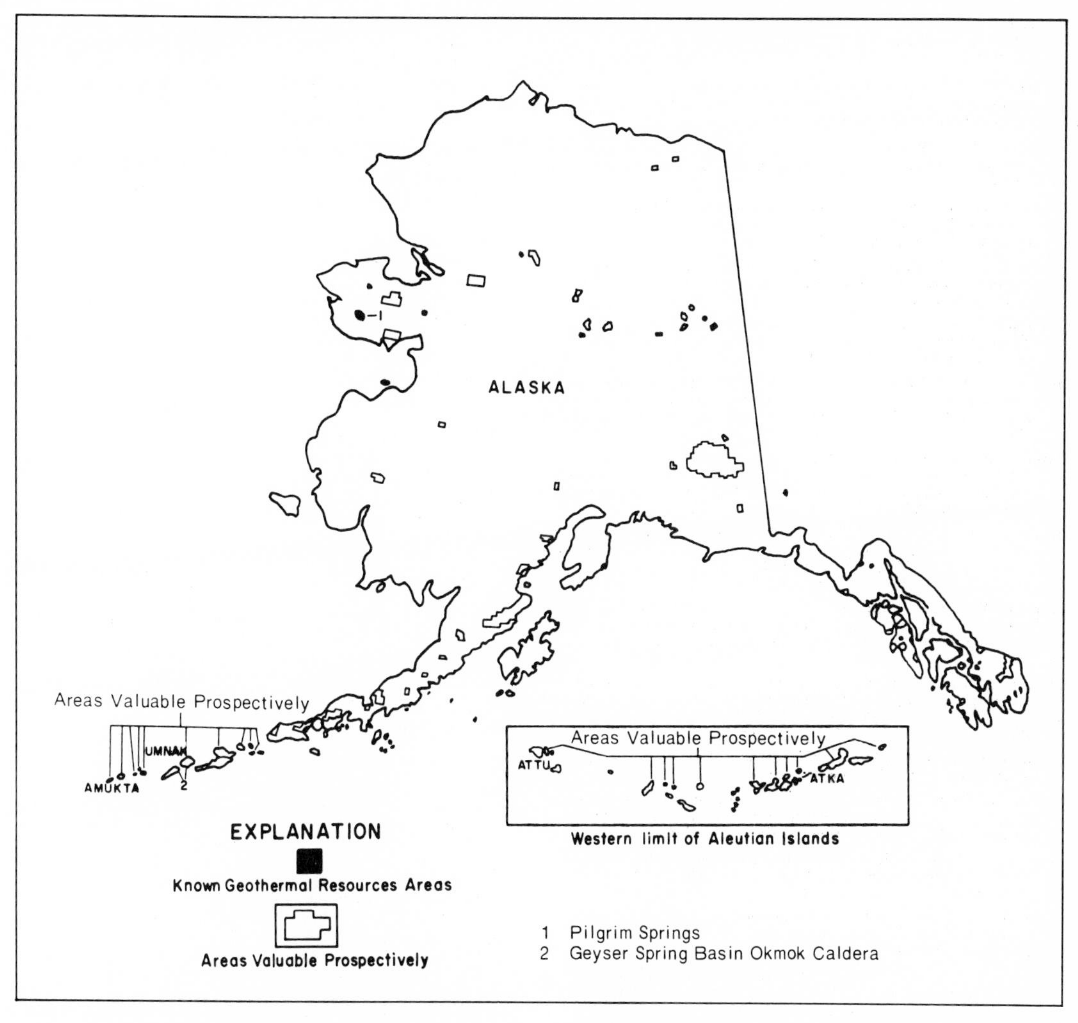

FIGURE 24.—Map of Alaska showing lands classified for geothermal resources effective December 24, 1970 (from Godwin and others, 1971).

Battini, F. and Menut, P., 1964, Etude structurale de la zone de Roccastrada en vue de la recherche de vapeur par les methodes geophysiques, gravimetriques et electriques: U. N. Conf. New Sources of Energy, Rome, 1961, Proc., v. 2, Geothermal Energy: I, p. 73-81.

Berry, F. A. F., 1966, Proposed origin of subsurface thermal brines, Imperial Valley, California (abs.): Am. Assoc. Petroleum Geologists Bull. v. 50, p. 644-645.

Biehler, S., Kovach, R. L., and Allen, C. R., 1964, Geophysical framework of the northern end of the Gulf of California structural province, *in* Marine geology of Gulf of California, a symposium: Am. Assoc. Petroleum Geologists Mem. 3, p. 126-143.

Biehler, S. and Rex, R. W., 1971, Structural geology and tectonics of the Salton trough, southern California, *in* Geol. Soc. America Cord. Mtg., Riverside, March, Field Trip Guidebook: Geol. Excur. in So. California, p. 30-42.

Birch, F., 1950, Flow of heat in the Front Range, Colorado: Geol. Soc. America Bull., v. 61, p. 567-630.

Blackwell, D. D., 1969, Heat flow determinations in the northwestern United States: Jour. Geophys. Research, v. 74, p. 992-1007.

——— 1971, Heat flow: EOS (Trans. Am. Geophys. Union), v. 52, no. 5, p. 135-139.

Bodvarsson, Gunnar, 1964, Physical characteristics of natural heat resources in Iceland: U. N. Conf. New Sources of Energy, Rome, 1961, Proc., v. 2, Geothermal Energy: I, p. 82-89.

——— 1970a, Evaluation of geothermal prospects and the objectives of geothermal exploration: Geoexploration, v. 8, p. 7-17.

——— 1970b, An estimate of the natural heat resources in a thermal area in Iceland: U. N. Symp. on Dev. and Util. of Geothermal Resources, Pisa, Italy.

Bodvarsson, G. and Palmason, G., 1964, Exploration of subsurface temperature in Iceland: U. N. Conf. New Sources of Energy, Rome, 1961, Proc., v. 2, Geothermal Energy: I, p. 91-98.

Boldizar, T., 1970, Geothermal energy production from porous sediments in Hungary: U. N. Symp. on Dev. and Util. of Geothermal Resources, Pisa, Italy.

Bolton, R. S., 1970, The behavior of the Wairakei geothermal field during exploitation: U. N. Symp. on Dev. and Util. of Geothermal Resources, Pisa, Italy .

Bowen, R. G. and Peterson, N. V., 1970, Thermal springs and wells in Oregon: Oregon Dept. Geol. and Mineral Industries, Misc. Paper 14.

Bradbury, J. J. C., 1970, The economics of geothermal power (Rapporteur's Report): U. N. Symp. of Dev. and Util. of Geothermal Resources, Pisa, Italy.

Brune, J. N. and Allen, C. R., 1967, A microearthquake survey of the San Andreas fault system in southern California: Seis. Soc. America Bull., v. 57, p. 277-296.

Buachidse, I. M., Buachidse, G. I., and Shaorshadse, M. P., 1970, Thermal waters of Georgia: U. N. Symp. on Dev. and Util. of Geothermal Resources, Pisa, Italy.

Burgassi, R., 1964, Prospecting for geothermal fields and exploration necessary for their adequate exploitation, performed in various regions of Italy: U. N. Conf. New Sources of Energy, Rome, 1961, Proc., v. 2, Geothermal Energy: I, p. 117-133.

Burgassi, R., Battini, F., and Mouton, J., 1964, Prospection geothermique pour la recherche des forces endogenes: U. N. Conf. New Sources of Energy, Rome, 1961, Proc., v. 2, Geothermal Energy: I, p. 134-140.

Burgassi, P. D., Ceron, P., Ferrara, G. C., Sestini, G., and Toro, B., 1970, Geothermal gradient and heat flow in the Radicofani region (east of Monte Amiata, Italy): U. N. Symp. on Dev. and Util. of Geothermal Resources, Pisa, Italy.

Burrows, W., 1970, Geothermal energy resources for heating and associated applications in Rotorua and surrounding areas: U. N. Symp. on Dev. and Util. of Geothermal Resources, Pisa, Italy.

Calamai, A., Cataldi, R., Squarci, P. and Taffi, L., 1970, Geology, geophysics, and hydrogeology of the Monte Amiata geothermal fields: Geothermics, Pisa, Italy, Instituto Internazionale per le Ricerche Geotermiche (CNR), v. 1 (Spec. issue).

Campbell, K. V. and others, 1970, A survey of thermal springs in Washington State: Northwest Science, v. 44, no. 1, p. 1-11.

Cataldi, R., Ceron, P., Di Mario, P., and Leardini, T., 1970, Progress report on geothermal development in Italy: U. N. Symp. on Dev. and Util. of Geothermal Resources, Pisa, Italy.

Chen, C. H., 1970, Geology and geothermal power potential of the Tatun volcanic region: U. N. Symp. on Dev. and Util. of Geothermal Resources, Pisa, Italy.

Cheng, W. T., 1970, Geophysical exploration in the Tatun volcanic region, Taiwan: U. N. Symp. on Dev. and Util. of Geothermal Resources, Pisa, Italy.

Clacy, G. R. T., 1968, Geothermal ground noise amplitude and frequency spectra in the New Zealand volcanic region: Jour. Geophys. Research, v. 73, p. 5377-5383.

Clayton, R. N., Muffler, L. J. P. and White, D. E., 1968, Oxygen isotope study of calcite and silicates of the River Ranch No. 1 well, Salton Sea geothermal field, California: Am. Jour. Sci., v. 266, p. 968-979.

Compendium of Papers, Imperial Valley — Salton Sea Area Geothermal Hearing, 1970, Geothermal Resources Board, California, Oct. 22 and 23.

Cormy, G., Demians d'Archimbaud, J., and Surcin, J., 1970, Prospection geothermique aux Antilles francaises Guadeloupe et Martinique: U. N. Symp. on Dev. and Util. of Geothermal Resources, Pisa, Italy.

Crosby, J. W., III, 1971, New developments in geothermal exploration: 1st Northwest Conf. on Geothermal Power, Dept. of Natural Resources, Olympia, Washington.

Dawson, G. B., 1964, The nature and assessment of heat flow from hydrothermal areas: New Zealand Jour. Geol. and Geophys., v. 7, p. 155-171.

Dawson, G. B. and Dickinson, D. J., 1970, Heat flow studies in thermal areas of the North Island of New Zealand: U. N. Symp. on Dev. and Util. of Geothermal Resources, Pisa, Italy.

de Anda, L. F., Septien, J. I., and Elizondo, J. R., 1964, Geothermal energy in Mexico: U. N. Conf. New Sources of Energy, Rome, 1961, Proc., v. 2, Geothermal Energy: I, p. 149-164.

De Grys, A., Vera, J., and Goossens, P., 1970, A note on the hot springs of Ecuador: U. N. Symp. on Dev. and Util. of Geothermal Resources, Pisa, Italy.

Decius, L. C., 1964, Geological environment of hyperthermal areas in continental United States and suggested methods of propecting them for geothermal power: U. N. Conf. New Sources of Energy, Rome, 1961, Proc., v. 2, Geothermal Energy: I, p. 166-177.

Decker, E. R., 1969, Heat flow in Colorado and New Mexico: Jour. Geophys. Research, v. 74, p. 550-559.

Dibblee, T. W., Jr., 1954, Geology of the Imperial Valley region, California, *in* Geology of Southern California: California Div. Mines Bull. 170, Chap. 2, p. 21-28.

Dowgiallo, J., 1970, Occurrence and utilization of thermal waters of Poland: U. N. Symp. on Dev. and Util. of Geothermal Resources, Pisa, Italy.

Economic Potential of Geothermal Resources in California, 1971, Geothermal Resources Board, California, 35 p.

Ellis, A. J., 1970, Quantitative interpretation of chemical characteristics of hydrothermal systems: U. N. Symp. on Dev. and Util. of Geothermal Resources, Pisa, Italy.

Epis, R. C., ed., 1968, Cenozoic volcanism in the Southern Rocky Mountains: Colorado School Mines Quart., v. 63, no. 3, 287 p.

Facca, G., 1970, General report of the status of world geothermal development (Rapporteur's Report): U. N. Symp. on Dev. and Util. of Geothermal Resources, Pisa, Italy.

Facca, G. and Ten Dam, A., 1964, Geothermal power economics, rev. ed.: Los Angeles, Worldwide Geothermal Exploration Co., 45 p.

Feng, T. T. and Huang, K. K., 1970, Exploration of geothermal resources in the Tatun volcanic region, Taiwan, Republic of China: U. N. Symp. on Dev. and Util. of Geothermal Resources, Pisa, Italy.

Fournier, R. O. and Rowe, J. J., 1966, Estimation of underground temperatures from the silica content of water from hot springs and wet-steam wells: Am. Jour. Sci., v. 264, p. 685-697.

Fournier, R. O. and Truesdell, A. H., 1970, Chemical indicators of subsurface temperature applied to hot spring waters of Yellowstone National Park, Wyoming, U. S. A.: U. N. Symp. on Dev. and Util. of Geothermal Resources, Pisa, Italy.

Frolov, N. M. and Vartanyan, G. S., 1970, Types of commercial deposits of thermal subterranean waters and some aspects relating to the assessment of their reserves: U. N. Symp. on Dev. and Util. of Geothermal Resources, Pisa, Italy.

Fujii, Y., and Akeno, T., 1970, Chemical prospecting of steam and hot water in the Matsukawa geothermal area: U. N. Symp. on Dev. and Util. of Geothermal Resources, Pisa, Italy.

Geological Society of America, Rocky Mountain Section, 1968, Igneous and hydrothermal geology of Yellowstone National Park: Geol. Soc. America Rocky Mtn. Sec., Bozeman, Mont., May 1968, Field Trip No. 6, 36 p.

George, R. D., Curtis, H. A., Lester, O. C., Crook, J. K., Yeo, J. B. and others, 1920, Mineral waters of Colorado: Colorado Geol. Survey Bull. 11, 474 p.

Geothermal Resources, 1967, 4th Prog. Rept. to the Legislature (California), Senate Permanent Factfinding Committee on Natural Resources, Sec. I, 66 p.

Gilbert, C. M., Christensen, M. N., Al-Rawi, Y., and Lajoie, K. L., 1968, Structural and volcanic history of Mono Basin, California-Nevada, *in* Coats, R. R., and others, eds., Studies in volcanology (Williams vol.): Geol. Soc. America Mem. 116, p. 275-329.

Godwin, L. H., Haigler, L. B., Rioux, R. L., White, D. E., Muffler, L. J. P., and Wayland, R. G., 1971, Classification of public lands valuable for geothermal steam and associated geothermal resources: U. S. Geol. Survey Circ. 647, 18 p.

Gomez Valle, R. G., Friedman, J. D., Gawarecki, S. J., and Banwell, C. J., 1970, Photogeologic and thermal infrared reconnaissance surveys of the Los Negritos — Ixtlan de los Hervores geothermal area, Michoacan, Mexico: U. N. Symp. on Dev. and Util. of Geothermal Resources, Pisa, Italy.

Grindley, G. W., 1964, Geology of the New Zealand geothermal steam fields: U. N. Conf. New Sources of Energy, Rome, 1961, Proc., v. 2, Geothermal Energy: I, p. 237-245.

——— 1965, The geology, structure, and exploitation of the Wairakei geothermal field, Taupo, New Zealand: New Zealand Geol. Survey Bull. n.s. 75, 131 p.

——— 1966, Geological structure of hydrothermal fields in the Taupo Volcanic Zone, New Zealand (abs.): Bull. volcanol., v. 29, p. 573-574.

——— 1970, Sub-surface structures and relation to steam production in the Broadlands geothermal fields, New Zealand: U. N. Symp. on Dev. and Util. of Geothermal Resources, Pisa, Italy.

Griscom, A. and Muffler, L. J. P., 1971, Aeromagnetic map and interpretation of the Salton Sea geothermal area, California: U. S. Geol. Survey, Geophys. Inv. Map GP-754.

Groh, E. A., 1966, Geothermal energy potential in Oregon: Oregon Dept. Geol. and Mineral Industries, Ore Bin, v. 28, p. 125-135.

Haigler, L. B., 1969, Mineral and water resources of Arizona: Arizona Bur. Mines Bull. 180.

Hamilton, W., 1960, Late Cenozoic tectonics and volcanism of the Yellowstone region, Wyoming, Montana, and Idaho, *in* Billings Geol. Soc. 11th Ann. Field Conf., West Yellowstone — Earthquake Area, 1960: p. 92-105.

Hatherton, T., MacDonald, W. J. P., and Thompson, G. E. K., 1966, Geophysical methods in geothermal prospecting in New Zealand: Bull. volcanol., ser. 2, v. 29, p. 487-498.

Hatton, J. W., 1970, Ground subsidence of a geothermal field during exploitation: U. N. Symp. on Dev. and Util. of Geothermal Resources, Pisa, Italy.

Hayakawa, M., 1970, The study of underground structure and geophysical state in geothermal areas by seismic exploration: U. N. Symp. on Dev. and Util. of Geothermal Resources, Pisa, Italy.

Hayakawa, M., Takaki, S., and Baba, K., 1967, Geophysical study of Matsukawa geothermal area, Northeast Japan: Japan Geol. Survey Bull. 18, 73.

Healy, J., 1964, Geology and geothermal energy in the Taupo Volcanic Zone, New Zealand: U. N. Conf. New Sources of Energy, Rome, 1961, Proc., v. 2, Geothermal Energy: I, p. 250-257.

——— 1970, Pre-investigation geological appraisal of geothermal fields: U. N. Symp. on Dev. and Util. of Geothermal Resources, Pisa, Italy.

Helgeson, H. C., 1968, Geologic and thermodynamic characteristics of the Salton Sea geothermal system: Am. Jour. Sci., v. 266, p. 129-166.

Heylmun, E. B., 1966, Geothermal power potential in Utah: Utah Geol. and Mineralog. Survey Spec. Studies 14, 28 p.

Hochstein, M. P. and Dickinson, D. J., 1970, Infrared remote sensing of thermal ground in the Taupo region, New Zealand: U. N. Symp. on Dev. and Util. of Geothermal Resources, Pisa, Italy.

Hochstein, M. P. and Hunt, M. T., 1970, Seismic, gravity, and magnetic studies, Broadlands geothermal field, New Zealand: U. N. Symp. on Dev. and Util. of Geothermal Resources, Pisa, Italy.

Hochstein, M. P., Innes, D. G., and Carrington, L., 1967, Seismic studies in the Broadlands geothermal area, p. 7-14, *in* Banwell and others, Broadlands geothermal area geophysical survey, New Zealand Dept. Sci. Indus. Res., Geophys. Div., unpubl. rept. 48.

Hodder, D. T., 1970, Application of remote sensing to geothermal prospecting: U. N. Symp. on Dev. and Util. of Geothermal Resources, Pisa, Italy.

Horton, R .C., 1964, Hot springs, sinter deposits, and volcanic cinder cones in Nevada: Nevada Bur. Mines Map 25.

Hunt, M. T., 1970, Net mass loss from the Wairakei geothermal field, New Zealand: U. N. Symp. on Dev. and Util. of Geothermal Resources, Pisa, Italy.

Ishikawa, T., 1970, Geothermal fields in Japan considered from the geological and petrological viewpoints: U. N. Symp. on Dev. and Util. of Geothermal Resources, Pisa, Italy.

James, R., 1968, Wairakei and Larderello: geothermal power systems compared: New Zealand Jour. Sci. and Tech. v. 11, p. 706.

Kaufman, A., 1970, The economics of geothermal power in the United States: U. N. Symp. on Dev. and Util. of Geothermal Resources, Pisa, Italy.

Keller, G .V., 1970, Induction methods in prospecting for hot water: U. N. Symp. on Dev. and Util. of Geothermal Resources, Pisa, Italy.

Kiersch, G. A., 1964, Geothermal steam: origin, characteristics, occurrence and exploitation: U. S. Air Force Cambridge Research Labs. (AFCRL-64-898), Washington, D. C., U.S. Dept. of Commerce.

Koenig, J. B., 1963, Geologic map of California, Santa Rosa sheet: California Div. Mines and Geol.

——— 1967, The Salton-Mexicali geothermal province: Calif. Div. Mines and Geol. Mineral Inf. Service, v. 20, no. 7, p. 75-81.

——— 1969, The Geysers geothermal field: California Div. Mines and Geol., Mineral Inf. Service, v. 22, no. 8, p. 123-128.

——— 1970, Geothermal exploration in the Western United States: U. N. Symp. on Dev. and Util. of Geothermal Resources, Pisa, Italy.

——— 1971, Geothermal development: Geotimes, v. 16, no. 3, p. 10-12.

Koga, A., 1970, Geochemistry of the waters discharged from drill hole at Otake and Hatchobaru areas (Japan): U. N. Symp. on Dev. and Util. of Geothermal Resources, Pisa, Italy.

Komagata, S., Iga., H., Nakamura, H., and Minohara, Y., 1970, The status of geothermal utilization in Japan: U. N. Symp. on Dev. and Util. of Geothermal Resources, Pisa, Italy.

Kottlowski, F. E., Weber, R. H., and Willard, M. E., 1969, Tertiary intrusive-volcanic-mineralization episodes in the New Mexico region: Geol. Soc. America Ann Mtg., Atlantic City, N.J., (abs.), p. 278-280.

Kremar, B. and Milanovic, B., 1970, Geothermal exploration of hot water sources in Carpathians of Yugoslavia and Czechoslovakia: U. N. Symp. on Dev. and Util. of Geothermal Resources, Pisa, Italy.

Lange, A. L. and Westphal, W. H., 1969, Microearthquakes near The Geysers, Sonoma County, California: Jour. Geophys. Research, v. 74, no. 17, p. 4377-4378.

Larderello and Monte Amiata, 1970, Electric power by endogenous steam: Rome, Ente Nazionale per L'Eneggia Elettrica (ENEL), 44 p.

Lee, K., 1969, Infrared exploration for shoreline springs of Mono Lake, California test site: Stanford Univ. Remote Sensing Lab. Tech. rept. no. 69-7, 196 p.

Lee, W. H. K., ed., 1965, Terrestrial heat flow: Am. Geophys. Union Geophys. Mon. Ser. no. 8, 276 p.

Lee, W. H. K. and Uyeda, Seija, 1965, Review of heat flow data, *in* Lee, W. H. K., ed., Terrestrial heat flow: Am. Geophys. Union Geophys. Mon. Ser. no. 8, p. 87-190.

Lovering, T. S., 1965, Some problems in geothermal exploration: Soc. Mining Engr. Trans., v. 232, p. 274-281.

Lovering, T. S. and Goode, H. D., 1963, Measuring geothermal gradients in drill holes less than 60 feet deep, East Tintic district, Utah: U. S. Geol. Survey Bull. 1172, 48 p.

Lumb, J. T., and MacDonald, W. J. P., 1970, Near-surface resistivity surveys of geothermal areas using the electromagnetic method: U. N. Symp. on Dev. and Util. of Geothermal Resources, Pisa, Italy.

Lyon, R. J. P. and Lee, K., 1970, Remote sensing in exploration for mineral deposits: Econ. Geol., v. 65, p. 785-800.

Mahon, W. A. J., 1970, Chemistry in the exploration and exploitation of hydrothermal systems: U. N. Symp. on Dev. and Util. of Geothermal Resources, Pisa, Italy.

Makarenko, F. A., Mavritsky, B. F., Lokchine, B. A. ,and Kononov, V. I., 1970, Geothermal resources of the USSR and prospects for their practical use: U. N. Symp. on Dev. and Util. of Geothermal Resources, Pisa, Italy.

Marinelli, G., 1964, Les anomalies thermiques et les champs geothermiques dans le cadre des intrusions recentes en Toscane: U. N. Conf. New Sources of Energy, Rome, Italy, Proc., v. 2, Geothermal Energy: I, p. 288-291.

McEuen, R. B., 1970, Delineation of geothermal deposits by means of long-spacing resistivity and airborne magnetics: U. N. Symp. on Dev. and Util. of Geothermal Resources, Pisa, Italy.

McKenzie, G. R., and Smith, J. H., 1968, Progress of geothermal energy development in New Zealand: Proc. 8th World Power Conf., Moscow.

McMillan, D. A., 1970, Economics of The Geysers geothermal field, California: U. N. Symp. on Dev. and Util. of Geothermal Resources, Pisa, Italy.

McNitt, J. R., 1963, Exploration and development of geothermal power in California: California Div. Mines and Geology Spec. Rept. 75, 45 p.

——— 1964, Geology of The Geysers thermal area, California: U. N. Conf. New Sources of Energy, Rome, Proc., v. 2, Geothermal Energy: I, p. 292-302.

——— 1965, Review of geothermal resources, *in* Lee, W.H.K., ed., Terrestrial heat flow: Am. Geophys. Union Geophys. Mon. Ser. no. 8, p. 240-266.

——— 1968, Geology of the Kelseyville quadrangle, Sonoma, Lake, and Mendocino counties, California: California Div. Mines and Geol. Map Sheet 9.

——— 1970, The geological environment of geothermal fields as a guide to exploration (Rapporteur's Report): U. N. Symp. on Dev. and Util. of Geothermal Resources, Pisa, Italy.

Meidav, T., 1970, Application of electrical resistivity and gravimetry in deep geothermal exploration: U. N. Symp. on Dev. and Util. of Geothermal Resources, Pisa, Italy.

Mercado, S., 1969, Chemical changes in geothermal well M-20, Cerro Prieto, Mexico: Geol. Soc. America Buil., v. 80, p. 2623-2630.

——— 1970, High-activity hydrothermal zones detected by Na/K, Cerro Prieto, Mexico: U. N. Symp. on Dev. and Util. of Geothermal Resources, Pisa, Italy.

Mooser, F., 1965, Progress report on recent developments of geothermal energy and volcanology in Mexico: Bull. volcanol., ser. 2, v. 28, p. 69-73.

Mori, Y., 1970, Exploitation of Matsukawa geothermal area: U. N. Symp. on Dev. and Util. of Geothermal Resources, Pisa, Italy.

Moxham, R. M., 1969, Aerial infrared surveys at The Geysers geothermal steam field, California: U. S. Geol. Survey Prof. Paper 650-C, p. C106-122.

Muffler, L. J. P., 1971, Evaluation of initial investigations, Dieng geothermal area, central Java, Indonesia: U. S. Geol. Survey, open file rept. 21 p.

Muffler, L. J. P. and White, D. E., 1969, Active metamorphism of Upper Cenozoic sediments in the Salton Sea geothermal field and the Salton trough, southeastern California: Geol. Soc. America Bull., v. 80, p. 157-182.

Mundorff, J. C., 1970, Major thermal springs of Utah: Utah Geol. and Minealog. Survey, Water Resources Bull. 13, 60 p.

Nakamura, H., Sumi, K., Katagiri, K., and Iwata, T., 1970, The geological environment of Matsukawa geothermal area, Japan: U. N. Symp. on Dev. and Util. of Geothermal Resources, Pisa, Italy.

Nicholls, H. R. and Rinehart, J. S., 1967, Geophysical study of geyser action in Yellowstone National Park: Jour. Geophys. Research, v. 72, no. 18, p. 4651-4663.

Noguchi, T., 1970, An attempted evaluation on geothermal energy in Japan: U. N. Symp. on Dev. and Util. of Geothermal Resources, Pisa, Italy.

Noguchi, K., Goto, T., Ueno, S., and Imahashi, M., 1970, Geochemical investigation of the strong acid water from the bored wells Hakone, Japan: U. N. Symp. on Dev. and Util. of Geothermal Resources, Pisa, Italy.

Nolan, T. B. and Anderson, G. H., 1934, The geysers area near Beowawe, Eureka County, Nevada: Am. Jour. Sci., v. 27, p. 215-229.

Oki, Y., and Hirano, T., 1970, Geothermal system of Hakone Volcano: U. N. Symp. on Dev. and Util. of Geothermal Resources, Pisa, Italy.

Oxburgh, E. R. and Turcotte, D. L., 1970, Thermal structure of island arcs: Geol. Soc. America Bull., v. 81, p. 1665-1688.

Pakiser, L. C., 1961, Gravity, volcanism, and crustal deformation in Long Valley, California: U. S. Geol. Survey Prof. Paper 424-B, p. 250-253.

Palmason, G., Friedman, J. D., Williams, R. S. Jr., Jonsson, J., and Saemundsson, K., 1970, Aerial infrared surveys of Reykjanes and Torfajokull thermal areas, Iceland, with a section on the cost of exploration surveys: U. N. Symp. on Dev. and Util. of Geothermal Resources, Pisa, Italy.

Peterson, N. V., and Groh, E. A., 1967, Geothermal potential of the Klamath Falls area, Oregon: a preliminary study: Oregon Dept. Geol. and Mineral Industries, Ore Bin, v. 29, p. 209-231.

Peterson, N. V., and McIntyre, J. R., 1970, The reconnaissance geology and mineral resources of eastern Klamath County and western Lake County, Oregon: Oregon Dept. Geol. and Mineral Industries Bull. 66, 70 p.

Phillips, K. N., 1941, Fumaroles of Mount St. Helens and Mount Adams: Mazama (Portland, Ore.), v. 23, no. 12, p. 37-42.

Piip, B. I., Ivanov, V. V., and Averiev, V. V., 1964, The hyperthermal water of Pauzhetsk, Kamchatka, as a source of geothermal energy: U. N. Conf. New Sources of Energy, Rome, 1961, Proc., v. 2, Geothermal Energy: I, p. 339-345.

Rex, R. W., 1966, Heat flow in the Imperial Valley of California (abs.): Am. Geophys. Union Trans., v. 47, p. 181.

——— 1970, Investigation of geothermal resources in the Imperial Valley and their potential value for desalination of water and electricity production: Riverside, California Univ. Inst. Geophys. and Planetary Phys., 14 p.

——— 1971, Geothermal energy — the neglected energy option: Bull. Atomic Scientists, v. 27, no. 8, p. 52-56.

Rinehart, C. D. and Ross, D. C., 1964, Geology and mineral deposits of the Mount Morrison Quadrangle, Sierra Nevada, California: U. S. Geol. Survey Prof. Paper 385, 106 p.

Rinehart, J. S., 1968a, Seismic signatures of some Icelandic geysers: Jour. Geophys. Research, v. 73, p. 4609.

——— 1968b, Geyser activity at Beowawe, Eureka County, Nevada: Jour. Geophys. Research, v. 73, p. 7703.

Risk, G. F., MacDonald, W. J. P., and Dawson, G. B., 1970, D.C. resistivity surveys of the Broadlands geothermal region, New Zealand: U. N. Symp. on Dev. and Util. of Geothermal Resources, Pisa, Italy.

Ross, S. H., 1970, Geothermal potential of Idaho: U. N. Symp. on Dev. and Util. of Geothermal Resources, Pisa, Italy.

Roy, R. F., Decker, E. R., Blackwell, D. D. and Birch, F., 1968, Heat flow in the United States: Jour. Geophys. Research, v. 73, p. 5207-5221.

Saito, M., 1964, Known geothermal fields in Japan: U. N. Conf. New Sources of Energy, Rome, 1961, Proc., v. 2, Geothermal Energy: I, p. 367-404.

Sass, J. H., Lachenbruch, A. H., Munroe, R. J., Greene, G. W., and Moses, T. H., Jr., 1971, Heat flow in the Western United States: Jour. Geophys. Research, v. 76, no. 26, p. 6376-6413.

Sato, H., 1970, The present state of geothermal development in Japan: U. N. Symp. on Dev. and Util. of Geothermal Resources, Pisa, Italy.

Sigvaldsson, G. E. and Cuellar, G., 1970, Geochemistry of the Ahuachapan thermal area, El Salvador, Central America: U.N. Symp. on Dev. and Util. of Geothermal Resources, Pisa, Italy.

Smith, J. H., 1970, Geothermal development in New Zealand: U.N. Symp. on Dev. and Util. of Geothermal Resources, Pisa, Italy.

Smith, R. L., Bailey, R. A., and Ross, C. S., 1961, Structural evolution of the Valles caldera, New Mexico, and its bearing on the emplacement of ring dikes: U. S. Geol. Survey Prof. Paper 424-D, p. D145-149.

——— 1970, Geologic map of the Jemez Mountains, New Mexico: U. S. Geol. Survey Misc. Geol. Inv. Map I-571.

Stone, R. T., 1971, Implementing the Federal Geothermal Steam Act of 1970: Northwest Conf. on Geothermal Power, 1st, Dept. of Natural Resources, Olympia, Washington.

Strangway, D. W., 1970, Geophysical exploration through geologic cover: U. N. Symp. on Dev. and Util. of Geothermal Resources, Pisa, Italy.

Studt, F. E., 1964, Geophysical prospecting in New Zealand's hydrothermal fields: U. N. Conf. New Sources of Energy, Rome, 1961, Proc., v. 2, Geothermal Energy: I, p. 380-384.

Studt, F. E. and Thompson, G. E. K., 1969, Geothermal heat flow in North Island of New Zealand: New Zealand Jour. Geol. and Geophys., v. 12, p. 673.

Sukharev, G. M., Vlasova, S., and Taranukha, Y. K., 1970, The utilization of thermal waters of the developed oil deposits of the Caucasus: U. N. Symp. on Dev. and Util. of Geothermal Resources, Pisa, Italy.

Summers, W. K., 1965a, A preliminary report of New Mexico's geothermal energy resources: New Mexico Bur. Mines and Mineral Resources Circ. 80, 41 p.

——— 1965b, Chemical characteristics of New Mexico's thermal waters — a critique: New Mexico Bur. Mines and Mineral Resources Circ. 83, 27 p.

——— 1970a, Geothermal prospects of New Mexico: U. N. Symp. on Dev. and Util. of Geothermal Resources, Pisa, Italy.

——— 1970b, New Mexico's thermal waters: New Mexico Bur. Mines and Mineral Resources microfilm.

——— 1971, Annotated and indexed bibliography of geothermal phenomena: New Mexico Bur. Mines and Mineral Resources.

Summers, W. K. and Ross, S. H., 1971, Geothermics in North America: present and future: Wyoming Geol. Assoc. Earth Sci. Bull., March, p. 7-22.

Tamrazyan, G. P., 1970, Continental drift and thermal fields: U. N. Symp. on Dev. and Util. of Geothermal Resources, Pisa, Italy.

Ten Dam, A. and Erentoz, C., 1970, Kizildere geothermal field, Western Anatolia: U. N. Symp. on Dev. and Util. of Geothermal Resources, Pisa, Italy.

Ten Dam, A. and Khrebtov, A. I., 1970, The Menderes Massif geothermal province: U. N. Symp. on Dev. and Util. of Geothermal Resources, Pisa, Italy.

Thompson, G. A. and White, D. E., 1964, Regional geology of the Steamboat Springs area, Washoe County, Nevada: U. S. Geol. Survey Prof. Paper 458-A, 52 p.

Thompson, G. A., Meister, L. J., Herring, A. T., Smith, T. E., Burke, D. B., Kovach, R. L., Burford, R. O., Salehi, I. A., and Wood, M. D., 1967, Geophysical study of Basin-Range structure, Dixie Valley region, Nevada: U. S. Air Force Cambridge Research Labs. (AFCRL-66-848), Washington, U. S. Dept. of Commerce.

Thompson, G. E. K., Banwell, C. J., Dawson, G. B., and Dickinson, D. J., 1964, Prospecting of hydrothermal areas by surface thermal surveys: U. N. Conf. New Sources of Energy, Rome, 1961, Proc., v. 2, Geothermal Energy: I, p. 386-400.

Thorsen, G. W., 1971, Prospects for geothermal energy in Washington: Northwest Conf. of Geothermal Power, 1st, Dept. of Natural Resources, Olympia, Washington.

Thorsteinsson, T. and Eliasson, J., 1970, Geohydrology of the Laugarnes hydrothermal system in Reykjavik, Iceland: U. N. Symp. on Dev. and Util. of Geothermal Resources, Pisa, Italy.

Tikhonov, A. N. and Dvorov, I. M., 1970, Development of research and utilization of geothermal resources in the USSR: U. N. Symp. on Dev. and Util. of Geothermal Resources, Pisa, Italy.

Tonani, F., 1967, Geothermal exploration in El Salvador: U. N. Rept., Central America.

——— 1970, Geochemical methods of exploration for geothermal energy: U. N. Symp. on Dev. and Util. of Geothermal Resources, Pisa, Italy.

Vakin, E. A., Polak, B. G., Sugrobov, V. M., Erlik, E. N., Belousov, V. I., and Pilipenko, G. F., 1970, Recent hydrothermal systems of Kamchatka: U. N. Symp. on Dev. and Util. of Geothermal Resources, Pisa, Italy.

Ward, P. L. and Jacob, K. H., 1971, Microearthquakes in the Ahuachapan geothermal field, El Salvador, Central America: Science, v. 173, p. 328-330.

Ward, P. L., Palmason, G., and Drake, C., 1969, Microearthquake survey and the mid-Atlantic ridge in Iceland: Jour. Geophys. Research, v. 74, no. 2, p. 665-684.

Waring, G. A., 1965, Thermal springs of the United States and other countries of the world — a summary: U. S. Geol. Survey Prof. Paper 492, 383 p.

White, D. E., 1955, Thermal springs and epithermal ore deposits, *in* Pt. I of Econ. Geol., 50th Anniversary vol., 1905-55, p. 99-154.

——— 1964, Preliminary evaluation of geothermal areas by geochemistry, geology, and shallow drilling: U. N. Conf. New Sources of Energy, Rome, 1961, Proc., v. 2, Geothermal Energy: I, p. 402-408.

——— 1965, Geothermal energy: U. S. Geol. Survey Circ. 519, 17 p.

——— 1967, Some principles of geyser activity, mainly from Steamboat Springs, Nevada: Am. Jour. Sci., v. 265, p. 641-684.

——— 1968a, Hydrology, activity, and heat flow of the Steamboat Springs thermal system, Washoe County, Nevada: U. S. Geol. Survey Prof. Paper 458-C, 109 p.

——— 1968b, Environments of generation of some base-metal ore deposits: Econ. Geol., v. 63, p. 301-335.

——— 1969a, Rapid heat-flow surveying of geothermal areas utilizing individual snowfalls as calorimeters: Jour. Geophys. Research, v. 74, p. 5191-5201.

——— 1969b, Thermal and mineral waters of the United States —a brief review of possible origins: Internat. Geol. Cong., 23rd, Prague 1969, v. 19, p. 269-286.

——— 1970, Geochemistry applied to the discovery, evaluation, and exploitation of geothermal energy resources (Rapporteur's Report): U. N. Symp. on Dev. and Util. of Geothermal Resources, Pisa, Italy.

White, D. E., Muffler, L. J. P., and Truesdell, A. H., 1971, Vapor-dominated hydrothermal systems compared with hot-water systems: Econ. Geol., v. 66, p. 75-97.

White, D. E., Muffler, L. J. P., Truesdell, A. H., and Fournier, R. O., 1968, Preliminary results of research drilling in Yellowstone thermal areas: Geol. Soc. America; Rocky Mtn. Sec. Program (abs.), p. 84.

White, D. E. and Roberson, C. E., 1962, Sulphur Bank, California, a major hot-spring quicksilver deposit, *in* Petrologic Studies (Buddington vol.): Geol. Soc. America, N. Y., p. 397-428.

White, D. E., Thompson, G. A., and Sandberg, C. H., 1964, Rocks, structure, and geologic history of Steamboat Springs thermal area, Washoe County, Nevada: U. S. Geol. Survey Prof. Paper 458-B, 63 p.

Whiteford, P. C., 1970, Ground movement in the Waiotapu geothermal region, New Zealand: U. N. Symp. on Dev. and Util. of Geothermal Resources, Pisa, Italy.

Wilson, S. H., 1970, Statistical interpretation of chemical results from drillholes as an aid to geothermal prospecting and exploitation: U. N. Symp. on Dev. and Util. of Geothermal Resources, Pisa, Italy.

Yamasaki, T., Matsumoto, Y., and Hayashi, M., 1970, The geology and hydrothermal alterations at Otake geothermal area, Kujya Volcano Group, Kyushu, Japan: U. N. Symp. on Dev. and Util. of Geothermal Resources, Pisa, Italy.

Zen, M. T. and Radja, V. T., 1970, Result of the preliminary geological investigation of natural steam fields in Indonesia: U. N. Symp. on Dev. and Util. of Geothermal Resources, Pisa, Italy.

12

Reprinted from National Research Council-National Academy of Sciences, Committee on Geological Sciences, *The Earth and Human Affairs,* 1972, 112-124; by permission of Harper & Row, Publishers, Inc.

THE ALASKA OIL PIPELINE

Committee on Geological Sciences
Division of Earth Sciences
National Research Council-National Academy of Sciences

"A large hot oil pipeline buried for hundreds of miles in permafrost has no precedent."[1]

Perhaps the best way to integrate and emphasize some of the principles and concerns that we have outlined thus far is to look at a major environmental issue like the proposed construction of the Alaskan oil pipeline. Briefly, the proposal is to build a pipeline, 789 miles (1270 kilometers) long, from the edge of the Arctic Ocean on the north coast of Alaska, through the center of the state, across coastal plains, mountain ranges, rivers and streams, to the port of Valdez on the southern coast where the oil will be transported by ocean tankers to refineries in other regions of the United States.

---

[1]Final Environmental Impact Statement, Proposed Trans-Alaska Pipeline, Vol. 1, p. 93 (March, 1972).

Numerous questions are currently being debated, pro and con, regarding the impact of the pipeline and the importance of the oil. These questions include such things as the disturbance or possible destruction of the landscape and the flora and fauna; the vulnerability of the pipeline to damage or failure from floods and earthquakes; the problems of building the pipeline itself, as well as access roads, pumping stations, and airstrips, on permanently frozen ground; the claims of the native population—Eskimos, Indians, and Aleuts—to most of the land over which the pipeline would go; the significance of the oil in meeting future national petroleum needs; and the economic value of the oil in helping to balance our international accounts. Many of these issues are beyond the scope of this book. However, one particular aspect, that of the geologic nature of the proposed pipeline route, does interest us here. If the pipeline is to be built, what special aspects of Alaskan geology ought to be considered in its design, construction, and maintenance?

the route

In 1968, oil and gas were discovered in Triassic sandstones, 1 to 2 miles (1.5 to 3 kilometers) below ground, on the north coast of Alaska near Prudhoe Bay, several miles inland from the Arctic Ocean. An estimated 10 billion barrels of oil in these subsurface reservoirs cover an area the size of Massachusetts, and perhaps many billions more are yet to be discovered. The Arctic Ocean is frozen most of the year, so it is impractical to ship the oil by tanker from the north coast. Instead, a pipeline would have to be built to transport the crude oil from Prudhoe Bay southward across Alaska to the port of Valdez where tankers can operate throughout the year.

As shown in Figure 5-1, the pipeline will start near Prudhoe Bay and first cross about 125 miles (210 kilometers) of flat, treeless tundra to the foothills of the Brooks Range. This region of Alaska is referred to as the "North Slope" because the land slopes gently northward from the Brooks Range to the shores of the frigid Arctic Ocean. Along much of this segment the pipeline will follow the flood plain of the Sagavanirktok and Atigun rivers that flow northward from the Brooks Range into the Arctic Ocean.

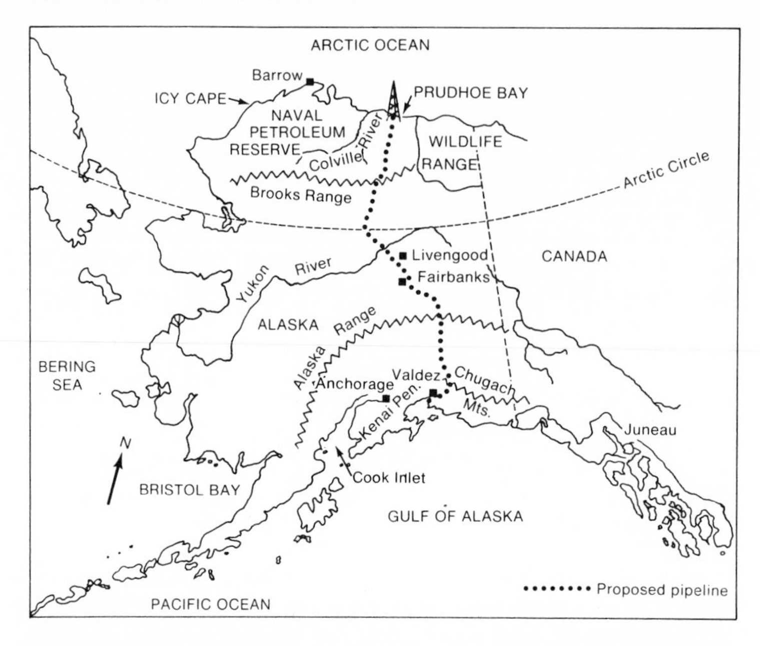

Figure 5-1. Map showing the route of the proposed Alaskan pipeline. The oil would be transported from oil wells on the north coast 789 miles (1270 kilometers) to the terminal facilities at the port of Valdez on the south coast.

The next segment of the pipeline, about 90 miles (145 kilometers) long, will cross the rugged Brooks Range, which is the Alaskan continuation of the Rocky Mountains. The pipeline will climb to an elevation of 4800 feet (1463 meters) at the Dietrich Pass. To this point the oil will be raised by three pumping stations, and then flow by gravity for the next 340 miles (547 kilometers).

On the south side of the Brooks Range the pipeline will cross a 350-mile (563-kilometer) wide interior plateau, a region of forests and streams. This section of the route includes the westward-flowing Yukon River and the city of Fairbanks.

The next segment of the route, about 75 miles (120 kilometers), crosses the Alaska Range, at close to 3000 feet (more than 900 meters) and then goes down toward the south coast. On the north side of the Alaska Range the pipeline will cross a major fault, the Denali fault.

After the Alaska Range, the route traverses the Copper River Basin for about 100 miles (160 kilometers) to the Chugach Mountains near the south coast. Rising and falling across this range for some 40 miles (64 kilometers), the pipeline will at last reach Valdez, about 10 miles (16 kilometers) farther on, located at the head of Prince William Sound on the south coast of the state.

the land

The proposed route of the pipeline lies within several major physical environments. From north to south they are the North Slope, the Brooks Range, the interior Central Plateau, the Alaska and Chugach Ranges with the Copper River Basin lying between them, and the Port Valdez area. Each of these environments has peculiar geologic characteristics affecting construction and maintenance of the proposed pipeline.

Permafrost. Both the world's polar regions are underlain by perennially frozen ground, or permafrost. Because of the below-freezing temperatures that prevail in these regions, moisture within the soil and bedrock exists in the frozen rather than liquid state. Approximately 85 percent of Alaska is underlain by permafrost that varies in thickness from 1300 feet (396 meters) near Barrow in the north to less than 1 foot (30 centimeters) in the southern part of the state.

In summer, when air temperatures are above freezing, the uppermost layers of the ground above the permafrost become warm enough for the ice to melt, but because of the permafrost layer below, the melted water collects at the ground surface, making a soggy, poorly drained landscape (Figure 5-2). The thickness of this uppermost, or "active," layer that freezes in the winter and thaws in the summer is controlled by the soil characteristics, the insulating effect of the vegetation mat, and the mean annual temperature.

In most areas of Alaska, rock and soil material have been frozen for thousands of years, extending well back into the glacial ages. In some places permafrost is more recent where new sediment has been deposited, where lakes and streams have changed location, or where humans have disturbed the natural terrain. The addition or removal of surface materials, such as sediment, water, or vegeta-

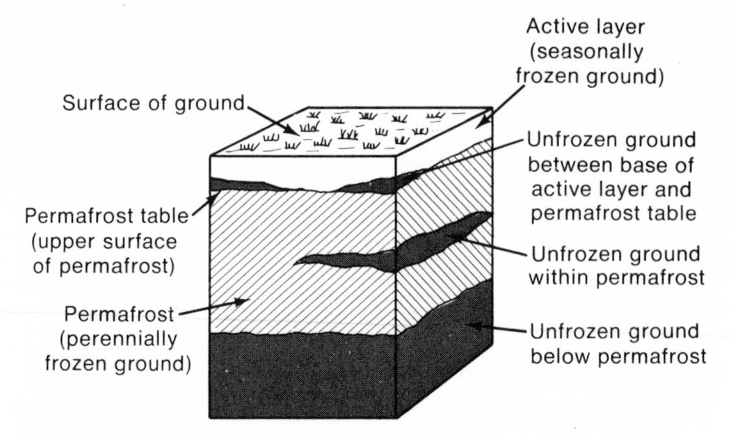

Figure 5-2. Block diagram showing permafrost lying below active layer and above unfrozen ground. The active layer may be inches to feet in thickness; the permafrost may be up to a thousand feet or more in thickness. The permafrost may include pockets of unfrozen material.

tion, changes the insulation of the ground surface so that the upper surface of the permafrost may rise or fall.

In a strict sense, permafrost refers to permanently frozen rock or soil, with or without ice. What concerns us here, however, is not icefree or dry permafrost but permafrost that has high ice content. The reason for this is that natural or man-induced thawing of ice-rich permafrost can result in dramatic changes of the physical properties of the rock or soil. The melting of ice within ice-rich permafrost causes the soil particles or rock fragments to collapse together because of the decreased volume occupied by the water as compared to that of the ice. Moreover, depending upon the slope of the land, the water may flow away from the thawed site by gravity. The change in volume and water flow often cause irregular subsidence of the ground, liquefaction and flowage of soils, and alterations in surface water drainage (Figure 5-3).

In building the proposed pipeline complex the insulating layers of the ground above the permafrost inevitably will be modified. Even the simple disruption of the vegetative mat of the ground by a vehicle destroys the insulating layer, and could cause melting and subsidence of the land. What in winter may be a hard, frozen roadway, becomes in summer an impassable muddy track (Figure 5-4).

Figure 5-3. Gravel road in Alaska showing severe irregular settling of the ground resulting from permafrost thawing below. Uneven distribution of ice within the permafrost caused the differential subsidence of the land along the roadway. SOURCE: U.S. Geological Survey, Professional Paper 678)

Figure 5-4. A road on the North Slope formed originally in winter on the frozen ground surface. After the spring thaw, the road has become a muddy ditch. Disruption of the insulating active layer has caused thawing of the underlying ice-rich permafrost resulting in a water-logged, poorly drained surface. SOURCE: U.S. Geological Survey.

Earthquakes and faults. Alaska, with its rugged mountain ranges, active fault zones, and earthquake activity, is geologically the northwestern extension of the western United States. As noted earlier, the pipeline route will cross the Denali fault, a zone of fracture where ground dislocation occurred in Holocene time with total horizontal movements of about 1000 feet. Within the estimated 30-year lifetime of the pipeline, significant movement may again occur. But when such movement might occur, and how great the ground displacement might be, is impossible to predict.

The Chugach Mountains along the southern part of the state are extremely tectonically active, and faults are undoubtedly present there. Detailed geologic field mapping and studies of microearthquake activity will be necessary in this area to delineate these faults. Other faults have been mapped in the area between the Yukon River and Fairbanks in the Central Plateau, but we do not know whether they actually intersect the pipeline route or what their activity might be.

Ground shaking from earthquakes also occurs in Alaska. Such shaking causes landslides as well as liquefaction of water-rich soils and sediments. Geologic studies indicate that any point along the southern two-thirds of the pipeline route could experience an earthquake with a Richter magnitude of 7 or more. (In 1937 the area southeast of Fairbanks experienced a 7.3 magnitude earthquake. The 1964 Alaskan earthquake was magnitude 8.5, whereas the 1971 San Fernando Valley earthquake in California was magnitude 6.6.) One or more such earthquakes is almost a certainty during the 30-year lifetime of the pipeline.

Rivers and streams. The route proposed for the pipeline crosses more than 50 major streams, the largest of which is the 0.5-mile (0.8-kilometer) wide Yukon River. In some areas the pipeline will follow river valleys, as in the north along the Sagavanirktok and Atigun rivers, usually being buried in the river alluvium. During the winter, Alaskan rivers and streams are frozen solid. At the annual spring thaw, these waterways flood, carrying away the accumulated ice and snow of the previous winter. Not only do water levels rise but the stream and river beds are deeply scoured by the floods. Where the pipeline is above ground, it must be sufficiently high to be protected from the flood waters. Where the pipeline is buried within the river and stream channels, it must be below the depth of maximal erosion. Pipeline plans call for construction across rivers and streams that could withstand the extreme sort of flood that might occur once in 50 years.

Construction of the pipeline also will alter surface water drainage in places where the pipeline is above ground, lying on a gravel pad. There will have to be culverts through the gravel pad to allow passage of surface water. Otherwise, the gravel pad will act as a dam and the ponds that could form would accelerate thawing of the underlying permafrost.

Sand and gravel. Approximately 83 million cubic yards of construction materials, particularly sand and gravel, will have to be quarried along the pipeline route. A total of 288 potential sites have been designated to supply these materials, which will be mainly taken from stream and river deposits and upland sources such as alluvial fans and glacial sediments. Quarrying in watercourses will alter channel flow and cause siltation. Excavation of upland areas will change the slope of the terrain, result in thaw areas underlain by permafrost, and increase local erosion. Special precautions will have to be taken to leave the land after quarrying in a condition that would minimize these effects.

Marine environment. Terminal facilities at the south end of the pipeline at Valdez will have to be built to handle the 600,000 barrels of oil arriving daily (at peak production this will increase to 2,000,000 barrels per day). Valdez lies at the northern end of Prince William Sound. This deep-water, icefree port is part of a larger coastal zone that runs from the tip of the Aleutian Islands in the west to California in the southeast. This area, which rims the Gulf of Alaska, supports a great variety and abundance of marine life including fish, especially salmon and herring, shellfish, seabirds, and marine mammals such as sea otters, whales, and seals.

The harbor where tankers will take on the oil is relatively sheltered; thus, no special precautions for storms will be necessary. The threat of tsunami, however, is more serious because of the susceptibility of the area to earthquakes. The enormous waves generated by earthquakes could seriously damage the port's large oil tankers and petroleum handling facilities. However, the storage tanks will be above the level of the tsunami generated by the 1964 earthquake.

We have touched on only the highlights of the physical environments along the route of the proposed pipeline and suggested some general problems that might result from the impact of the pipeline complex on the natural environment. Some of the problems we have posed become clearer after we describe the construction of the pipeline.

the structure

The proposed pipeline will be constructed from special steel pipe, 48 inches (122 centimeters) in diameter and with walls ½ inch (1.2 centimeters) thick. Pipe lengths 40–60 feet (12–18 meters) long will be specially welded and coated on the outside to prevent corrosion when buried. Where placed above ground, the pipe will be insulated with plastic foam and covered with thin metal sheathing.

The temperature of the oil as it comes out of the ground will be 57°C and will be kept at about that value because pre-cooling would be costly, the oil would become very viscous at lower temperatures and more difficult to pump, and at too low temperatures thick waxy deposits would form inside the pipe.

The preferred method of construction is to bury the pipeline in a trench to protect it from vandalism, keep it out of the way of animals, maintain an even temperature within the flowing oil, and minimize construction costs. However, because of the presence of ice-rich permafrost that becomes unstable when thawed, the pipeline will have to be above ground almost half the route, where it could be carried either on a gravel pad or on vertical steel supports.

Underground construction. For roughly 312 miles (502 kilometers) of the route (39 percent) the pipe will be buried as most oil pipelines are. Conventional burial, as this type of construction is called, is within a trench, 6 feet (1.8 meters) wide and 8 feet (2.4 meters) deep, with the pipe surrounded by packing material to prevent damage from bedrock, loose debris, and backfill. A cap or crown of fill at the top of the trench is added to compensate for any anticipated settlement of the backfill (Figure 5-5).

Conventional burial will be used where the ground is bedrock or well drained sediment. Burial will also occur where thawing of

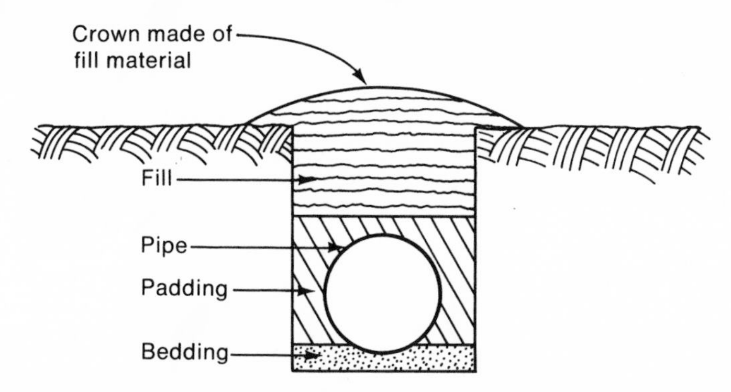

Figure 5-5. Diagram showing suggested design of conventional underground burial of pipeline.

permafrost would not result in too great subsidence or flowage of materials away from the pipe, which would leave the pipe unsupported and liable to rupture.

Special pipe burial will be required along about 70 miles (113 kilometers) or 9 percent of the route in areas where thawing of the permafrost must be minimized or prevented altogether. In this case the pipe may have an insulating layer surrounded by a special fluid that absorbs the excess heat from the oil. In the north, this technique is expected to prevent permafrost thaw. In the south, it has been suggested that special, refrigerator-like condensers be added that would extract extra heat from the soil and rock around the pipe and transfer it to the atmosphere (Figure 5-6).

Surface construction. In many sections of the proposed route, burial below ground would make the ground unstable by the thawing of permafrost. In these areas the pipe will be laid on a gravel pad, up to 6 feet (1.8 meters) thick, that will act as an insulating layer (Figure 5-7). The pipe will be set out in a slight zig-zag pattern to permit thermal expansion and contraction without unduly straining the pipe. The pipe itself will be heavily insulated to prevent too much cooling that would make the oil too viscous to move. About 154 miles (248 kilometers) or 19 percent of the pipeline will be supported on gravel.

In some areas where gravel is unavailable, where slopes are too steep, or where there might be too much settling of the ground

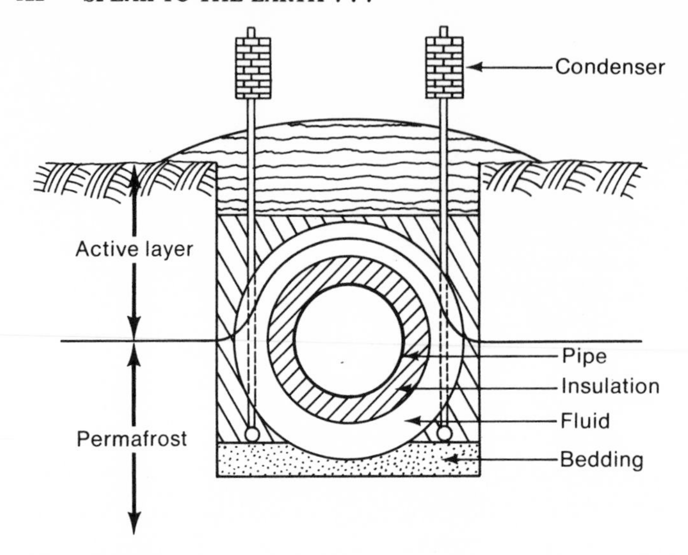

Figure 5-6. Diagram showing suggested design of special underground burial of pipeline where ice-rich permafrost must not be allowed to thaw. Pipe is insulated and has a special fluid around it to absorb heat. In the south, condensers will be added to withdraw heat and transfer it to the atmosphere.

below the pad, the pipe would be supported on steel piles 2–8 feet (0.6–2.4 meters) high and 60–70 feet (18–21 meters) apart (Figure 5-8). This technique would be used along 197 miles (317 kilometers) or 25 percent of the proposed route.

At river crossings the pipeline would be either buried below the river bed for 53 miles (85 kilometers) or 7 percent or carried overhead by simple spanning for 3 miles (5 kilometers) or less than 1 percent. The pipeline would cross the Yukon River, about 0.5 mile (0.8 kilometers) wide at this point, via a highway bridge. Where it is buried in river beds the pipe will have to be weighted down with concrete to prevent possible floating. The pipeline will cross 44 roads and will be buried at least 4 feet (1.2 kilometers) below the roadway. Of the total pipeline length, 435 miles (700 kilometers) or 55 percent will be buried and 354 miles (570 kilometers) or 45 percent will be elevated.

Additional construction. Besides the pipeline itself, additional facilities will have to be built. These include a 361-mile (581-

kilometer) gravel road, 28 feet (9 meters) wide and 3–6 feet (1–2 meters) high, from the Yukon River north to Prudhoe Bay; oil field structures such as pumps, feeder lines, storage tanks, and crew quarters; 5 pumping stations along the route with 7 added later on;

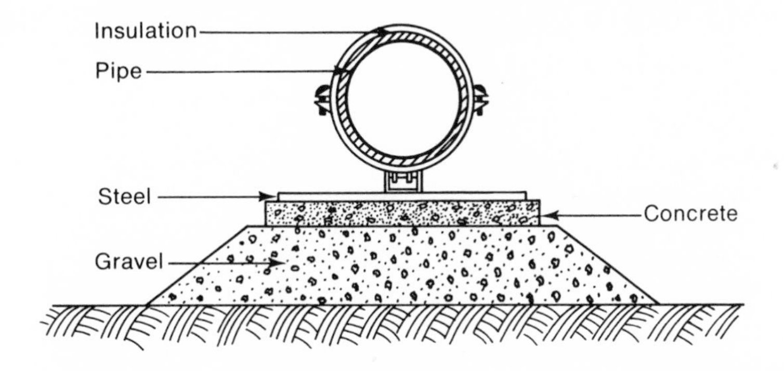

Figure 5-7. Diagram showing suggested design of surface construction of pipeline on a gravel pad.

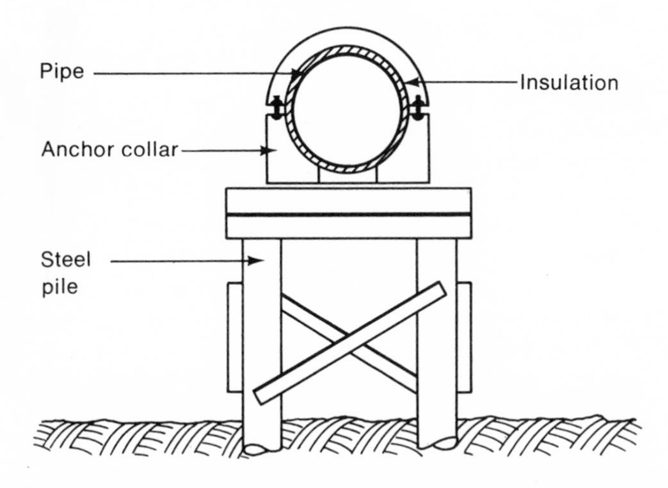

Figure 5-8. Diagram showing suggested design of surface construction of pipeline on steel piles. This mode of construction will be necessary when gravel is unavailable, when slopes are steep, or when too much settling would occur under a gravel pad.

7 airfields during construction, 3 of which will be permanent for pipeline maintenance; 900 acres (364 hectares) of terminal facilities at Valdez including storage tanks, pumps, oil-tanker berths, and oil-loading equipment; and a telecommunications network, including 26 microwave towers, 30 feet (9 meters) high.

The oil fields at Prudhoe Bay will also produce large quantities of natural gas. A small fraction of this can be used for heating and energy to drive equipment. Initially, the excess gas can be reinjected into the oil reservoir. Eventually, however, it will have to be transported out because at peak well production there will be 1.5 billion cubic feet (43 million cubic meters) of gas per day, and this will be too much to reinject. (State law prohibits burning the gas at the wells to get rid of it because this wastes a valuable natural resource.)

A gas pipeline using the same corridor as the oil pipeline is not planned because of the high cost of liquefying the gas at Valdez for tanker shipment. Also, tanker activity in the port would have to increase by 50 to 100 percent and therefore require an appropriate increase in port facilities.

Although no formal proposal has been made, a gas pipeline across Canada—possibly up the Mackenzie River Valley—to Edmonton, Canada, would probably have to be constructed in the future if the proposed oil pipeline is built. From Edmonton the gas pipeline could join the existing one that comes into the midwestern United States. Even though a gas pipeline would run at a much lower temperature and the pollution resulting from a rupture would be far less than an equivalent oil spill, construction of a gas pipeline would still have a significant impact on the Alaskan landscape. Because no formal proposal has been made yet for a gas pipeline, it is not possible now to evaluate how great that impact might be.

[*Editor's Note:* Material has been omitted at this point.]

13

Reprinted from *U.S. Geol. Surv. Circ. 699,* 13-14, 16-28 (1972)

MINERAL RESOURCES OFF THE NORTHEASTERN COAST OF THE UNITED STATES

Frank T. Manheim

[*Editor's Note:* In the original, material precedes this excerpt.]

SAND, GRAVEL, AND MUD

As pointed out by Emery (1968), Rexworthy,[4] Schlee (1968), and Schlee and Pratt (1970), sand and gravel represent not only a very large, but probably the most immediate, resource likely to be exploited on the offshore Atlantic margin of the United States. Offshore sand and gravel production around the United Kingdom now amounts to about 14 million tons per year or 13 percent of total production, and demand and production continue to grow (Hess, 1971).

In the United States, sand and gravel production makes up about one-fifth of the total nonmetallic industry in value, amounting to about 1 billion tons and about $1.1 billion value in 1970. In addition, somewhat less than 1 billion tons of crushed stone and other aggregate competed with sand and gravel for use in building and paving of roads (tables 2 and 3).

In spite of the current standstill in offshore leasing or exploration activities preparatory to exploitation (see "Environmental Factors"), several factors combine to exert pressure toward the mining of building aggregates offshore in the near future. Spreading urban areas continue to encroach on or exhaust sand and gravel quarries, which must normally be located close to the major-use sites. This is so because transportation is the principal factor in the cost of sand and gravel. Moreover, restrictions arising from enviromental and esthetic concern, rural zoning, and rise in land values decrease the availability and increase the cost of quarries. This is even more true of hard-rock quarries, which utilize blasting and are regarded as locally undesirable.

Most sand and gravel is used for construction (building and road) purposes, and value and production have increased over the last few decades at about 4 to 5 percent per year, on the average. Cement, prestressed concrete, ready-mixed concrete, and other special products all depend on constant and convenient supplies of aggregate materials.

[*Editor's Note:* Certain figures, tables, and some text material have been omitted at this point.]

[4] S. R. Rexworthy, 1968, The sand and gravel industry of the United States of America with special reference to exploiting the deposits offshore the eastern seaboard: Unpub. rept. for Ocean Mining Aktiengesell., 48 p.

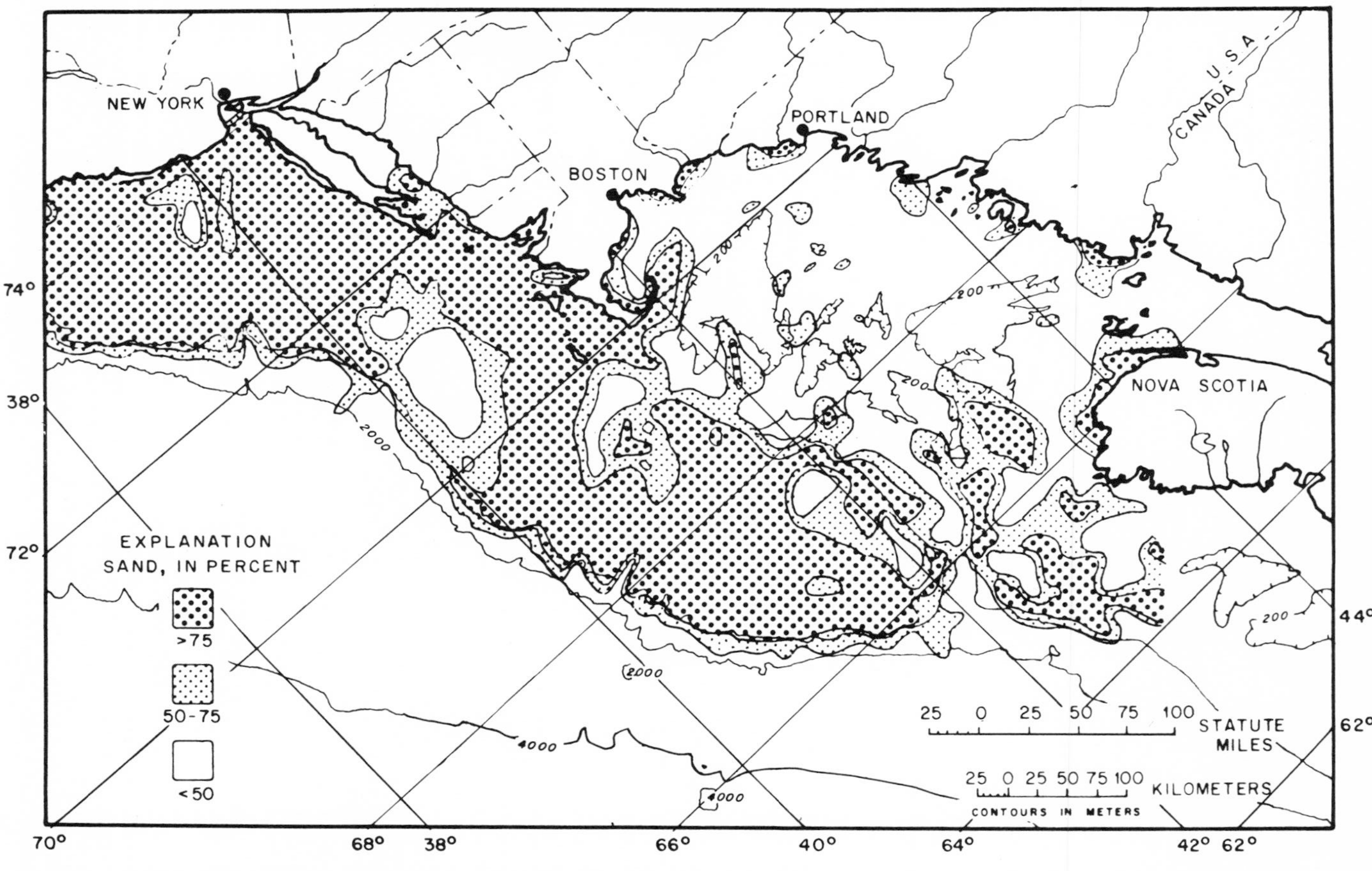

FIGURE 11.—Distribution of sand on the continental margin off Northeastern United States (redrawn from Schlee and Pratt, 1970, pl. 5A).

Existing sand deposits on the shelf are very large (fig. 11). In the area shown in figure 11, surficial sand deposits grading 75 percent or more cover 112,000 km². One can get an idea of the sand resources contained in this area by assuming a dredging depth of 3 m (10 ft) and 50 percent porosity (1.35 tons of dry sand per cubic meter of wet sediment). The total of about 450 billion tons of dry sand obtained is enough at current consumption rates to meet the needs of all the northeastern coastal states of the United States and the Maritime Provinces of Canada for several thousand years. Though poorly known at present, actual thicknesses are probably at least 10 times greater. Drill cores will be required to establish depths and characteristics of the sand.

A few sand deposits, such as those off New Jersey (fig. 12) (Schlee, 1964), have been

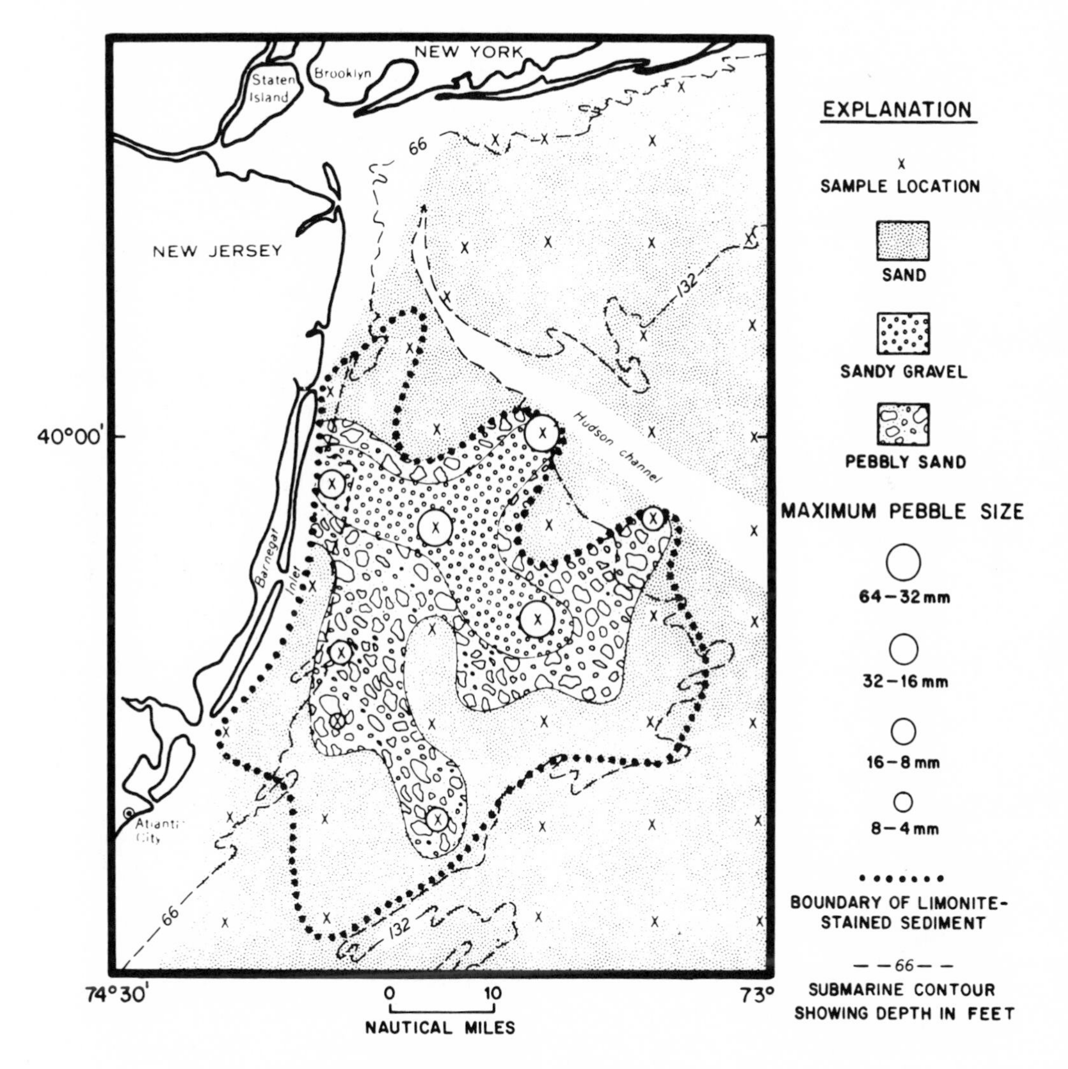

FIGURE 12.—Distribution of sand and gravel off New Jersey (from Schlee, 1964).

sufficiently well delineated to indicate their direct suitability for immediate exploitation. Leases have been requested, but activities await promulgation of clear regulatory guidelines by State and Federal agencies. Other sand supplies north of Cape Cod also offer clear potential (fig. 13). Another prime target for dredging, though farther offshore, is Nantucket Shoals, which has been a major hazard to navigation, as attested to by the hundreds of wrecks still located there.

Gravels are much less abundant than sands on the shelf and generally would require longer transport to consumer areas (figs. 14 and 15). However, gravels have greater value and, as has been pointed out by recent lively discussions on the question of offshore sand and gravel mining (Davenport, 1971; Taney, 1971), distance is a much less important factor at sea than on land. Whereas the capital costs to initiate viable commercial dredging operations offshore are much higher than on land, say on the order of $5 million dollars, the barging and delivery of materials to coastal urban centers is much less costly than trucking (main land means) or even rail. With efficient technology and good organization, there seems little doubt that major centers could begin to utilize seafloor aggregates within the next 5 or 10 years.

Sediments containing more than 50 percent gravel cover about 15,000 km^2 in the area shown in figure 14. In the smaller area in figure 15, presumably a potential supplier to the Greater Boston and Massachusetts shore communities, sediments containing more than 60 percent gravel cover 345 km^2, which, under assumptions previously made for sand dredging, would yield 1.4 billion tons.

In the United Kingdom, subsidies of about 20 percent of costs were offered earlier to stimulate building of offshore dredging equipment, but these have been discontinued as no longer necessary (Hess, 1971). Private surveys of conditions in the United States also indicate that subsidy of offshore sand and gravel production would not be needed, given a stable and appropriate legal climate.

Sand and gravel are not the only potentially valuable sediments in the offshore bottoms. The natural grading of materials by grain size and mineralogic and chemical compositions offers a wide range of earth materials which might find uses through technologic innovations. A recent discovery (patent pending) by D. C. Rhoads, R. B. Gordon, and M. A. Ruggiero (Bingham Laboratory, Yale University, New Haven, Conn.) is that ceramic material of superior quality can be made by firing fine marine sediment that contains less than 10 percent calcium carbonate and between 1 and 3 percent organic matter. Advantages of the marine material are that it need not be subjected to complex blending processes involving different raw material sources, it can be obtained in very large volumes by simple surface dredging of deposits in Long Island Sound, Buzzards Bay, and other areas along the northeast continental margin, and it has much lighter finished weight than brick or other comparable ceramics. Channel dredging materials may be suitable, as may be muds in contaminated bays or tidal water bodies, for the excess organic carbon attributable to pollution products actually promotes favorable attributes in the finished ceramic; the gases released on firing create high porosity and therefore lightness. The sea salt incorporated in the raw mud also aids in the fluxing of the ceramic to create desirable strength and other properties. One may point out that muddy basins where the indicated materials occur (fig. 16) are generally poor fishing grounds and otherwise play a subordinate role in the marine biologic cycle offshore, insofar as present information permits assessment. Under certain conditions, removal and use of the mud from polluted areas, such as inner Long Island Sound (Gross and others, 1971), may actually improve the local aquatic habitat much more rapidly than could be done by the slow process of natural deeutrophication after the input of pollutants is stopped. This kind of innovative and coordinated use of offshore resources needs especial encouragement and stimulation.

ENVIRONMENTAL FACTORS

Shelf sands and gravels are metastable deposits on the northeastern continental margin and are undergoing erosion and winnowing by natural forces such as bottom currents. In contrast to widespread impressions to the contrary,

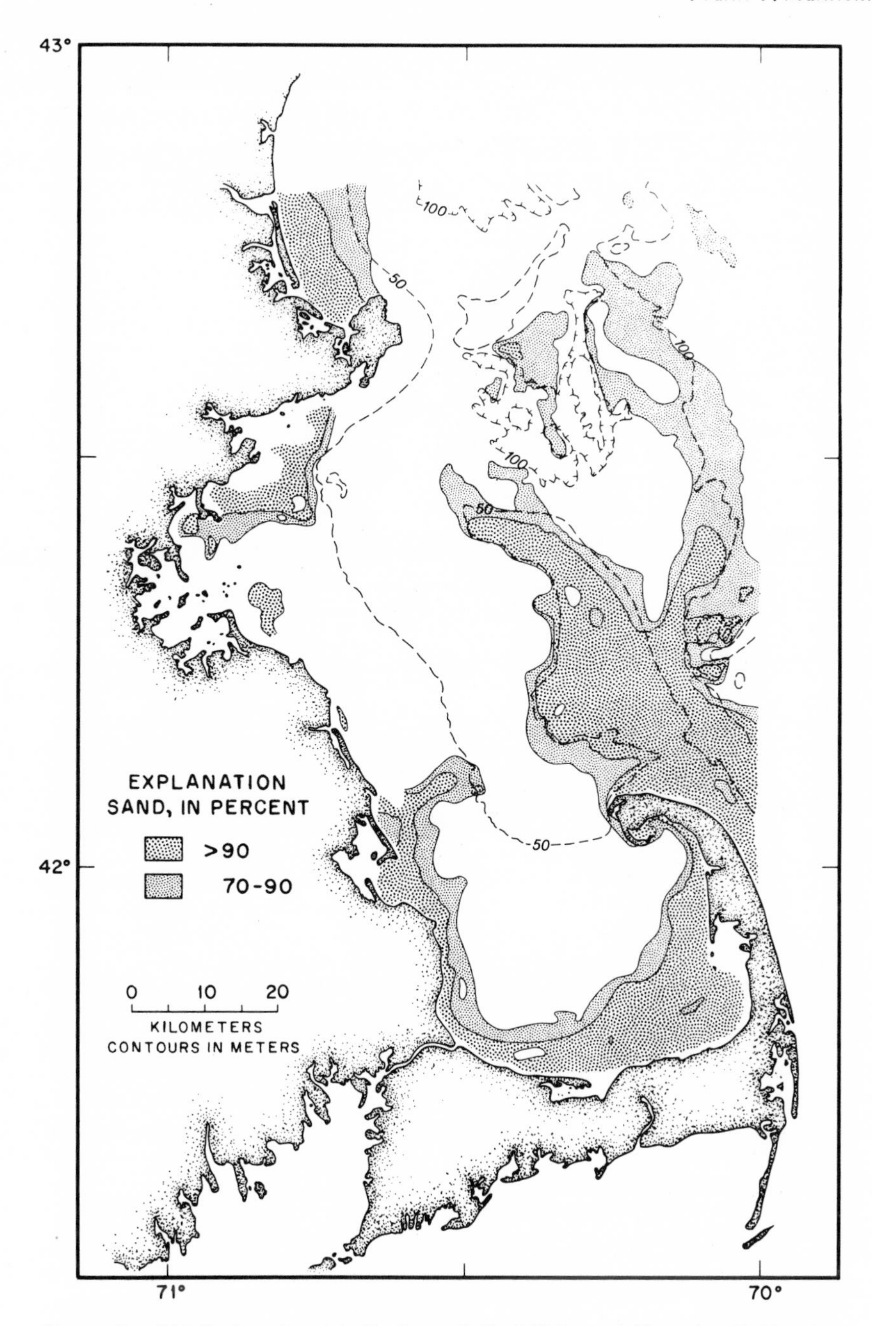

Figure 13.—Distribution of sand in the inner Gulf of Maine and Massachusetts Bay (redrawn from Schlee and others, 1971).

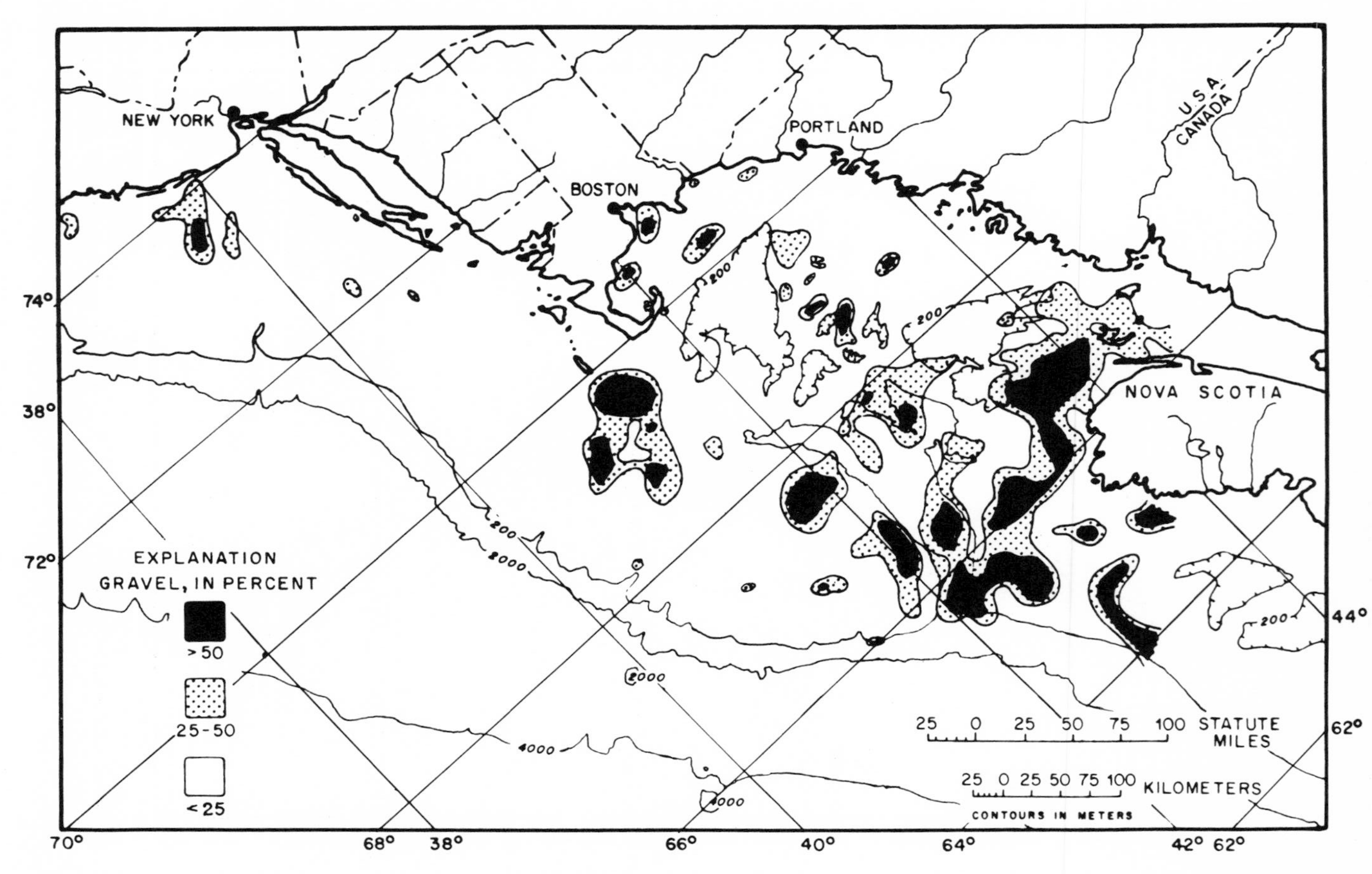

FIGURE 14.—Distribution of gravel on the continental margin off Northeastern United States (redrawn from Schlee and Pratt, 1970, pl. 5*B*).

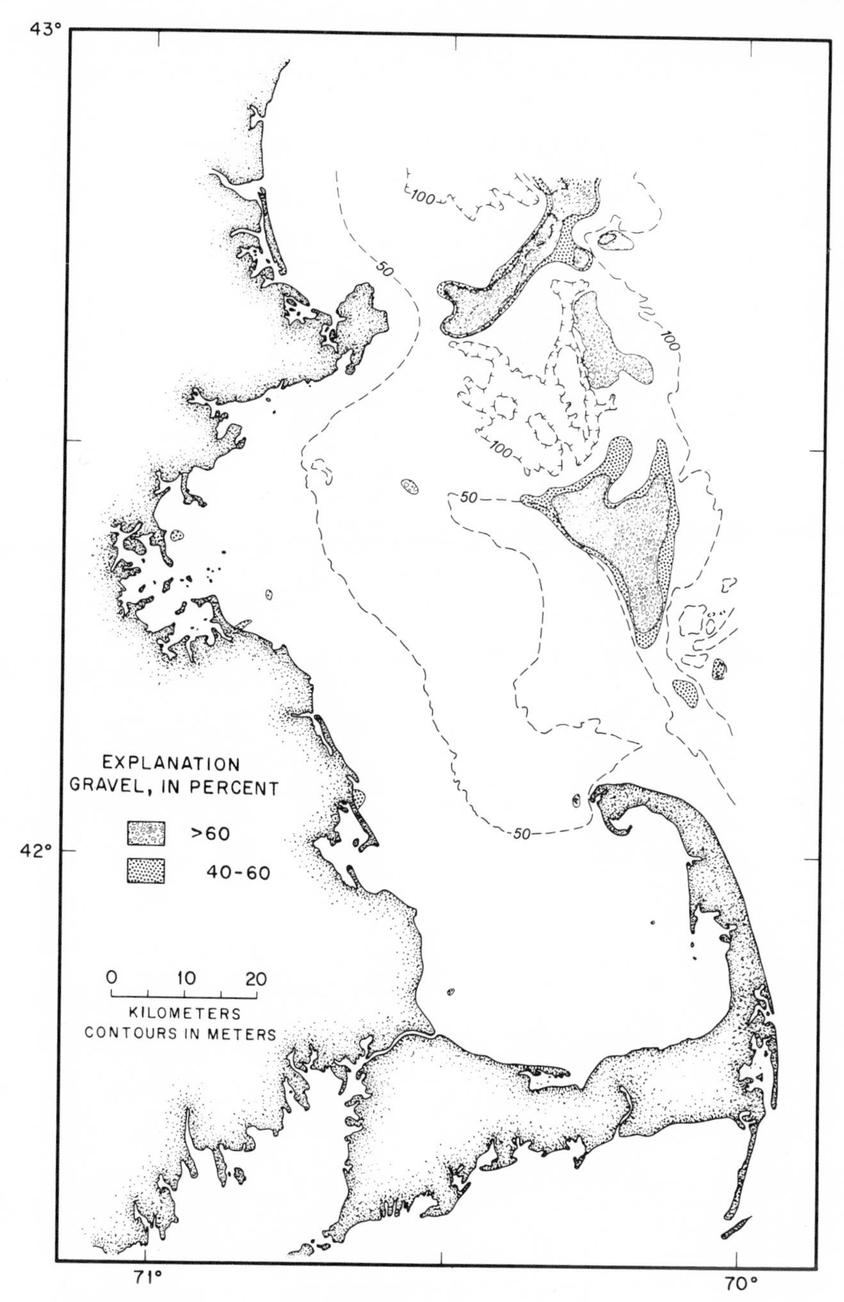

FIGURE 15.—Distribution of gravel in the inner Gulf of Maine and Massachusetts Bay (redrawn from Schlee and others, 1971).

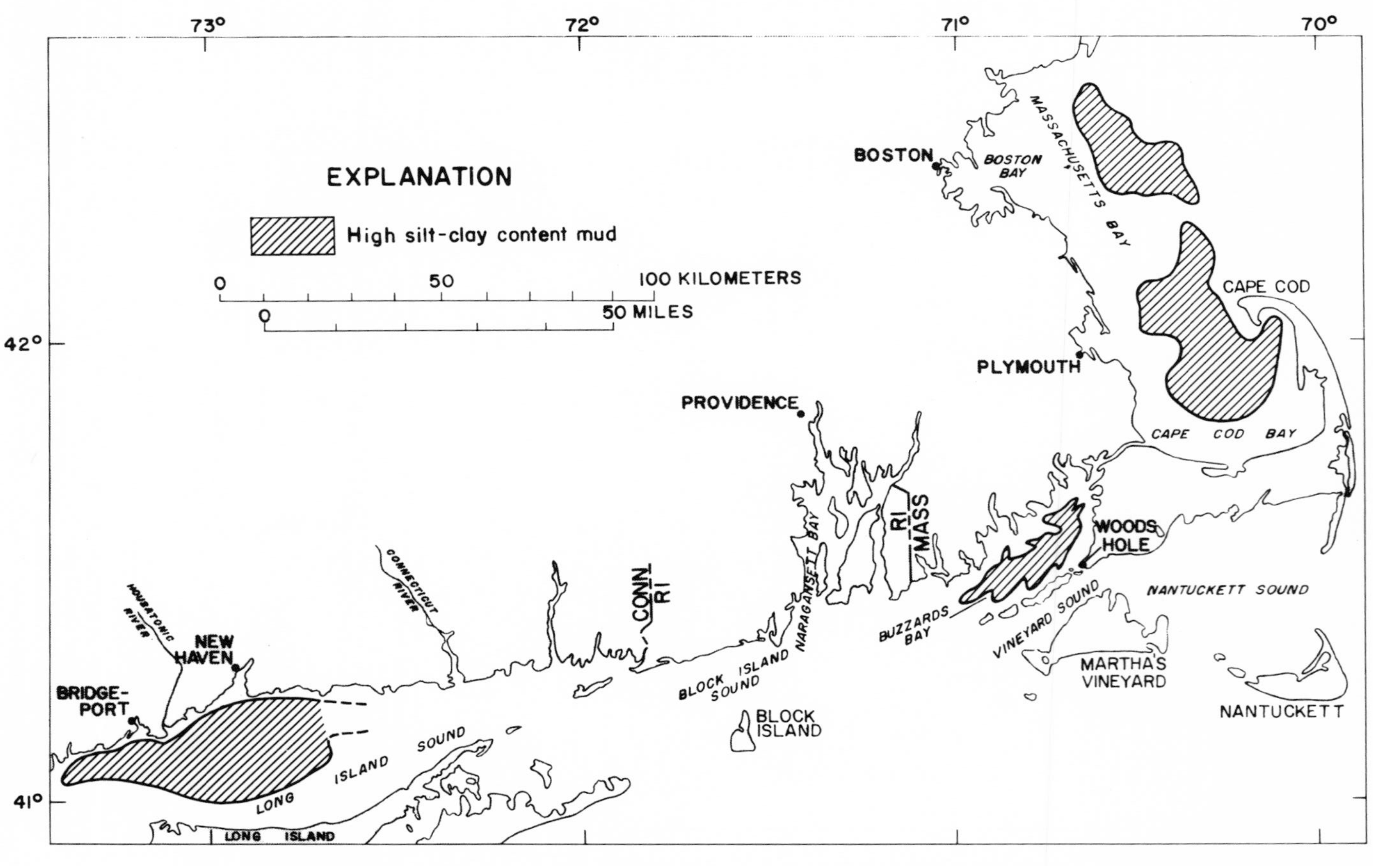

FIGURE 16.—Offshore distribution of sediments that can be used in an innovative ceramic manufacturing process.

most estuaries on the continental margin may not have a net loss of sediment to the shelf and open ocean; rather, sediment moving into the estuaries by bottom transport from the shelf is tending to fill them up on a net basis (fig. 17A) even though outward transport of river sediments does take place at the same time (Meade, 1969). Thus, the dredging of a harbor mouth, as in the case of Savannah Harbor, has not increased outward loss of sediment, but rather has accelerated the silting up to the harbor, because movement of bottom material inward is facilitated (H. B. Simmons, cited in Meade, 1969). From this point of view, mining of offshore sands may be taking materials which are tending to move into the continent in any event (fig. 17B).

A different problem is posed by beach and shore erosion. Whether or not the taking of sand from one area might promote erosion in another coastal zone must be assessed separately for each case. On the one hand, experience in the United Kingdom, where offshore sand and gravel have been mined since World War II, has brought to light only a few cases where coastal erosion has been attributed to sand mining, and these occurred where the mining was in relatively close proximity to the affected shore. On the other hand, studies of actual movement of bottom materials in the Atlantic offshore region are still very limited. Moreover, though the data are still inadequate, it can be shown qualitatively that storms and hurricanes have a much greater influence on beach erosion and coastal modification in many areas than day to day action (Hayes, 1967; Coastal Plains Center for Marine Development Services, 1970, and publications cited therein). Thus, there will continue to be coastal erosion with or without mining operations.

Marine life is potentially affected by marine mining and other activities that modify sediments. For example, Massachusetts oyster production has declined from 24 million pounds (meat) annual production in 1910 to only a few tens of thousands of pounds per year in recent years. Discussion with officials of the Massachusetts Department of Natural Resources and other observers indicates that the major causes of the destruction of oyster fisheries include dredge and fill operations and no or inadequate management of oyster beds in the public (that is, town) domain: For example, in some places oysters are taken without returning the shell to provide cultch or other hard surfaces needed by oyster spat as substrates. Marine biologists familiar with shelf fauna and flora indicate that mining operations on the shelf are not inherently incompatible with proper maintenance of marine life and commercial and sports fisheries. However, they indicate that proper design of operations and coordination with fisheries authorities are important factors in assuring ecologically judicious development. For example, mining operations might be shifted to different areas according to seasons of the year, in order to avoid the spawning grounds of given fish species.

A serious problem at present is the expanding pressure on existing sites for dumping of dredge spoils. For example, dredgings in Delaware Bay amounted to nearly 20 million cubic yards per year until a few years ago, when damming off of the spoils reduced the need for dredging to about 10 million cubic yards per year. However, in the foreseeable future all available dumping space will be exhausted if dumping is not to be done where environmental damage will result or where exorbitant transport costs are incurred. Extracts from the earlier U.S. Commission on Marine Science, Engineering and Resources (Coastal Research Notes, 1971) indicated that the draft requirements of the new super tankers (up to 70 ft) are so great that in some areas channels extending as much as 60 miles from shore or deep-water terminals would be needed to accommodate the vessels. Even medium-sized super tankers cannot be accommodated by any present-day Atlantic ports. Alternatives to dredging may not be simple or cheap because they involve the possibility of declining port usage and the necessity of supplying energy and essential fuels by other means. According to the reports, dredgings even at current rates would fill most currently available disposal sites in 8 to 10 years.

If innovative and efficient technological means could be found to utilize dredge spoils to supply aggregate needs (other than fill), then a dual advantage would be gained; the destruction of land areas for sand and gravel mining purposes could be reduced, at least for coastal communities, and a troublesome waste product could be turned to productive uses. The ceramic process mentioned earlier is an example of innovation

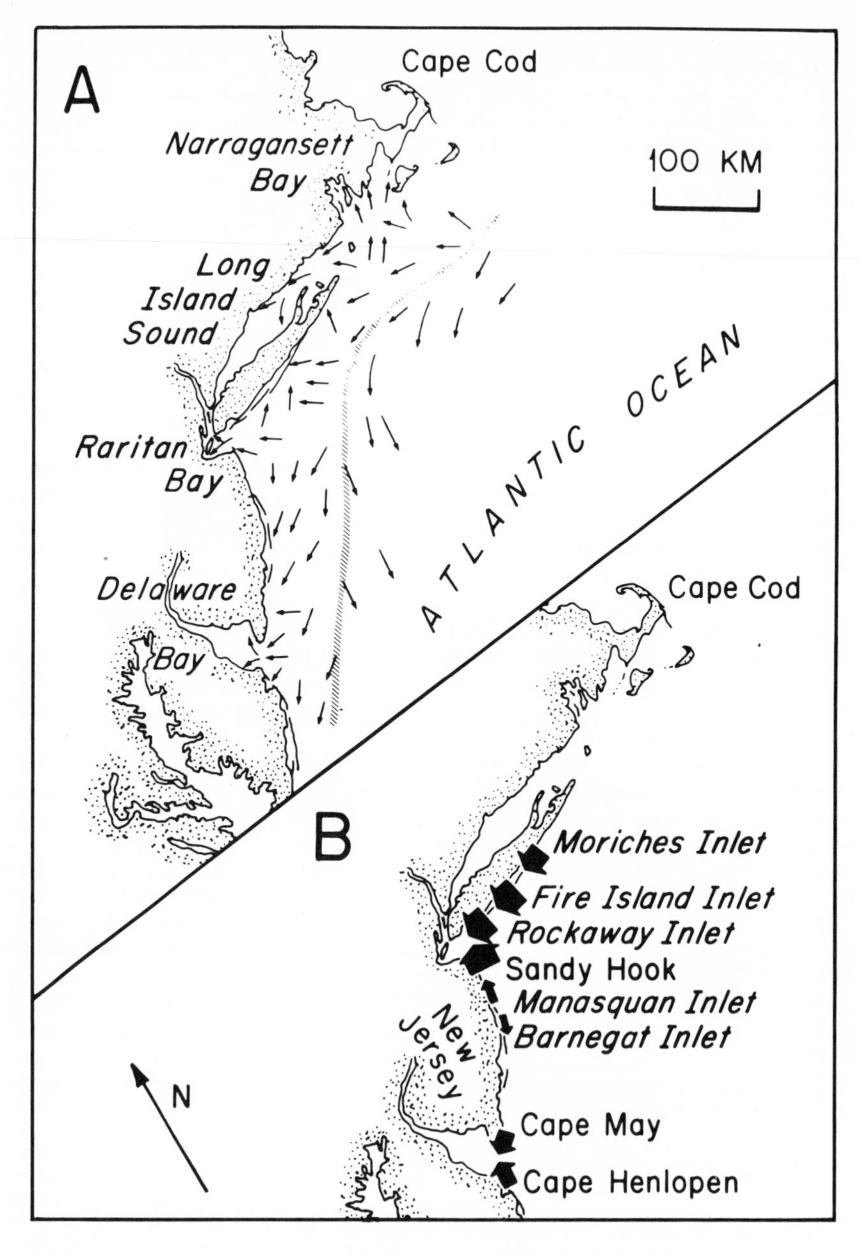

which has promise in this direction. The approximate regional balance of dredge spoils and sand and gravel consumption is shown in figure 18.

Innovative use of offshore bottoms might even improve the aquatic or biologic environment under certain conditions. An example of the "surgical removal" and utilization of polluted mud has already been mentioned. D. C. Rhoads (oral commun., 1971) has also pointed out that mud bottoms (especially if they are highly organic owing to deposition of sewage or other organic matter of primary or secondary origin) might serve as sources of food to nourish an intensive above-bottom aquaculture (for example, mussels, clams, oysters). Freed from predators, and with good substrates provided, the organisms could use excess organic matter and nutrients not only to produce marketable protein, but help deeutrophicate previously contaminated basins. Pilot studies for analogous types of aquaculture are being carried on by J. H. Ryther and his coworkers at the Woods Hole Oceanographic Institution.

The history of the Atlantic offshore region shows that we are now at a crossroads. Before, when mining leases could have been obtained at nominal cost, there was little incentive to prospect offshore. Now, when the growth of coastal populations and increased demand for products potentially available from the sea bottom renders such exploration and exploitation attractive, there is virtually a complete moratorium on leasing while State, Federal, and local regulatory agencies reassess their operations procedures in response to the growth in environmental concern, as well as improved knowledge of the natural environment itself.

Massachusetts, for example, has recently enacted a moratorium on all offshore leasing within the State boundaries (Mass. Sen. 737, Ch. 567, 1971), and another bill (Sen. 1445, Ch. 742, 1971) provides a natural sanctuary around the entire Cape Cod region, stipulating that no dredging or dumping of whatever kind may be performed, except for beach restoration purposes. The legislature has also allocated $200,000 to the Department of Natural Resources to study the environmental aspects of the Massachusetts offshore region with a view to preparing guidelines for leasing when and if the Department deems that offshore exploitation can safely be carried out.

Keys to proper and environmentally judicious management of offshore resources lie in improving presently uncertain or overlapping jurisdiction and coordination and effectiveness of investigatory and regulatory bodies. With limited resources, regulatory bodies must have broad knowledge or access to expertise in many overlapping fields: marine biology, geology, engineering practice, economics, law, politics, and psychology. A short account and bibliography of problems potentially affecting marine mining is provided by McKelvey, Tracey, Stoertz, and Vedder (1969) and Battelle Memorial Institute Staff (1971), respectively.

If regulation is inadequate, damage or public loss of confidence leading to prohibitory legislation or crippling litigation may result. If regulation is unnecessarily rigid or if its inadequacy leads to blanket prohibition, needed and useful goods and services may be denied the public, or unfavorable environmental developments may occur by default. It is often pointed out that offshore work requires adequate capital, technology, and organization; responsible organizations possessing these prerequisites will hesitate to take political risks over and above the usual economic risks involved in new ventures. They need dependable guidelines under which to work.

FIGURE 17 (left).—Directions of general residual current (A) along the bottom on the continental shelf (from Bumpus, 1965), and (B) directions and magnitudes of sand movement on the beaches (from Meade, 1969, fig. 9; reprinted from "Journal of Sedimentary Petrology" with permission of Society of Economic Paleontologists and Mineralogists). Width of arrow shaft in part B indicates estimated amounts of sand; the widest represents about 370,000 m^3 per year.

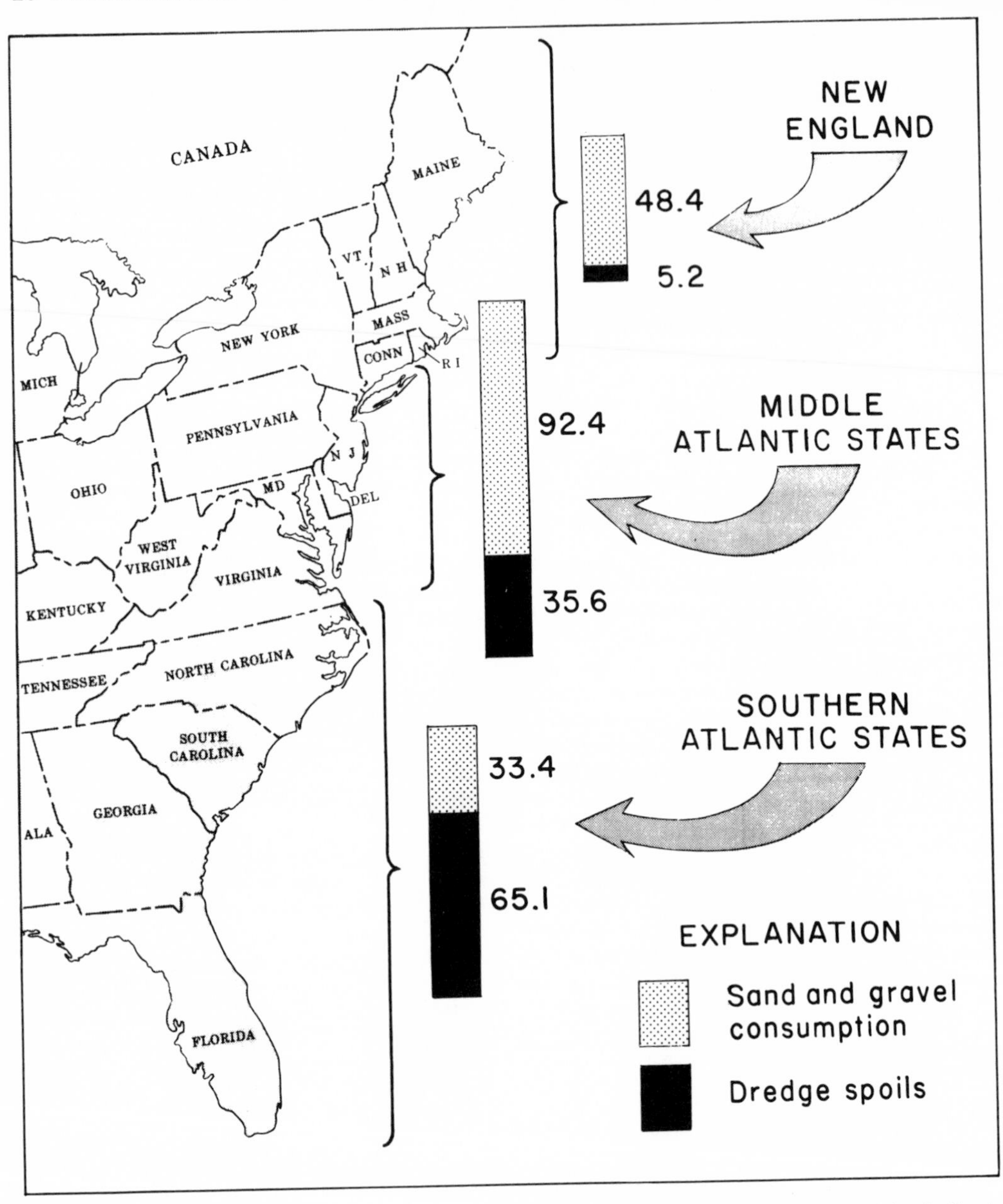

FIGURE 18.—Balance of dredge spoil and consumption of sand and gravel (in millions of tons) in Atlantic Coast States. Dredge spoil data from V. L. Andreliunas, U.S. Army Corps of Engineers (oral commun., 1972) refers to 1971 dredging in and around navigable waterways. Sand and gravel consumption from U.S. Bureau of Mines (1969, p. 986–987).

REFERENCES CITED

Battelle Memorial Institute Staff, 1971, Environmental disturbances of concern to marine mining research, a selected annotated bibliography: U.S. Natl. Oceanic and Atmospheric Adm. Tech. Memo. ERL MMTC-3, 72 p.

Bumpus, D. F., 1965, Residual drift along the bottom on the continental shelf in the Middle Atlantic Bight area: Limnology and Oceanography, v. 10, supp. (Redfield Volume), p. R50–R53.

Canada Geological Survey, 1949, Geologic map of the Maritime Provinces (New Brunswick, Nova Scotia, and Prince Edward Island): Canada Geol. Survey Map 910A.

Cathcart, J. B., 1968, Phosphate in the Atlantic and Gulf Coastal Plains, *in* Forum on geology of industrial minerals, 4th, Austin, Tex., 1968, Proceedings: Austin, Texas Univ. Bur. Econ. Geology, p. 23–24.

Coastal Plains Center for Marine Development Services, Washington, D.C., 1970, Bibliography on hurricanes and severe storms of the coastal plains region: Coastal Plains Center Marine Devel. Services Pub. 70–2, 71 p.

Coastal Research Notes, 1971, [On port-dredging problems]: Coastal Research Notes, v. 3, no. 6, p. 3.

Davenport, J. M., 1971, Incentives for ocean mining: a case study of sand and gravel: Marine Technology Soc. Jour., v. 5, p. 35–40.

Duane, D. B., 1969, A study of New Jersey and northern New England coastal waters: Shore and Beach, v. 37, no. 2, p. 12–16.

Emery, K. O., 1966, Geological methods for locating mineral deposits on the ocean floor, *in* Exploiting the ocean: Marine Technology Soc. Conf. and Exhibit, 2d Ann., 1966, Trans., p. 24–43.

——— 1968, The continental shelf and its mineral resources, *in* Selected papers from the Governor's Conference on Oceanography, New York, 1967: [Albany, New York State Dept. Commerce] p. 36–51.

Emery, K. O., and Noakes, L. C., 1968, Economic placer deposits of the continental shelf: U.N. Econ. Comm. Asia Far East Comm. Coord. Joint Prosp. Mineral Resources Asian Offshore Areas Tech. Bull., v. 1, p. 95–111.

Emery, K. O., Uchupi, Elazar, Phillips, J. D., Bowin, C. O., Bunce, E. T., and Knott, S. T., 1970, Continental rise off eastern North America: Am. Assoc. Petroleum Geologists Bull., v. 54, p. 44–108.

Goldsmith, Richard, 1964, Geologic map of New England: U.S. Geol. Survey open-file rept., 3 sheets, scale 1:1,000,000.

Gross, M. G., Black, J. A., Schramel, R. J., and Smith, R. N., 1971, Survey of marine waste deposits, New York Metropolitan region: New York State Univ., Stony Brook, Marine Sci. Research Center Tech. Rept. 8, 72 p.

Hathaway, J. C., 1971, Data File, Continental Margin Program, Atlantic Coast of the United States, Volume 2, Sample collection and analytical data: Woods Hole Oceanog. Inst. Ref. 71–15, 496 p.

Hawkes, H. E., and Webb, J. S., 1962, Geochemistry in mineral exploration: New York, Harper and Row, Publishers, 415 p.

Hayes, M. O., 1967, Hurricanes as geological, agents, south Texas coast: Am. Assoc. Petroleum Geologists Bull., v. 51, no. 6, p. 937–942.

Hess, H. D., 1971, Marine sand and gravel mining industry of the United Kingdom: U.S. Natl. Oceanic and Atmospheric Adm. Tech. Rept. ERL 213-MMTC 1, 176 p.

Hülsemann, Jobst, 1967, The continental margin off the Atlantic coast of the United States; carbonate in sediments, Nova Scotia to Hudson Canyon: Sedimentology, v. 8, p. 121–145.

Libby, F., 1969, Searching for alluvial gold deposits off Nova Scotia: Ocean Industry, v. 4, no. 1, p. 43–47.

Maher, J. C., 1971, Geologic framework and petroleum potential of the Atlantic Coastal Plain and continental margin: U.S. Geol. Survey Prof. Paper 659, 98 p., 17 pls.

Manheim, F. T., and Pratt, R. M., 1968, Geochemistry of manganese-phosphorite deposits on the Blake Plateau, *in* Woods Hole Oceanographic Institution Summary of investigations in 1967: Woods Hole Oceanog. Inst. Ref. 68–32, p. 49.

McKelvey, V. E., Tracey, J. I., Jr., Stoertz, G. E., and Vedder, J. G., 1969, Subsea mineral resources and problems related to their development: U.S. Geol. Survey Circ. 619, 26 p.

McKelvey, V. E., and Wang, F. F. W., 1969, World subsea mineral resources: U.S. Geol. Survey Misc. Geol. Inv. Map I-632.

Meade, R. M., 1969, Landward transport of bottom sediments in estuaries of the Atlantic Coastal Plain: Jour. Sedimentary Petrology, v. 39, no. 1, p. 222–234.

Pratt, R. M., 1971, Lithology of rocks dredged from the Blake Plateau: Southeastern Geology, v. 13, p. 19–38.

Pratt, R. M., and McFarlin, P. F., 1966, Manganese pavements on the Blake Plateau: Science, v. 151, p. 1080–1082.

Ross, D. A., 1970, Atlantic continental shelf and slope of the United States—Heavy minerals of the continental margin from southern Nova Scotia to northern New Jersey: U.S. Geol. Survey Prof. Paper 529-G, 40 p.

Schlee, John, 1964, New Jersey offshore gravel deposit: Pit and Quarry, v. 57, p. 80–81.

——— 1968, Sand and gravel on the continental shelf off the northeastern United States: U.S. Geol. Survey Circ. 602, 9 p.

Schlee, John, Folger, D. W., and O'Hara, C. J., 1971, Bottom sediments on the continental shelf of the northeastern United States, Cape Cod to Cape Ann, Massachusetts: U.S. Geol. Survey open-file rept.

Schlee, John, and Pratt, R. M., 1970, Atlantic continental shelf and slope of the United States—Gravels of the northeastern part: U.S. Geol. Survey Prof. Paper 529–H, 39 p.

Stanley, D. J., Swift, D. J. P., and Richards, H. G., 1967, Fossiliferous concretions on Georges Bank: Jour. Sed. Petrology, v. 37, p. 1070–1083.

Taney, N. E., 1971, Comments on "Incentives for Ocean Mining": Marine Technology Soc. Jour., v. 5, p. 41–43.

Trumbull, J. V. A., and Hathaway, J. C., 1968, Dark-mineral accumulations in beach and dune sands of Cape Cod and vicinity, *in* Geological Survey research 1968: U.S. Geol. Survey Prof. Paper 600–B, p. B178–B184.

Uchupi, Elazar, 1966, Structural framework of the Gulf of Maine: Jour. Geophys. Research, v. 71, p. 3013–3028.

——— 1968, Long lost Mytilus: Oceanus, v. 14, no. 3, p. 2–7.

U.S. Bureau of Mines, 1969, Minerals yearbook, 1968, Volumes 1–2, Metals, minerals, and fuels: Washington, D.C., U.S. Govt. Printing Office, 1208 p.

——— 1971, Minerals yearbook, 1969, Volumes 1–2, Metals, minerals and fuels: Washington, D.C., U.S. Govt. Printing Office, 1194 p.

14

Reprinted from *Science,* 179(4076), 859-864 (1973)

Hydrological and Ecological Problems of Karst Regions

Hydrological actions on limestone regions cause distinctive ecological problems.

H. E. LeGrand

The ecological patterns of the land regions of the world are classified in many different ways. They are seldom classified according to their relationships to the underlying rocks because generally the ecological zones or belts are so extensive that they encompass a wide variety of rocks. Moreover, in most regions the influence of climate is so great that it tends to mask the more subtle influence of the underlying rocks except for some special situations. Limestone terranes form one distinctive exception. It is my purpose in this article to point out the relationships of hydrology to the ecology in limestone regions that have been altered by processes of karstification. The results of karstification, such as scarce soils and water at the land surface, so obviously control the existence and behavior of organisms that a gross understanding of the ecology can be assumed and need not be explicit for this purpose.

The term "karst" has long been used to define the sum of the phenomena characterizing regions where carbonate rocks, chiefly limestone, are exposed. "Karstification" refers to the many processes that result in the development of surface and subsurface features that are distinctive in many carbonate rock terranes. Karst features develop where water containing carbon dioxide has been able to move on and through carbonate rocks and to remove some of the rock in solution.

The development of karst features, particularly caverns and other large openings, depends on (i) the presence of soluble rocks, (ii) the presence of carbonic acid, (iii) ample precipitation, (iv) openings in the rocks, and (v) a favorable topographical and structural setting.

The ecology of carbonate rock regions ranges greatly, depending on climate, of course, but also on several gross geologic factors. There are delicate balances in the development of karst areas that are related to the character of the rock: whether it is essentially pure, whether it contains insoluble material throughout or only in thin discrete beds, and whether it has primary permeability or secondary permeability through fractures and caverns. Geologic structure and history are everywhere involved in karst development because carbonate rocks must be uplifted from below sea level to the landscape environment where fresh water can infiltrate and move through them. Such structural conditions as folding and faulting may cause the rock to lie below or within the local water circulation system. Topographic relief gives impetus to circulation of water. Only within the freshwater system can the full interplay of climate, topography, and geology exist to develop karst features.

The interplay between carbonate rocks and climatic, geologic, topographic, and hydrologic factors creates a wide variety of environments, ranging from the subsurface to the surface and particularly determining soil patterns and water distribution. These environments, in turn, have had their effect on the local development of plants and animals and on the culture and history of man. Many problems of carbonate terranes are related directly to the influence of carbonate rocks on the local hydrologic regimen. By defining the problems and putting them into perspective it may be possible to improve man's understanding of the economic and ecologic balances in regions where carbonate rocks occur.

Some of the characteristic problems of carbonate rock terranes, such as those related to scarcity of soils, scarcity of surface streams, and rugged topography, are obvious and somewhat distinctive; these characteristic features are developed by natural processes. In fact, natural processes in some carbonate regions may have caused a greater restriction in the development of biota than man can ever be suspected of causing. There are complex insidious problems that develop as man disturbs the natural balance of geologic and hydrologic conditions in carbonate rock terranes. It is necessary to understand environmental relationships—especially those involving hydrology—to determine whether actions by man affecting one part of the carbonate rock system will have a direct or indirect detrimental effect somewhere else in the total carbonate rock system. It is conventionally considered that man inherits a good natural environment but that he tends to degrade it. A contrary consideration may be posed in the following questions. Has man inherited a rather harsh and poor environment in carbonate rock terranes? If so, can he improve it?

Ecological Problems

The ecology of any region is inherently related to climate, topography, soils, and of course, to the complex of characteristic biota. An additional factor that may be called "terrane permeability" is especially significant to the ecology of carbonate rock terranes. Features of climate, topography, and soils are observable or discernible and, therefore, do not need amplification for this discussion. Permeability, on the other hand, is less readily discernible and has a bearing on many human problems in karst terranes.

Carbonate rocks are exposed and karstified in glacial, temperate, and tropical settings that range from wet to very dry. Karst in arid regions, such as that in the Western Desert of Egypt and in the Nullarbor Plain of Australia, is mostly a relic of development during an earlier and more humid period.

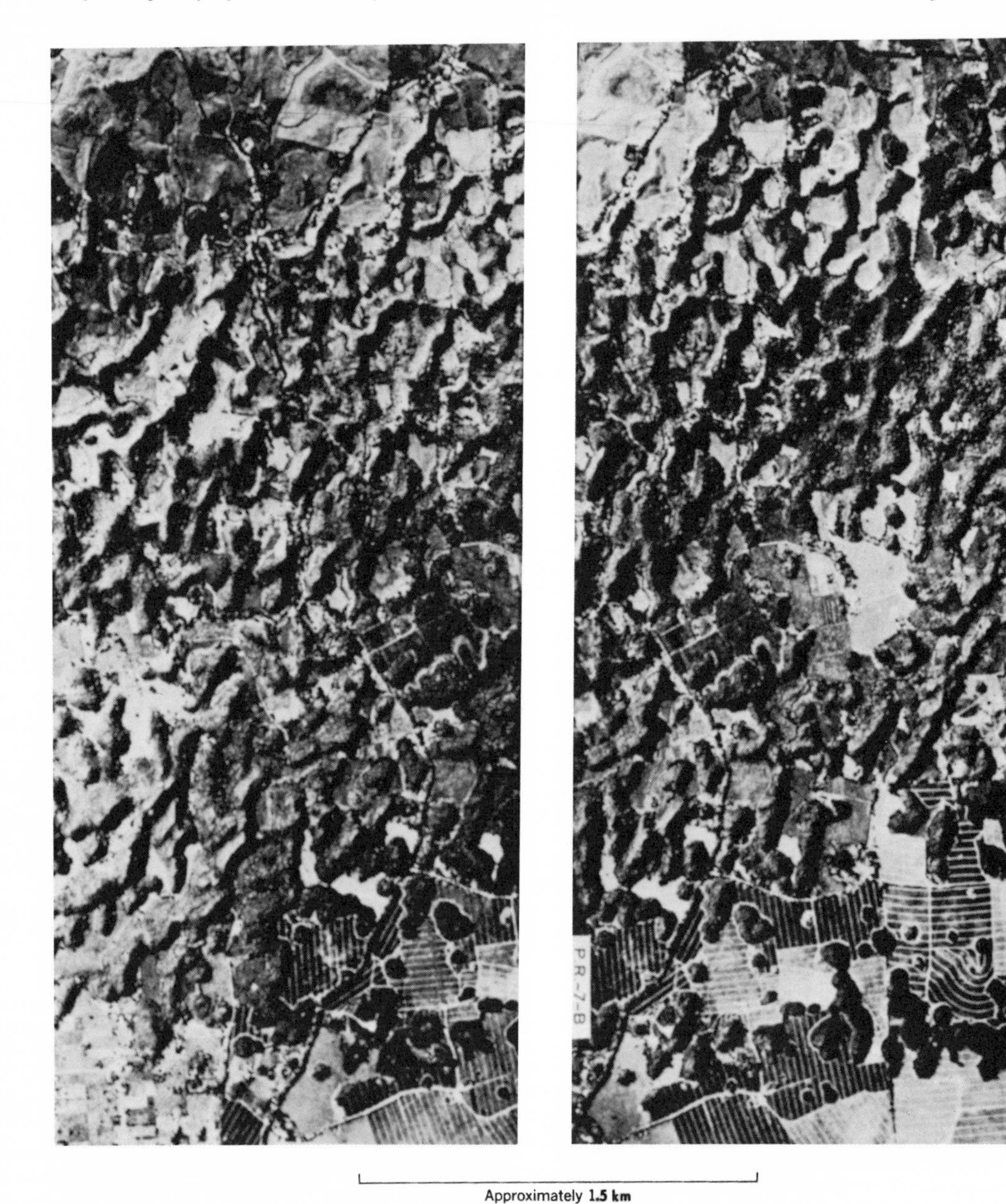

Fig. 1. Stereoscopic view of a karst terrane in northern Puerto Rico (U.S. Geological Survey, Manatí quadrangle, 7.5′). The rugged topography consists of sinkholes and towers.

Karst topography is uneven. The unevenness ranges from subdued topographic relief in some sinkhole areas to precipitous sinks and conical hills, typical of the Cockpit Country of Jamaica, West Indies, and parts of northern Puerto Rico (Fig. 1). Differential erosion is common where soluble carbonate rocks are adjacent to nonsoluble rocks and, depending on imbalances of several factors, the carbonate rocks stand either higher or lower than the adjacent rocks. In the warm humid Appalachian region of the eastern United States the carbonate rocks tend to occupy elongated valleys, but in many other regions, such as in the arid Southwest, the carbonate rocks make prominent ridges. In some areas, the terrane is so rugged that human habitation and communication are limited, as in parts of Yugoslavia and Greece.

While undergoing karstification, many carbonate rocks leave little insoluble residue, and their soils are regenerated more slowly than those on nonsoluble rocks. Once carbonate rocks on upland slopes have been stripped of soil, they tend to remain denuded, even in humid regions where soil-forming processes are favorable. Soils of upland karst regions are washed into sinkholes and other karst lowlands. In some low-lying areas, the soils are protected from further removal long enough to form laterite and even bauxite.

The permeability of the entire carbonate rock system is important because while caverns and other openings are enlarging and while permeability is increasing, the water table is progressively lowering itself to greater depths below the land surface. Permeability does not at first glance seem significantly related to ecology, but many practical problems in carbonate areas are related to permeability. These include (i) scarcity and poor predictability of groundwater supplies, (ii) scarcity of surface streams, (iii) instability of the cavernous ground, (iv) leakage of surface reservoirs, and (v) an unreliable waste-disposal environment.

1) *Scarcity and poor predictability of groundwater supplies.* Although the extremely high permeability of some carbonate rocks results in high yields of wells, in other areas the high permeability causes the water to drain quickly out of the region or to great depths. In Bermuda, for example, the permeability of the near-surface limestone is so great that water quickly moves into the ground and laterally to the sea, and only a thin sheet of fresh water lies on the basal salt water. The Miami area of Florida has a comparably high permeability, but the quartz sand in the solution openings reduces the permeability slightly and slows down the outflow to the sea. The limestone in Bermuda can hardly be called an aquifer, whereas that in Miami is one of the most prolific water producers known. In most karst regions of the world the permeability is too high or too low for water supply needs.

The uneven distribution of permeability in most karst aquifers leads to several types of problems. The preferential circulation of water along some fractures and the enlargement of these fractures by solution causes an arterial system in which water tends to collect in large openings and to discharge as widely spaced large springs. Wells penetrating the large openings have yields much greater than those that end in rock with smaller openings. The yield of wells in such an arterial system cannot be predicted consistently because the common techniques of interpolation and extrapolation, so useful in more homogeneous aquifers having more uniform flow, are not applicable in simple form. The permeability tends to

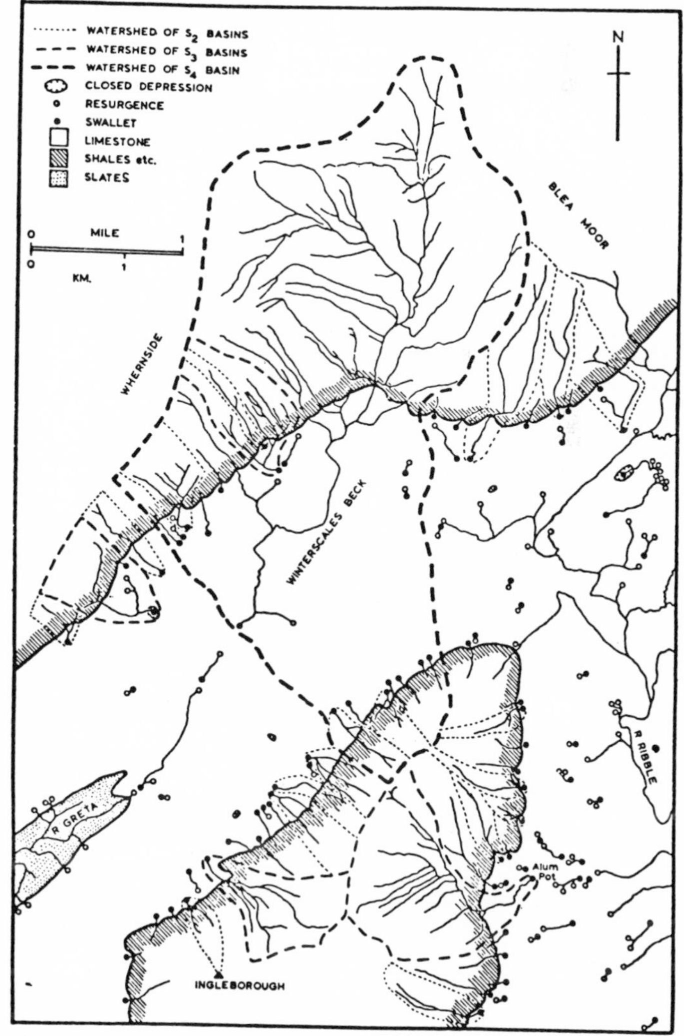

Fig. 2. Some karst features of the Ingleborough District, Yorkshire, England, showing the scarcity of streams in a karst region. [From (*1*)]

Fig. 3. Two new sinkholes recently appeared in a farm in Lebanon Valley, Pennsylvania. The farm is underlain by limestone. [Courtesy of the Pennsylvania Geological Survey]

decrease downward in most carbonate rock terranes, so that the freshwater zone is thin. Old and nearly stagnant brackish water underlies most of the carbonate rock aquifers of the world, and this salty water may present problems in the production of water to individual wells and in the overall development of the aquifers.

2) *Scarcity of surface streams.* Fully developed karst regions are characterized by the absence, or scarcity, of surface streams (Fig. 2) (*1*). The dependence of some biota on streams and the convenient uses that man has made of streams are well known and need not be reviewed here. Where surface streams are scarce or absent, many people must turn to the ground environment for water supplies from wells and for disposal of wastes. The difficulties inherent in these arrangements are discussed at greater length below. The centralization of water discharge as large springs poses subtle management problems because competition arises between users of well water and users of spring water; these problems are not confined to carbonate rock terranes but are likely to be especially acute in them because their ecology and environment are more restrictive or specialized.

3) *Instability of the ground.* The solution of carbonate rocks and removal of dissolved mineral matter by circulating groundwater produce underground cavities that weaken the structure of the rock above them. Although collapse may be a purely natural event, activities of man such as heavy structural loading and excessive pumping of water from wells accelerate subsidence of the ground in many areas (*2*) (Fig. 3).

4) *Leakage of surface reservoirs.* Where dams are proposed to store surface water in regions underlain by carbonate rocks there is concern as to whether the reservoirs will hold water. The leakage is due to the great permeability of the rocks above the water table. As the water attempts to rise in a surface reservoir it enters various caves and other solution openings which had been dry in the previously unsaturated zone. In such areas it is necessary to inject great volumes of grout in a curtain along the dam so that the particular reservoir will hold water, and this increases the cost of the project. Not only are the test drilling and grouting expensive, but in some fully developed karst areas there is a risk that even grouting will not be adequate to ensure the success of the reservoir.

It is possible to take advantage of the situation, however. In Yugoslavia, for example, a new dam with a grouting curtain about 9 kilometers long was designed for a reservoir capacity that included the groundwater backed up behind the curtain. Not only was the reservoir volume thus greatly increased, but the water stored in the solution openings was far less subject to losses by evaporation.

5) *Unreliable waste-disposal environment.* In general, carbonate terranes are not considered to be satisfactory areas for disposal of wastes into the ground. If the permeability of the carbonate rocks is low the rocks do not accept waste at sufficient rates, and if it is high they permit the waste to move downward quickly and to be transported rapidly to some point of discharge without time for the waste to be oxidized or otherwise purified. The poor capabilities of karst regions for waste disposal are worsened by the thinness of the soil zone. Thick but moderately permeable soils tend to sorb many pollutants chemically or physically and to slow the movement of polluted water so that oxidation or some decay mechanism can allow the pollutant to be attenuated. Pollution of karst groundwater from disposal of wastes in the ground and from undesigned leaching of materials at the land surface is common. In many places such pollution seriously threatens the health of the polluting areas and minimizes the development of groundwater in lower places. The large springs that concentrate the discharge of polluted groundwater become themselves the focus of water problems to the people who depended on them for potable supplies.

Hydrogeologic Conditions

The following are brief sketches of some hydrogeologic settings of a few carbonate rock regions. The relations of the hydrogeologic settings to the present ecology can be inferred.

1) *Central United States.* In Tennessee, Kentucky, and Missouri there are separate but somewhat similar karst settings. Each is characterized by a slight structural dome which has brought to the land surface Middle Paleozoic carbonate rocks, concentrically surrounded by younger Paleozoic noncarbonate rocks. Topographic relief locally or regionally is not as great as in many karst regions. Sinkholes abound and cuestas tend to define the boundaries between carbonate and noncarbonate rocks. The climate is temperate and humid. The permeability is unevenly distributed, a large percentage of the wells having low yields and a smaller percentage having high yields. Streams are fairly scarce, most of the streams being fed by large springs. Soils are generally thin. Many wells are polluted, chiefly because of effluent from

septic tanks. Locally, the collapse of land into subsurface caverns forms sinkholes, which in the more populated areas may create serious problems for houses, roads, and public utilities.

2) *Puerto Rico and Jamaica.* Tertiary limestones are widespread on both Puerto Rico and Jamaica. Both islands were uplifted many hundreds of meters above sea level during Miocene time, and the limestones have been continuously karstified since then. Because the topographic relief is great and the climate is humid and tropical, solutional erosion and attendant land collapse have resulted in deep sinks and pronounced conical hills, which combine to make the terrane difficult to cross. Soils are thin, especially on upland slopes. The water table is deep below upland areas, and large caverns in the unsaturated zone transmit storm waters to openings, such as sinkholes, in the lowlands, where flooding may sometimes occur. Because the permeability of the carbonate rocks is unevenly distributed, the yields of wells range greatly. Some springs are polluted at times. Large amounts of fresh groundwater are discharged to the sea, while some inland areas are short of water because the runoff quickly infiltrates to the subsurface by way of the intricately dissected surface.

3) *Nullarbor Plain of southern Australia.* The most extensive continuous karst region of the world lies as a raised, but nearly level plain along the south-central coast of Australia. Having been subjected to slow karstification through almost continuously arid conditions since Miocene time, this region has almost no soil and vegetation (*3*). Scattered sinks occur over much of the region. The water table lies only slightly above sea level for thousands of square kilometers, and the groundwater is almost everywhere salty. Among the few living things are low shrubs on the land surface and bats, spiders, and bugs in caves.

4) *Central Florida.* Limestone of Tertiary age in central Florida is near the land surface, where it is on a structural arch. The Tertiary limestone is covered by a veneer of Pleistocene quartz sand and clay and, radially from the arch, the limestone is buried under increasingly thickening deposits. Sinkholes are common, but topographic relief is subdued because surficial sands and clays drape into the sinks. The humid climate, coupled with the low topography, keeps the water table near the land surface. Lakes are common. Solution cavities abound in the limestone, but almost all are below the water table. Problems of land subsidence and stability of lake levels are of increasing concern. The quartz sands and clays on the limestone prevent some problems characteristic of most karst terranes. For example, the soils cannot be removed easily, and the filling of solution cavities with sand has caused a slight reduction in the permeability of the limestone. The slight reduction in permeability is favorable because it permits the aquifer to store more water for local use seasonally and to recharge the artesian part of the system, which extends out radially from the structural arch in central Florida.

5) *Northern Yucatan Peninsula, Mexico.* The northern third of the Yucatan Peninsula, as much as 150 kilometers wide, is an almost level karst plain underlain by nearly horizontal Tertiary limestone. The plain ranges in elevation from sea level on the coast to about 40 meters in the interior. The climate is not arid, but precipitation is seasonal. There are no surface streams. Bare, fluted limestone is exposed over large areas; the limestone is almost everywhere pitted and scarred by solution depressions and small ridges. Because of the high permeability of the limestone, the water table is only slightly above sea level (*4*), and storage of fresh water is limited. The scarcity of soils has retarded farming and may lead to water supply problems in villages. The absence of streams caused the ancient Mayans, as well as the present civilization, to rely on "cenotes," or steep-walled natural wells, in which the water table is exposed at the bottom as a small and often lovely lake.

6) *Northern and eastern Mediterranean region.* The karstic areas of Yugoslavia and Greece are classic examples of conditions in complex topographic and structural terranes. Limestones in the Adriatic region, deposited through much of Mesozoic time, were uplifted, folded, and faulted during the Cenozoic. The upland areas are practically bare, and what little soil there is tends to be concentrated in the valleys. Although the rocks are intricately channeled, the spring rains and runoff are so great that many of these valleys are temporarily flooded in the wet seasons. Streams are scarce in the fully developed karst areas. Groundwater lies deep below the upland areas and tends to be funneled away along large subsurface channels to large springs. These springs may resurge above or below sea level. Enormous quantities of fresh groundwate are wasted as they mix with sea water at submarine springs.

Maintaining Ecological Balance

We often assume that there was a fair degree of ecological balance in all regions of the world before man significantly changed the landscape. Yet in karst regions, more than in many other specialized environments, ecological conditions were already skewed and the biota were developed in special and sometimes erratic ways. The scarcity of soils, the scarcity of water at the land surface, and the rugged terrane are not conducive to a flourishing and expansive environment. Man has added to the already skewed situation by deforesting hilly karst, which in turn has accelerated the erosion of the scanty soil, and by polluting the groundwater in many places. The question is whether nan will continue to compound the problems of the karst environment or whether he will adjust to and improve it.

Some of the problems relating to climate seem almost insurmountable. For example, the immense karst desert of Nullarbor Plain is simply waterless and uninhabitable. To expand the development of organisms significantly seems impossible.

The difficulties of rugged topography, such as that in the Cockpit Country in Jamaica and that in northern Guatemala, can be partly overcome—insofar as man's access and habitation are concerned—by building modern roads through the regions. It would be inappropriate here to debate the question of building roads.

Some restoration of soils in less arid karst regions can be made through reforestation, such as that attempted in parts of Yugoslavia. Locally in some karst areas there is an "inversion of soils," more soil being trapped in sinkholes and solution openings below the land surface than occurs on the land surface. Recovery of some of the subsurface soils to distribute them again on the land surface may be possible if satisfactory recovery techniques can be developed.

The hydrologic problems in a karst area need to be considered both separately and in conjunction with other aspects of the environment. For example, heavy local pumping of groundwater in one area may cause local catastrophic land subsidence elsewhere. A problem that man can partly solve is the elusiveness of groundwater and the difficulty of capturing it in karst regions. In interior karst regions groundwater tends to discharge as large springs that lead to freshwater rivers. In coastal karst regions, however, freshwater springs and rivers may be scarce; large volumes of fresh water move directly into the sea. The high and uneven distribution of permeability in karst regions makes this waste to the sea a serious problem because it deprives man and other life of the nourishment that water in the right places brings.

Attempting to salvage this fresh water that is wasting to the sea is not likely to have offsetting ecological problems. Yet, the delicate natural hydrological balances within a karst region can easily be disturbed, and advantages gained by altering the hydrology may be offset by various detrimental effects.

Changes in hydrological balances are not unique to karst regions, but karst regions are more sensitive than other regions and the counteracting problems may be especially severe. It is necessary to understand environmental relationships, particularly those involving hydrology, to determine whether some actions by man are warranted. As a result of increased studies evaluation of these hydrological problems is coming into better focus. There is a need for improved knowledge of this important subject, which can be applied to the development of carbonate rock terranes for human use.

Fortunately, we can take specific action to maintain and upgrade the ecology of carbonate rock regions without waiting for additional field data to be collected. Our review of some hydrological relationships indicates that the ecological problems have generic solutions. These relationships have patterns that can be cast in terms of rules and principles. For example, good circulation of subsurface water leads to appreciable solution of the rock, solution leads to overall high permeability, and high permeability tends to lead to a water table well below most of the land surface and to scarcity of water on the land surface. Such general relations and tendencies need to be combined with pertinent specific data for a particular region or area. For ecological evaluations, it would be foolhardy to treat each carbonate rock terrane solely on its own merits, and to collect masses of data before decisions on the ecology are made. The data available are sufficient to give good first-round approximations to the solutions needed. Successively better approximations can be made as more data are considered, but in no case do we need to wait until new data are collected.

Summary

Climate exerts a universal dominant influence on ecology, but processes of karstification have an equally high ecological influence in carbonate rock regions. Development of karst features depends greatly on the degree to which water containing carbon dioxide has been able to move on and through carbonate rocks and to remove some of the rock in solution. Distinctive features of many karst terranes include scarcity of soils, scarcity of surface streams, and rugged topography; less distinctive are the highly permeable and cavernous rocks, especially at the shallow depths. This high permeability gives rise to many practical problems, including (i) scarcity and poor predictability of groundwater supplies, (ii) scarcity of surface streams, (iii) instability of the ground, (iv) leakage of surface reservoirs, and (v) an unreliable waste-disposal environment.

Natural karst processes in some carbonate rock regions have caused a greater restriction in the development of biota than man can ever be suspected of causing.

References

1. P W Williams, *Ir. Speleol.* **1**, 23 (1966).
2. V T. Stringfield and H. E. LeGrand, *J. Hydrol.* **8** (Nos. 3 and 4), pp. 349–417 (1969).
3. J. N. Jennings, *Trans. Roy. Soc. S. Aust.* **87**, 41 (1963).
4. W Back and B. B. Hanshaw, *J. Hydrol.* **10** (No 4), 330 (1970).

15

Reprinted from *Univ. Missouri-Kansas City Geograph. Publ.*, 1, 13-15, 17, 21, 24, 26, 28-29, 31, 33-36, 47 (1972)

OCCUPANCE AND USE OF UNDERGROUND SPACE IN THE GREATER KANSAS CITY AREA

Truman P. Stauffer, Sr.
University of Missouri-Kansas City
Dept. of Geology-Geography

Introduction

Secondary use of mined-out space in underground limestone quarries is emerging in the Kansas City area where abandoned quarry sites are being renovated into warehouses, factories, and offices. Mining techniques, ceiling heights, and pillar distribution are being altered as planning for the secondary use of mined-out space is combined with the quarrying process. Even the historic urban-quarry land use conflict invites a reappraisal in that this development provides a potential resolution of the problem. Mutual subsidization by dual rock and space interests is changing the economics of rock production.

A national survey conducted by the author early in 1971 revealed that Missouri led the contiguous United States in the secondary use of space mined out of a limestone matrix with the greatest concentration occurring in the Kansas City area (Figure 1).

The economic impact stemming from revitalization of once dormant abandoned quarries is felt in Kansas City as 1,641 employees earn $13,000,000 annually some 50-100 feet below ground and property values are increased both adjacent to and including the mine sites.

The secondary use of an otherwise abandoned mine is a dynamic factor in the neighborhood. The use of the subsurface is often accompanied by a plan of total development which includes intensive use of the surface. A more permanent geographic relationship with the community replaces the temporary perspective usually associated with the quarry operation. The economic input from secondary use commonly exceeds that of its rock production.

Land use problems of an urban-quarry nature are being resolved as quarry operators sit with developers and plan the future of their mining site as a permanent industrial park with storage facilities, factories, and offices below ground, and regular use of the surface for stores, parks, and common urban development. Long range plans are laid for the use of the mined-out space which is more valuable than the once-over exploitation of the rock. The neighborhood accepts

[*Editor's Note:* A row of asterisks indicates omission of material.]

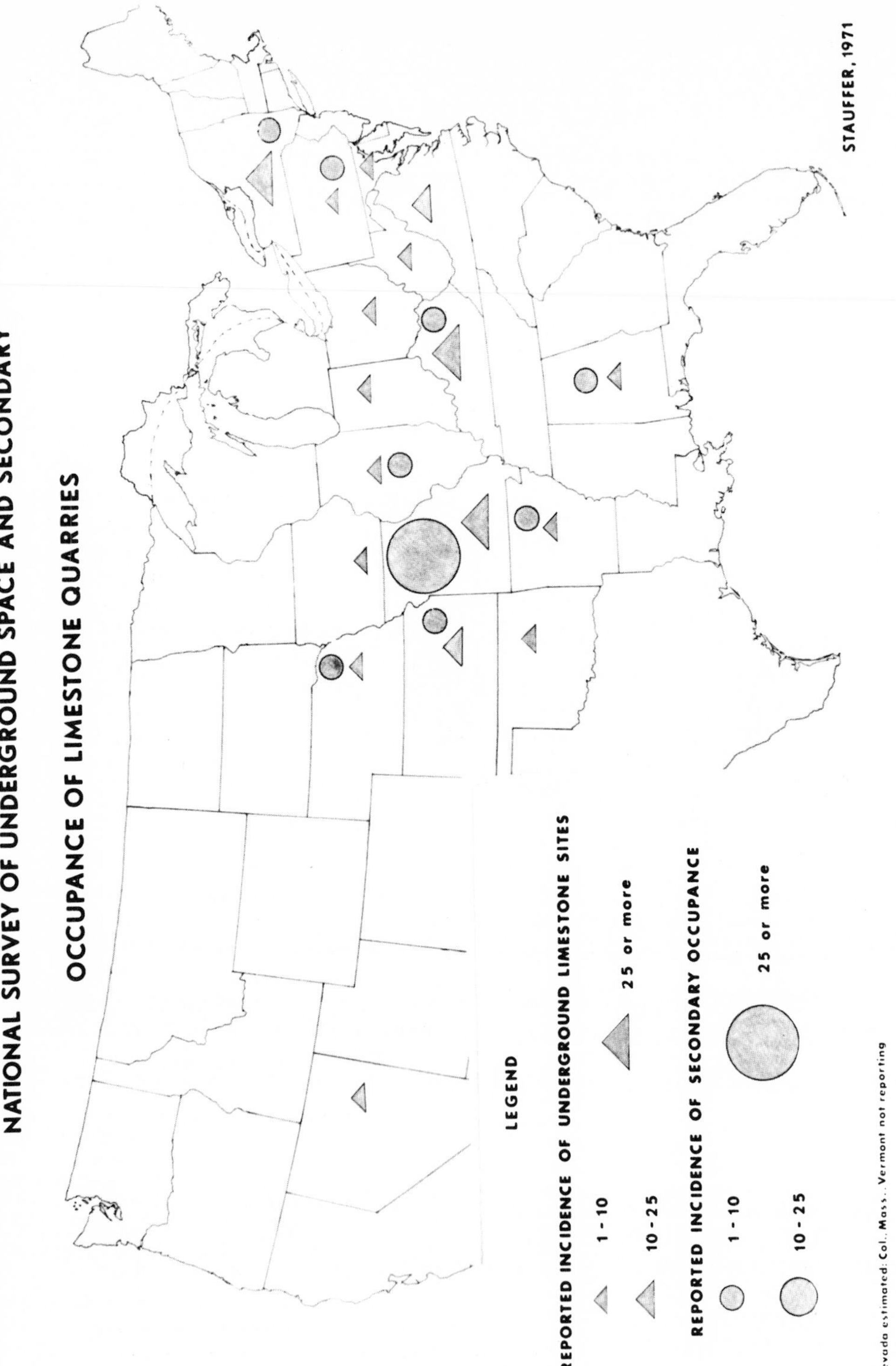

NATIONAL SURVEY OF UNDERGROUND SPACE AND SECONDARY OCCUPANCE OF LIMESTONE QUARRIES

STAUFFER, 1971

LEGEND

REPORTED INCIDENCE OF UNDERGROUND LIMESTONE SITES

1 - 10

10 - 25

25 or more

REPORTED INCIDENCE OF SECONDARY OCCUPANCE

1 - 10

10 - 25

25 or more

Nevada estimated: Col., Mass., Vermont not reporting

Fig. 1. MAP OF NATIONAL SURVEY RESULTS

the planned secondary use of a mine with far greater approval than the spectre of fenced wasteland often left by mining and later engulfed in the path of urban growth. The continued use of the land assures tax potential and economic support for the city.

Major secondary uses are warehousing, factories, and offices in that order. One-seventh of Kansas City's warehousing is now underground. The Department of Agriculture reports that Kansas City has 34,473,000 cubic feet capacity for frozen food storage about one-tenth of the total such capacity of the entire nation most of which is underground. A lawnmower parts factory makes use of the unlimited carrying capacity of the floors for its heavy industrial tooling and milling, while a sailboat factory utilizes the easily controlled humidity for setting its lacquers and glues. Printing shops take advantage of the controlled humidity for maintaining quality control.

Currently there are 28 subsurface sites in varying stages of development in the Kansas City area, encompassing nearly 45 million square feet of space, 30-200 feet below the surface, with floors that dip less than ten feet per mile

Long term occupance of the subsurface for a variety of functions has clearly demonstrated several distinct advantages in use of the underground environment, for example:

1) mined-out space can be purchased or leased at a fraction of the cost for comparable surface facilities

2) roof and foundation problems are reduced or eliminated

3) floors are capable of supporting unlimited installations or storage weight

4) complete noise and vibration control is possible

5) the areas are fireproof and command the lowest insurance rates in the area

6) greatly reduced cost in heating, air conditioning, or freezing as the consequence of insulation from the rock, and

7) vastly improved security for equipment, records, and personal protection

SOCIAL AND ECONOMIC IMPACT

Greater Kansas City has a new concept in space use, becoming, in the last twelve years, the center for secondary use of underground mined-out limestone quarries. The location of warehousing, manufacturing, businesses, and offices in an otherwise subterranean desert of quarried wasteland is especially significant, since this means the redemption of blighted areas within the boundaries of an urbanized region where both land values and land usage are at a premium.

The use, re-use, constancy of use, and multiple use of space within an urbanized area is of vital importance. It would seem appropriate that a study of the character, nature, and distribution of a particular multiple use of space within an urbanized setting should properly fall within the domain of geography which by tradition addresses itself to problems of description and analysis of areal differentiation.

One use of the land commonly allied with the growth of cities is the quarrying of limestone for building materials. Nearness to limestone deposits has historically been viewed as an asset by both user and supplier with the development of mine and city within close proximity. Today, the boundaries of many cities, and especially their urbanized fringes, have expanded to include their mining areas. Quarries, once viewed as assets, have become thorny problems in the sides of those who must decide what to do with a "hole in the ground" within the limits of the enlarged city.

Uses such as public warehousing, manufacturing, salesrooms, and offices, which are commonly associated with surface locations, are currently being located in mined-out quarries. In this development, Kansas City has a new concept in the use of otherwise abandoned space.

The use of underground space for storage, offices, and industry is no longer restricted to the use of abandoned mines, but more recently planning for secondary use of space is combined with quarrying to the mutual advantage of both. Spatial variations in mining methods are currently developing as an adjunct to planning for the secondary use of mined-out space.

* * * * * * *

The significant trend toward use of underground space for storage, offices, and industry began with the use of abandoned mined-out space but has now advanced to the point that secondary use of space is of greater importance than the rock products.

* * * * * * *

In 1966 there was 35 million square feet of distribution warehousing space reported, which was an increase of 10 million square feet over

1961. Five million square feet of the total warehousing space was underground in mined-out limestone caves making the Kansas City area first nationally in underground warehousing space as of that date (Kansas Citian, 1967, p. 62). According to my current research, there is approximately 13.5 million square feet of warehousing underground in the area.

* * * * * * *

These sites have both truck and rail loading facilities, and can distribute products overnight to Oklahoma, Arkansas, western Kansas, southeast Nebraska, and the rest of Missouri. These facilities for overnight delivery allow grocery stores to turn over their inventories as many as 50 times a year.

The geographic location of Kansas City in the great "food basket" of the United States and midway between the western area of the United States, which produces some 50 percent of the nation's processed and frozen foods, and the eastern region, which buys two-thirds of the production, is ideally suited for a storage-in-transit point. The first underground freezer storage room was developed in 1953 and Inland Cold Storage took the lead as the world's largest refrigerated warehouse handling 8 million pounds daily. Over a pound of food for each person in the United States can be stored in this facility at a given time (Jewett, 1967, p. 12).

* * * * * * *

Tunneled into one of the numerous limestone bluffs that characterize the terrain around Kansas City is the Brunson Instrument Company which manufactured the surveying instruments used on the moon. Personnel in its 140,000 square feet factory 77 feet below ground ranges as high as 435 among whom are technicians in optical tooling. Precision settings are made at any hour in this vibration free environment whereas only the low traffic hours of 2-4 a.m. could be used in a former surface location (Harbinger, 1970, pp. 3-9).

Amber Brunson was the first to quarry rock as a secondary process with the thought of obtaining the underground space for a factory being his primary objective. Occupation of the underground factory began in 1960 and his facilities have since been an object of interest, both nationally and internationally. He is recognized as the "father" of planned secondary use since the previous uses of underground space were afterthoughts.

* * * * * * *

Brunson led in pre-planned mining and pillar arrangement which served the purpose of secondary occupance.

This impact of secondary usage is of greater magnitude when one considers that each year some 115 acres of land are consumed in primary usage whereas the secondary usage of the underground space is continuous year after year. Secondary usage has now exceeded stone production as an economic factor in the Kansas City area by some 4.7 times in numbers employed and 4.3 times in annual wages.

The economic impact of increased land values has not been felt in its entirety. It is highly speculative and difficult to accurately ascertain since much depends on pending legislative action with regard to split fee title. At this time the surface area and the subsurface is owned and developed by the same individual or corporation.

Unmined land with stone potential at the edge of Kansas City can be bought for about $1,500 per acre. This is in a large quantity and along rough bluffs of outcropping limestone. Mined land where the subsurface is suitable for development can sell for $28,000 per acre and when developed may be as high as $60,000. These figures could change with title and taxation clarification. However, it should be pointed out that land where mined and abandoned has remained close to the $1,500 per acre level if not lower and quoted as high as $60,000 where the subsurface is developed. The economic impact of secondary use of mined sites can therefore be as high as 4,000 percent.

* * * * * * *

Analysis of the types of uses made of underground space shows warehousing and storage as the principal usage making up 89% of the total; manufacturing accounting for an additional 7%, and the rest composed of offices. The ability to create freezer space from part of the storage total at a low cost due to the natural insulation of the surrounding rock is reflected in the location of the world's largest refrigerated storage in Kansas City with some 3,000,000 square feet available. Major reasons for the location of public warehouses in Kansas City are Kansas City's geographical position and excellent transportation facilities.

In a recent article, "A Plea for Storage Geography," the importance of storage as a critical leveler between production and consumption was stressed. The vital place of storage pits in the preservation of the sweet potato and the storage cities of Egypt are cited as examples of early significance (Lewthwaite, 1969, pp. 1-4). Crop specialization in today's world make storage of even greater importance. One of the major industries in the use of underground space in the Kansas City area is the storage of fruits and vegetables from the time of harvest until the market is in demand. Fresh vegetables are brought from California and held until demand in New York creates the desired market. One of the more fortunate aspects of this trend toward use of low cost underground storage is that of freeing both production and market from total dependence on climatic conditions.

The underground sites with secondary usage show a close affinity with railroads and highways. The nearest railroad averaged less than 1 mile from the site while the mean for major highways was 0.5. The mean distance to downtown Kansas City for the sites studied was 7.3 miles.

Whereas Jean Gottmann viewed the skyscraper as a vertical extension of a horizontal street pattern underground development seems to be a lateral or horizontal extension of the valleyed industrial pattern (Gottmann, 1966, p. 196).

The importance of the Central Business District was not indicated in this study. The regional market dominated the Central Business District in importance as a locational factor.

* * * * * * *

GEOLOGIC SETTING

Erosive processes in the Mesozoic and Tertiary eras removed a considerable portion of the Pennsylvanian rock from the area, leading to some of the topographic features existing today (McCourt, 1917, pp. 80-1). Kansas City, stratigraphically, is in the middle of a 150-mile-wide belt of such exposed Pennsylvanian rocks that extend around the Ozark Uplift in a north-south direction through western Missouri and eastern Kansas. The beds slope west-northwest away from the Ozarks and toward the Forest City Basin with dips that seldom exceed 20 feet per mile. The surface beds in the Kansas City region are part of the Missourian Series, one of five that make up the Pennsylvanian System in the northern Midcontinent (Greene & Howe, 1952, p. 6).

* * * * * * *

The major feature which pervades this study area is the outcrops of limestone that form bluffs. Wherever extensive erosion has occurred it gives rise to scarped edges of Paleozoic rocks as bluffs around the valleys. Such exposures are so common that roads have historically been built by portable crushers moving along valley roads during construction, which machines work the convenient ledges with minimal hauling. Relics of cirque-like quarries can be seen along many highways, some having degraded into dump sites.

The exploitation of large reserves of readily accessible beds of high quality limestones of the Kansas City Group is an important factor in the growth of Kansas City into an important industrial center (Gentile, 1965, p. 50).

* * * * * * *

PHYSICAL FACTORS LIMITING UNDERGROUND DEVELOPMENT

To the geographer interested in the ways in which man has utilized the subsurface in the Kansas City area, a number of questions come to mind regarding the feasibility of extending underground space development to other regions. It is reasonable to suspect that more than mere coincidence produced the use of secondary space in mined-out areas. The analysis of those factors which limit or control such development are vital to understanding the potential for extension of space use (Dean et al, 1969, pp. 35-55).

A combination of six major physical factors have provided the opportunity for this development. They are:

(1) A massively bedded limestone

The unit must be massively bedded in order to support its own weight when undermined. Thin to medium bedding, persistent jointing and interlayering of clay or shale are prohibitive factors.

(2) Sufficient thickness

Economically viable room height and an adequate roof are the two factors determining the thickness of the potential rock unit. According to this study, 12 - 13 foot room height suffices for economic rock production as well as most secondary uses, and sufficient roof thickness seems to require a minimum of 8 feet in order to include at least two massive beds of rock. It appears, therefore, that rock units with potential for secondary use should be at least 20 feet thick. Modifications from this limitation may possibly be accomplished by a different room height or by lowering the floor into the underlying strata.

(3) Overlying impermeable shale

The fact that karst topography does not occur in the Kansas City area can be partially attributed to the sealing effect of the overlying shales. Where the shales have been dissected or removed by erosion, seepage, roof fall, and surface caving can result. The existence of an intact overlying impermeable shale member is necessary to avoid the danger of mine subsidence.

(4) Nearly level stratigraphy

Tilted strata obviously defies secondary development within competitive economic range. Slight inclination also introduces problems. A 4 percent grade presents difficulty in efficient railroad car switching, and benching of the floor creates steps and ramps which add to the problems of handling of goods by trucks, carts, and forklifts. Leveling the floor by filling robs the ceiling height and leveling by added excavation involves diversified matrix materials. Drainage can be an expensive maintenance problem where any slope occurs opposite to the underground extension. Whereas problems begin with even small bedding inclination, a 4 percent grade appears to be the point beyond which development is not feasible. The average dip of the Bethany Falls limestone in the Kansas City area is 10 - 20 feet per mile, a dip of approximately one-fifth of a degree.

(5) Competent overburden

For minimum roof stability and surface preservation above underground space a sufficient thickness of overburden should be present which includes impermeable shales and a buffer against erosion and mass wasting. In the Kansas City area this requires a minimum of 40 feet of overburden, thus indicating the need for the superjacent Winterset member. In this area, such a thickness of overburden generally provides sealant shales which protect the underground openings from vadose water and insure the strength of overlying rock by virtue of its mass. Exceptions occur where aeolian deposits constitute the overburden. These areas must be considered independently.

(6) Natural accessibility

Vertical shaft mining of limestone necessitates the use of elevators for secondary use of the resulting space. Where rent values are sufficiently high this may be economically feasible, but it is highly unlikely that underground space use on a grand scale is apt to develop except with a more economical means of accessibility. A dendritic drainage pattern which dissects the strata and exposes the potential rock matrix to accessibility by horizontal means appears to be a more reasonable prerequisite to extensive development of both the mine and the mined-out space. Valley walls provide natural accessibility and valley floors are the natural locus for rail and highway routes.

* * * * * * *

PLANNING

"The planner has an opportunity and an obligation to plan for the sequential use of mineral lands so that these lands may return to the community a value many times greater than what might have been had they been used for a single purpose rather than for multiple purposes in sequence." (Vineyard, 1968, p. 5)

The presence of acres of desecrated mined out land within the boundaries of greater Kansas City begs the attention of concerted thought and planning for their recovery.

The factors which created such wasted land and the necessary steps to prevent its escalation have too often been overlooked. Hewes cites the limited attention given by geographers to the general matter of destructive exploitation despite Jean Brunhe's early observation (Hewes, 1966, p. 95). Geologists have often been more interested in finding a mineral than reclaiming the site after extraction.

The successful reclamation of some mined areas by secondary use has invited an evaluation of former mining practices and their subsequent effect on later usage of the mined area. Where mining practices have been less exploitive the areas left have been partially recoverable by use of roof bolting to compensate for inadequate overburden. However, where mining is done in conjunction with planning for secondary use the entire mined site can be developed on both surface and subsurface levels. Public regulations should now require that all operators follow basic mining practices which assure a stable surface above the mines that can be used for residential and/or other purposes. Though multiple and continued use of any area should be self disciplining by virtue of its economic advantage avoidance of wasted areas within an urban setting should be a matter of public concern. The reconciliation of the traditional urban-quarry conflict by the development of industrial parks in mined out areas now attests the value of planning land use toward this goal.

* * * * * * *

REFERENCES

Dean, Thomas J., Jenkins, Gomer, Williams, James H., 1969, Underground Mining in the Kansas City Area, Missouri Mineral Industry News, Vol. 9, No. 4, Missouri Geological Survey, Rolla, Missouri, pp. 35-56.

Gentile, R. J., 1965, Economic Mineral Commodities of the Kansas City Area: Missouri Geological Survey and Water Resources, Report of Investigations, No. 31, pp. 50-57.

Gottmann, Jean, 1966, "Why The Skyscraper," Geographical Review, Vol. 56, pp. 190-212.

Greene, F. C., and Howe, W. B., 1952, Geologic Section of Pennsylvanian Rock Exposed in the Kansas City Area: Missouri Geological Survey and Water Resources, Inf. Circular No. 8.

Harbringer, 1970, Brunson. . . The Man and His Company, Vol. 11, No. 9, pp. 3-9.

Hewes, Leslie, 1966, Common Responsibility of Economic and Cultural Geography, Guest Editorial, Economic Geography, Vol. 42, p. 95.

Jewett, J. M., Hornbaker, A. L., and Press, J. E., 1967, A Guidebook for the Third Forum on the Geology of Industrial Minerals, Univ. of Kans., Lawrence, Kansas, 13 pp.

Lewthwaite, Gordon R., 1969, A Plea for Storage Geography, The Professional Geographer, Vol. XXI, No. 1, pp. 1-4.

Luce, William N., 1971, What's Ahead in Tunneling Technology, Paper Presented at the Annual Meeting of the Geological Society of America, Washington, 13 pp.

Parizek, E. J., 1965, Stratigraphy of the Kansas City Group: Missouri Geological Survey and Water Resources, Report of Investigations, No. 31, pp. 32-49.

Vineyard, Jerry D., 1968, Natural Resources, Recreational Geology and Planning, Annual Spring Conference of Missouri Planning Association, March 15-16, Rolla, Missouri.

Part V

HAZARDOUS EFFECTS OF GEOLOGIC ENVIRONMENTS

The papers offered in this part deal with mass movement, earthquakes, volcanoes, and health and disease. There is a broad band of overlap between this part and the one that follows, which will become evident in the papers. (At this point note is taken of the important recent book, *Cities and Geology* by Robert F. Legget, which includes a long and excellent section on geologic hazards. See Supplemental Readings.)

MASS MOVEMENT

Mass movement refers to a variety of dislocations of earth materials, identified by such terms as collapse, subsidence, landslides, rockslides, settlement, lowering, and mass waste. Specialists on these problems make many more distinctions, related to substances involved, angles of movement, and the factors that induce the movements. In brief, however, from an environmental perspective, the important point is the impact of a mass movement on persons and enterprises that depend on stability of the landscape. In Paper 16 there may not be a total reflection of the environmental problem, but it is evident that mass movement can be a startling and unwanted event. (Another Benchmark volume, by S. A. Schumm, treats slope morphology, and another is planned on landslides.)

Quick clays are a relic of Pleistocene glacial action. They are postglacial clays of mixed mineralogy, often mixed with marine salts, and are known especially from Scandinavia, the USSR, Canada, Alaska, and the northeastern United States. The characteristics of quick clay and the causes of slides in their areas of occurrence have been studied, but the problem of determining whether the clays can be controlled is still under investigation.

EARTHQUAKES

Earthquakes are discussed extensively in the literature, but usually in terms that are meaningful only to geologists and seismologists. Even the reporting in the public press, which mentions the rating of intensity, is more or less meaningless to the layman, except when translated into money figures expressing extent of damage.

The almost daily occurrence of earth tremors in California has tended to obscure the fact that severe earthquakes have been experienced in other parts of the United States. The major seismic events, called the New Madrid earthquakes, that occurred in the mid-United States during 1811 and 1812, and a catastrophe that hit in the area of Charleston, South Carolina, in 1886, are sometimes overlooked, no doubt because they have not been repeated. Unfortunately, the detailed literature on them is interesting more from a historical viewpoint to antiquarians who are familiar with old landmarks in these regions.

The study of seismic risk and evaluation of hazard reduction in further occurrences are being pursued actively by many scientists in different parts of the world. The question that most often arises is whether frequency or severity is the main factor in weighing the environmental hazard. A catastrophic event once in a century is a major risk, though, if spaced at 100-year intervals, it will not be so terrifying to most persons whose life span is shorter.

VOLCANIC ACTIVITY

Volcanic activity is a hazard over which humans may be presumed to have little control, if any. The literature on the subject drifts into accounts of interest only to subject specialists—volcanologists, mineralogists, petrologists, and climatologists. The violent effects of eruptions are well known, but because of the relative rarity of such events there is little continuing attention to the subject. The scientific literature is strangely unconcerned with the environmental impact of volcanic events. A main interest today relates to the role of volcanic dust and its long-term climatic impact, as opposed to the immediate human suffering that may result from eruptions.

HEALTH AND DISEASE

Health and disease has become a live subject of investigation for a small number of geologists, whose field of interest is appropriately named *medical geology.* Determination of relationships between the geologic environment and the incidence of health problems goes a step further than the field of *medical geography,* which deals with the distribution of chronic and scattered health problems related to general geographic factors and is for the most part quantified by statistical analysis. Although various factors of geography are examined, the geologic factor is not put in the foreground. It should be recognized that environmental effects are manifested in numerous ways. Trace-metal contamination of groundwater and the correlation between radiogenic rocks and cancer are cases in point. Another interesting effect is the pervasive influence of climate on human behavior, usually connected with seasonal or occasional phenomena. Thus, the erratic behavior of inhabitants of regions bordering the Alps both on the north and the south when subject to certain persistent winds (the *Foehn*) has been accepted as an extenuating circumstance in judicial decisions on illegal acts. Health hazards are not, however, exclusively the result of the environment; as with other hazards, the contribution of man is evident in the creation of air and water pollution. One insoluble problem, as with all natural hazards, is the virtual impossibility of regulating the human willingness to be exposed to an undesirable environment.

["Supplemental Readings" begin on the following page.]

SUPPLEMENTAL READINGS

Almagia, Roberto. 1924. Les Éboulements en Italie. Genève, Soc. de Géog., *Materiaux pour l'etude de calamités*, **2**, p. 99-121.

Anderson, Alan, Jr. 1973. Earthquake Prediction—The Art of the State. *Saturday Rev. Sci.*, Feb., p. 25-33.

Armstrong, R. W. 1971. Medical Geography and Its Geologic Substrate. In: *Environmental Geochemistry in Health and Disease* (H. L. Cannon and H. C. Hopps, eds.) Geol. Soc. America Mem. 123, p. 211-219.

——. 1972. Cancer Infectious Diseases Related to Geochemical Environment—Is There a Particular Kind of Soil or Geologic Environment That Predisposes to Cancer? *N.Y. Acad. Sci. Ann.*, **199**, p. 239-248.

Barany, Jean, and Louis Galacz. 1965. Fréquence du cancer et structure géologique. *J. Radiol. Electrologie Méd. Nucl.*, **46**(10), p. xxxii, xxxiv, 132-133.

Cannon, Helen, and D. F. Davidson (eds.). 1967. *Relation of Geology and Trace Elements to Nutrition.* Geol. Soc. America Spec. Paper 90, 68 p.

——, and H. C. Hopps (eds.). 1971. *Environmental Geochemistry in Health and Disease.* Geol. Soc. America Mem. 123, 230 p.

Clark, Blake. 1969. America's Greatest Earthquake. *Reader's Digest*, **94** (April), p. 110-114.

Crandell, D. R., and H. H. Waldron. 1969. Volcanic Hazards in the Cascade Range. In: *Geologic Hazards and Public Problems* (R. A. Olson and M. M. Wallace, eds.), p. 5-18. U.S. Office of Emergency Preparedness, Region 7, Santa Rosa, Calif.

Crawford, C. B. 1968. Quick Clays of Eastern Canada. *Eng. Geol.*, **2**(4), p. 239-265. (Natl. Res. Council Canada, Div. Bldg. Res., Tech. Paper 281.)

Crawford, M. D., M. J. Gardiner, and J. N. Morris. 1968. Mortality and Hardness of Local Water-Supplies. *The Lancet*, 7547, p. 827-831.

Curry, M. G., and G. G. Gigliotti (compilers). 1971. *Cycling and Control of Metals.* Washington, D.C., U.S. Environmental Protection Agency, National Environmental Research Center, 187 p. (Proc. Environ. Resources Conf., Columbus, Ohio, Oct. 31-Nov. 2, 1971.)

Dutton, C. E. 1890. *The Charleston Earthquake of August 31, 1886.* Washington, D.C., Government Printing Office, 528 p. (From 9th Ann. Rept., Director, U.S. Geol. Surv. 1887-1888.)

Fairbridge, R. W. (ed.). 1972. *Encyclopedia of Geochemistry and Environmental Science.* New York, Van Nostrand Reinhold Company.

Fleischer, Michael. 1972. An Overview of Distribution Patterns in Trace Elements of Rocks. *N.Y. Acad. Sci. Ann.*, **199**, p. 6-16.

——. 1973. Natural Sources of Some Trace Elements in the Environment. In: *Cycling and Control of Metals* (M. G. Curry and G. M. Gigliotti, compilers), p. 3-10. Cincinnati, Ohio, U.S. National Environmental Research Center. (Proc. Environ. Resources Conf., Columbus, Ohio, Oct. 31-Nov. 2, 1972.)

Fuller, M. L. 1912. *The New Madrid Earthquake.* U.S. Geol. Surv. Bull. 494, 119 p.

Glökler, Konrad. 1967. Zwei Bergrutsche an der oberen Mosel. *Z. Geomorph.*, (n.s.) **11**(1), p. 93-102.

Heim, Albert. 1932. *Bergsturz und Menschenleben.* Zürich, Fretz und Wasmuth, 218 p.

Henkel, G. 1965. Fluoraufkommen im Trinkwasser im Zusammenhang mit verschiedenen geologischen Formationen und Wetterverhältnissen Thüringens. *Deut. Stomatol.*, **15**(2), p. 121-133.

Holmsen, P. 1953. Landslips in Norwegian Quick-Clays. *Geotechnique*, 3.

Hopps, H. C., and H. L. Cannon (eds.). 1972. *Geochemical Environment in Relation to Health and Disease.* N.Y. Acad. Sci. Ann., **199,** 352 p.

Howe, Ernest. 1969. *Landslides in the San Juan Mountains, Colorado.* U.S. Geol. Surv., Prof. Paper 67, 58 p.

Humphreys, W. J. 1913. Volcanic Dust and Other Factors in the Production of Climatic Changes, and Their Possible Relation to Ice Ages. *J. Franklin Inst.,* **176,** p. 131-172.

Kates, R. W., J. E. Haas, and D. J. Amaral. 1973. Human Impact of the Managua Earthquake. *Science,* **182**(4116), p. 981-990.

Kerr, P. F., and I. M. Drew. 1968. Quick Clay Slides in the U.S.A. *Eng. Geol.,* **2**(4), p. 215-238.

Kingscote, B. F. 1970. Correlation of Bedrock Type with the Geography of Leptospirosis. *Can. J. Comp. Med.,* **34**(1), p. 31-37.

Knight, S. H. 1937. *The Rocks and Soils of Wyoming and Their Relations to the Selenium Problem.* Univ. Wyoming Agr. Expt. Sta., B. 221.

LaMoreaux, P. E., and W. M. Warren. 1973. Sinkhole. *Geotimes,* **18**(3), p. 15.

Lawson, A. C., et al. 1908-1910. *California Earthquakes of April 18, 1906.* Washington, D.C., Carnegie Institution, Publ. 87, 2 vols. and atlas. (Reissued, 1969.)

Legget, R. F. 1973. *Cities and Geology,* New York, McGraw-Hill Book Co., 624 p.

Leighton, F. B. 1966. Landslides and Hillside Development. In: *Engineering Geology in Southern California* (R. Lung and R. Proctor, eds.), p. 149-193. Assoc. Eng. Geol., Los Angeles Section.

Liebling, R. S., and P. F. Kerr. 1965. Observations on Quick Clay. *Geol. Soc. America Bull.,* **76**(8), p. 853-874.

McKinley, Carl. 1886. *A Descriptive Narrative of the Earthquake of August 31, 1886, with Notes of Special Investigations.* Charleston, S.C. Yearbook, 1886, p. 345-439. (With map prepared by Earl Sloane, U.S. Geol. Surv.)

Marsden, S. S., Jr., and S. N. Davis. 1967. Geological Subsidence. *Sci. Amer.,* **216**(6), p. 93-100.

Mayuga, M. N., and D. R. Allen, 1966. Long Beach Subsidence. In: *Engineering Geology in Southern California* (R. Lung and R. Proctor, eds.), p. 281-285. Assoc. Eng. Geol., Los Angeles Section.

Merke, F. 1967. Weitere Belege für die Eiszeit als primordiale Ursache des endemischen Kropfes: Eiszeit und Kropf in Wallis. *Schweiz. Med. Wochenschr.,* **97**(5), p. 131-140.

Milne, W. G., and A. G. Davenport. 1969. Distribution of Earthquakes in Canada. *Seismol. Soc. America Bull.,* **59,** p. 729-754.

Morse, W. C. 1935. *Geologic Conditions Governing Sites of Bridge and Other Structures.* Missouri Geol. Surv. Bull. 27, 19 p.

Morton, D. M., and Robert Streitz. 1967. Landslides. *Calif. Div. Mines Geol., Mineral Inform. Serv.* **20**(10), p. 123-139, (11) p. 135-140.

Oliveira, Ricardo. 1972. *An Example of the Influence of Lithology on Slope Stability.* 24th Intern. Geol. Congr. Proc., Sec. 13, p. 142-149.

Poland, J. F. 1969. Land Subsidence in the Western United States. In: *Geologic Hazards and Public Problems* (R. A. Olson and M. M. Wallace, eds.), p. 77-96. U.S. Office of Emergency Preparedness, Region 7, Santa Rosa, Calif.

——, and G. H. Davis. 1969. Land Subsidence Due to Withdrawal of Fluids. *Geol. Soc. America Rev. Eng. Geol.,* **2,** p. 187-269.

Powell, W. J., and P. E. LaMoreaux. 1969. *A Problem of Subsidence in a Limestone Terrane at Columbiana, Alabama.* Geol. Surv. Alabama Circ. 56, 30 p.

Reaves, Bill. 1974. Charleston Quake Recalled by Tremor. *Wilmington* (North Carolina) *Star-News,* Nov. 29, p. 8-E.

Reid, H. F. 1914. The Lisbon Earthquake of November 1, 1755. *Seismol. Soc. America Bull.*, **4**, p. 53-80.

Scholz, Christopher, et al. 1973. Earthquake Prediction: A Physical Basis. *Science*, **181**(4102), p. 803-810.

Sharpe, C. F. S. 1938. *Landslides and Related Phenomena.* New York, Columbia University Press, 138 p. (Reprinted by Pageant Books, Inc., Paterson, N.J., 1960.)

Skempton, A. W., and D. J. Henkel. 1955. A Landslide at Jackfield, Shropshire, in a Heavily Overconsolidated Clay. *Geotechnique*, **5**(2).

Stringfield, V. T., and R. C. Smith. 1956. *Relation of Geology to Drainage, Floods, and Landslides in Petersburg Area.* West Virginia Geol. Econ. Surv., Rept. Invest. (13), 19 p.

Terrosi, F. 1968. L'Influenza dei Fattori Geo-Fisici e Climataci sull' Accrescimento del Bambino nella Provincia di Grosseto. *Arch. Ital. Pediat. Puericolture (Pisa)*, **26**(3), p. 234-239.

Trelease, S. F. 1949. *Selenium: Its Geological Occurrence and Its Biological Effects in Relation to Botany, Chemistry, Agriculture, Nutrition, and Medicine.* New York, privately printed, 292 p.

UNESCO/International Association of Scientific Hydrology. 1970. *Land Subsidence—A Contribution to the International Hydrological Decade.* Gentbrugge, Belgium, and Paris, IASH and UNESCO, 2 vols.

Van Ness, G. B. 1967. Geologic Implications of Anthrax. In: *Relation of Geology and Trace Elements to Nutrition* (H. Cannon and D. F. Davidson, eds.). Geol. Soc. America Spec. Paper 90.

Varnes, D. J. 1958. Landslide Types and Processes. In: *Landslides and Engineering Practice*, p. 20-47. Natl. Res. Council, Highway Res. Board, Spec. Rept. 29.

Warren, H. V., and R. E. Delavault. 1965. Medical Geology. *Geotimes*, **10**(Sept.), p. 14-15.

Webb, J. S. 1971. Regional Geochemical Reconnaissance in Medical Geography. In: *Environmental Geochemistry in Health and Disease* (H. Cannon and H.C. Hopps, eds.), p. 32-42. Geol. Soc. America Mem. 123.

Wexler, Harry. 1952. Volcanoes and World Climate. *Sci. Amer.*, **186**(4), p. 74-80.

Young, Patrick. 1974. Seismic Sleuthing. *Saturday Rev./World*, **1**(12), p. 48-50.

Abstracts (1972-1974) from the Geological Society of America, Abstracts with Programs

Carter, B. R. 1973. Colluvium: A Geologic Hazard in the Southeast. **5**(5), p. 385.

Carter, C. H. 1973. Natural Geologic Hazards: Landslide, Shore Erosion, and Flooding Along the Ohio Shore of Lake Eric. **5**(7), p. 569.

Clark, J. R. 1973. Selenium Anomalies, Health Hazards, and Pyrite Deposits in Chambers County, Alabama. **5**(5), p. 387-388.

Fischer, P. J. 1973. Environmental Hazards of the Northern Santa Barbara Shelf, California. **5**(1), p. 42-43.

Klusman, R. W., and H. Sauer. 1972. Some Possible Relationships of Water and Soil Chemistry to Cardiovascular Diseases in Indiana. **4**(7), p. 563-564.

Lange, I. M., and J. C. Avent. 1973. Ground-Based Thermal Infra-Red Surveys as an Aid in Predicting Cascade Eruptions: A Preliminary Report. **5**(1), p. 69.

Nilsen, T. H., and E. E. Brabb. 1973. Environmental Effects of Landslides in the San Francisco Bay Region, California. **5**(7), p. 751.

Runnels, D. D. 1973. Molybdenum in the Environment—Establishing Background Levels. 5(6), p. 508.
Saucier, R. T. 1973. Effects of the New Madrid Earthquake Series in the Mississippi Alluvial Valley. 5(7), p. 793.
Sowers, G. F. 1972. Foundation Settlement in Limestone Terrane. 4(7), p. 673.
Stewart, D. M. 1973. The Role of Seismic Station CHC in Earthquake Research in Southeastern United States. 5(5), p. 438.
Szucs, F. K., and Samuel Jakab. 1973. A Comparative Survey of Gastric Cancer Occurrences in Eastern Europe and the United States. 5(4), p. 356.

Editor's Comments on Papers 16 Through 20

In Paper 16, Paul F. Kerr describes the extraordinary phenomenon of quick clay, "by far the most mobile of all the common solid materials on the earth's surface," which produces sudden, severe mass movement. The movement can be initiated by a natural or man-induced vibration or jar that converts the material from solid to liquid. "It will then flow rapidly along any slope or almost no slope at all."

Kerr was associated with Columbia University from 1924 to 1972, and occupied the Newberry Professorship of Geology from 1952 to 1965. Recently he has worked in a research capacity with Stanford University. He is the author of several outstanding textbooks. Among his diverse subjects of research, his work on clays is particularly well known.

Paper 17 is a concise discussion of cataclysmic earthquakes in the upper Mississippi valley. The author's remarks summarize the salient aspects of this long series of major occurrences and draw pertinent conclusions. Osterberg notes that the damage to life and property was slight, largely because the region was sparsely populated and the structures seem to have been quite resistant. The effects of similarly severe earthquakes today would be a disaster, since we have paid little attention to earthquake-resistant construction. A relatively minor earthquake occurred in the northeastern United States in 1973, but we should not ignore the possibility of extreme damage to metropolitan centers on the eastern coast of the United States.

The highly destructive earthquake that centered on Managua, Nicaragua, in December 1972 has been discussed in Paper 18, a comprehensive and well-prepared study by a team of U.S. Geological Survey members. Portions of this excellent publication, which should be read in its entirety, are reproduced.

The area is identified as one of unusually high risk in terms of geologic hazards. It is important to observe that the area is endangered by both seismic and volcanic activity. The authors stress the great importance of taking heed of these hazards in the reconstruction of the area, something that is almost invariably ignored in places that are dangerous.

The authors, Robert D. Brown, Peter L. Ward, and George Plafker, have combined their experience on studies of earthquakes with consideration of environmental impacts and other hazards in earthquake-prone regions.

Only one article on volcanic activity (Paper 19) has been chosen as representing a connection with environmental problems. This brief discussion is concerned with the climatic effects, impact on agriculture, and stratospheric particulate loading resulting from the eruptions of Tambora and Krakatau in the nineteenth century. Krakatau is better known, but the author, Kenneth E. F. Watt, refers to the Tambora event as "the largest known volcanic eruption in recorded history." An interesting point is the reflection by Watt on air pollution created by human action as compared with the effects produced by volcanic eruptions.

John S. Stevenson and Louise S. Stevenson are the authors of Paper 20, a concise overview of medical geology. They speak of the recognition that man is affected by the transfer of elements to his body, and that the method of transfer is extremely complex. Worthy of note in their discussion is the observation that "medical statistics are ordinarily given for political rather than geologically significant areas" and that most statistics have to do with death rather than illness. They focus attention on dental health, rock dust diseases of the lungs, and calculi in the body.

The slow development of medical geology is typical of the entire range of subjects coming under the heading of environmental geology. Once again, deplorably, the initial steps taken to bring geologists into conjunction with environmental problems was the force of war, which produced the forward-looking field of military geology.

The authors occupy positions at McGill University. John S. Stevenson occupies the chair of Sir William Dawson Professor of Geology. Louise S. Stevenson is curator of geology in the Redpath Museum.

Reprinted with permission from *Sci. Amer.*, **209**(5), 132–138, 140, 142 (1963)

QUICK CLAY

It is a water-soaked glacial deposit that sometimes changes suddenly from a solid to a rapidly flowing liquid, causing disastrous landslides in parts of Scandinavia and Canada

by Paul F. Kerr

A farm owned by a man named Borgen lies on a plain near the Norwegian city of Oslo. A shallow ravine runs through the farm. On the afternoon of December 23, 1953, Borgen was walking along the ravine and noticed a small landslide at one point in the bank. The slide was so insignificant that even an expert in soil mechanics would hardly have given it a second thought. Farmers in Norway, however, are aware that the clay of their country sometimes behaves peculiarly. Borgen hurried home and moved his family to a neighbor's house some distance away.

The next morning the site of his home had vanished—house, barn and all. In its place was a circular crater some 650 feet in diameter and 30 feet deep. Many of the things that had stood on the site were deposited against a highway bridge more than a mile down the ravine, where in the morning Borgen found a cow and a horse plastered with mud climbing out of a soupy lake.

During the night a section of the ravine bank had suddenly opened, and a river of mud about 75 feet wide had poured down the ravine. The land on which the farmhouse stood had suddenly become fluid and flowed through the channel formed by the collapse of the ravine bank. The avalanche of mud and buildings had stopped only after piling up against the bridge.

The Borgen incident, reported in detail by the Norwegian mineralogist I. Th. Rosenqvist, is a characteristic example of the behavior of the substance known as quick clay. This singular material acts like certain gels (for example iron hydroxide) that, when suddenly jarred, promptly turn into a liquid. Such substances are called thixotropic, a Greek-derived term meaning "turning by touch." Quick clay, a natural aggregate made up of very fine mineral particles and water, behaves in a similar fashion. A liquid thixotropic gel, however, in time reverts to its solid state, whereas a quick clay does not.

Quick clay can be changed from solid to fluid by an earthquake, an explosion or even the jar of a pile driver. It will then flow rapidly along any slope or almost no slope at all. Quick clay can slide like an avalanche over flat land with a slope of less than one degree, rafting along heavy structures in its path.

The areas where quick-clay slides are reported most frequently are Norway and Sweden, which have several each year, and the valleys of the St. Lawrence, Ottawa and Saguenay rivers in Canada. It appears that some have also occurred

QUICK-CLAY SLIDE of June 7, 1957, at Göta in Sweden was one of the largest of modern times. Clay beneath a pulp mill suddenly became liquid and rafted most of the buildings and huge stacks of logs up to 220 feet toward the Göta River. Three workers were killed,

in Maine, Vermont and northern New York. The most damaging slide on record was one that took place at Verdal in Norway in 1893. It wrecked a settled area of three and a half square miles and killed 120 persons. Two similar disasters, which have been studied in more detail, occurred during the 1950's.

At 11:40 A.M. on November 12, 1955, an area within the town of Nicolet in Quebec suddenly began to slide away. In less than seven minutes the rapidly departing clay left a hole 600 feet long, 400 feet wide and 20 to 30 feet deep in the heart of the town. It carried away a school (the Académie Commerciale), a garage, several other buildings and a bulldozer. The whole mass flowed into the Nicolet River, and the schoolhouse wound up on the riverbank near a bridge, where it caught fire. Fortunately it was empty, its students having been dismissed for a midday recess, but three persons in the town lost their lives in the landslide.

A considerably larger slide had devastated the south end of the town of Surte in southern Sweden, a community of several thousand people, on September 29, 1950. The town sits in a flat valley on glacial clay deposits at least 120 feet deep. It seems that a pile driver, preparing a foundation for a new building, may have started the slide. At 8:10 A.M. a large section of the town, most of it residential, began to slip away toward the nearby Göta River. The huge mass, estimated to contain some 106 million cubic feet of soil and gravel and bearing 31 houses, flowed rapidly over the flat terrain. It picked up a paved highway and railroad roadbed in its path and finally plunged into the river, almost completely choking the channel. According to witnesses the entire slide took place in less than three minutes. It killed one person, injured 50 and destroyed the homes of 300.

Clearly the quick-clay phenomenon has practical as well as scientific interest, particularly for those concerned with the location of towns, airfields and farms. The Air Force Cambridge Research Laboratories, the Royal Swedish Geotechnical Institute in Stockholm, the Norwegian Geotechnical Institute in Oslo, the Mineralogical Laboratory of Columbia University and the National Research Council of Canada have all conducted studies of quick clay, focusing in particular on the mechanism of its movement and on possible ways of preventing slides.

Broadly speaking, a material slides when it has the following three characteristics: a layered fine structure, a high content of particles less than two microns in diameter and a high content of water. Even clay can hold its place on a fairly steep slope if it is dry. But when a hillside with a high content of clay absorbs much water, it tends to become unstable. Thus rainstorms may be responsible for most landslides. Asbestos, an unusual material, illustrates the action of water particularly well. Near the town of Coalinga in the coastal mountains of California remarkable landslides have occurred in a soft, powdery rock made up largely of short-fiber asbestos. These landslides, draped over the mountainside on slopes ranging from five to 20 degrees, have made scars hundreds of feet across and about a mile long. When the asbestos is dry, it remains firmly in place, but when it is wet, it may start to slide.

some buildings were destroyed, many were badly damaged and some survived almost intact. Most of the stacks of logs remained standing. The slide also blocked a canal by the river. In the mile-long, 800-foot-wide area of the slide the ground sank as much as 25 feet. This panorama, from the Royal Swedish Geotechnical Institute in Stockholm, is a mosaic of seven aerial photographs.

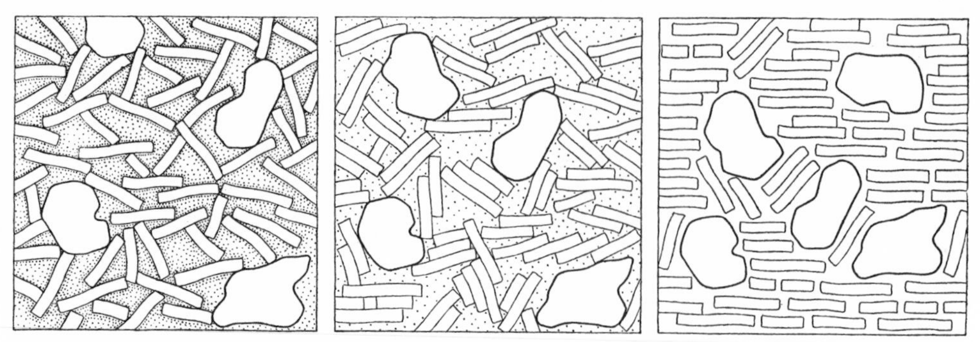

COLLAPSE OF QUICK CLAY is shown on microscopic scale in these schematic cross sections. Undisturbed clay (*left*) is thought to possess a "house of cards" structure. The "cards" are the flat bits of clay minerals. Irregular blobs represent sand grains and tiny dots are dissolved salt, which provides electrolytic "glue" for structure. Water (*color*) is being squeezed out as clay collapses (*middle*). Most of salt was leached out before collapse. Remolded or collapsed clay (*right*) contains very little water.

TWO SAMPLES OF SAME QUICK CLAY show startling contrast. Column of undisturbed clay (*left*) holds 11 kilograms (24 pounds). It can support 2,100 pounds per square foot of surface. Another piece of the same clay pours like a liquid after being stirred in a beaker. No water was added. Demonstration was made by Carl B. Crawford of the National Research Council of Canada.

Quick clay is an extreme case; it is by far the most mobile of all the common solid materials on the earth's surface. It has both a high water content and a mineral texture that allows it to flow with the utmost ease. A mass of quick clay that has lain undisturbed for thousands of years can be jarred into motion by any sudden shock.

Investigation with refined methods has elicited a good deal of information about the composition of quick clays and the history of their formation. With the electron microscope one can see that the clay is made up largely of flaky particles less than two microns in diameter. X-ray examination shows that these flakes are crystals of various silicate minerals. They have been identified mainly as illite, montmorillonite, chlorite and kaolinite.

The quick clays in Norway, Sweden and Canada were formed during the last advance of the great ice sheets of the Pleistocene epoch. They are essentially pulverized rock and other fine material that the glaciers ground off the land and deposited on what was then the sea floor around the Scandinavian peninsula, which was much smaller then than it is now, and in the Champlain Sea, which covered Canada's present St. Lawrence Valley. After the retreat of the ice some 10,000 years ago, the submerged areas gradually rose above water; in Scandinavia the uplift has raised some of the old sea-floor deposits to as much as 650 feet above sea level. Evidence of the elevation of the land can be seen plainly along the Norwegian coast, where rock quarries of early man that once were at the water's edge now stand well above sea level.

Thus much of the land around the Scandinavian peninsula and in eastern Canada contains strata of clay that were originally laid down on the sea bottom. The quick clays are recognizable by their several distinctive properties. They are generally dark blue-gray when wet. Their sensitivity—that is, the ease with which they can be triggered to flow—depends on four main physical features: (1) usually more than 50 per cent of the solid matter in the clay mass consists of particles less than two microns in diameter; (2) the fine particles are not coagulated but are loosely dispersed through the mass; (3) the water content of the mass is often higher than 50 per cent by weight; (4) the salt content of this water is comparatively low (usually less than five grams per liter, whereas that of sea water is about 35 grams per liter). The amount of salt is

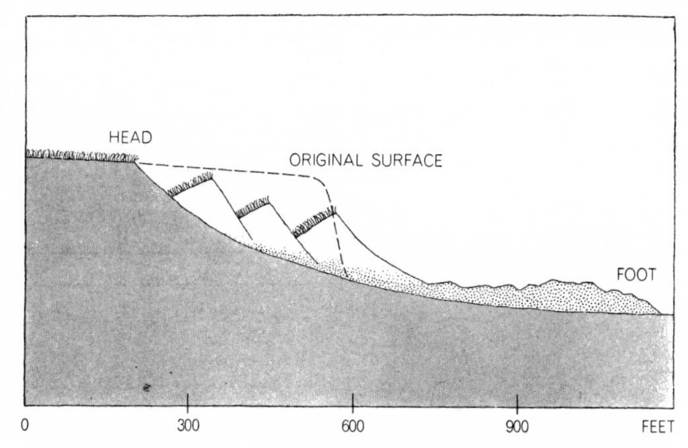

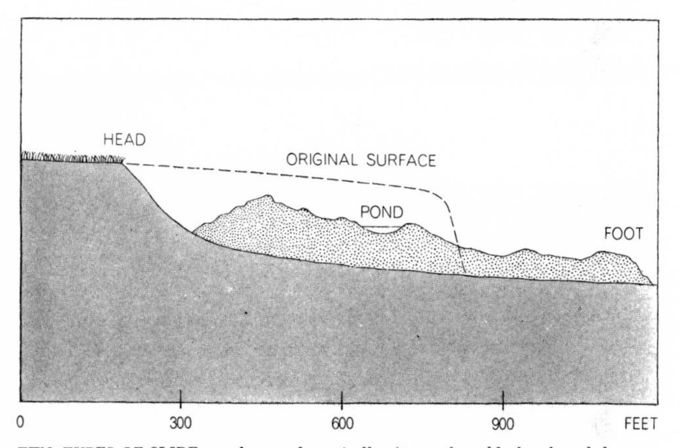

TWO TYPES OF SLIDE are shown schematically. At top three blocks of earth have rotated in a clay-water slurry. In the other slide total liquefaction has occurred and water forced out of the clay has formed a pond. The vertical dimension is exaggerated five times.

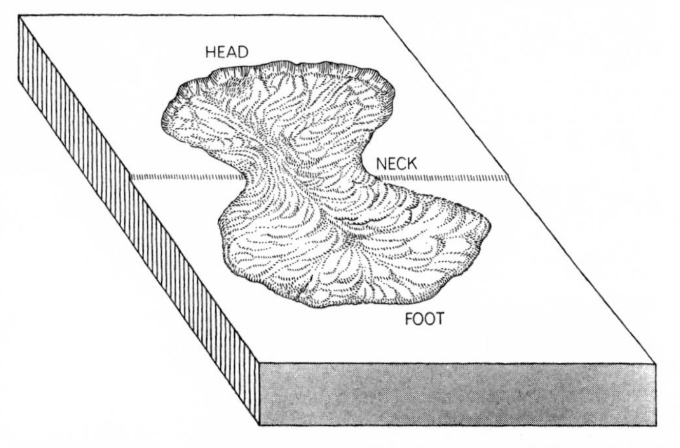

TYPICAL SLIDE has a figure-eight appearance. Material that leaves a hollow at the head region piles up at the foot. The low cliff at the neck is often the bank of a river.

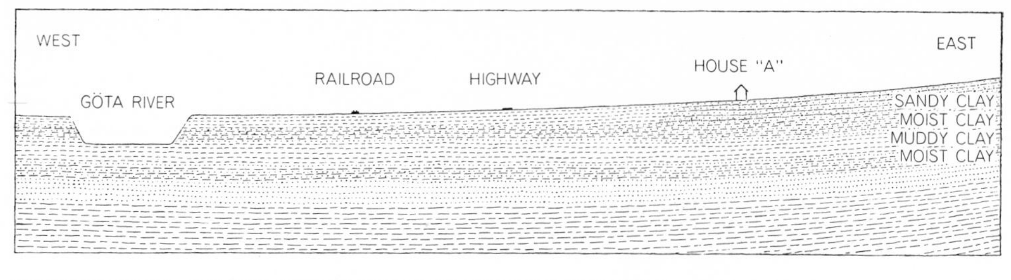

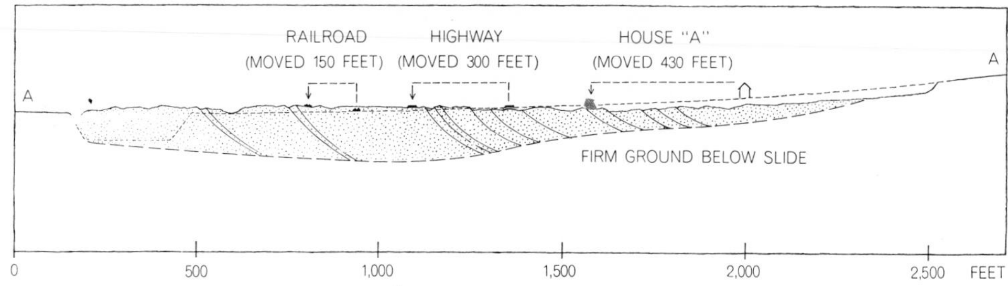

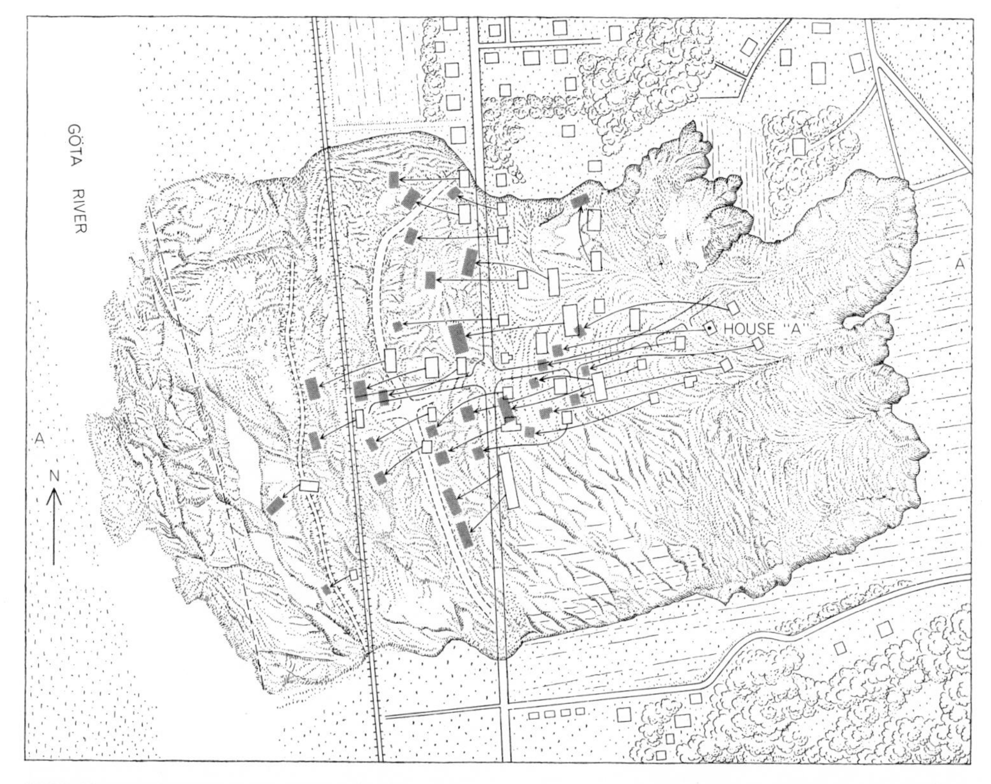

GREAT SLIDE AT SURTE on the Göta River in Sweden took place in 1950. Cross sections along broken colored line *A-A* in vertical view at bottom show low slope and other conditions before slide (*top*) and after slide (*middle*). Slide involved both liquefaction and rotation of soil. Houses were rafted as shown by arrows in bottom diagram. Highway and railroad also moved. Occupants of houses were aware of rocking, rising and falling motions but not of sliding. As the structures came to rest at crazy angles, one house split in half and another overturned. The entire slide took less than three minutes, moved 106 million cubic feet of soil and gravel, killed one person, injured 50 and destroyed the homes of 300. Water from the clay formed several large ponds.

important because electrolytes in a mass of clay tend to bind the clay particles together; consequently, as salt is leached out of the clay, it becomes more thixotropic.

Each of these features has been examined closely in the laboratory. Electron micrographs of samples of Norwegian quick clay have shown that in sensitive clay the fine particles tend to be separated and dispersed, whereas in clay that has flowed and become remolded the particles are packed closely together in aggregates. Justus Osterman and his co-workers at the Royal Swedish Geotechnical Institute have pointed out that the particles in clay tend to be dispersed by the action of certain natural chemicals, such as the organic acids in peat bogs and some alkaline earths, in much the same way that the mud used in oil-well drilling is kept loose and fluid by the addition of tannic acid.

Salt, as we have noted, has the opposite effect: it acts to hold the clay together. The quick clays laid down in ocean bottoms started with a high salt content, but in the millenniums since they were lifted above water much of the salt has been leached out of them, particularly at high elevations, as Rosenqvist has found in Norway. A few quick-clay bodies were originally deposited in fresh water; these, of course, contain little salt. It has been shown that quick clays with a low salt content rank high in sensitivity to thixotropic disturbance.

Apparently the most critical of all the factors is water content; the more water in the mass, the higher its sensitivity. (The index of sensitivity is technically taken to be the difference in shear strength before and after the clay has flowed; this ratio is sometimes as high as 100 to one.) A quick clay commonly contains so much water that after a slide it often leaves small ponds in its wake.

The area of the Nicolet slide in Quebec was found afterward to be heavily charged with water. A sewer in this area had broken down a few months before, and the subsequent accumulation of sewer water in the clay may well have been responsible for the slide. A heavy accumulation of water was also found in the slide area at Surte in Sweden, one of the most thoroughly studied of all quick-clay avalanches.

The Surte terrain is a classic example of a build-up of marine glacial clay deposits. In late glacial times, when the old valley of the Göta River in southwestern Sweden was covered by ocean, the valley became filled with an extremely fine flour of rock particles scoured down from the hills by the melting ice. Geological processes later raised the accumulated strata above sea level. Over the centuries river water permeated the clay strata and leached out much of the sea salt. The combination of high water content and removal of the salt left the clay mass in an unstable condition, and it would appear that the hammering of a pile driver, used in Surte for the first time in 1950, in addition to the normal train and highway traffic, were sufficient to start the clay sliding.

Can quick-clay slides be prevented? So far this problem has received much less attention than the causes of slides and the nature of the quick clays themselves. Some investigators believe that control of these slides would be economically impracticable; they point out that the cost of exploring and applying preventive treatment to a farm area, for example, might be greater than the value of the land. Studies of the problem of preventing slides are nevertheless being conducted, particularly by Osterman's group at the Royal Swedish Geotechnical Institute.

At least two obvious possibilities invite exploration. One is control of the water content of the quick clay. If percolation of water into the clay can be prevented, this may keep the water content below the critical level and also

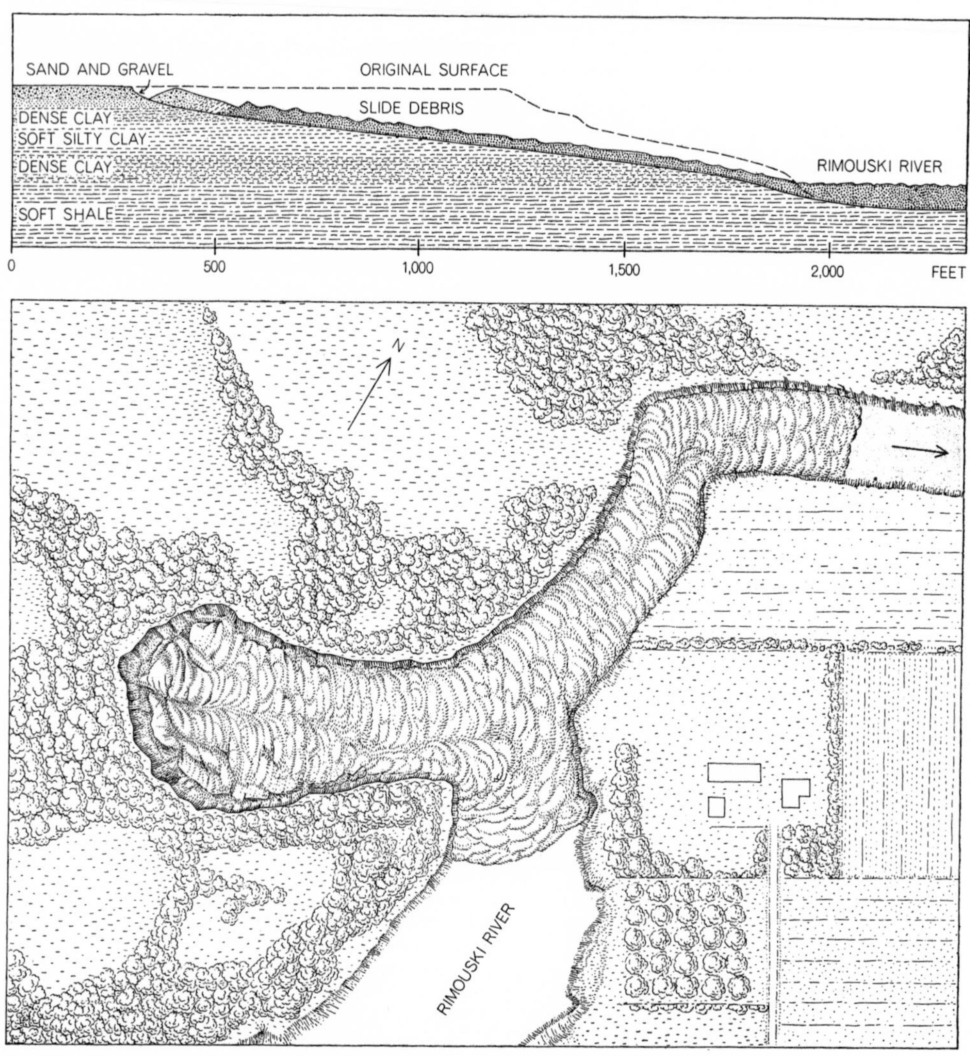

SLIDE IN QUEBEC on the Rimouski River was caused partly by the 14-degree slope (*cross section at top*), much steeper than slopes usually associated with quick-clay slides. A layer of soft, silty clay heavily charged with water formed a slip surface. The debris filled the river for several thousand feet and created a temporary lake. Hummocky surface is typical of clay slide.

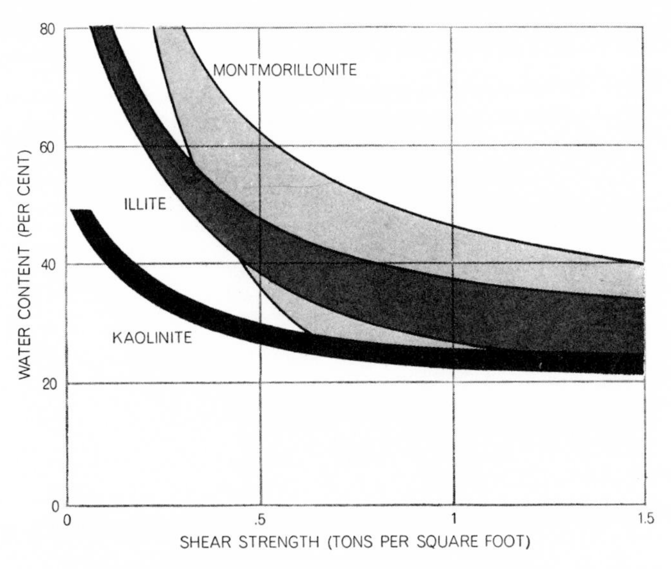

SHEAR STRENGTH of minerals common to quick clay declines rapidly as water content rises. Bands indicate the range of strengths found at the various per cents of water.

stop the leaching out of salt and other binding electrolytes. Therefore drainage of both the surface and the underground water in quick-clay areas might be effective as a slide-preventing measure.

The other possibility that is being tested is artificial injection of electrolytes into the quick clay. Applying salt and lime by means of shallow-bore holes in test plots, experimenters have had some success in improving the cohesion and stability of the clay.

Whether or not effective and practical methods of control will eventually be worked out, it is important to know where the dangerous areas are. Considerable work in mapping them has already been done. Fortunately aerial photography is helpful, because it shows up the scars of quick-clay slides. The Geological Survey of Canada has made photographs and maps of more than 50 slide areas in the valleys of the St. Lawrence, Saguenay and Ottawa rivers. Geological agencies in Norway, Sweden and the U.S. also have made aerial photographs of slide areas. These are not, however, the only countries in the world where quick clay is a matter of concern. It appears that southern Chile and possibly the Peruvian Andes may also have such deposits.

Furthermore, some of the well-known "turbidity currents" on ocean floors seem to be akin to quick-clay slides [see "The Origin of Submarine Canyons," by Bruce C. Heezen; SCIENTIFIC AMERICAN, August, 1956]. These flows of mud, sand and rock debris, which have been studied extensively by Bruce C. Heezen, Maurice Ewing and David B. Ericson of the Lamont Geological Observatory of Columbia University, have a great deal in common with the clay flows on land. It was once believed that the ocean slides occurred only on the sea-floor slopes off the continents, but it now appears that they take place also in the flat ocean basins and even in some of the deepest oceanic troughs.

On November 18, 1929, the transatlantic telegraph cables between the U.S. and Europe began to go out one after another. Within a period of 13 hours 17 minutes, 12 submarine cables stopped transmitting. The interruption followed a sharp earthquake whose epicenter was placed at the Grand Banks south of Newfoundland. The cable breaks were attributed at the time to the shock of the quake. Many years later, however, the Lamont investigators found that the breaks must have been caused by a massive flow of mud started by the earthquake.

These studies have established reasonably well that the area of flow was some 230 miles wide and proceeded for about 400 miles—not merely down the slope of the continental shelf but far along the flat ocean floor. At the last cable severed by the slide the slope of the floor was less than one degree, yet the flow was rushing along at no less than 14 miles per hour (judging from the known time and distance between this cable break and the one before).

Apparently the main body of the flow was fine clay, bearing along on its surface a great quantity of sharp sand and fine gravel. Cores from a number of places on the deep sea floor show that it contains a high proportion of very fine clay, with the minerals illite, montmorillonite and chlorite predominating. These are the same materials that predominate in the quick clays on land. Furthermore, it seems more than a coincidence that the oceanic flows, like those of quick clay, can be triggered by an earthquake or a similar shock, can flow rapidly over almost flat terrain and are capable of carrying along immense masses of heavy material.

It seems likely that much of the clay of the ocean floor is in a sense thixotropic and is sometimes set in motion by the same mechanisms as the quick-clay slides on land. This is an interesting hypothesis that deserves exploration; the movement and deposition of material on the ocean floor is one of the main themes in the formation of the solid surface of the earth.

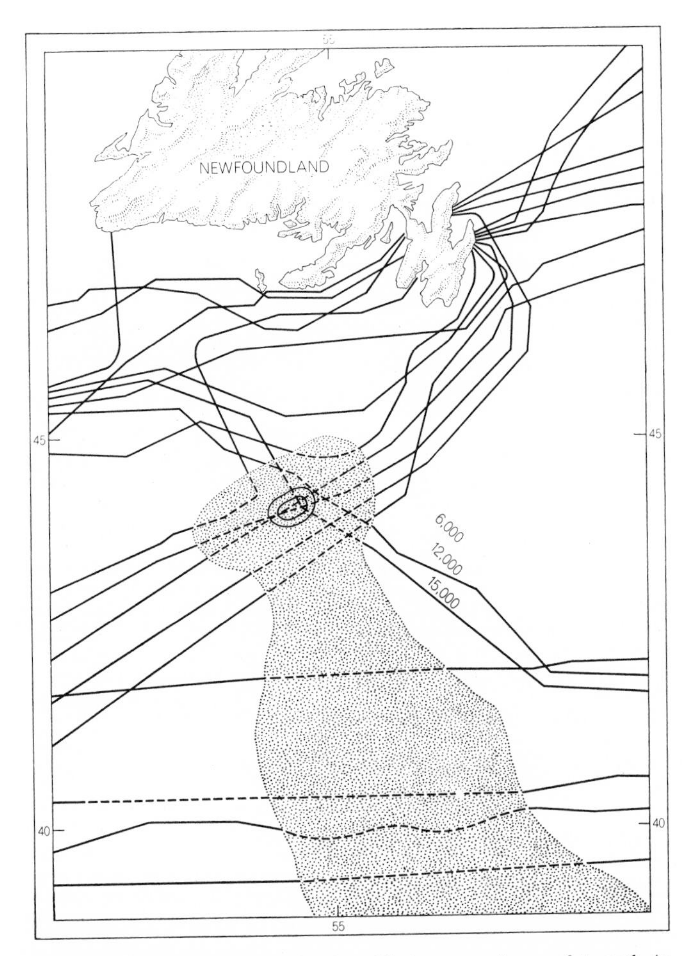

UNDERSEA SLIDE in 1929 cut 12 submarine cables in sequence from north to south. An earthquake triggered the slide. Its epicenter is marked by the concentric rings. Region of cable damage outlines probable slide area (*stippling*). Changes in color shading denote depth. Much of area had slope of less than one degree, but sliding material traveled rapidly.

ADDENDUM BY PAUL F. KERR

Paper 16 was printed by *Scientific American* in November 1963. The following April the Good Friday earthquake at Anchorage, Alaska, caused a number of devastating quick clay landslides. In the Turnagain area alone 75 homes were badly damaged or destroyed.

The strata of the Anchorage area are largely flat lying and of glacial origin. A sea cliff 50 to 70 feet high follows the south border of Knik arm of Cook Inlet. Quick clay crops out at the base of the sea cliff and extends inland. Overlying the quick clay are layers of silt, sand, and gravel. The underlying quick clay stratum is about 10 to 20 feet thick.

A large part of the material beneath the Turnagain area was water saturated. The coarse sands and gravels tended to cohere. However, the wet clays were thixotropic. When subjected to an earthquake shock, the clay liquified and flowed like soup.

The soupy clay mixture moved underground beneath the overlying glacial materials. It passed under the elevated flats of Turnagain and emerged along the shoreline. The shore was covered with a mass of sticky mud. At the same time large, block-like masses of sand and gravel broke away from the sea cliff causing the cliff to retreat inland. Choice houses from the land behind the sea cliff were caught in this movement and rafted hundreds of feet seaward.

A number of other slide areas caused by clay mobility formed but Turnagain was the most outstanding quick clay slide at Anchorage. Other quick clay areas are recognized in the United States and Canada. A publication of the U.S. Geological Survey calls attention to numerous slides along the Columbia River in Washington. Glacial clays occur along the river for a considerable distance both upstream and downstream from Coulee Dam. In excavating for a powerhouse extension at the dam a large clay mass was uncovered. Because of the tendency of such clay masses to slide and the importance of the construction, it was elected to remove the clay

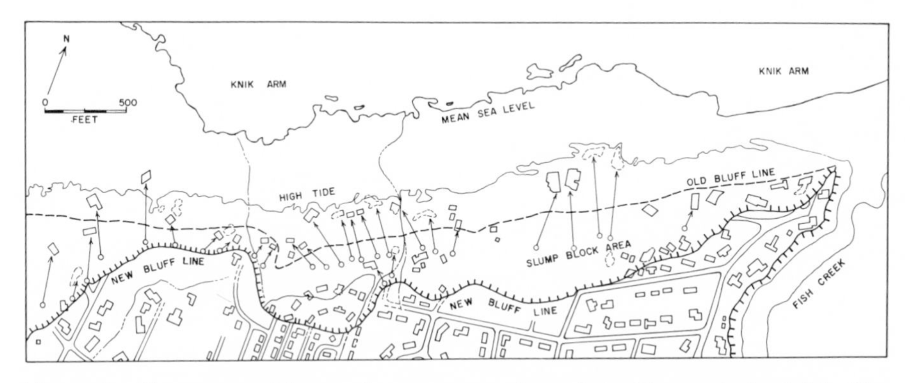

Figure 1 Movement of houses in eastern portion of the Turnagain landslide. The new bluff line marks the present north boundary of the residential area [*Eng. Geol.*, 2(4), 1968]. Copyright © 1968 by Elsevier Scientific Publishing Company.

entirely. Unfortunately, economic factors often prevent such a solution of a landslide problem.

It seems likely that quick clays may be found in northern latitudes around the world. Occurrences are known in the U.S.S.R., Scandinavia, Canada, and the northern United States. Thick clay strata in the St. Lawrence Valley in Canada have produced many slides.

Often clay masses causing landslides occur in open fields or uninhabited areas. Occasionally, however, as at Nicolet or along the Saguenay River in Quebec, Canada, major damage has resulted.

Mineralogical studies of the clay minerals in a number of quick clay slides have been made. Illite and chlorite are common constituents. These flaky minerals are usually less than 2 microns across. Fine grains of quartz, feldspar, and ferromagnesian silicates are likewise common but range widely in amount. Where abundant the clay is less sensitive or prone to slide action. In general, quick clays are of glacial origin. Some have accumulated on land, others in shallow waters along marine margins. Some are of freshwater deposition, others have formed under saline conditions. The fine clays are a product of glacial abrasion.

Many clay-bearing areas that yield landslides have formed under conditions differing from those that yield quick clays. These provide different problems. The clay minerals differ. Montmorillonite and illite are common, but other layer lattice minerals are also slide prone when wet. Slopes range from almost flat to steep mountainsides. Movements may be gradual or rapid. In such slides the role of clay minerals may become quite complex. Treatment of these slides is beyond the scope of this discussion.

17

Reprinted from *The Testing World,* No. 21, 11–12 (1967–1968), published by Soiltest, Inc.

The Earthquakes that Rearranged Mid-America

Jorj O. Osterberg
Professor, Civil Engineering,
The Technological Institute,
Northwestern University

The violence of the New Madrid earthquakes makes them rank with history's severest shocks in Lisbon, California, Tokyo, and Alaska, but few seismic records were kept on the U.S. Western frontier in 1811 and 1812.

Despite the great notoriety of the San Francisco earthquake of 1906 and the other San Francisco, Long Beach, and Los Angeles earthquakes of 1933 and '34, the greatest earthquake in North American history did not occur in California and was not one earthquake at all but a series of quakes.

These great shocks, 2,000 miles from the famed San Andreas fault of the West Coast and far from any great mountain system, rumbled through the rolling wooded country in the heart of the United States in late 1811 and early 1812. They were the New Madrid earthquakes, so named because their epicenter was near New Madrid, Mo., a Mississippi River town in Southeastern Missouri.

A Rare Recurrence of Quakes

There were three major shocks in the New Madrid series, the first on December 16, 1811, and the other two on January 23 and February 7, 1812. But as is usual in major earthquakes, there were many intervening shocks. These were less severe, but only by contrast to the three biggest quakes. What was quite unusual about the New Madrid earthquakes was the repetition of major shocks. The recurrence of quakes in a period of months or even years is quite rare.

The New Madrid earthquakes, gigantic disturbances, had a magnitude of XII, as expressed on the Mercalli Intensity Scale of I to XII, meaning that they were the largest, most severe type of earthquakes on record.

These quakes were so great that they caused major topographic changes over an area of 50,000 square miles, and badly shook a total area of more than 1,000,000 square miles. The earthquakes were felt over two-thirds of the U.S. Shaking was recorded in New Orleans, Boston, New York, Colorado, Montana, and as far north as northern Canada. Repeated strong shocks rocked Louisville, Ky., 200 miles away.

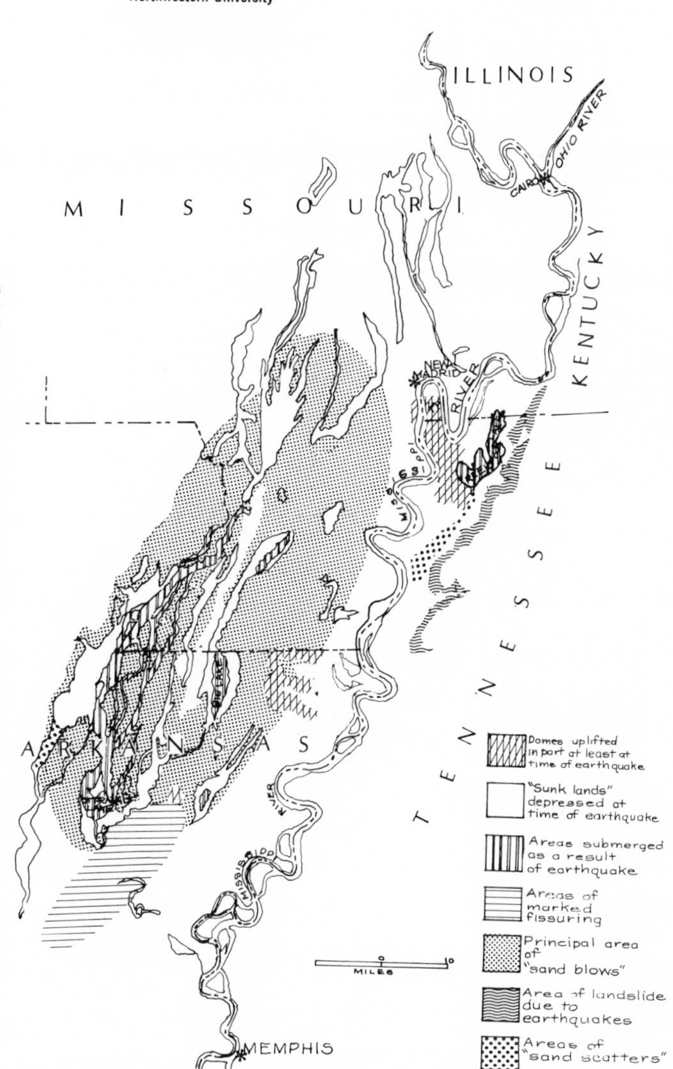

Center of New Madrid earthquakes, the region of marked earth disturbances, may have covered as much as 50,000 sq. mi.—the size of New York State—and ranged 100 mi. north to south from near Cairo, Ill. to Memphis, and 50 mi. east to west, from the Chickasaw Bluffs in Tennessee to Crowley Ridge in Arkansas. Map key shows various earthquake phenomena in area.

If New Madrid earthquakes had struck populated areas such as Anchorage, Alaska, shown here smashed by the 1964 Good Friday quake, loss of life and financial losses would have been huge.

In Cincinnati, Ohio, 400 miles distant, the shocks toppled chimneys. In Washington, D.C., 800 miles away, the shocks registered IV on the Mercalli Intensity Scale, that is, strong enough to awaken sleepers; rattle dishes and windows; stress walls till they emit cracking sounds; and rock buggies and wagons in the street.

In the area of greatest destruction, south of New Madrid, a whole region 150 miles long and 40 miles wide in Tennessee and Arkansas sank three to nine feet. Mississippi River water rushed into the depression.

The Earth Waved Like the Sea

The New Madrid earthquakes began suddenly without any warning at 2:00 a.m. December 16, 1811.

The ground rose and fell as earth waves, like swells of the sea, passed across the surface tilting trees, interlocking branches, and opening up the soil in deep cracks. Landslides swept down the steeper slopes. Along the Mississippi the shocks crumbled banks at many places, and what the tremors did not dislodge, the waves sweeping the river did.

The great waves on the Mississippi swamped many boats and washed others ashore. Whole islands disappeared, and new islands appeared where none existed before.

Shocks for a Year and a Half

During the entire days of December 16 and December 17, 1811, the shocks continued at rather short intervals, but gradually diminished in intensity. They occurred sporadically until January 23, when there was another severe shock, almost as intense as the first. Then, on February 7, after about two weeks of virtually no shocks, there were several tremendous upheavals that equaled or even surpassed the initial shock of December 16. For several days the earth was in constant tremor, and for fully a year from that date small shocks occurred at intervals, then gradually died out.

Record-keeping at that time was so unsystematic that the total number of shocks is not known. One observer in Louisville actually counted and recorded more than 250 shocks, but then gave up, apparently because he couldn't be bothered to continue. A geologist who has made an exhaustive study of the available eyewitness reports and written records of the New Madrid quake has estimated that there were approximately 1,900 shocks strong enough to be felt 200 miles away in Louisville during that total 1½-year period of the New Madrid earthquakes. Eight of these are considered to have been violent shocks of magnitude X.

After the Fissures, Fishing

Probably the most spectacular physical changes caused by the earthquakes were in that huge 150-mile-long area that caved in south of New Madrid. Local depressions 20 or 30 miles long by four or five miles wide were formed and some of them flooded. Today two of these are large lakes popular with fishermen, Reelfoot Lake in Tennessee and St. Francis Lake in Arkansas.

The flexing and waving of the earth's surface during the quake broke the earth open in great fissures, some of these also filling with water.

Rumbling, Whistling, Eerie Noises

This was a noisy earthquake. Not only was there the rumbling of the shifting earth and rock, but as the water and sand gushed up through cracks and holes in the ground they pushed out air and other gas that whistled and made eerie noises.

Along the Mississippi and its tributaries, as the cracks opened in the earth they quickly flooded, and as they closed again this water too geysered high into the air with large volumes of sand, mud, and gas.

A horseback-riding witness to the phenomena of the sand blows and geysers reported:

"There was a blowing up of the earth with loud explosions. It rushed out in all quarters bringing with it an enormous quantity of carbonized wood, reduced mostly to dust which was ejected to a height of from 10 to 15 ft. and fell in a black shower mixed with the sand which its rapid motion had forced along."

150 Miles of Pollution and Sand Floods

Another observer noted:

"The sulphurated gases that were discharged during the shock tainted the air with their noxious effluvia and so strongly impregnated the water of the river to the distance of 150 miles below that it could hardly be used for any purpose for a number of days. Many hundreds if not thousands of acres of land were made useless by deposits of sand spewed out that flowed over the land, thus destroying the trees and covering the fertile topsoil. These are referred to as sand scatters. The sand is in some places up to five feet in thickness but generally only one or two feet."

Not only did the earth cave in, it also heaved up in domed areas. These domes, some of them 15 miles wide and 20 ft. high, formed islands in some instances or uplifted land formerly in the flood plains to new high and dry levels.

A Huge Burden of Sand and Gravel

What caused the New Madrid earthquakes? The truest answer is that the cause is not known.

Probably what happened in 1811-1812 is that the huge quantities of sand and gravel that had been deposited during the last ice age and washed down the Mississippi valley resulted in such a heavy overburden that it warped the underlying limestone bedrock. Then, perhaps through a preexisting fault in the bedrock or the creation of a new fault, the adjustments of the crust to this heavy overburden caused one or more of the quakes.

The Dead? A Half Dozen, A hundred, A . . .

It is difficult to know how many lives were lost in the New Madrid earthquakes. The population in the area was small and widely dispersed. Records show that at least half a dozen were known dead, but this figure very well could have mounted to 100 or more. Their deaths could have gone unnoticed because of the catastrophic nature of the quakes.

Undoubtedly there would have been more fatalities if the structures in this part of the country had been more "advanced," that is, built of stone, brick, and plaster. Fortunately they were largely log and frame houses that rolled with the punch of the shocks rather than crumbling around their inhabitants and crushing them.

In this perhaps there is a lesson to be learned.

We should consider what would happen if a "New Madrid" earthquake occurred today—in one of the many parts of the world that is thought, by popular conception, to be earthquake-free.

In many of these areas there are none of the building-code provisions for earthquake-resistant building construction that are found in regions frequently affected by earthquakes, such as California and Japan. The clear possibility of serious quakes is such supposedly earthquake-free areas makes such provisions highly desirable.

18

Reprinted from *U.S. Geol. Surv. Prof. Paper 838,* 1-2, 26-34 (1973)

GEOLOGIC AND SEISMOLOGIC ASPECTS OF THE MANAGUA, NICARAGUA, EARTHQUAKES OF DECEMBER 23, 1972

By R. D. Brown, Jr., P. L. Ward, and George Plafker

ABSTRACT

The Managua, Nicaragua, earthquake of December 23, 1972 (Richter magnitude of 5.6, surface-wave magnitude of 6.2), and its aftershocks strongly affected an area of about 27 square kilometers centered on Managua. Within this area, over 11,000 people were killed and 20,000 were injured. About 75 percent of the city's housing units were destroyed or rendered uninhabitable leaving between 200,000 and 250,000 people homeless, and property damage exceeded half a billion dollars. As a consequence, the economy and government of the city, and to a large extent the entire country, were severely disrupted.

Surface geology shows that there are at least four subparallel strike-slip faults spaced 270 to 1,150 meters apart in the Managua area that slipped in a predominantly sinistral (left-lateral) sense during the earthquake. Aftershock studies show that at least one of these northeast-trending faults extends from the surface to a depth of 8 to 10 km (kilometers) over a maximum length of about 15 km. The faults are mappable on land for 1.6 km, 5.1 km, 5.9 km, and 2.7 km; aftershock data indicate that faulting extends at least 6 km northeast of the city beneath Lake Managua. Horizontal displacements vary, with the maximum aggregate sinistral slip ranging from 2.0 to 38.0 centimeters. There is also a local small down-to-the-southeast vertical component of slip on three of the four faults. The nature and distribution of the surface faulting are consistent with a tectonic origin for the earthquake.

The extensive destruction and loss of life in the Managua area were caused by a combination of the following factors: (1) occurrence of the earthquake on faults directly beneath the city, (2) poor behavior of structures, chiefly tarquezal (wood frame and adobe) and masonry, during strong seismic shaking, and (3) direct displacement of structures, streets, and utilities by faulting. The historic record of seismicity and geologic evidence of active Holocene faulting and volcanism together show that Managua is an unusually high risk area in terms of geologic hazards and that these hazards should be a primary consideration in evaluating reconstruction of Managua.

INTRODUCTION

Managua, Nicaragua's political capital, its business and industrial center, and by far its largest city, was struck by three moderate-sized earthquakes within less than an hour in the early morning of December 23, 1972. The earthquakes and related surface faulting severely damaged the central part of the city, interrupted essential services, and, by their effect on Managua, severely disrupted the entire Nicaraguan economy. The first and largest earthquake was felt at 12:30 a.m., local time. It was assigned a Richter magnitude, M_b, of 5.6 (surface-wave magnitude, M_s, of 6.2) by seismologists of the U.S. National Oceanic and Atmospheric Administration (National Oceanic and Atmospheric Administration, 1973). The two largest aftershocks were felt at about 1:18 a.m. and 1:20 a.m. Both were smaller (M_b, 5.0 and 5.2) than the main shock, but were large enough to cause substantial additional damage. According to eyewitness accounts, many buildings that were structurally weakened but still standing after the main earthquake suffered additional damage or collapsed during these aftershocks.

The earthquake sequence killed over 11,000 people and injured another 20,000, caused more than half a billion dollars property damage, and destroyed or rendered uninhabitable 75 percent of the city's housing units leaving between 200,000 and 250,000 people homeless out of a total Managua population of around 500,000. Interviews with residents of Managua indicate that many left their homes and moved into the streets as the shaking from the first earthquake subsided. Many of these people were still in open areas when the aftershocks were felt and thereby escaped possible injury or death in the further collapse of buildings. Aftershock activity continued for weeks after the initial earthquake, with the frequency and magnitude of aftershocks progressively diminishing with time. All of the significant damage resulted either from the first three shocks or from fires that followed shortly thereafter. The earthquakes were of moderate size but caused extensive damage because (1) they occurred at shallow depth under the city, (2) at least four surface faults broke in and near Managua, and (3) most buildings had little resistance to seismic shaking.

An accurate evaluation of the geologic hazards and the possibility of future earthquakes like those of December 23 is critical to future development plans. Such evaluations have obvious applications in formal planning and in plan implementation by governmental bodies. Less obvious perhaps is the degree to which such evaluations are used by financial institutions, insurance companies, and by business and industry. In recent years, in various parts of the world, geologic knowledge concerning recognized active faults and other clearly identifiable geologic hazards has been increasingly applied by private industry to decisions on site selection, mortgage loan evaluation, and the setting of insurance rates. These nongovernmental decisions can profoundly affect the pattern of growth and development simply by directing or influencing the flow of investment capital away from high-risk sites and towards those where the level of risk is deemed more acceptable.

Much current planning, both at governmental and private levels, reflects the viewpoint that earthquake safety in modern cities involves designing for the interaction of two complex systems: the manmade system that is the city itself, and the natural system consisting of the geologic processes that cause or accompany a major earthquake. Successful planning for earthquake safety involves far more than the prevention of structural failure in buildings. It should include, as well, ensuring the integrity of communication lines, water service, sanitation facilities, and emergency services such as police, fire, and hospital facilities. Such planning should also recognize that massive economic loss will recur in accordance with the recurrence rates of catastrophic geologic processes. Such losses are largely independent of structural design and construction practices, which are directed primarily to the safety of human lives, at least insofar as earthquake-resistant characteristics are concerned. Comprehensive urban planning for earthquake safety depends first of all on a clear understanding of the processes that accompany earthquakes and how these processes may affect the works of man.

[*Editor's Note:* Certain figures, tables, and some text material have been omitted at this point.]

SETTING OF THE EARTHQUAKES

REGIONAL TECTONIC RELATIONS

Managua lies within the trend of volcanic and earthquake activity that girdles the Pacific Ocean basin and that popularly is referred to as the "Pacific Ring of Fire." According to modern geologic theory, the earthquakes and volcanic activity around the Pacific result from relative movement between large plates of the earth's crust. Certain boundaries between such mobile plates are defined by long, linear trenches in the seafloor, well-defined zones of earthquake activity that are shallow near the trench and deepen toward adjoining continental areas, and linear chains of volcanoes that parallel both the trench and the trend of the zone of earthquakes. All of these characteristic features occur in Central America and have been active there for several million years (Dengo, 1968; McBirney and Williams, 1965). Clearly, the historic volcanism and earthquakes are natural and continuing processes that man must understand and plan for if he wishes to live and prosper here.

Major geologic features in Central America are the Middle America Trench, a pronounced linear feature 4 to 5 km deep along the Pacific Coast from central Mexico to Costa Rica (shown on index map, fig. 1), and the chain of young andesitic stratovolcanoes extending from western Guatemala to Panama. Most earthquake activity in Central America is in a belt about 200 km wide that parallels the trench. Where the focal depths of these earthquakes can be well determined, they exhibit a systematic distribution—shallow near the trench and deeper with increasing distance towards the northeast (Molnar and Sykes, 1969). The zone of earthquake activity thus dips about 45° NE. and extends from very near the surface at the Middle America Trench to more than 170 km deep at points farthest from the trench. In Nicaragua, earthquake activity related to this dipping zone extends as far inland as Lake Managua and Lake Nicaragua. The line of volcanoes that extends through most of Nicaragua approximately follows the northeasternmost limit of earthquake activity. Earthquakes along this zone since 1963, when the data are most complete, have ranged up to magnitude 6 (National Oceanic and Atmospheric Administration, 1973), but Gutenberg and Richter (1954) report some events as large as magnitude 7.7 in the period since 1913. Because Managua lies 100 to 200 km above this zone, even large earthquakes are unlikely to cause severe damage, although shallow earthquakes in this zone could cause damage in the Pacific coastal areas of Nicaragua.

The northeast-dipping zone of earthquake activity marks the boundary between two crustal plates. The Caribbean plate on the northeast includes most of Central America and extends northeast into the Caribbean. The Cocos plate on the southwest extends into the Pacific Ocean from the Middle America Trench. Geologic and geophysical evidence suggests that the Pacific, or Cocos plate, is moving relatively towards the northeast and is slowly being driven beneath the Caribbean plate along the plate boundary.

The Managua earthquakes of December 23, 1972, were at much shallower depths than the inferred crustal boundary between the Cocos and Caribbean plates, and the observed surface faulting, described in this report, exhibits a much different geometry than that of the plate boundary. For these and other reasons, it is unlikely that the December 23 earthquakes are a simple and direct result of relative plate movement between these two major crustal blocks. More likely they are caused by relatively shallow adjustment to accumulating crustal strain within the southwesternmost part of the Caribbean plate. This interpretation is favored both by the historic record of shallow-focus earthquakes in the Managua area and by the surface trend of the volcanic chain which passes through the Pacific coastal part of Nicaragua. The line of recent volcanoes in Nicaragua exhibits a marked bend or offset to the south in the segment between the volcano Momotombo on the northwest shore of Lake Managua and Masaya Caldera to the southeast of Managua. Detailed crustal structure and geology are not known well enough in the Managua area to specify the relations between the plate boundary, the line of volcanic activity offset to the south in a dextral sense, and shallow-focus earthquakes like those of December 23 with sinistral offset of the ground. A close relationship between all three, although still unproven, is an attractive hypothesis for testing and studying.

PHYSIOGRAPHY

The nature of the land surface in and around Managua provides important clues both to the geologic history of the area and also to the kinds of damage that may be expected in future earthquakes. Many of the surface effects of the December 23 earthquakes are likewise related directly to easily observed topographic features.

Much of the city of Managua and most of the surrounding areas affected by the earthquakes are on a surface that dips a few degrees towards the north. A few north-flowing washes drain this surface and feed into Lake Managua, but all are small and none are incised more than a few tens of meters into the surface. More deeply incised ravines are common further south, however, in the upland area lying west of Masaya Caldera. Except near the Chiltepe Peninsula, similar low relief is also found along the shoreline of Lake Managua, and at most places near Managua the lake appears to be very shallow for considerable distances offshore.

This gently north-dipping surface is interrupted in several places by low hills, most of which are clearly of relatively recent volcanic origin. Examples include Tiscapa near the south edge of the city, the hill enclosing Lake Asososca on the west, and the ridgeline on which the Nejapa pits southwest of Managua are located. Few of the hilly areas rise more than about 100 meters above the general surface, and few exhibit steep slopes. Steep slopes are found, however, in the crater walls at Tiscapa, Asososca, and in most of the other interior depressions of volcanic origin.

Several lines of evidence suggest that the gentle, relatively undissected surface at Managua and extending generally to the southeast is very young. This surface appears to be graded to Masaya Caldera, and locally perhaps to other nearby volcanic centers. Its essentially planar form has not yet been modified greatly by erosion, sedimentation, or other geologic processes, and the rock materials that underlie it exhibit generally the same inclination as does the surface. Most of these near-surface rocks are lapilli or ash derived from nearby volcanic sources such as Masaya.

If, as appears likely, the surface in and near Managua is a relatively young constructional feature, the task of evaluating earthquake risk becomes more difficult. Geologists commonly recognize and evaluate active faults, those which are capable of generating destructive earthquakes, by their surface topographic expression. Recurrent movement on faults produces well-defined scarps, trenches, alined stream courses, and other linear topographic features that not only mark the fault trend, but provide clear evidence of repeated activity along the same lines. These identifying characteristics, however, can be destroyed by other geologic processes, and if such other processes are operative, the record of faulting is apt to be blurred or completely obliterated. However, a very young surface provides a useful means of dating fault-formed features that clearly cut or offset it. Hence, if the young surface near Managua does locally show evidence of fault displacement, such displacement must be very young indeed.

The general low relief and absence of steep slopes in and near Managua also had an important bearing on the kinds of damage that resulted from the earthquake. Landslides and other kinds of slope failure are often among the most important causes of property damage in large and moderate earthquakes. Although many slope failures of different kinds could be observed after the earthquake, most of these were small; there were far fewer than are usually seen in areas with even moderate slopes. Other factors probably also contributed to the low incidence of slope failures, but the low relief and the relatively small area covered by steep slopes were major ones.

NEAR-SURFACE ROCK UNITS

The severity and distribution of damage resulting from destructive earthquakes depend to a large extent upon the nature of the near-surface geology. Different kinds of rock units respond to shaking in quite different ways, and in many well-observed earthquake areas, a very close correlation has been noted between the geology and the intensity of damage. Although the relation between damage from shaking and geology is far from simple, damage is commonly greatest over thick accumulations of poorly compacted water-saturated deposits and is least over relatively dense well-consolidated rocks.

Our knowledge of the geologic units that underlie Managua comes from published geologic maps of the area, from published scientific papers, from our own observations of scattered exposures of bedrock units, and from a few unpublished records from water wells. The data are inadequate for a detailed analysis of the geology, and they allow us to "see" only about 200 m beneath the surface. Nevertheless, the different lines of evidence are consistent, and they indicate that the city is underlain by a relatively homogeneous sequence of rocks, predominantly volcanic but with many interbeds of water-worked volcanic debris.

Exposures in and near Managua show that most of the volcanic debris is composed of lapilli-sized (4 to 32 mm) angular basaltic scoria. The scoria, or cinder deposits, contains almost no fine-grained ash except as thin beds a few centimeters thick. Both the scoria and the thin beds of ash are pyroclastic debris and appear to be derived either from Masaya or from the line of volcanic vents immediately to the west of Managua. Locally, these beds contain interbeds of more compact fine-grained rocks that are the products of volcanic mudflows. Unlike the scoria, the mudflow deposits are firm and relatively well lithified. They are thick and firm enough to be quarried for building stone west and southwest of Managua, and Williams (1952) has described quarried localities at which the imprints of human feet can be seen on exposed bedding surfaces.

The sequence of interbedded scoria, ash, and mudflow deposits appears to underlie nearly all of Managua, or at least those parts of the city that exhibited the greatest damage (fig. 1). The relative proportions of each rock type vary somewhat in different exposures and in the logs of wells, and the sequence is characterized by lensing and by channeling where water-worked deposits are evident. Despite these variations, lapilli-sized scoria appears to be the dominant lithology at least to the depths known from drilling, about 200 m.

Some confidence in extrapolating units for considerable distances from outcrops, wells, or artificial exposures is gained from the structural attitude of the rocks. In spite of the several faults described in this report, the rocks are little deformed and generally dip about 4° N. They are more steeply inclined, however, within a few hundred feet of the faults.

The lack of interstitial fine-grained matrix in the scoria, the rough exterior and vesicularity of individual granules, and the angularity of the granules together contribute to form a rock unit that is extremely porous and permeable and that has a low bulk density. Largely because of the angular, rough surface of the lapilli-sized fragments, this rock is fairly stable under static loads, and where it is undisturbed it will stand in near-vertical slopes. It is clearly much less stable under dynamic load conditions, such as the shaking that accompanies earthquakes. This was well shown by the numerous small debris-falls (fig. 19) that accompanied the earthquakes of December 23.

Somewhat different geologic relations are evident west of Managua along the line of volcanic centers that extends south from Lake Jiloa through Lake Asososca. There, relatively dense lava flows and vent debris are associated with pyroclastic deposits (fig. 1). Damage in this area was much less intense than in the central city, and although a major part of the difference in intensity is due to distance from the epicenter of the main shocks, some of the difference may be related to the differences in geologic conditions between the two areas.

Despite the general uniformity of ground response in the damaged area, it is likely that shaking was more intense than it would have been in an area underlain by well-consolidated, relatively dense bedrock. This conclusion, however, is based more on knowledge of other earthquakes and research results than on direct observation of ground effects at Managua.

GROUND-WATER RELATIONS

A major factor controlling damage in many earthquakes is ground water. Ground water in permeable zones can result in liquefaction and loss of strength in foundation materials. A near-surface water table, even if unconfined, can lead to slope failures, lateral spreading on low slopes, and to other kinds of failure.

Ground-water levels in the Managua area appear to be well below the surface except in the northernmost part of the city, where they are at or near the level of Lake Managua. An unpublished map of the ground-water surface prepared by Hazen and Sawyer, Engineers, New York-Managua (1964), shows that the surface of the ground water is from 10 to 30 m beneath the ground surface in most of the area that was damaged, and that the piezometric surface slopes northward somewhat more gently than the land surface. The high porosity and permeability of the rock units that contain the ground water, and the lenticular nature of most of the impermeable interbeds, are considered by us as evidence that the ground-water system is relatively open and that confined aquifers are relatively unimportant in the part of the geologic section penetrated by wells.

VOLCANIC RISK AT MANAGUA

In addition to geologic hazards related to earthquakes, the Managua area has had a long and active history of volcanism, and the future risk from destructive volcanic eruptions should be considered in reconstruction planning. A thorough discussion of the volcanic risk is far beyond the scope of this fieldwork and report but nevertheless, we feel that the seriousness of this risk warrants a brief outline and evaluation of the available data.

There are three types of recent volcanoes in Nicaragua. According to McBirney (1955),

> the first, and by far the most common group is the Strombolian type, characterized by a steep sided structure of ash, cinders, and vesicular lava. This group includes the volcanoes El Viejo, Telica, Cerro Negro, Asososca de Leon, Santa Clara, Momotombo, Chiltepe, Concepcion, Madera, and a host of minor cinder cones. The activity of these volcanoes, which is often intermittent over many years, is normally solfataric, the volume of solid ejecta being subordinate to that of steam and other gaseous elements.
>
> The second group is of the Krakatoan type usually characterized by a low, shield-like structure composed of successive layers of massive lava flows and a large, steep-walled collapse crater. These volcanoes have been notable for sudden, paroxysmal eruptions, usually culminating long periods of dormancy, during which enormous quantities of gas and pumice are ejected in the short period of a few days. At the final stage of such eruptions a cylindrical portion of the dome has usually collapsed into the vacated magma chamber forming a large, vertical-walled caldera. In this class we find Cosequina, Apoyeque, * * * and Apoyo.

The third type is the Masaya type, of which Masaya is the only example in Nicaragua. Masaya is quite similar to those Hawaiian volcanoes that consist of a caldera formed by repeated collapses of vents within the summits of a flattish basaltic shield volcano as magma migrates upward from great depth. It has been the most consistently active volcano in Central America in historic times (McBirney, 1956). There is "no trace of the characteristic pumice beds, which are so voluminous about the other [more explosive] calderas * * *" (McBirney,

1956), and McBirney concludes that while gas emission may damage crops, as happened in the period around 1927 and 1954, "* * * little is to be feared from lava eruptions because of the large volume that must be filled before any of the existing craters overflow. Even an eruption of lava from the flanks of the Nindiri-Masaya group or from any other vents on the caldera floor would not be likely to endanger any center of population." No lava flow has covered the Managua area in historic time. One flow, however, believed to have erupted in 1670 (McBirney, 1956), did run 9 km northward, within 3½ km of the present site of the international airport. Masaya could pose a threat to substantial development in the region between Managua and the City of Masaya.

Two of the three most explosive and potentially devastating volcanoes in Nicaragua, Apoyeque and Apoyo, are within 35 km of the center of Managua. The vent occupied by Lake Jiloa on the flank of Apoyeque also "appears to be the source of thick pumice beds typical of an explosive eruption" (McBirney, 1955). Furthermore, Lakes Tiscapa, Asososca, and Nejapa are collapse craters from recent volcanic activity (McBirney, 1955). These calderas and craters appear to be dormant. McBirney (1955), however, reports that the temperatures in Lake Nejapa are abnormally high and that the chemical content of the water implies that the lake is fed by hot springs. On June 8, 1852, the first indication of a new eruption of Masaya was "when Lake Masaya, together with Lakes Tiscapa, Asososca, Apoyo, and others began to 'boil.' Most likely this 'boiling' was actually an emission of gases from the lake bottom" (McBirney, 1956). This observation shows, however, that these features are merely dormant and not dead.

Thus there is significant volcanic risk in the Managua area not only from lava flows but from the possibility of a truly devastating eruption. What makes evaluation of volcanic risk particularly difficult is the question of time scale. There has been no historic Krakatoan-type eruption in the Managua area, but there may have been one large eruption since human habitation of the area (Williams, 1952). Another eruption may be thousands of years away. Devastating eruptions typically occur, however, only hours to weeks after the first visible signs of a reawakening of activity at previously dormant vents. The volcanic risk needs to be carefully evaluated and taken into account in the reconstruction of Managua.

SEISMIC RISK AT MANAGUA

HISTORIC SEISMICITY

Damaging earthquakes have occurred frequently in Nicaragua. Montessus De Ballore (1888) lists earthquakes in 1528, 1663, 1844, 1849, 1858, 1862, 1881 and 1885, but from his descriptions, it is difficult to tell where these events occurred. The earthquakes of 1844, 1858, and 1881, however, caused damage in the region of Managua. Earthquakes in 1898, 1913, 1918, 1928, and 1931 also caused damage in Managua (list compiled by Ken Jorgensen, Panama Canal Company, written commun., 1966).

All accounts of the earthquake of March 31, 1931, indicate that it was remarkably similar to the 1972 earthquake in most respects. The event was of magnitude 5.3 to 5.9 (Gutenberg and Richter, 1954) and caused ground fracturing along a northeast-trending fault in the western part of Managua. The downtown area was heavily burned. About 1,000 people (Sultan, 1931) were killed out of a population of about 40,000 (Durham, 1931). Most homes were destroyed, and utilities were seriously damaged.

A small earthquake (magnitude 4.6) occurred in Managua on January 4, 1968. It caused the heaviest damage in the Colonia Centroamerica, but no loss of life occurred (Brown, 1968).

These few data on historic seismicity show how common earthquakes are in the Managua area. From these data and the regional tectonic relations discussed above, it seems certain that damaging earthquakes will occur again in the Managua area.

The data are inadequate for determining a statistical recurrence rate of earthquakes, but it seems reasonable to expect an earthquake in Managua similar to that of December 23, 1972, within the next 50 years.

A COMPARISON

No method has been developed to quantify the earthquake risk in one area as compared to another. Too many factors, many of them as yet poorly understood, must be taken into account. Considerable research is being done and needs to be done in the future to find methods for defining comparative risk. Some qualitative comparisons, however, can be made on the basis of existing data.

All three of the earthquakes that shook Managua between 12:30 and 1:30 a.m. on December 23 were of moderate magnitude. The greatest of these, at magnitude 5.6, was smaller than the San Fernando, Calif., earthquake (6.6) of February 9, 1971, and much smaller than such great earthquakes as the Alaskan earthquake of March 27, 1964 (8.4), the

San Francisco earthquake of April 18, 1906 (8.3), the Niigata, Japan, earthquake of June 16, 1964 (7.5), or the Peruvian earthquake of May 31, 1970 (7.7). Because the magnitude scale is exponential, each integer step—for example, from 6.0 to 7.0—represents an increase in released energy of about 30 times. Accordingly, a magnitude 8 earthquake releases nearly 1,000 times the energy of a magnitude 6 earthquake. The area of the fault that slipped in the Managua earthquake is on the order of 100 km^2, whereas faults that slip during events of magnitude 6.5 and 7.5 typically have areas on the order of 500 km^2 (Hamilton, 1972) and 2,000 km^2 (Aki, 1966), respectively. In view of the complex regional tectonics in the Managua area, we would guess that it is unlikely that there are faults with areas much larger than 500 km^2. On this basis, there appears little likelihood that earthquakes much greater than magnitude 6.5 will occur in the immediate vicinity of Managua. Of course, an earthquake with magnitude larger than 8.0 might easily occur on the large faults associated with the Middle America Trench and the zone of underthrusting of the Cocos plate, but the energy source from such earthquakes would be 100 to 200 km distant from Managua.

Maximum expected magnitude is, however, not the only consideration. Damage caused directly by an earthquake is primarily related to the amount that the ground accelerates during the event, the duration of the shaking, the number of fractures going through buildings and other structures, and the amount of displacement on these fractures. Acceleration is attenuated logarithmically with distance. The data in figure 23 show that the peak acceleration of 0.31g (F. Matthiesen, oral commun., 1973) recorded at the ESSO refinery during the main Managua earthquake is about the same as might be expected somewhere between 30 and 50 km from an earthquake of magnitude 7.7. The duration of shaking also is attenuated with distance in a roughly similar way (Page and others, 1972). Thus the intensity and duration of ground shaking in Managua were large compared with that observed in many cities shaken by larger earthquakes because the Managua earthquake occurred almost directly below the central part of the city. The acceleration would probably have been 10 times less if the earthquake had occurred only 20 to 40 km distant. For instance, there was no noteworthy damage at Masaya, Tipitapa, or other nearby cities.

Statistically, seismologists find that in a region where there is one earthquake of magnitude 8 in a given period of time, there are approximately 10 earthquakes greater than magnitude 7, 100 greater than magnitude 6, 1,000 greater than magnitude 5, and so forth. Although it is dangerous to extrapolate this relation from region to region, a city that is so close to a fault and is built in such a way that it can be destroyed by a magnitude 6.0 earthquake might be destroyed much more often than a city that could sustain an earthquake of magnitude 7.5 with little damage.

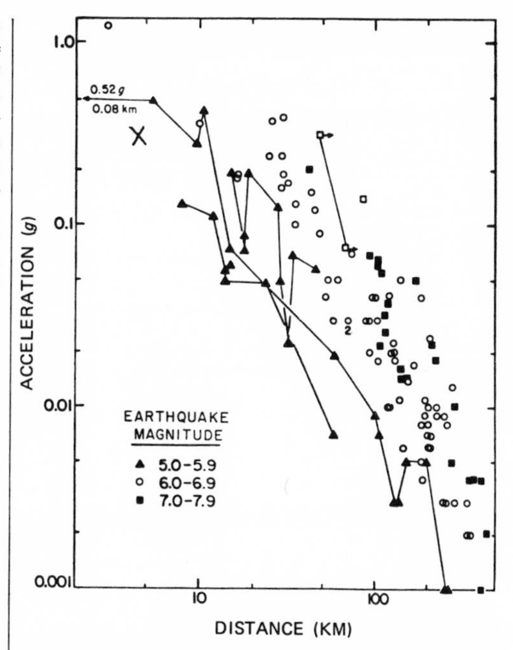

FIGURE 23.—Peak horizontal acceleration versus distance to the slipped fault as a function of magnitude (after Page and others, 1972). The X is the peak acceleration of 0.31g observed at the ESSO refinery for the Managua earthquake.

Proximity to faults and ground displacement beneath structures can significantly increase damage. No place in the central two-thirds of Managua is more than one-half kilometer from one of the four faults that moved during this earthquake sequence or the fault that moved during the 1931 earthquake. Within the approximately 15-km^2 city limits of Managua there are 11 km of faults active within the last 42 years—a fault density of roughly 0.73 km/km^2. We are not aware of a similar density of faults in any other city. Even in the entire 50 km^2 area included on the 1:10,000-scale topographic map of Ma-

TABLE 4.—*Comparison of fault density at Managua and vicinity with other urban areas in seismically active zones*

[Faults included are only those with known Holocene or historic displacement]

Community	Approximate population	Fault length (km)	Area (km^2)	Length per km^2 (km)	Data source
Managua, Nicaragua	400,000	11.0	15.0	0.73	This report.
Berkeley, Calif	116,716	11.1	25.8	.43	Radbruch (1967).
Oakland, Calif	362,100	55.9	138.0	.41	Radbruch (1967).
Managua and vicinity, Nicaragua	500,000	18.0	50.0	.36	This report.
Fukui, Japan	744,230	1.5	6.2	.24	Collins and Foster (1949).
Hayward, Calif	100,000	23.4	96.7	.24	Radbruch (1967).
San Bruno, Calif	36,254	3.2	14.4	.22	Brown (1970).
San Leandro, Calif	70,300	7.9	38.7	.21	Radbruch (1967).
Woodside, Calif	4,875	6.4	36.1	.18	Schlocker and others (1965).
Fremont, Calif	123,273	34.9	246.4	.14	Radbruch (1967).
Greater Los Angeles area, Calif	6,755,000	46.2	590.8	.08	Wentworth and others (1970).

nagua and vicinity (part of which is shown as plate 1), there are at least 18 km of active faults with a density of 0.36 km km^2. In table 4, fault density at Managua is compared with the density of faults along which there has been late Quaternary movement in other seismically active urbanized areas elsewhere. Clearly, the hazard from active faults is as great, if not greater, at Managua than at any other large city for which data are available.

The pattern of active faults in Managua differs from those in most other urban areas crossed by faults, and it differs in such a way as to increase the hazard. In most urban areas crossed by active faults, the fault breaks are simple—either a single continuous break or a narrow band of subparallel or en echelon breaks a few tens of meters to several hundred meters wide—so that the hazard from surface displacements can be well defined. The four faults recognized and described in this report, and a fifth which moved in 1931, together constitute a wide band of active faults which trends northeastward across the central part of the city. Together these five active fault traces pose a major threat to much of the urbanized area and to yet undeveloped land lying on their trend and immediately south of the city. New displacements may occur on any or all of these faults during future earthquakes, for at least two of them show clear evidence of repeated movement in the past. This pattern of faulting, which defines a band 3 km wide, suggests also that future surface displacements may not be confined only to those faults which are now known. New branch faults and subsidiary faults may occur within the zone or outside of it.

CONCLUSIONS

The extensive destruction and loss of life in the Managua earthquakes of December 23, 1972, were caused almost entirely by the following:

1. Occurrence of the earthquakes directly beneath the city.
2. Poor construction of the buildings, chiefly of tarquezal and masonry, which had very little shear resistance to lateral forces imposed by the strong seismic shaking. (These effects are being studied and reported in detail by other investigators.)
3. Direct displacement on four subparallel surface faults through the Managua area.

From the standpoint of risk from earthquakes, and possibly also volcanism, Managua is situated in an exceptionally hazardous location.

On the basis of available geologic and seismologic data the following conclusions appear warranted:

1. Earthquakes comparable in magnitude to those of 1931 and 1972 can reasonably be expected within the next 50 years.
2. Some of these earthquakes will be accompanied by surface faulting like that in 1931 and 1972.
3. Maximum hazard from surface faulting is along the trace of known active faults, five of which have been recognized.
4. New surface faulting is possible, and even likely, within a broad zone that includes all of the present area of Managua.
5. Other conditions of foundation materials, design, and construction being equal, maximum damage from shaking will be controlled largely by the proximity of structures to the surface ruptures and, in the case of a dipping fault, to the fault plane at depth.
6. In terms of the damage they cause, secondary geologic effects such as slope failure, liquefaction, and compaction will be far less significant than shaking and fault displacement.
7. The nature and distribution of the surface faulting are consistent with a tectonic origin for

the 1931 and 1972 earthquakes.
8. Catastrophic eruptions from nearby volcanic centers pose a hazard that may be as great as that from earthquakes, but one that is as yet largely unevaluated.

RECOMMENDATIONS

A reconstruction and redevelopment plan for Managua that is sound and economically feasible should be based on informed evaluations by experts from a number of disciplines. Key roles in the long-range decisions that will govern future development should be played by earth scientists, engineers, city planners, economists, and political scientists. The required action can take several routes simultaneously, among the most critical of which are:

1. Evaluation of the present and potential sites for development so that the seismologic-volcanologic hazards can be minimized.
2. Development of adequate emergency facilities and response systems to reduce the impact of natural or other disasters.
3. Adoption and strict enforcement of building codes and zoning ordinances that would ensure the integrity of vital utilities and emergency services such as communications, water, police, fire, and hospital facilities.

Comprehensive planning for the future of Managua depends first of all on an understanding of the geologic hazards and how these hazards may affect the works of man. The problems of emergency response systems as well as construction and zoning practices are beyond the scope of this report and require the expertise of others. However, some of the specific recommendations that can be made regarding the geologic and seismologic problems are:

1. A full evaluation of the hazard from earthquakes is required as a basis for local zoning and structural design criteria. This would involve detailed geologic and seismologic studies primarily directed towards delineating active faults and predicting the level of shaking and acceleration that can be expected in future earthquakes. Other potential geologic hazards such as the possibility of landslide damage to existing and planned critical facilities, such as the Lake Asososca water intake and pumping facility, should also be considered.
2. The hazard from catastrophic volcanic eruptions should be evaluated. This would entail detailed geologic studies to deduce the eruptive histories of volcanoes in the Managua area and geophysical monitoring to determine their present state of activity.
3. To the extent possible, essential underground service facilities, such as sewer and waterlines, electric power and telephone lines, should be routed so that they cross known active fault zones in the fewest possible places. Where crossings are unavoidable, design provisions should be made for fault displacements of at least the amounts reported here.
4. Emergency and critical facilities, such as hospitals, fire stations, police stations, powerplants, schools, and important government buildings, should be sited well away from known active faults and, to the extent possible, outside of the zone in which surface faulting is prevalent.
5. Disaster relief planning for future destructive earthquakes should be undertaken and periodically reviewed; the 1931 and 1972 earthquakes provide patterns that should be incorporated into such plans. Especially important are the fault trends, amount and nature of displacement, the rupture of waterlines at fault crossings, and the effects of such ruptures on postearthquake fire hazard.
6. Regional earth science studies should be undertaken on a long-range basis to evaluate safe sites in Nicaragua for future growth and development. Such studies should include both geological field investigations and monitoring of seismic and volcanic processes.

REFERENCES CITED

Aki, K., 1966, Generation and propagation of G waves from from the Niigata earthquake of June 16, 1964. Part 2. Estimation of earthquake moment, released energy, and stress-strain drop from the G wave spectrum: Earthquake Research Inst. Bull., v. 44, p. 73–88.

Allen, C. R., and Nordquist, A. R., 1972, Foreshock, main shock, and larger aftershocks of the Borrego Mountain Earthquake: U.S. Geol. Survey Prof. Paper 787, p. 55–86.

Brown, R. D., Jr., Vedder, J. G., Wallace, R. E., Roth, E. F., Yerkes, R. F., Castle, R. O., Waananen, A. O., Page, R. W., and Eaton, J. P., 1967, The Parkfield-Cholame earthquakes of June to August 1966: U.S. Geol. Survey Prof. Paper 579, 66 p.

Brown, R. D., 1968, Managua, Nicaragua earthquake of January 4, 1968: Proj. Rept., Nicaragua Inv. (IR) NI-1, 16 p.

Brown, Robert D., Jr., 1970, Faults that are historically active or that show evidence of geologically young surface displacement, San Francisco Bay Region—A progress report: Oct. 1970: U.S. Geol. Survey open-file map.

Chinnery, M. A., 1963, The stress changes that accompany strike-slip faulting: Seismol. Soc. America Bull., v. 53, p. 921–932.

Clark, M. M., 1972, Surface rupture along the Coyote Creek fault: U.S. Geol. Survey Prof. Paper 787, p. 55–86.

Collins, J. J., and Foster, H. L., 1949, The Fukui earthquake, Hokuriku region, Japan: Office of the Engineer, General Headquarters, Far East Command, 81 p.

Dengo, G., 1968, Estructura geologia, historia tectonica, y morfologia de America central: Instituto Centroamericano de Investigacion y Technologia Industrial (ICAITI), Guatemala, 50 p.

Durham, H. W., 1931, Managua—Its construction and utilities: Eng. News Record, April 23, p. 696–700.

Eaton, J. P., O'Neill, M. E., and Murdock, J. N., 1970, Aftershocks of the 1966 Parkfield-Cholame, California, Earthquake.: A detailed study: Seismol. Soc. America Bull., v. 60, p. 1151–1197.

Greensfelder, R., 1968, Aftershocks of the Truckee, California earthquake of September 12, 1966: Seismol. Soc. America Bull., v. 58, p. 1607–1620.

Gutenberg, Beno, and Richter, C. F., 1954, Seismicity of the earth and associated phenomena: Princeton, N.J., Princeton Univ. Press, 310 p.

Hamilton, R. M., 1972, Aftershocks of the Borrego Mountain Earthquake from April 12 to June 12, 1968: U.S. Geol. Survey Prof. Paper 787, p. 31–54.

Kachadoorian, Reuben, Yerkes, R. F., and Waananen, A. O., 1967, Effects of the Truckee, California, earthquake of September 12, 1966: U.S. Geol. Survey Circ. 537, 14 p.

Kuang, J., and Williams, R. L., 1971, Mapa Geologico de Managua, Nicaragua, sheet 2952 III, 1:50,000, CATASTRO.

McBirney, A. R., 1955, The origin of the Nejapa Pits near Managua, Nicaragua: Bull. Volcanol., Serie II, Tome XVII, p. 145–154.

——— 1956, The Nicaraguan volcano Masaya and its caldera: Am. Geophys. Union Trans., v. 37, no. 1, p. 83–96.

McBirney, A. R., and Williams, H., 1965, Volcanic history of Nicaragua: Calif. Univ. Pubs. Geol. Sci., v. 55, 73 p.

Molnar, P., and Sykes, L. R., 1969, Tectonics of the Caribbean and Middle America Regions from focal mechanisms and seismicity: Geol. Soc. America Bull., v. 80, p. 1639–1684.

Montessus De Ballore, F., 1888, Tremblements de terre et eruptions volcaniques au Centre Amérique: Dijon, 281 p.

National Oceanic and Atmospheric Administration, 1973, Preliminary determination of epicenters, No. 76–72.

Page, R. A., Boore, D. M., Joyner, W. B., and Coulter, H. W., 1972, Ground motion values for use in the seismic design of the trans-Alaska pipeline system: U.S. Geol. Survey Circular 672, 23 p.

Radbruch, D. H., 1967, Map showing recently active breaks along the Hayward fault zone and the southern part of the Calaveras fault zone, California: U.S. Geol. Survey open-file map.

Ryall, A., Van Wormer, J. D., and Jones, A. E., 1968, Triggering of microearthquakes by earth tides and other features of the Truckee, California, earthquake sequence of September, 1966: Seismol. Soc. America Bull., v. 58, p. 215–248.

Santos, C., 1972, La hipotética probabilidad de occurrencia de temblores en la ciudad de Managua durante el verano de 1973: Unpub. rept., Santos and Heilemann, Consultores de Ingeneria, Managua, Nicaragua, 26 p.

Schlocker, Julius, Pampeyan, E. H., and Bonilla, M. G., 1965, Approximate trace of the main surface fault zone between Pacifica and the vicinity of Saratoga, California, formed during the earthquake of April 18, 1906: U.S. Geol. Survey open-file map.

Sultan, D. I., 1931, The Managua Earthquake: Military Engineer, v. 23, p. 354–361.

Thatcher, Wayne, and Hamilton, R. M., 1973, Aftershocks and source characteristics of the 1969 Coyote Mountain earthquake, San Jacinto fault zone, California: Seismol. Soc. America Bull. (In press.)

Tsai, Y. B., and Aki, K., 1970, Source mechanism of the Truckee, California, earthquake of September 12, 1966: Seismol. Soc. America Bull., v. 60, p. 1199–1208.

Ward, P. L., and Gregersen, Soren, 1973, Comparison of earthquake locations determined with data from a network of stations and small tripartite arrays on Kilauea Volcano, Hawaii: Seismol. Soc. America Bull., v. 63, p. 719–751.

Wentworth, C. M., Ziony, J. I., and Buchanan, J. M., 1970, Preliminary geologic environmental map of the greater Los Angeles area, California—TID–25363: U.S. Geol. Survey; available from Clearinghouse Federal Sci. and Tech. Inf., Springfield, Va., 41 p.

Williams, Howell, 1952, Geologic observations on the ancient human footprints near Managua, Nicaragua: Carnegie Inst. Washington Pub. 596, pt. 1, 31 p., 11 figs.

Wyss, Max, and Brune, J. N., 1968, Seismic moment, stress, and source dimensions for earthquakes in the California-Nevada region: Jour. Geophys. Research, v. 73, p. 4681–4694.

Wyss, Max, and Hanks, T. C., 1972, Source parameters of the Borrego Mountain Earthquake: U.S. Geol. Survey Prof. Paper 787, p. 24–30.

19

Reprinted with permission from *Saturday Rev./World (Sci.),* 55(52), 43–44 (1973)

TAMBORA AND KRAKATAU: VOLCANOES AND THE COOLING OF THE WORLD

BY KENNETH E. F. WATT

You have every reason to be confused as to the possible effects of pollution on the weather. Some scientists have predicted that increasing global air pollution will reduce the amount of solar energy penetrating the atmosphere. This, they say, will cause a decline in the global average air temperature. Other scientists say the earth's weather will warm up. This, it is explained, will come about because of the increase in carbon dioxide concentration in the atmosphere. Since carbon dioxide is a strong absorber of infrared radiation, it will act like the glass roof that prevents loss of heat energy from inside a greenhouse—the infamous "greenhouse effect"—and trap radiation in the atmosphere.

Three types of arguments can be put forth to settle the matter as to which mechanism will be most important. One approach is to argue from mathematical theories of climate determination, using computer-simulation studies. This line of argument is not completely trustworthy, because of the embryonic state of such theories. A second line of argument, pursued by Reid Bryson and Wayne Wendland at the University of Wisconsin, is to discover by statistical analysis the relative strengths of the two processes during recent decades. They found that the impact of the temperature-lowering mechanism will override the impact of the temperature-raising mechanism. Since the temperature has in fact been dropping for more than two decades, their argument is compelling.

Kenneth Watt, who writes regularly for SR, *is a systems ecologist at the University of California, Davis.*

However, a third line of argument opens up a most fascinating area of research, both in interpretation of history and in climatological prediction. This argument asserts that the most realistic means of assessing the effect of air pollution is to find some gigantic natural event in the past that operated in a way analogous to modern pollution. Fortunately, we are provided with such historical "experiments" by records of a few particularly gigantic and explosive volcanic eruptions.

A close reading of old newspapers gives evidence that people in the early nineteenth century were not aware of any relationship between unusual weather and volcanoes. To my knowledge, most ancient civilizations thought that the earth's weather was almost totally determined by astronomical phenomena, and this belief explains the spectacular development of observational astronomy in those civilizations. The first man known to have recognized that volcanoes could affect the weather was Benjamin Franklin in 1784, following a series of enormous eruptions in Japan and Iceland the previous year. During the twentieth century many climatologists and geophysicists have come to suspect that volcanoes influenced weather and climate and, more recently, that they had effects similar to the global increase in atmospheric pollution.

There are two obvious sources of information on the impact volcanic eruptions could have on weather and hence on crop production. One is the tables of past weather data recently constructed by historical climatologists such as Gordon Manley and H. H. Lamb. The other is old newspapers, which yield detailed information on local weather, crop growing, planting and harvesting conditions, and agricultural commodity prices.

Consider the effect of the eruption of Tambora in Sumbawa, the Dutch East Indies, in April of 1815. This was the largest known volcanic eruption in recorded history: over the period 1811 to 1818 an estimated 220 million metric tons of fine ash were ejected into the stratosphere. Of this, 150 million metric tons were added to the stratospheric load in 1815 alone, mostly in April. Krakatau, which is much better known because it erupted more recently (1883), ejected only 50 million metric tons of ash into the global stratosphere. It is now known that the ash from certain of these very large and explosive volcanic eruptions spreads worldwide in a few weeks and does not sink out of the atmosphere totally until a few years have elapsed. During all of this time the ash is back-scattering incoming solar radiation outward into space and consequently chilling the earth.

The magnitude of the chilling can be ascertained from Manley's tables for central England. The Tambora volcano erupted most explosively in April of 1815; by November the average temperature of central England had dropped 4.5° F. The following twenty-four months was one of the coldest times in English history. Specifically, 1816 was one of the four coldest years in the period 1698 to 1957; the coldest July in the 259-year period was in 1816; October of 1817 was the second coldest October; May of 1817 was the third coldest May. However, these figures convey little sense of the impact of such chilling on society. For that we must turn to the newspapers of the time. All of the following quotations are from *Evans and Ruffy's Farmers' Journal and Agricultural Advertiser*.

"We had fine mild weather until about the 20th, when it set in cold, with winds at East and North-East, with partial frosts; these together have greatly retarded the operations in Agriculture, and very many cannot purchase seed corn, so that thousands of acres will pass over untilled, and sales of farming stock, and other processes in law, drive many of this useful class in society into a state of despondency. . . . The wheats, late sown . . . have been partially injured by the frosty mornings. . . . Sheep and lambs have suffered from the severity and variableness of the weather. . . . The doing away the Income Tax, and the war duty on Malt, will afford some relief, but are wholly insufficient in themselves to restore this country to its former state of happiness and prosperity." (April 8, 1816)

"From about the 9th or 10th of this month, we have never had a day without rain more or less, sometimes two or three days of successive rain with thunder storms. The hay is very much injured; a considerable part of it must have laid on the ground upwards of a fortnight. . . . Wheat is looking as well as can be expected, considering the deficiency of plants in the ground, and those very weakly . . . but still far short of an average crop." (August 12, 1816)

"Throughout the whole month the air has been extremely cold; there has not been more than two or three warm days, being at other times rather cloudy and dark, and the sun seldom seen. The Oats . . . on high situated ground . . . are the most backward and miserable crop ever seen . . . for the greater part of the Wheat, where the mildew did not strike, has been very much affected by the rust or canker in the head. . . . There have been many seizures for rent this month, and many a farmer brought to nothing, and we hear of very few gentlemen who are inclined to lower their farms as yet; it seems they are determined to see the end." (September 9, 1816)

The preceding quotations all refer to agricultural conditions in England. To indicate that this was a worldwide rather than a local phenomenon, the following quotation, from a letter printed in the same newspaper, suggests the state of affairs in other countries.

"Last year was an uncommon one, both in America and Europe: We had frosts in Pennsylvania every month the year through, a circumstance altogether without example. The crops were generally scant, the Indian Corn particularly bad, and frost bitten; the crops, in the fall and in the spring, greatly injured by a grub, called the cutworm . . . ." (November 10, 1817).

Thus, in the case of a volcano, which puts an immense load of pollution into the atmosphere suddenly—unlike modern pollution, which builds up gradually—we have a gigantic experiment, the effects of which can be clearly traced throughout all social and economic systems. For example, the volcano of 1815 had a clear-cut influence on world agricultural-commodity markets, as one would expect from the preceding descriptions of the consequences of weather deterioration on crops. Many measures of market conditions could be used to make this point, but the one I have selected is the highest asking price for best-quality flour within a month of trading on the London commodities market. The following table shows how this price changed, from the period before the volcano, to the peak price in June of 1817, to the normal price, which was finally reached again by the end of December 1818. The table indicates how volcanic

HIGHEST ASKING PRICE FOR A SACK OF FLOUR (IN SHILLINGS)

Year	June	December
1814	65	65
1815	65	58
1816	75	105
1817	120	80
1818	70	70*

(*65 by end of month)

eruptions can serve as the basis for interdisciplinary research, in which a pulse due to a physical event can be tracked through biological, social, economic, and political phenomena.

For example, an interesting feature of this table is the long lag between the time the pollution was introduced into the upper atmosphere and the time that the price elevation was at its peak (April 1815 to June 1817—twenty-six months). These time lags are characteristic of complex systems and indicate why cause and effect are often not connected in peoples' minds: by the time the effect has occurred, everyone has forgotten the cause.

Could the gradual increase in worldwide air pollution concentrations become serious enough to bring crop production to a halt in high latitudes? To answer this question, we must consider the rate of build-up in pollution now and the likelihood that political power to enforce adequate pollution control will materialize.

The worldwide stratospheric particulate loading due to Krakatau and Tambora was 50 and 220 million metric tons respectively. If worldwide man-caused stratospheric particulate loading continues to build up at recently prevailing rates, the following table indicates the likely outcome.

WORLDWIDE STRATOSPHERIC PARTICULATE LOADING DUE TO MAN (MILLIONS OF METRIC TONS)

YEAR	
1970	2
2000	15
2010	30
2020	59
2030	116
2040	228

Thus, we see that by about 2018, if present trends continue, the permanent particulate load in the stratosphere would be equal to that produced temporarily by Krakatau, and by about 2039 the permanent load would be equal to that produced by Tambora.

Will they continue? The reader must make his own judgments. There is ample printed evidence testifying to the difficulty of requiring automobile manufacturers to conform to the limits required by the Clean Air Act of 1975. Also, a casual reading of industry journals does not suggest that industry would like to arrest the rate of increase in sales of oil, gas, coal, or other polluting substances. Computer-simulation studies of trends in the use of fossil fuels do not indicate any significant lessening of the rate of increase in pollution prior to about 2004 unless stringent controls are introduced. Thus, without a really massive political and social change of a type that does not seem likely, judging from present attitudes, the quotations from English farm newspapers in 1816 and 1817 may well be read as a scenario for the future.

There are further complications if these are not enough. What if man continues to build up the pollution load in the atmosphere, and then a volcano of the order of Tambora adds still more pollution to the atmosphere? How likely is this to happen? The answer is: very likely. The period since 1835 has been remarkably free of major volcanic eruptions of the explosive, Vesuvian type. But over the historical record an average of five very large volcanoes has occurred each century. Luckily for us, the modern period of great technological activity has been free from major volcanic eruptions, to an almost historically unique extent. But at some time our luck may run out.

Another complication is what is going on in the minds of people. The world population of 1816 had no idea that there was any relationship between the phenomenally bad weather they were experiencing and air pollution. Will we be any wiser? Can we change in time? Will we be able to take the necessary action to ensure a brighter end to the scenario for civilization? Simple extrapolation of present trends does not lead to an encouraging answer, but history teaches us that sudden, surprising changes in political and social attitudes do occur. Perhaps it will occur once more—in time. □

20

Reprinted from *The Encyclopedia of Geochemistry and Environmental Sciences*, Rhodes W. Fairbridge, ed., Van Nostrand Reinhold Company, New York, 1972, pp. 696-699

MEDICAL GEOLOGY

John S. Stevenson and Louise S. Stevenson

There are two main aspects of medical geology: the study of the total geological environment and its relation to the health of man; and the study of petrographic and mineralogical methods of parts of the human body or foreign substances therein.

The Geological Environment and Health

That environmental factors are important and, at times, crucial in human health and disease has long been recognized by students of human geography (see *Medical Geography*). But whereas geographic studies comprise the entire range of environmental factors, including such variables as climate and sanitation, the geologist confines his attention to the soils and underlying bedrock. Because the relationship between man and his geological environment is frequently less direct than between man and, for example, his climate, it is not surprising that he has been slow in recognizing this relationship. Hence, most significant work in this field has been quite recent.

Warren and Delavault (1949, 1960), and others working in biogeochemical prospecting, drew attention to the large and selective concentrations of metals found in certain plants. This metal concentration appeared to be related to the underlying bedrock even in some cases where the soil cover was different in mineral content from that of the underlying rock. Attention was then focused on the fact that important trace elements found their way to the human body from both soils and the underlying bedrock, and the geological environment was seen to be significant in human health and disease.

Man depends ultimately upon the rocks of the earth's crust for his trace elements, but unfortunately the route that these trace elements may take, in getting from the earth's crust into human organs, is extremely complex. The relationships between trace elements in rocks and in soils, and between soils, plants and animals, are only now beginning to be understood (see *Trace Elements in Plants*).

Many important variations in rock composition point to the need for extremely careful and detailed study if significant applications of geology to health problems are to be made. Limestones and dolomites may be either high or low in lead content, two monzonites closely resembling one another mineralogically may vary greatly in trace elements, and even slates of the same geological age in one small area have been found to differ in content of some trace elements by a factor of as much as 20:1 (Warren, 1964b).

Although soils largely, but by no means entirely, derive their trace elements from the geological formations to which they ultimately owe their origin, there is always the problem of whether the soil is residual or transported. In any event, the total amounts of trace elements, as well as their distribution in soil, is often significantly altered by the soil-forming process. Again, plants may be variable and selective in their trace element content, and plant-animal relationships introduce further variables even before man enters the picture. It has been found that cattle can eat fodder that contains sufficient copper, yet suffer from copper deficiency because there is too much molybdenum in this fodder. Also, some animals can eat without harm a food containing materials that would be poisonous to most other animals.

The need for exact and detailed geological and mineralogical work is underlined by the small amounts of materials often involved in human health. Small deficiencies may lead to serious ailments and, on the other hand, quite modest amounts of materials may be highly poisonous.

Epidemiological studies are shedding some light on these problems, but correlations with trace element studies are still uncommon. Special difficulties arise because medical statistics are ordinarily given for political rather than geologically significant areas, and because most medical statistics concern death rather than illness.

The work of Helen Cannon, who has combined experience in study of uranium mineralization with geobotany, is pointing the way to significant advances in field correlation of geology and health. She has been interested in relating geochemical environments to incidence of cancers and cardiovascular diseases, and this involves careful and detailed geological field studies as a basis for a review of health in the area. Similarly, Warren (1964a) in cooperation with a number of physicians, is making important studies of copper, lead, zinc and molybdenum content of some rocks, soils and vegetal matter from areas reporting a high prevalence of multiple sclerosis.

That dental health is affected by trace elements is also generally acknowledged, with most of the attention centered on the reduction of caries by the addition of fluorine in suitable form and amount to drinking water. This program, now rather extensive, developed from observations that inhabitants of areas with high fluorine content in waters and soils had stronger and healthier teeth than the general population. Widespread research is continuing in the matter of optimum intake of fluorine for dental as well as for general health. It is possible that the keen interest and the positive results from this research have tended to obscure the role played by other trace elements in dental health, and there is need for further information.

Petrographic and Mineralogical Methods in Medicine

Rock Dust Diseases of the Lungs. In cases of rock dust exposure, it is important to know the type of dust and its origin, and this requires a detailed study of the rock being worked and of the minerals that go to make it up. Since it has been found that dust samples taken from different working faces, even in the same mine, show significant mineralogical differences from place to place, very careful petrographic and mineralogical studies are required (Stevenson, 1959).

Most of the dusts usually encountered by man, such as the sands of the desert and the dusts of the street, are too large to reach the lungs and are trapped in the nose and upper respiratory passages. However, particles 5μ in diameter or less can reach the alveoli of the lungs, and many dusts resulting from mining and quarrying are in this size range. In the ashed lung specimens of men who have died from silicosis, Hunter (1959) reports that the most representative particle measured 1μ in diameter.

Considerable confusion has arisen because, in medical literature, silicosis is usually cited as being caused by the inhalation of "free or uncombined silica," and the great danger of many silicate minerals was long overlooked. However, it is true that many rock dusts, such as those from limestones, dolomites, and iron oxide ores, which do not contain siliceous accessory minerals, seem to be inert when taken into the lung and can be inhaled without serious effect.

Careful mineralogical study of ashed lung tissue from patients suffering from silicosis has shown that the most dangerous rocks contain siliceous minerals which have a tendency to shatter into fibrous forms. Thus pure quartz, unless it is very finely ground, is far less dangerous than sericite or chrysotile. The ashed lung tissue of silicotic patients generally reveals a characteristic felting produced by minute, usually fibrous, silicates. Although originally such studies could be made only in autopsies, these contributed greatly to knowledge of the disease and to proof of industrial injury valuable to the heirs in legal compensation cases. Now, however, most ashed lung studies are based on biopsies, and are helpful to the physician in interpreting the patient's symptoms and advising further treatment. They have also been valuable in alerting management to dangerous areas of rocks rich in fibrogenic siliceous minerals, so that special precautions could be taken, particularly in "hard rock" metal mines and asbestos mines.

Coal-miners' pneumoconiosis, rather than classical silicosis, is the most widespread disease in coal mines, and because it does not develop as rapidly as silicosis, its danger was not recognized for many centuries. However, since dust at a coal face is nearly always a mixture of coal and siliceous material, continued inhalation of coal dust generally produces a slowly progressive fatal disease.

The usual dust suppression methods in mines and quarries include ventilation, wet cutting, and extensive spraying of dusty areas. The prevention of silicosis by metallic aluminum has been accomplished in recent years, based largely on research coordinated by the McIntyre Research Foundation in Ontario. It was found that inhalation of small amounts of aluminum dust with the rock dust rendered the siliceous material insoluble and hence benign, and that miners who inhaled aluminum dust before going on shift were protected from the dust hazard. Aluminum prophylaxis is now used extensively in many parts of the world (Irwin, 1964).

Calculi in the Human Body. Medical problems of the cause and recurrence of calculi have been greatly aided by petrographic and mineralogical studies of these stones, using the methods ordinarily employed in the study of true rocks and minerals. This is especially true of such concretions as kidney stones, which are generally composed of materials identical to minerals found in the earth's crust.

Prien and Frondel (1947) have applied the oil immersion method of optical crystallography with great success in the identification of the component minerals in calculi, and they

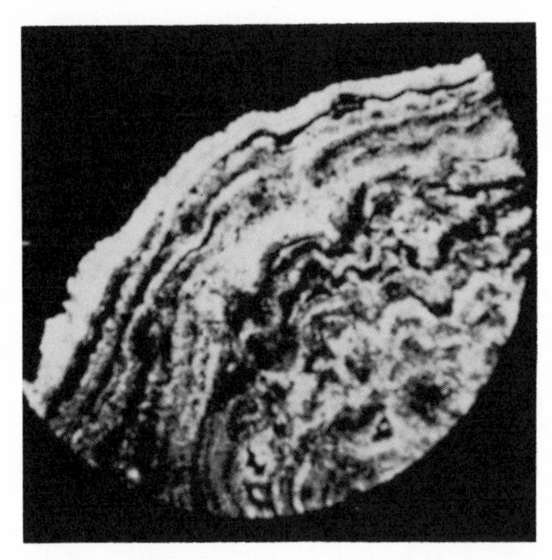

FIG. 1. A portion of a urinary calculus showing colloform structure. Light gray rings are clear, crystalline whewellite; darker grey rings are yellow whewellite; white outer ring is weddellite. The dark lines between some of the rings are organic matter, probably blood (crossed nicols, 20×).

have made significant statistical studies of the various types of stones. Whewellite, weddellite, struvite, apatite (both carbonate-apatite and hydroxyapatite), and brushite were found to be common constituents.

Lucas, Stevenson and Stevenson (1950) have emphasized the value of augmenting oil immersion identification with petrologic study in thin sections (Figs. 1 and 2). The structure of the calculus may then be seen, and the formation of the stone, with the paragenesis of the constituent minerals, traced from the centers of crystallization to the outer encrustation. With sufficient knowledge of the case history of the patient, it may be possible to correlate pathogenesis and petrogenesis.

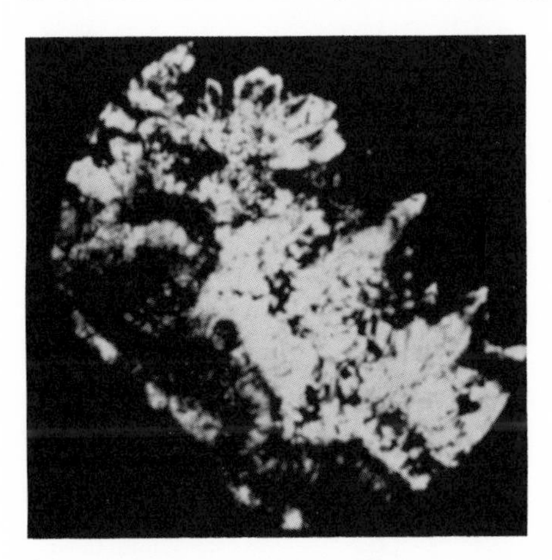

FIG. 2. The rim of the same calculus shown in Fig. 1. Weddellite crystals are encrusted on whewellite "tree-rings"; darker material between the crystals is struvite (crossed nicols, 40×).

Mineralogical Studies of Human Bones and Teeth. The "mineral" composition of bones and teeth is similar to that found in rocks of the earth and, hence, may be studied in the same way; indeed it is the mineralogist who has been able to explain to the physiologist the true nature of the "inorganic" component of bones and teeth (McConnell, 1962).

It has been found that dental enamel is composed of a single "mineral" phase which is dahllite (carbonate hydroxyapatite). Dentin and bone have a composition very similar to dental enamel, and their crystalline structure must be essentially isostructural with that of dental enamel. Fossil studies have shown that the usual composition of fossil teeth and bone is also dahllite, although fossilized material may become francolite (carbonate fluorapatite) by fluorine enrichment during metamorphism.

Studies of fluoride in bones and teeth by mineralogical methods suggest that fluoride, when ingested, may change the crystal texture of bone and tooth apatites and stabilize them by increasing crystal size and decreasing crystal strain.

References

Hunter, Donald, 1959, "Health in Industry," Pelican Medical Series, Pelican Books, pp. 176–215.

Irwin, D. A., 1964, "The Use of Aluminum in the Prevention of Silicosis." Publ. McIntyre Research Foundation, Toronto, Ont., pp. 1–9.

Lucas, O. C., Stevenson, John S., and Stevenson, Louise S., 1950, "Petrologic study of a urinary calculus," *Trans. Roy. Soc. Can., Ser. III, Section IV,* **44,** 35–40.

McConnell, Duncan, 1962, "The crystal structure of bone," *Clinical Orthopaedics.* No. 23, 253–268, J. B. Lippincott Co.

Prien, Edwin L., and Frondel, Clifford, 1947, "Studies in Urolithiasis: I. The composition of urinary calculi," *J. Urol.* **57,** No. 6, 949–991.

Stevenson, John S., 1959, "Mineralogy and the field geologist," *Can. Mineralogist,* **6,** Part 3, 303–306.

Warren, Harry V., 1964a, "Geology, trace elements and epidemiology," *Geograph. J.* **130,** Part 4, 525–528.

Warren, Harry V., 1964b, Introductory remarks, Medical Geology and Geography Symposium, A.A.A.S., Montreal, Dec. 28, 1964 (Report in *Science,* 1965).

Warren, Harry V., and Delavault, R. E., 1949, "Further studies in biogeochemistry," *Bull. Geol. Soc. Am.,* **60,** 531–559.

Warren, Harry V., and Delavault, R. E., 1960, "Observations on the biogeochemistry of lead in Canada," *Trans. Roy. Soc. Can., Ser. III, Section IV,* **54,** 11–20.

Part VI

HUMAN INFLUENCES: BENEFICIAL AND DELETERIOUS

The difficulty of distinguishing between natural and man-made effects on the physical environment is that the direct cause of some events could be attributed to either source. In the opening remarks for Part V, this "gray zone" is mentioned: it is evident that more than half the papers in that part have clear relations to this. It is evident also that in all the papers in Part VI the natural tendencies that are "triggered" (stimulated into action) by human interference are displayed. In other words, man has become increasingly a significant geologic agent; formerly regarded as having only local influence, he is now, with the aid of modern technology, extending his influence over vastly broader areas. Using modern American idiom, that is what physical environment is all about, man either *against* or *with* nature.

In Part V only the hazardous situations were considered. Obviously, at times nature serves man, and indications of this more pleasing situation are given in somewhat extended comment on aesthetic quality of the physical environment in Part IV. So much attention has been directed to man's less desirable actions that any civilized and sensible ones in assisting nature are often overlooked by some members of modern society. Their concern is understandable, but they fail to realize that even without human interference nature is also not faultless. For example, people disturbed by brown-colored streams seem to think that this is always pollution caused by humans. This is, of course, not correct; it is not usually an example of pollution, although excess sediment may be a man-exacerbated problem. Normal processes of erosion and drainage produce the effect, often seasonally, and we must accept these natural effects. As in the case of the historic Nile floods, the deposited sediments had the favorable effect of enriching the land. The same floods, however, were associated with slope wash in parts of Ethiopia, which produced deleterious soil erosion.

Flooding is a natural function of many fluvial regimes and represents a way for nature to clean out valleys clogged with debris of dead vegetation and rock materials that could not be moved under conditions of normal stream flow. It should be remembered that natural processes appear to be primarily destructive, but they are balanced with constructive effects. For example, the continual reconstitution of low-relief coasts, while destroying some beaches, creates new ones without interference of man.

Persons fighting the onslaughts of developers usually fail to convey the ultimate implications for mankind in their arguments, which are intensely real; they speak mainly of their worry over the fate of the fauna and flora when deprived of their habitat. We must be concerned with our collective fate as humans as well as with the welfare of other occupants of the earth. What is required is proper balance in thinking about the respective roles of man and other phenomena of the earth he occupies. The whole meaning of this book is that we should find the relationship between man and nature. Geoscientists must realize that they have a mission to perform, along with everyone else in our society.

SUPPLEMENTAL READINGS

Adams, W. M. (ed.). 1970. *Engineering Seismology: The Works of Man.* Geol. Soc. America, Eng. Geol. Div., Eng. Geol. Case Hist. 8.

Banat, K., Ulrich Förstner, and German Müller. 1972. Schwermetalle in Sedimenten von Donau, Rhein, Ems, Weser und Elbe im Bereich der Bundesrepublik Deutschland. *Naturwissenschaften,* **59**(12), p. 525-528.

——. 1972. Schwermetalle in den Sedimenten des Rheins. *Umschau,* **72**(6), p. 192-193.

Bauer, A. M. (ed.). 1968. *Mining in an Urban Landscape.* University of Guelph, School of Landscape Architecture Proc. Symp., Nov. 19-20, 1968, 86 p.

Bodner, R. M., and W. T. Helmsley. 1972. *Evaluation of Abandoned Strip Mines as Sanitary Landfills.* 3rd Mineral Waste Utilization Symp. (Chicago) Proc., p. 130-138. (Sponsors: U.S. Bureau of Mines and IIT Research Institute.)

Born, S. M., and D. A. Stevenson. 1969. Hydrologic Considerations in Liquid Waste Disposal. *J. Soil Water Conserv.,* **24**(2).

Bräuner, Gerhard. 1973. *Subsidence Due to Underground Mining.* U.S. Bur. Mines, IC-8571-8572.

Brown, J. M., Jr. 1969. *Planning a Sanitary Landfill.* Univ. Kansas Bull. Eng. Architecture, 60, p. 15-19.

Bülow, Kurd von. 1953. An-aktualistische Wesenszüge der Gegenwart. *Deut. Geol. Ges. Z.,* **105**(2), p. 183-196.

Carder, D. S. 1970. *Reservoir Loading and Local Earthquakes.* Geol. Soc. America Eng. Geol. Case Hist. 8, p. 51-61. (Engineering Seismology: The Works of Man.)

Cronin, L. E. 1967. The Role of Man in Estuarine Processes. In: *Estuaries* (G. A. Lauff, ed.), p. 667-689. Amer. Assoc. Advan. Sci. Publ. 83.

Davis, J. H. 1956. Influences of Man on Coast Lines. In: *Man's Role in Changing the Face of the Earth* (W. L. Thomas, Jr., ed.), p. 504-521. Chicago, University of Chicago Press.

Demek, Jaromir. 1968. Beschleunigung der geomorphologischen Prozesse durch die Wirkung der Menschen. *Geol. Rundschau,* **58**(1), p. 111-121.

Dittmer, Ernst. 1954. Der Mensch als geologischer Faktor an der Nordseeküste. *Eiszeitalter Gegenwart,* 4(5), p. 210-215.

Doehring, D. O., and J. H. Butler. 1974. Hydrogeologic Constraints in Yucatán's Development. *Science,* **186**(4164), p. 591-595.

Doerr, Arthur, and L. Guernsey. 1956. Man as a Geomorphological Agent: The Example of Coal Mining. *Assoc. Amer. Geog. Ann.,* **46**, p. 197-210.

Evans, D. M. 1966. Man-Made Earthquakes in Denver. *Geotimes,* **10**(9), p. 11-18.

——. 1966. The Denver Area Earthquakes and the Rocky Mountain Arsenal Disposal Well. *The Mountain Geologist,* 3(1), p. 23-36. (Also in Geol. Soc. America Eng. Geol. Case Hist. 8, p. 25-32.)

Fels, E. 1954. *Der wirtschaftende Mensch als Gestalter der Erde.* Stuttgart, 282 p.

——. 1965. Nochmals: Anthropogene Geomorphologie. *Petermanns Mitt.,* **109**, p. 9-15.

Fischer, Ernst. 1915. Der Mensch als geologischer Faktor. *Deut. Geol. Ges. Z.,* **67**, p. 106-148.

Flawn, P. T., L. J. Turk, and C. H. Leach. 1970. *Geologic Considerations in Disposal of Solid Wastes in Texas.* Texas Bur. Econ. Geol. Circ. 70-2.

France, Centre National pour l'Exploitation des Océans. 1974. Etudes ecologiques des effets de l'exploitation des gisements marins de sables et graviers. *Bull. Inform.,* 62, p. 8-9.

Golomb, Berl, and H. M. Eder. 1964. Landforms Made by Man. *Landscape,* **14**(1), p. 4-7.

Guy, H. P. 1970. *Sediment Problems in Urban Areas.* U.S. Geol. Surv. Circ. 601-E, 8 p.

Häusler, H. 1959. *Das Wirken des Menschens im geologischen Geschehen.* Naturkundliches Jahrb. Stadt Linz (Austria).

Healy, J. H., et al. 1968. The Denver Earthquakes. *Science,* **161**(3848), p. 1301-1310.

Holz, R. K. 1969. Man-made Landforms in the Nile Delta. *Geograph. Rev.,* **59**(2), p. 253-269.

Jäckli, Heinrich. 1962. Die Beziehungen zwischen Mensch und Geologie. *Wasser- und Energiewirtschaft (Linth-Limmat),* Sonderheft.

——. 1964. Der Mensch als geologischer Faktor. *Helvetica,* **19**(2), p. 87-93.

Jennings, J. N. 1965. Man as a Geological Agent. *Australian J. Sci.,* **28**(4), p. 210-215.

Jolliffe, I. P. 1968. *Planning and Research Problems in the Exploitation of Coastal Areas.* 23rd Intern. Geol. Congr. Proc., Sec. 12, p. 96-103.

Judson, Sheldon. 1968. Erosion of the Land, or What's Happening to Our Continents? *Amer. Scientist,* **56**, p. 356-374.

——. 1968. Erosion Rates near Rome, Italy. *Science,* **160**(3835), p. 1444-1445.

Kiersch. G. A. 1964. The Vaiont Reservoir Disaster. *Civil Eng.,* 34(3). [Revised: California Division of Mines and Geology, *Mineral Inform. Serv.,* **18**(1965), p. 129-138.]

Lachenbruch, A. H. 1970. *Some Estimates of the Thermal Effects of a Heated Pipeline in Permafrost.* U.S. Geol. Surv. Circ. 632, 23 p.

Landon, R. A. 1969. Application of Hydrogeology to the Selection of Refuse Disposal Sites. *Ground Water,* **7**(6), p. 9-13.

LeGrand, H. E. 1957. Role of Ground Water Contamination in Water Management. *Amer. Water Works Assoc. Jour.,* **59**(5), p. 557-565.

——. 1965. Patterns of Contaminated Zones of Water in the Ground. *Water Resources Res.* **1**(1), p. 83-95.

——. 1968. Urban Geology and Waste Disposal. *Geotimes,* **13**(5), p. 23.

Leopold, L. B. 1968. *Hydrology for Urban Planning—A Guidebook on the Hydrologic Effects of Urban Land Use.* U.S. Geol. Surv. Circ. 554, 18 p.

Lucas, C. P. 1915. *Man as a Geographical Agency.* Brit. Assoc. Advan. Sci., 84th Ann. Mtg., Sec. E, Rept., p. 426-439.

McGauhey, P. H. 1968. Earth's Tolerance for Wastes. *Texas Quart.,* **11**(2), p. 36-42.

——. 1968. Manmade Contamination Hazards to Ground Water. *Ground Water,* **6**(3), p. 10-13.

Maneval, D. R. 1972. Coal Mining vs. Environment: A Reconciliation in Pennsylvania. *Appalachia,* **5**(4), p. 10-40.

Müller, German, and Ulrich Förstner. 1973. Cadmium-Anreicherungen in Neckar-Fischen. *Naturwissenschaften,* **60**.

Müller, Stephan. 1970. Man-Made Earthquakes: Ein Weg zum Verständnis natürlicher seismische Aktivität. *Geol. Rundschau,* **59**(2), p. 792-805.

Newton, J. G., and L. W. Hyde. 1971. *Sinkhole Problems in and near Roberts Industrial Subdivision, Birmingham, Alabama—A Reconnaissance.* Geol. Surv. Alabama Circ. 68, 42 p.

Piper, A. M. 1969. *Disposal of Liquid Wastes by Injection Underground.* U.S. Geol. Surv. Circ. 631, 15 p.

Schäfer, Wilhelm. 1973. Der Oberrhein, sterbende Landschaft. *Natur. Museum,* **103**, p. 1-29, 73-81, 110-123, 137-153.

Scheidt, M. E. 1967. Environmental Effects of Highways. *J. Sanit. Eng. Div. Proc. ASCE,* **93**, SA-5, p. 17-25.

Sherlock, R. L. 1922. *Man as a Geological* Agent. London, Witherby, 372 p. (Abridged edition: *Man's Influence on the Earth,* London, Butterworth & Company Ltd., 256 p., 1931.)

U.S. Department of the Interior, Strip and Surface Mine Study Policy Committee. 1967. *Surface Mining and Our Environment.* 124 p. Washington, D.C.

Weyl, Richard. 1967. *Der Mensch im Spiel der geologischen Kräfte.* Giessen, Justus-Liebig-Univ. Schriften, 6, 22 p.

Wilmoth, B. M. 1970. Environmental Geologic Hazards of West Virginia. *West Virginia Acad. Sci. Proc.,* **41**(1969), p. 174-183. (West Virginia University B., ser. 70, no. 11-13.)

Wixson, B. G., et al. 1972. The Lead Industry as a Source of Trace Elements in the Environment. In: *Cycling and Control of Metals* (M. G. Curry and G. M. Gigliotti, compilers), p. 11-19. U.S. Environmental Protection Agency, Natl. Environ. Res. Center (Conf. Proc.).

Woeikof, A. 1901. De l'Influence de l'Homme sur la Terre. *Ann. Geog.,* **10**, p. 97-114, 193-215.

Wolman, M. G. 1964. *Problems Posed by Sediment Derived from Construction Activities in Maryland.* Annapolis, Md.: Maryland Water Pollution Commission, 125 p.

Zwartendyk, Jan. 1971. *Economic Aspects of Surface Subsidence Resulting From Underground Mineral Exploitation.* Washington, D.C., U.S. Bureau of Mines and Pennsylvania State University, USBM Open-File Rept. 7-72. 412 p. (Available from NTIS, PB 207-512.)

Abstracts (1972-1974) from the Geological Society of America, Abstracts with Programs

Elmer, R. E., W. E. Cutcliffe, and J. R. Dunn. 1974. Geologic Planning Provides a Basis for Minimizing the Social Impact in Suburban Areas. 6(1), p. 22.

Hammer, T. R. 1972. Stream Channel Enlargement Due to Urbanization. 4(7), p. 526.

Hayes, M. O., P. K. Ray, et al. 1973. Urbanization as a Cause of Accelerated Beach Erosion. 5(7), p. 659.

Hoge, H. P. 1974. Report on an Evaluation of Environmental Hazards Related to Deep-Mining in Eastern Kentucky. 6(4), p. 365-366.

Leamon, A. R. 1973. Surface Mining for Coal in Tennessee—Its Status and Effects. 5(5), p. 411-412.

Merrill, R. D., and E. S. Bliss. 1973. Sierra Nevada Foothill Subdivisions in Fresno County, California: A Study in Environmental Mayhem. 5(1), p. 80.

Szucs, F. K. 1973. Urban Geological Zoning in the Greater Liege Area. 5(4), p. 355-357.

Editor's Comments on Papers 21 Through 25

Paper 21 is an attempt at an overview of human influences on the physical environment, which the author terms *anthropogeology*. Thus, we place it at the beginning of the part because it is concerned with the whole concept of this extremely vital continuing relationship, and not just with man creating local earthquakes or land collapse.

The examples mainly illustrate what has occurred in Switzerland, and can be said to deal with both the good and bad. On the whole, the well-ordered landscape of Switzerland attests to careful planning, and thus to beneficial effects. However, Jäckli does note that the demand for natural resources in Switzerland is a matter of real concern. Depletion of what is there and the side effects of exploitation are both decidedly troublesome.

Looking beyond the confines of Switzerland, Jäckli discusses the action taken by man to make coastal readjustments. His example is the Netherlands, which brings us face to face with a problem that cannot be ignored. With its high population density and need for land to conduct supporting activities, the very existence of that country depends on regulating nature. The conversion of areas in that country which were constantly under water to land (*polders*), or dike building to prevent the invasion of seawater into farmland is hardly a frivolous program. The problems of the Netherlands may not impress themselves too forcibly on some people in countries such as the United States, in which a critical land shortage has not yet become a reality. With the increasing concentration of population in coastal areas of the United States, however, this will become a problem of no minor proportions in the future.

Jäckli notes that the effects of some human interventions are tolerated and others are uncontrolled. For geologists the principle of actualism is a basic axiom, but the entry of man into the development of the environment tends to confuse the issue. That man upsets natural processes is plain to see. The question is: How can the geololist best aid the direction of human interventions for predictable and desirable ends?

Jäckli is a consulting geologist in Zürich. He also lectures in the well-known Eidgenössische Technische Hochschule (ETH) and the University of Zürich.

Paper 22 is included here because it deals specifically with environmental pollution of rivers and lakes in a defined geographic region, much of which is heavily populated and highly industrialized. Thus, the pollution "is due to a strong increase of industrial waste, domestic sewage and agricultural runoff." The authors portray the hazards with excellent graphics, which leave little doubt that man-made pollution far outweighs the contribution of the immediate natural background.

The heavy-metal concentrations in sediments of the Rhine, Danube, and other central European rivers are sizable. Förstner and Müller make an important point concerning these accumulations; independently of the pollution problem, they represent wasted economic resources, and the authors ask: "Since the natural resources of most heavy metals are very limited, the question arises, how long can we afford the extravagance of wasting vital materials (bound to the suspended matter and dissolved in the water, which are lost each year to the sea. . .)." Therefore, it becomes necessary to cope with man-made pollution, not simply by denying the introduction of undesirable elements into surface water and the ground, but by seeing to it that they are retrieved for useful purposes.

Förstner has taught in the University of Kabul (Afghanistan) and is now a member of the faculty of the Sedimentology Laboratory, University of Heidelberg. For several years Müller headed the sedimentology laboratory of a major oil company; he then joined the faculty of the University of Tübingen and since 1964 has been at the University of Heidelberg, where he is director of the sedimentology laboratory.

William J. Schneider has provided in Paper 23 a brief, very informative, and well-organized report on the hydrologic implications of solid-waste disposal. He deals mainly with the situation in the United States, but remarks also on situations in foreign regions. Obviously, there is a distinctly undesirable level of pollution. The leachate that reaches the groundwater is "generally both biologically and chemically contaminated." Schneider appeals for "full consideration of the water resources in selection of sites for solid-waste disposal."

Schneider has had many years of experience with U.S. Geological Survey. His water-resource investigations have been carried out in the northeastern United States, Appalachia, and Florida.

In Paper 24 Nikola P. Prokopovich concentrates on exogenic subsidence, which, he states, "was barely known at the turn of the century." It is most common in densely populated areas. However, although exogenic subsidence is particularly attributed to human activity, it is not at all limited to instances of human interference (e.g., karst solution, cliff collapse). The observation that "the susceptibility to subsidence is the direct result of the geologic past of the area" is important. Prokopovich considers exogenic subsidence to be a form of pollution because it is "usually an undesirable alteration of natural conditions." This is a rather unique interpretation of the term pollution, which, by and large, is questionable. It does not detract from the merit of the paper.

Prokopovich is a geologist with the U.S. Bureau of Reclamation.

Sinkholes occur typically but not exclusively in areas underlain by carbonate rocks. The occurrence of these features and, in general, of areas of subsidence, is a striking, unwanted situation from various obvious standpoints.

From Alabama, it is reported that in Birmingham more than 200 collapses occurred in an area less than half a mile square from 1963 to 1970, and in Greenwood more than 150 were noted, which formed from 1950 to 1972.

The authors of Paper 25, on the Greenwood area, explain that "sinkholes result from a variety of causes, but the occurrence of numerous sinkholes in a restricted area is commonly associated with a substantial lowering of the water table which may be natural, man-created, or a combination of both." The additional information given is that here the general lowering of the water table resulted from large withdrawals of groundwater from wells and mines compounded with a prolonged drought during the 1950s, which made the area prone to the development of sinkholes.

Newton is a hydrologist with the U.S. Geological Survey. Scarbrough is chief, Geological Division, Geological Survey of Alabama, and Copeland is chief, Programs and Plans, of the same organization.

21

This abridged translation was prepared expressly for this Benchmark volume by Elisabeth W. Betz from "Elemente einer Anthropogeologie," in Eclogae Geol. Helv., *65(1), 1–19 (1972).*

ELEMENTS OF ANTHROPOGEOLOGY

Heinrich Jäckli

INTRODUCTION

The term *anthropogeology* means the relationships between man and geologic phenomena, including those of the past, present, and future (Häusler, 1959).

The fundamental question is: What are the relationships between man and geologic phenomena? In the following an attempt is made to order these relationships systematically.

I. UTILIZATION OF MINERAL RESOURCES

The building contractor requests the geologist to tell him where he should open a gravel pit from which he can obtain clean, sandy, loam-free gravel of consistent quality in the largest possible quantity at the lowest possible cost, without causing undesirable effects, such as encroaching on the groundwater, disturbing cased springs, and leaving ugly scars in the landscape. A pottery manufacturer would ask the same kinds of questions about clay, a cement producer about limestone and marl, and a salt producer or soda manufacturer about rock salt. Similar advice is sought on ore deposits and energy resources. In every case the stratigraphic and lithologic conditions are studied and restrictions imposed by morphologic and tectonic factors are determined.

When resources are depleted and are not renewable in the same quantity, man is guilty of despoliation. An example from Switzerland of a resource that will become a rare commodity in the future is gravel. If for reasons of protecting the groundwater, we avoid excavating gravel from areas around well casings and below the water table, and work only in the dry zone above the water table and in the areas designated for present and future construction, at the present rate of depletion of usable gravel the resource will have been exhausted in only a few centuries. If the rate of removal increases in the same degree as in past years, the supply will be exhausted that much sooner.

In Figure 1 these relationships are shown graphically for Aargau, one of the most gravel-rich cantons in Switzerland. On the abscissa is time in years, and on the ordinate the gravel supply still existing in 1970. Since gravel excavation in 1970 was at the rate of about 3 million cubic meters per year, constant excavation at this rate would use up the 1970 reserves in 466 years (as shown by the diagonal line beginning at the lower left). In the last 4 years the annual increase in excavation was about 0.25 million cubic meters; if this rate of increase were maintained in the future, the excavation curve would become a parabolic line instead of a

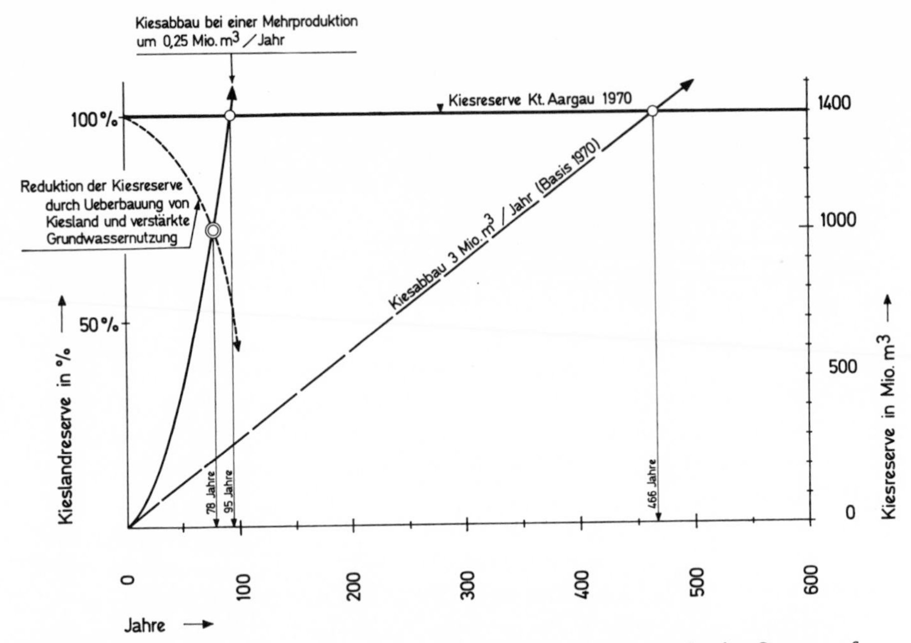

Figure 1 Relationship of gravel reserves to gravel excavation in the Canton of Aargau as a function of time. Reproduced from *Eclogae Geol. Helv.*, 65(1), 4 (1972); copyright © 1972 by the Swiss Geological Society.

straight line, and the present reserves would be used up in 95 years. At the same time, however, there is a decrease in the workable reserves through construction on gravel deposits and through greater use of groundwater (shown in Fig. 1 as a broken line beginning at the upper left). Where the two curves intersect, in 78 years, the remaining workable gravel supply will have been used up.

Estimation of the supply of gravel and the extent of future reduction is obviously not precise; nevertheless, the diagram clearly shows that we cannot continue the increase in exploitation that has occurred in recent years.

Groundwater, too should be regarded as a useful raw material. Thus, more must not be taken from the ground than is continuously replenished either by seepage of precipitation or infiltration of surface water. The worldwide phenomenon of a sinking natural water table as a result of modern groundwater use shows, however, that this rule is not heeded everywhere and that in very large areas groundwater supplies are being radically exploited, a practice that must eventually lead to a disturbance of natural balances.

II. ADAPTATION TO GEOLOGIC CONDITIONS

A. Static Elements: Foundation Soil Conditions

Construction activity is adapted as much as possible to the existing geologic conditions. The careful engineer, therefore, asks geologists about the composition of the foundation soil on which he wants to build, how to assess the susceptibility to settling, what pressure is admissible on the soil at a specific depth, and where the water table may be expected. He asks the geologist to judge the stability of a natural embankment, how to tell the limits of a landslide, and how a projected tunnel, road, or power line can avoid the slide. He asks the geologist about the geologically favorable location for a dam, how to judge the imperviousness of a projected reservoir, what kind of geologic formations the pressure tunnel will go through, and what tunneling difficulties must be anticipated.

The geologist must react to these questions so that the engineer can either adapt his structures practically to the geologic conditions or avoid problematic geologic conditions altogether. The geologist does not necessarily make the decision; but he does have to inform the decision maker objectively.

The geologist appraising a road-building project may recognize that a certain slope segment is in the process of a slow creeping movement. There may be two ways of avoiding this sliding slope: either by going around it or traversing it. Often more important factors than the geologic ones are brought to bear on the decision, such as the construction time or technical transportation requirements.

Frequently, the geologist has the unpleasant experience of finding that his geologic warnings are not fully considered; later, if a too optimistic cost estimate has been exceeded, it is stated, directly or indirectly, that the geologic advice has failed. The consulting geologist may have to accept blame, sometimes justifiably but very often not.

B. Dynamic Elements

1. Regions of Active Sedimentation or Erosion

For human adaptation we must not consider only static elements (e.g., the foundation soil) but also dynamic elements. Man generally avoids regions that are especially active geologically. He avoids areas of sedimentation, which are marked by constantly recurring floods, and areas of erosion, where spontaneous slides or torrent damage of all kinds are to be expected. He looks instead for areas of least geologic activity for his settlements and traffic routes. That this desire for geologic quiet exists in large parts of the world where it cannot actually be realized, is evident.

2. Regions of Volcanic and Earthquake Activity

Man regards volcanoes as danger spots to be avoided, but his fear of them is only of short duration. The region of Vesuvius serves as a representative example of countless volcanic areas: the fertile volcanic ash that buried Herculaneum and Pompeii continues to attract farmers again and again for cultivation and settlement.

One would expect man to avoid active earthquake zones as well, but his fear of these dangers is empirically short lived. The famous San Andreas Fault in California, which destroyed San Francisco in 1906 and claimed hundreds of thousands of lives, has been overrun by new suburbs in recent years—although the geologist can predict with certainty that the fault movements will continue and that heavy earthquake damage will result in the newly developed residential areas.

C. Hydrogeologic Elements

Where water is rare, man looks for places to settle with reliable springs or groundwater that can be obtained by means of wells. Many original settlements in all climatic zones can easily be explained by these two conditions. Interesting cases are in desert regions, where single isolated springs or groundwater occurrences make life possible in an otherwise perilous environment.

The task of the geologist today is not only the utilization, but also the protection, of springs and groundwater. All too often he must exercise a restraining effect on engineers and administrators when the latter make plans that would jeopardize current or future usage.

III. ACTIVE INFLUENCES ON GEOLOGIC CONDITIONS AND PROCESSES

A. Deliberate Changes

1. River Diversions

River diversions probably belong to the oldest human influences on geologic processes.

a. Old interventions. In central Switzerland, diversion of the Rengbach at Littau is known from the sixteenth century. Part of the rock bed was artifically deepened, especially during 1572 and 1577, to prevent overflow at high-water stages toward Kriens and Lucerne (Roesli, 1965).

Furthermore, it is known that in 1591 and 1592 near Zug the outlet of the Lake of Zug (northeast of Lucerne) into the Lorze was deepened to lower the water level of the lake and to hinder shore flooding. Similar projects to lower lake levels were tried subsequently at other lakes northwest of Lucerne.

b. The Kander Tunnel. Compared to the late-medieval river diversions and deepenings, which were rather modest human interventions in the natural hydrological process, the diversion of the Kander River into the Lake of Thun appears to have been a measure of striking audacity. Until 1714 the Kander flowed across the Aare plain to the Aare River. The Kander and Zulg rivers, both gravel-rich torrents, used the Aare plain to the northwest of Thun as an area for sedimentation and simultaneously caused a backwash of the Aare, which created frequent flood conditions on the shores of the lake. Nonetheless, the inhabitants of the Aare plain felt that these geologic conditions were unalterable and had accommodated to them, locating their settlements and roads on the safe flanks of the plain and otherwise avoiding the actual sedimentation area in the valley bottom.

Diversion of the Kander River to relieve high-water conditions was begun on 1 April 1711. An open cut was planned to divert the Kander to the lake across a rise, but, because of the cost factor, a tunnel was driven in the morainic material with an incline of 6½ percent. When in 1714 the Kander was directed through this artificial bed to the lake, the tunnel collapsed. The stream, owing to headward erosion, deepened its bed (as much as 21 meters) so rapidly that none of the water could enter the old channel to the Aare River, but instead all poured into the lake. Simultaneously, a river delta developed on the shore of the lake (Beck, 1943; Grosjean, 1962).

c. The Linth Works. Around 100 years later another artificial interference in natural sedimentation in an alluvial plain took place, the correction of the Linth River (central eastern Switzerland).

In the Linth plain between Lake Walen and upper Lake Zurich, sedimentation from the meandering Linth produced a continual backwash into Lake Walen. There was increasingly heavy flood damage in towns at both ends of the lake. This condition could not be altered by raising the dams along the old Linth and other streams that were the main gravel suppliers for the lowlands southeast of Lake Zurich. The broad plain became marshy because the water table rose and the tributaries overflowed. Endemic malaria is said to have prevailed at that time in all the lowland villages and along the lake, causing a higher death rate and lowering the average life span in these villages to around 10 years less than in the rest of Switzerland.

Reorganization was accomplished by the Linth Works. First, the Mollis (now called Escher) Canal conducted the Glarus-Linth directly into Lake Walen. It was opened on 8 May 1811. The Linth could then deposit its sediments directly in the lake, and the Linth plain would no longer serve as a sedimentation area. Furthermore, the course of the stream could be straightened and its gradient in the lower Glarus region could be increased; finally, Lake Walen could peak high waters of the Linth from the upland. The second important construction was the Linth Canal, which permitted the water from Lake Walen to flow to Lake Zurich by the shortest possible route. It was opened on 17 April 1816.

Hans Conrad Escher (1767–1823), the first engineering geologist in Switzerland, is closely connected with the Linth Works. He was honored by the Swiss Diet with the added surname on "von der Linth," which all his male descendants were allowed to carry.

d. The Jura Waters Correction. In the second half of the last century the alluvial plain between Lakes Murten, Neuchatel, and Biel was suffering from increased flooding and marshiness similar to that of the Linth plain. A fundamental reorganization was carried out: the Aare River was artificially connected to Lake of Biel in August 1878 by means of the Hagneck Canal and in this way given a new sedimentation area like that of the Linth. In addition, the Nidau-Büren Canal was built as an outlet for the lake, replacing an older canal. Some of the deepening process of the Hagneck Canal was optimistically left to the power of erosion of the Aare itself. This great work was completed in 1891 (Peter, 1922).

The second Jura Waters Correction has been in process for years and will soon be finished. It adds only hydraulic changes whereby the most critical canal and river lengths are being widened and deepened so that the levels of the three Jura border lakes can be kept more constant than before.

2. *Stream Embankments*

a. Suppression of artificial accumulation of gravel. To prevent flooding of the valley bottom by a main river or tributaries, man has long made use of embankments along both sides of streams, which are meant to prevent them from spilling over the banks. Connected with the construction of embankments are straightening of the stream bends and cutting of the meanders, to shorten the stream course and increase the gradient. Thereby a former tendency toward sedimentation is transformed frequently to an erosional tendency, which is desirable in many cases.

If the main stream is now prevented by the construction of lateral embankments from depositing gravel on the valley floor, it is forced to form a delta at the stream mouth. The growth of the delta is accelerated, which necessarily leads to a more rapid raising of the stream bed and then to a further heightening of the embankments.

Once man has begun to canalize rivers with lateral embankments, he is committed to the unending task of subsequent raising of these walls. It would be more compatible with geologic processes for the river to deposit laterally and so be able to raise the valley floor continually over its entire width. Today, however, such a geologic happening is unthinkable because the valley bottom has meanwhile been occupied by valuable settlements, factories, and roads, and natural deposition can never again be permitted. This is true in the Rhine, Rhone, and Po valleys.

b. Artificial deposition in the valley bottom. The Integrated Improvement of the Domleschg in Graubünden (southeastern Switzerland) is an interesting and successful intervention in the natural sedimentation process. The Nolla, formerly one of the most feared torrents of the canton, empties near Thusis from the steep erosional ravines of Piz Beverin into the Rhine plain. The Rhine valley bottom consisted of coarse Rhine gravel of recent age with little humus cover; thus the area was quite unproductive for agriculture. As part of the Integrated Improvement of the Domleschg, a water-control system was built in 1940 in the Nolla at Thusis and the mud-filled Nolla water was conducted to the fine gravelly valley floor of the Domleschg by means of specially constructed canals. In individual fields separated by low dikes black clay mud was deposited until the fine-grained sediment had attained a thickness of about 80 centimeters. Then that area was made available for agricultural use. In this case, man created a flood which deposited the nutrient-rich, fine-grained sediment. Thus, human activity relieved the main river, the Hinter-Rhein, of this sediment on the one hand, and, on the other, raised the valley bottom areally; connected with that action the fertility of the soil was substantially increased.

3. Control of Floods

Man can hinder undesirable downcutting in streams by erecting transverse barriers; on steep inclines these take the form of actual barrier steps. Simultaneously, the bed can be raised and widened. The stream, which is no longer engaging in downcutting, usually reacts with lateral erosion. If this is also undesirable, man must combat it by means of jetties.

4. Coastal Construction

Since early times, inhabitants of low-relief coasts have demonstrated active intervention of striking proportions into the natural geologic phenomena on the border between mainland and the sea (Dittmer, 1955).

For thousands of years man has observed the transgressive tendency of the North Sea against the mainland of northwestern Europe. The inhabitants of those coasts first defended themselves against the advance of the transgressing sea by erecting *wurten* (artificial hills of one to two meters in height), on which they built their settlements and churches in safety from storm flooding. Later, connected dikes were built to replace the *wurten* as further protection against the advancing North

Sea. The monstrous dike breaks in the Late Middle Ages resulted in catastrophic flooding of the hinterland but, geologically, it was a perfectly natural transgression of the North Sea on the subsiding low-relief coast. Since then people have wrested that land back from the sea in a systematic fashion.

Work during the last 40 years has resulted in extensive construction on the Zuyder Zee in Holland with its retaining dike, reclamation of the Wieringermeer, of the Noordoostpolder, and the eastern and southern Flevoland, totaling about 160,000 hectares of land.

Similar projects under the name Delta Plan have been started in the estuaries of the Rhine and Scheldt. Connecting dikes are being built from island to island to keep the North Sea from penetrating the coastal area. Seven hundred kilometers of old dikes, which would otherwise have to be maintained and raised again so as to really protect the hinterland, are being replaced by 23-kilometer-long dams containing locks for the Rhine water to flow through into the sea. These large constructions are supposed to protect the southwestern portion of the Netherlands from flooding. Simultaneously, they serve the hydroeconomy by retaining the Rhine's fresh water in the interior of the land while keeping the salt water out.

The Wadden Plan is also under discussion. With this plan the Dutch hope to link the West Frisian Islands and connect them with the mainland, and to drain the Wadden Zee that lies between them and the mainland.

Since the twelfth century, dike construction has withdrawn the former flood areas on both sides of the river mouths from the geologic effect of natural sedimentation and erosion, but the actual consolidation settling of still very loose, young sediments continues. Thus, the once inhabited and agriculturally useful marshes settle continually closer to the high-tide level of the sea. The movement of sea level itself has demonstrated a slight opposite tendency, which is generally interpreted as the eustatic effect of ice melting at the polar caps and in the glaciated mountains.

In harbor areas man has intervened in the natural silting-up process. Owing to navigational needs, he has tried to strengthen the natural erosional processes by artificially dredging ship channels. The coastal population, however, would rather see increased sedimentation by means of breakwaters and other coastal constructions. Here we can observe a true conflict of interests: the interests of shipping are diametrically opposed to those of the farmers on the coast (Simon, 1963, 1965).

B. Weak, Ineffective Intervention

If human intervention, running counter to the natural tendency of geologic processes, is not carried out forcefully enough or is not purposefully pursued, natural geologic processes can break through again. Man anthropocentrically identifies this phenomenon as natural catastrophe.

When a river levee bursts, flooding of the hinterland can lead to sudden sedimentation over wide areas of previously dry valley bottom. On 1 February 1953, the southwestern coast of Holland was overrun by a storm tide, which caused dike breaks and subsequently more flooding than had occurred since 1421. One hundred and sixty thousand hectares of land were submerged under water and 1,783 deaths were attributed to these events, which were witnessed by the most experienced hydraulic engineers. This land continues to sink because it still consists of young, unconsolidated alluvium; contrarily, the sea level is rising eustatically under the present climatic conditions. Thus, the difference between the level of the ocean and occupied land surface grows to the disadvantage of the latter. Such a region requires constant raising of the dikes against the sea and strengthening to protect the land against increased flood danger. It is most dangerous to fail to carry out this work regularly, once begun.

C. Unintentional, But Tolerated, Interventions

1. Artificial Depressions

Man intervenes in natural phenomena on the earth's surface without consciously wanting to cause geologic changes; however, he may also take this factor into account and accept the consequences.

For example, man creates new depressions by excavating for roads, railways tracks, buildings, mines, quarries, and so on, and uses this material for various purposes (without consciously wanting to bring about geologic changes).

2. Artificial Elevations

Corresponding to artificial depressions are artificial elevations in the form of deposits. These consist of dumps resulting from ore and coal mining, the waste from rock quarries, clay pits, lime and marl quarries, and excavated material from drifts and tunnels.

One special problem for our generation is the proper disposal of refuse, which annually amounts to about 200 kilograms per person. All too often the geologist must make the authorities aware that gravel pits above groundwater that is being used are not suitable refuse dumping sites. Thus, as all suitable clay and marl pits are filled, more and more natural depressions with loamy, impermeable foundation soil will be selected for the deposit of refuse. Farmers encourage the leveling of too steep landforms so that their agricultural use is simplified; but the geologist regrets the loss of numerous topographic features, which are of geologic origin and which help him to decipher the geologist history of the landscape.

3. Consequences of Man-made Reservoirs

With reservoirs man creates artificial sedimentation areas where erosion had prevailed and allows limnic sediments to be deposited instead of letting this material be transported by the river to its mouth. Peak flood stages are broken by a reservoir, and the transport, erosional power, and sediment load of its discharge are lessened.

D. Man-made Interventions with Unwanted, Uncontrolled Effects

1. Rock Slides

The most spectacular intervention of this kind must surely be artificially and unintentionally released rock slides. I will mention three in Switzerland: the slide at Plurs in Bergell, released by improper quarrying of serpentine on the southern flank above Plurs, on 4 September 1618, killing about 2,430 people; the slide at Elm, released by improper quarrying of roofing slate above Elm in Canton Glarus, on 11 September 1881, killing 116 people (Heim, 1932); the slide on Monte Toc at Longarone in the southern Tirolean Dolomites, released by fluctuations in the water level in the Vaiont reservoir, on 9 October 1963, killing about 2,000 (L. Müller, 1964, 1968).

2. *Soil Slides*

Included in the category of undesirable events are the numerous man-made soil slides, whether they were caused by cutting into a slope for an excavation or road, or by an inadmissible amount of filling, or by increased water content in the slope because of careless manipulations that decrease its shearing strength. In any case, human intervention may reduce the natural stability to such an extent that slow or sudden creeping will begin.

When the level of a lake is artificially changed, as is the rule with reservoirs, the stability of the bordering slopes is inevitably decreased and, in precarious spots, a spontaneous slide results.

Part of the daily activity of the consulting geologist is to make the engineer and building contractor aware of such dangers beforehand, or afterwards to explain to the builder, authorities, and insurance companies whether or not a slide had been caused by human intervention (thus representing a liability case) or whether it was to be considered a natural occurrence (for instance, damage from a sudden storm).

3. *Erosion*

The play of erosion and accumulation, which usually oscillates in a highly unstable balance in rivers, is occasionally influenced by man in an undesired direction. When, for example, construction for torrent control or gravel dredging is done in the watershed of a river, the effect (downstream) can be an uncontrolled abrupt change from gravelling to erosion or to an accelerated downcutting and, in the worst instances, even to collapse of bridges.

4. *Change in Facies*

Sedimentation in Switzerland's lakes occurred without much change for thousands of years in postglacial times, corresponding to the unchanged chemical composition of the lake waters. Through the introduction of domestic and industrial sewage into these lakes in the last decades there has been a process of fertilization with nutrients, among which phosphate has become a primary regulator. This has led to a fundamental change in facies in lake sediment; layers of sapropelites have recently been deposited on top of older and oxidized sediments, such as lake marl and inorganic silts and clays. Nipkow (1920) dates this abrupt facies change in the lower basin of Lake Zurich, on the basis of annual undisturbed varves, at 1886. Thomas (1971) has proved that the presumed eutrophy of Lake Zurich has been retrogressive since 1967 and, through oligotrophication, an increased oxygen content and decreased phosphate content have resulted. Thomas attributes this facies change to the effect of the recently built sewage-treatment plants in the drainage basin of the lake.

The ocean, which for millions of years remained uninfluenced by man, has become a rubbish pit for regions near the coast (Böhlmann, 1971). Domestic and industrial sewage, waste of all kinds, hydrocarbon and pesticides, are poured into rivers and, lately, poured directly into the ocean through pipelines or by transport ships. The unchanged facies of littoral sediments are beginning to exhibit anthropogene effects. Analogous to the changes in facies and organic sediments in inland waters is the overfertilizing of coastal regions and the poisoning of autochthonous flora and fauna.

5. *Changes in Groundwater*

The use of artificial fertilizers causes an uncontrolled artificial increase in potassium, nitrogen, and phosphorus compounds in upper soil layers, which are leached out again by precipitation and transported downward into the ground water.

Pyritiferous waste from bituminous coal mines dumped into pits will raise the sulfate content in the groundwater because precipitation flows through the exposed pits and causes pyrite oxidation (Siebert and Werner, 1969).

A similar situation exists for the deposit of other chemically active industrial and domestic wastes. Where not only solid but liquid wastes are transferred downward, there is the danger of uncontrollable groundwater pollution. Potash washings are allowed to sink into permeable karst water conductors that are not being used for drinking water needs at the moment in potash mining areas. For the time being, this practice is considered to be better than dumping the washings into surface water (Mayrhofer, 1965; Finkenwirth, 1967). It cannot be denied, however, that the future fate of such industrial wastes cannot be controlled at great depths, and it is entirely possible that, contrary to expectations, these wastes will again participate in the general water circulation and reappear suddenly in groundwater being used.

Infiltration from anthropogene-loaded rivers into groundwater can eventually lead to complete oxygen consumption and great harm to water quality. This phenomenon has been frequently observed along the sluices of river power plants. Characteristics of such groundwater are a very unstable chemical balance, low oxygen content, reduced nitrate, formation of ammonia, and a great rise in the content of dissolved iron and manganese.

6. *Artificial Seismism*

It has been recognized in recent times that damming great masses of water in reservoirs or sewage depressions under great pressure can release a spontaneous artificial seismism, which dies away in the course of months or years. An earthquake on 10 December 1967 with its epicenter at the Konya Reservoir, southeast of Bombay (water height of approximately 110 meters), reached a magnitude of M 6.4 and took about 20 lives (Müller, 1970). Whether an earthquake will occur and at what amplitude cannot be predicted. Man does not have command over these events.

7. *Terrain Settlement*

a. Through lowering of the groundwater level. Wherever the groundwater level is artificially lowered, be it in areal water drainage systems or in mining, it inevitably produces a settling movement on the surface. The terrain sinking in peat areas is about 10 to 30 percent of the total groundwater settlement; in settle-resistant sands and gravels it is on the order of 0.1 to 0.2 percent.

Sayma and Momikura (1971) report land sinking on the approximately 200-square-kilometer coastal plain around Saga, Japan. This process is undoubtedly connected with the very intensive use of groundwater for irrigation. The rate of depression in the last few years has averaged about 27 millimeters per year, with a maximum of 60 millimeters, and has resulted in much severe building damage.

According to Varnhagen (1967), in the lower Rhine basin, the most important lignite region of Germany, where Middle Miocene lignite is being mined from open pits to depth of 250 to 300 meters, a corresponding depression of the groundwater level is resulting. From hundreds of wells at depths of 150 to 350 meters over an area of about 500 square kilometers, some 1 billion cubic meters of groundwater is obtained annually. The groundwater depression results not only in the draining of many wells and groundwater systems, but forces the lignite mine operators to seek secondary water for their operations; it also produces measurable settling of the surface, which differs from place to place along the faults and can cause damage to railways, roads, pipelines, and cables. The terrain settling in the Erft area is observed periodically at about 1,200 leveled points. Settlement is increasing progressively with groundwater depression according to a logarithmic function, rather than a linear.

b. Through production of petroleum and natural gas. Where reservoir rocks are not well consolidated—as can be the case especially in coastal regions—there may be uncontrolled terrain settlement on the magnitude of decimeters, meters, or decameters, which in overbuilt regions inevitably leads to building damage (Lofgren, 1961, 1968).

In his use of both groundwater and petroleum, man interferes with the natural circulation of the interstitial liquid phase, changes pressure relationships, and replaces the original liquid with another liquid or leaves pore space.

c. Through mining. Extensive, uneven, and more damaging terrain settlement occurs in underground mining regions; for years this has been treated as "mining damage." Whether mineral resources are mined in drifts and large openings or extracted through leaching as brine, one must always expect uneven settlement on the surface and damage to real estate and man-made structures (Niemczyk 1949; Lütkens 1967).

8. *Eustatic Variations of Sea Level*

It need not always be geologists and engineers who interfere in natural geologic phenomena and release completely uncontrolled effects. This can be demonstrated by an example cited by Egli (1970: the carbon dioxide content of the air, as a result of man's burning processes in industry for heating, producing energy, and in motor vehicles, rose by about 12 percent between 1900 and 1950, according to Bolin and

Erikkson (1959). The resulting increased reflection has resulted in higher temperatures, amounting to 1.1 degrees per century, according to Plass (1956).

Such a general rise in temperature over the entire earth must lead to ice melting in glaciated mountains and especially at both polar caps, and, as a result, to glacial-eustatic rise in sea level and unavoidable flooding of low-lying coasts, for example in Holland, Germany, the Po delta at Venice, and other deltas.

When we try to classify the different kinds of human interventions, we may have a problem in determining whether certain effects are tolerated or uncontrollable.

The greater the experience of engineers and geologists, the less room there actually is for unexpected, uncontrolled effects. More often it becomes a question of a technical error when an unforeseen consequence arises.

Every engineering geologist is constantly under the pressure of his professional responsibility and his employer. Directly stated or tacitly, the engineering geologist must assume the responsibility for keeping costs down.

IV. LIMITATIONS OF THE PRINCIPLE OF ACTUALISM

Extrapolation of current geologic processes into the past and the future is a logical application of actualism, which was first introduced as a research method in geology by Hoff about 150 years ago and then by Lyell and others.

An example of this method is the periodic survey of the deltas of the Alpine rivers in the Pre-Alpine lakes, from which we can record growth and correlate numerically the rate of erosion of the rivers in their Alpine drainage areas (Eidgen. Amt für Wasserwirtschaft, 1939; Waibel, 1962). Using the delta surveys of the last decades, we can extrapolate from recent erosion in the Alps to the past. The result is—presuming an even climate and a quiet crust—that in the last 600,000 years the drainage areas of Swiss Alpine rivers have been lowered on the average about 200 to 300 meters by erosion. We can also extrapolate today's erosional yield into the future and predict that present drainage areas, based on erosion rates in Alpine border lakes, will be reduced by half in about 3 to 4 million years (Jäckli, 1958).

This actualistic thought process must, however, be circumscribed in view of the anthropogene influences. Man's influence has accelerated the erosion process where forest cutting and clear-cutting from the Middle Ages into the past century exposed the slopes, which had been sheltered by the forest, and subjected them to increased erosion. On the other hand, all artificial drainage, afforestation, torrent control, artificial sedimentation (as in the case of Domleschg), constructed reservoirs, and gravel dredging in rivers have a restraining effect on the growth of deltas.

If the geologist wishes to maintain the spatial and temporal geologic frame of reference, he will recognize that "man," as a factor in the continuing chain of natural processes, has played a decidedly locally limited role, and in time during only the past 3,000 years, However, the previous relationship between human and natural influence will indeed change fundamentally if nuclear explosives are used in mining and if nuclear devices are used to construct navigational channels, divert streams, create harbors, and make space for piers (Bülow, 1954).

The present generation of earth scientists has a responsibility toward the earth, and anthropocentrically speaking, it is in their care. When man interferes geologically in natural phenomena, it is necessary that he do so with moderation and reflection. Nongeologists require and expect the geologist to carry out only those interventions that have effects and consequences which they can understand and control in spatial and temporal respects.

REFERENCES

BASLER, E. (1971): *Umweltprobleme aus der Sicht der technischen Entwicklung.* Schweiz. Bauztg. *89*/13.

BECK, P. (1943): *Die Natur des Amtes Thun.* Druck- und Verlagsanstalt Ad. Schaer, Thun.

BENTZ, A., und Martini, H. J. (1968): *Lehrbuch der angewandten Geologie.* Ferdinand-Enke-Verlag, Stuttgart.

BÖHLMANN, D. (1971): *Müllgrube Meer?* Kosmos 7/71.

BÜLOW, K. VON (1954): *Anaktualistische Wesenszüge der Gegenwart.* Z. d. geol. Ges. *105.*

BUSINGER, A. (1836): *Der Kanton Unterwalden.*

COLLET, L. W. (1916): *Le charriage des alluvions.* Annalen Schweiz. Landeshydrographie, Bern.

DITTMER, E. (1955): *Der Mensch als geologischer Faktor an der Nordseeküste.* Eiszeitalter und Gegenwart *8.*

Eidg. Amt für Wasserwirtschaft (1939): *Deltaaufnahmen.*

EGLI, E. (1970): *Natur in Not. Gefahren der Zivilisationslandschaft.* Hallwag-Verlag, Bern und Stuttgart.

ENGLER, A. (1919): *Untersuchungen über den Einfluss des Waldes auf den Stand der Gewässer.* Mitt. schweiz. Anst. forstl. Versuchswesen *12.*

FINKENWIRTH, A. (1967): *Deep Well Disposal of Waste Brine in the Werra Potash Region. AIH-Mémoires VII. Hannover.*

GERBER, Ed. (1967): *Die Flussauen in der schweizerischen Kulturlandschaft.* Geogr. Helv. *22*/1.

GIERLOFF-EMDEN, H. G. (1954): *Die morphologischen Wirkungen der Sturmflut vom 1. Februar 1953 in den Westniederlanden.* Hamburger geographische Studien *4*, Hamburg.

GRAFTDIJK, K. (1960): *Holland bezwingt das Meer.* Verlag Wereldvenster, Baarn, Holland.

GROSJEAN, G. (1962): *Die Ableitung der Kander in den Thunersee vor 250 Jahren.* Jb. vom Thuner- und Brienzersee, Interlaken.

HÄUSLER, H. (1959): *Das Wirken des Menschen im geologischen Geschehen.* Naturk. Jb. der Stadt Linz.

HEIM, ALB. (1932): *Bergsturz und Menschenleben.* Vjschr. natf. Ges. Zürich, *77*/20.

JÄCKLI, H. (1957): *Gegenwartsgeologie des bündnerischen Rheingebietes. Ein Beitrag zur exogenen Dynamik alpiner Gebirgslandschaften.* Beitr. Geol. Schweiz, geotechnische Serie, *36.*

– (1958): *Der rezente Abtrag der Alpen im Spiegel der Vorlandsedimentation.* Eclogae geol. Helv. *51*/2.

– (1964): *Der Mensch als geologischer Faktor. 250 Jahre Kanderdurchstich.* Geogr. Helv. *19*/2.

KOLB, A. (1962): *Sturmflut 17. Februar 1962. Morphologie der Deich- und Flurbeschädigungen zwischen Moorburg und Kranz.* Hamburger geogr. Studien *16*, Hamburg.

LOFGREN, B. E. (1961): *Measurement of Compaction of Aquifer Systems in Areas of Land Subsidence.* U.S. Geol. Survey Research.

– (1968): *Analysis of Stresses Causing Land Subsidence.* U.S. Geol. Survey Prof. Paper 600-B.

LÜTKENS, O. (1967): *Bauen im Bergbaugebiet.* Springer-Verlag, Berlin.

MAYRHOFER, H. (1965): *Die Kaliabwässerversenkung in den Plattendolomit des Werra-Gebietes.* Exkursionsführer AIH-Kongress Hannover.

MEYER, R. (1970): *Die Beanspruchung der Umwelt durch die Besiedlung.* Aus: «Schutz unseres Lebensraumes», Verlag Huber, Frauenfeld.

MÜLLER, L. (1964): *The Rock Slide in the Vajont Valley.* Felsmechanik u. Ingenieurgeol. *2.*

– (1968): *New Considerations on the Vajont Slide.* Felsmechanik u. Ingenieurgeol. *6.*

MÜLLER, St. (1970): *Man-Made Earthquakes. Ein Weg zum Verständnis natürlicher seismischer Aktivität.* Geol. Rundschau *59*/2.

MORLOT, V. (1916): *Flusskorrektionen der Schweiz.* Schweiz. Oberbauinspektorat *5*, Bern.

NIEMCZYK, H. (1949): *Bergschadenkunde.* Glückauf, Essen.

NIPKOW, F. (1920): *Vorläufige Mitteilungen über Untersuchungen des Schlammabsatzes im Zürichsee.* Z. Hydrologie *1*, *100*–122.

PETER, A. (1922): *Die Juragewässerkorrektion.* Bern.

ROESLI, F. (1965): *Das Renggloch als geologisches Phänomen und als Beispiel einer frühen Wildbachkorrektion.* Eclogae geol. Helv. *58*/1.

SAYAMA, M., und MOMIKURA, Y. (1971): *Groundwater in the Shiroishi Plane (Japan).* International Association of Hydrogeologists, Asian Regional Conference, Guidebook for Kyushu Tour.

SIBERT, G., und WERNER, H. (1969): *Bergeverkippung und Grundwasserbeeinflussung am Niederrhein.* Fortschr. Geol. Rheinld. Westf. *17.*

SIMON, W. G. (1963): *Sturmfluten in der Elbe und bei Hamburg in historischer und aktuogeologischer Sicht.* Abh. Verh. nat. Ver. Hamburg *7.*

– (1965): *Geschichte des Elbe-Aestuars.* Abh. Verh. nat. Ver. Hamburg *9.*

SPITS, A. (1959): *Neues Land. Zuiderseewerke und Deltaplan.* Amsterdam.

– (1960): *Die Deltawerke.* Vereniging Nederland in den Vreemde. Amsterdam.

SCHMID, W. (1962): *Wildbachverbauungen und Flusskorrektionen im Einzugsgebiet der Linth/Limmat.* Wasser- und Energiewirtschaft *8.*

SCHULTZ, K. (1965): *Der Generalplan für das Land Schleswig-Holstein.* NZZ, Bl. 5, vom 10. 11. 1965.

Schweiz. Vereinigung für Innenkolonisation und industrielle Landwirtschaft (1945): *Die Integralmelioration in der Talebene Domleschg.*

STUMM, W. (1971): *Manipulation der Umwelt durch den Menschen.* NZZ Nr. 441 vom 22. 9. 1971.

THOMAS, E. A. (1971): *Oligotrophierung des Zürichsees.* Vjschr. naturf. Ges. Zürich *116*/1.

VARNHAGEN, B. (1967): *Untersuchung über den Zusammenhang zwischen Grundwasserabsenkung und Bodensetzung im Rheinischen Braunkohlenrevier.* Diss. T. H. Aachen.

VOGT, J. (1958): *Zur historischen Bodenerosion in Mitteldeutschland.* Petermanns Mitt. *3.*

WAIBEL, F. (1962): *Das Rheindelta im Bodensee. Seegrundaufnahme im Jahre 1961.* Bericht des österreichischen Rheinbauleiters, Bregenz.

ZWITTIG, L. (1964): *Die Beeinflussung des Grundwassers durch Mülldeponien.* Steirische Beitr. Hydrogeol.

22

Reprinted from *Geoforum*, **14**, 53–61 (1973)

Heavy Metal Accumulation in River Sediments: A Response to Environmental Pollution

Ulrich FÖRSTNER and German MÜLLER, Heidelberg*

Abstract: As evidenced by catastrophic cadmium and mercury poisonings in Japan, heavy metals belong to the most toxic environmental pollutants. Through the investigation of sediments, the extent, distribution and provenance of heavy metal contamination in rivers and lakes can be determined and traced. Eight heavy metals from the clay fraction of sediments from major rivers within the Federal Republic of Germany were determined by means of atomic adsorption spectrometry. Heavy metals especially known for their high toxicity are enriched most: mercury, lead and zinc by a factor of 10; cadmium by a factor of 50, as compared with the natural background of these elements. A mobilisation of heavy metals from the suspended load and from the sediments, as to be observed in rivers approaching the marine enviromment, could endanger marine organisms, thus negatively influencing the acquatic food chain. With a further increase of heavy metal pollution, a threat to the drinking water supplied by rivers and lakes cannot be excluded.

1. Introduction

Environmental pollution is clearly evidenced by the present situation of rivers and lakes in highly industrialized regions. Most of these surface waters are endangered in their biological existence; some have already reached conditions under which higher organisms can no longer survive. This development is due to a strong increase of industrial waste, domestic sewage and agricultural runoff which are accumulated in the very limited lake and river systems and then discharged into the oceans. Man himself is threatened in two main ways:

First of all, surface waters are providing more and more of the drinking water supply since the natural ground water resources are rapidly diminishing and further exploitation is becoming increasingly expensive and difficult. For example, two thirds of the total water supply of the Federal Republic of Germany originates from river or lake water; nearly 10 % of this water is used for drinking water purposes. A further increase of the surface water pollution would, therefore, seriously affect the quality of the drinking water.

Secondly, most of the contaminants are released into the sea. In the shallower parts, particularly at the margins of highly industrialized regions, p.ex. in the North Sea and in the Baltic Sea, the rate of pollution is highest. Some pollutants become further enriched in the aquatic food chain and may then represent a serious danger for man.

2. Heavy Metal Pollution

Spectacular accidents in Japan caused by mercury and cadmium have sparked general interest in heavy metals as potential hazards for man: 46 fishermen who had eaten fish from the Minamata Bay died from mercury poisoning. Villagers in the Jintsu River Basin were stricken with a skeletal disease known as osteomalacia (the Japanese called

* Doz. Dr. Ulrich FÖRSTNER, Prof. Dr. German MÜLLER, Laboratorium für Sedimentforschung, Universität Heidelberg, Postfach 840, D-69 Heidelberg 1, Germany (W)

it "itai-itai byo" -- the "ouch-ouch disease") – caused by the elevated cadmium content of the drinking water whose source was a river polluted by waste material from a closed zinc mine.

The particular threat of the heavy metals lies in the fact that they, in contrast to many organic pollutants, are not decomposed by microbiological activity. On the contrary, heavy metals can be enriched by organisms and the type of bonding can be converted to more poisonous metal-organic complexes, as observed in the transformation of elementary and divalent ionic mercury into methylmercury.

Heavy metals, which in normal concentrations are essential components of biochemical functions (Cu, Zn, Fe, Mn, Cr, Co), are toxic when present in higer concentrations: "The most important mechanism of toxic action is thought to be the poisoning of enzymes" (BOWEN 1966).The total extent of the heavy metal pollution is unknown, neither for the atmosphere nor for the aquatic environment. A first attempt to tackle this problem involves a comparison of the consumption of heavy metal with the natural concentration of these elements in the different spheres (lithosphere, pedosphere, hydrosphere, atmosphere). The ratio

$$\frac{\text{metal consumption (in tons/a)}}{\text{average metal content in a specific sphere (in g/ton)}}$$

is introduced as a measure of the relative pollution potential of each element in a certain sphere.

Table 1

- World's consumption of heavy metals in 1968 (after SAMES, 1971), metal contents in soils (after BOWEN, 1966) and "Index of relative pollution potential" (this work) for the pedosphere.

- Welt-Schwermetallverbrauch 1968 (nach SAMES, 1971), Metallgehalte von Böden (BOWEN, 1966) und „Index des relativen Verschmutzungspotentials" (diese Arbeit) für die Pedosphäre.

	Consumption x 1 000 t/y	Soils (ppm)	Index of relative pollution potential
Iron	400 000	38 000	1
Manganese	9 200	850	1
Copper	6 400	20	30
Zinc	4 600	50	10
Lead	3 500	10	35
Chromium	1 700	100	2
Nickel	493	40	1
Tin	232	10	2
Cobalt	19	8	0.2
Cadmium	15	0.06	25
Mercury	10	0.03	30

The „ Index of the Relative Pollution Potential"[1]) for some important industrial metals in the pedosphere is shown in Table 1. It becomes evident that the more „rare" (and more toxic) metals are enriched by an order of magnitude as compared with iron, manganese, chromium and nickel. Cobalt, on the other hand, is depleted.

It should be emphasized, however, that the recycling of the produced metals into the different spheres is neither uniform nor complete: a certain amount is „bound" in those objects which outlast longer periods of time (machines, instruments, construction material, coins, cutlery, just to mention a few examples); another portion is regained from scrap and waste.

In the long run, the bulk of the consumed metals is released into the environment. The various ecosystems are hereby affected in different ways as demonstrated by the following examples:

15 % of the world's lead production is consumed for gasoline additives; during combustion almost the entire lead content is emitted into the ambient *air.*

At least 20 % of the world's mercury production is employed in chlor-alkali-electrolysis; a large amount is released into surface *waters.*

3. Sediments as Pollution Indicators

Contaminants in aquatic systems can be investigated by analyzing either the water, the suspended material or the sediments, and by comparison of the obtained data with their natural background[2]).

Short-term measurements of pollutants in the *water* are (in most cases) not conclusive. Changes in water discharge, fluctuations in the predominance of certain source areas, and irregular local emissions – it is a well known fact, that certain pollutors empty out accumulated waste products during high water periods – are responsible for variations in orders of magnitude. The same holds true for the suspended material which has a high sorption capacity for metal ions and may also contain considerable amounts of particulate inorganic (metal compounds) and organic (p.ex. DDT, aminoacid-metal complexes) matter.

Only by long-term continuous measurements over several yearly cycles can a satisfactory result be obtained. Up to

[1]) The ratio was multiplied by the factor 10^{-4} in order to obtain whole numbers.

[2]) The problem of the „natural backround" is difficult to solve. Investigation often began when a world wide contamination was already present. With regard to heavy metals in aquatic sediments, we are in a more advantageous position: natural background values are „conserved" in sediments of the geological past. In this study, the average heavy metal composition of shales (TUREKIAN & WEDEPOHL, 1961) is used as a geochemical background.

now, these types of measurements have only been carried out at a very limited number of stations which do not permit an overall picture of the condition of a lake, river or sea.

In this situation, the study of the *sediments* seemed to be a possible solution to the problem. ZÜLLIG (1956) described the significance of „Sedimente als Ausdruck des Zustandes eines Gewässers" („sediments as a response to the condition of an aquatic system").

WEBB (1971) demonstrated that pollution reconnaissance can be carried out in stream sediments in the same way as in standard practice in mineral exploration.

Sediments are composed of numerous individual layers, each of which corresponds to a distinct condition of water flow: coarse layers represent bed load deposited during stronger currents, fine layers consist mainly in suspended load deposited during weaker (or absent) currents. Fine grained sediments built up of a large number of individual layers should therefore represent an average value for certain contaminants over a long period of time. This is especially true for heavy metals which are strongly adsorbed by the clay minerals and the organic fraction of the sediments.

4. Heavy Metal Concentrations in Sediments of Major Rivers in the Federal Republic of Germany

More than 100 sediment samples were collected during extremely low water levels in Winter 1971/72 from the otherwise flooded banks of the Rhine, Ems, Weser, Elbe and Danube and from their more important tributaries.

In order to obtain comparable data, the $< 2\ \mu m$ fraction („clay") of each sample was separated in settling tubes and analyzed for cadmium, mercury, cobalt, lead, copper, nickel, chromium and zinc.

An interim report (BANAT, FÖRSTNER & MÜLLER 1972a) containing both the results of this investigation and also of other research on heavy metals in the aquatic environment has been distributed privately and is now being published in revised and enlarged form (FÖRSTNER & MÜLLER 1973). Short summaries of the distribution pattern of certain elements in the different river systems have been recently published (BANAT, FÖRSTNER & MÜLLER 1972b, c).

The following chapters deal with (1) the distribution and accumulation of various heavy metals in sediments of a single river (the Rhine) from its source (Lake Constance) to its mouth (the North Sea); (2) the distribution and accumulation of two selected heavy metals (lead and cadmium) investigated in all major rivers of the Federal Republic of Germany and (3) a geochemical reconnaissance of the source of a highly toxic pollutant (cadmium) in the Neckar river.

4.1. Heavy Metal Accumulation in the Sediments of the Rhine (Fig. 1, Table 2)

Between Lake Constance and the North Sea, the Rhine flows through the most highly industrialized areas of Central Europe (Rhein-Main-Gebiet, Ruhr-Gebiet). This river is known as being extremely polluted in all respects and has not unjustly been designated the „Majestic Cesspool of Europe".

Heavy metal concentrations in the sediments of Lake Constance (Bodensee) do not considerably differ from the geochemical background. The Rhine between stations 1 and 8, however, is already significantly contaminated by mercury, lead and zinc; these metals are enriched by a factor of 7, 5 and 3, respectively. A strong increase of chromium in the sediments appears at station 9 where the Weschnitz, highly polluted by sewage from leather industry, flows into the Rhine. Between stations 9–19 (Dutch border) all elements investigated, except for nickel and cobalt, are further enriched by a factor of two to three.

In the Netherlands, an additional increase is found as determined by DE GROOT et al. (1971) at Biesbosch, at the beginning of the Rhine estuary. However, a direct comparison with our values is not possible since DE GROOT's data were obtained with the sediment fraction $< 16\ \mu m$, ours with the fraction $2\ \mu m$.

The extent of the heavy metal pollution in the Rhine sediments within the German borders becomes evident by subtracting the natural background values from those obtained in the three northernmost stations (17–19), Fig. 2. From cobalt to nickel to chromium the natural portion of these elements decreases considerably. Copper, zinc, lead, mercury and cadmium – metals with a high „index of relative pollution potential" are enriched by a factor of ten or more in the sediments by man's activity.

4.2. Heavy metals in sediments of major rivers in the Federal Republic of Germany

4.2.1. General distribution

Table 3 contains the minimum, maximum and average values for 8 heavy metals determined in the sediments of the Rhine , Weser, Elbe, Danube and in two important tributaries of the Rhine (Neckar, Main). The addition of these 8 elements results in values higher than 1000 ppm for each river; in the Elbe and the Weser over 2000 ppm are found. The zinc concentration in each river is higher than all other heavy metals together.

The arrangement of the rivers according to their average content of a certain heavy metal in the clay fraction of the sediments results in the following sequence presented in Table 4.

Highly polluted with respect to cadmium, mercury, lead and zinc, which are known for their toxicity, is the Elbe.

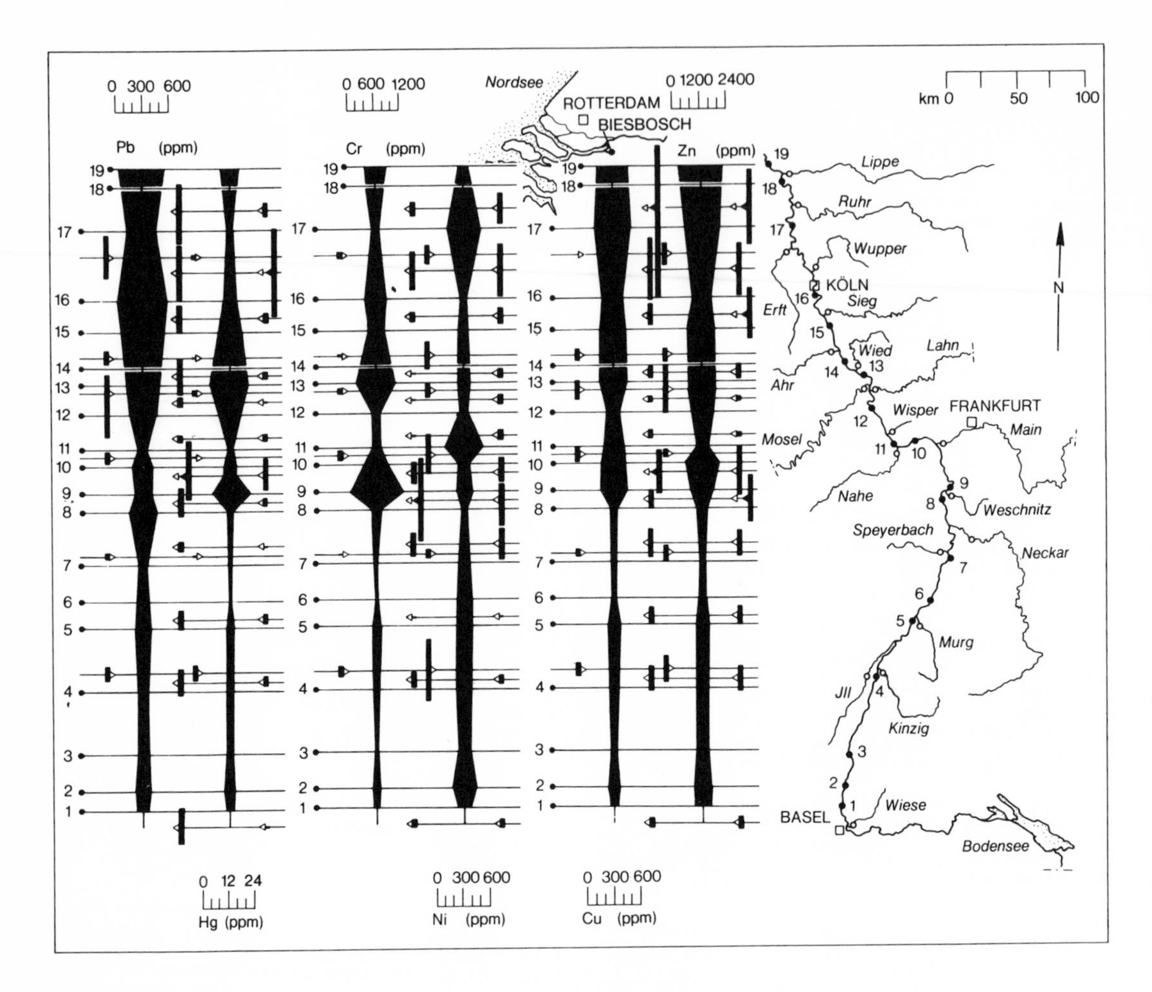

Fig. 1

- Heavy metal concentrations (in ppm) in the fraction < 2 μm of sediments of the Rhine and its tributaries (from BANAT, FÖRSTNER a. MÜLLER 1972b).
- Schwermetallgehalte (in ppm) in der Fraktion < 2 μm der Sedimente des Rheins und seiner Nebenflüsse (aus BANAT, FÖRSTNER u. MÜLLER 1972b).

The Neckar(Rhine) has the highest cadmium content. By comparison, the Danube and Ems have a low degree of pollution.

The highest content of each element was found in the tributaries: 38 ppm mercury and 830 ppm nickel in the Wupper (Rhine), 1200 ppm lead in the Aller (Weser), 1600 ppm copper and 340 ppm cobalt in the Diemel (Weser), 3 290 ppm zinc in the Ruhr (Rhine) and 6 400 ppm chromium in the Murr (Neckar).

4.2.2. Regional Distribution for Lead and Cadmium

As examples for the regional distribution of heavy metals in our river systems, lead and cadmium were chosen. Distribution charts for lead and cadmium are presented in Figs. 3 and 4.

Contents of *lead* are high in the lower and middle Rhine, Elbe and lower Main, whereas the Ems, Danube and upper Rhine are comparably less polluted.

Table 2

- Heavy metal concentrations (in ppm) in the < 2 μm fraction of sediments from Lake Constance and the Rhine river. Biesbosch data (after DE GROTT *et al.*, 1971) from the < 16 μm sediment fraction.

- Schwermetall-Konzentrationen (in ppm) in der < 2 μm-Fraktion der Sedimente vom Dosensee und vom Rhein. Die Werte von Biesbosch (nach DE GROOT *et al.*, 1971) beziehen sich auf die < 16 μm-Fraktion.

	Cd	Hg	Co	Pb	Cu	Ni	Cr	Zn
L. Constance	-	0.4	-	30	67	102	119	185
Rhine Station 1–8	4	3	19	155	86	152	121	520
Rhine Station 9–19	13	9	31.	369	286	175	493	1239
Biesbosch Netherl.	-	18	-	850	470	-	760	3900

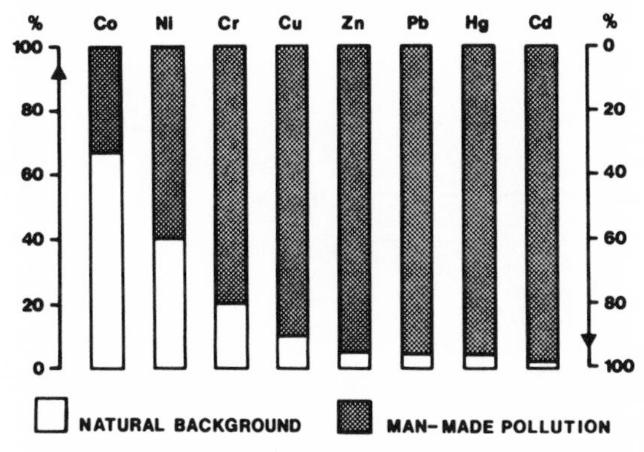

Fig. 2

- Relative amount of heavy metals derived from man-made pollution in Rhine sediments (stations 17–19 of Fig. 1).

- Relativer Anteil der zivilisatorischen Schwermetall-Belastung von Rhein-Sedimenten (Stationen 17–19, vgl. Fig. 1).

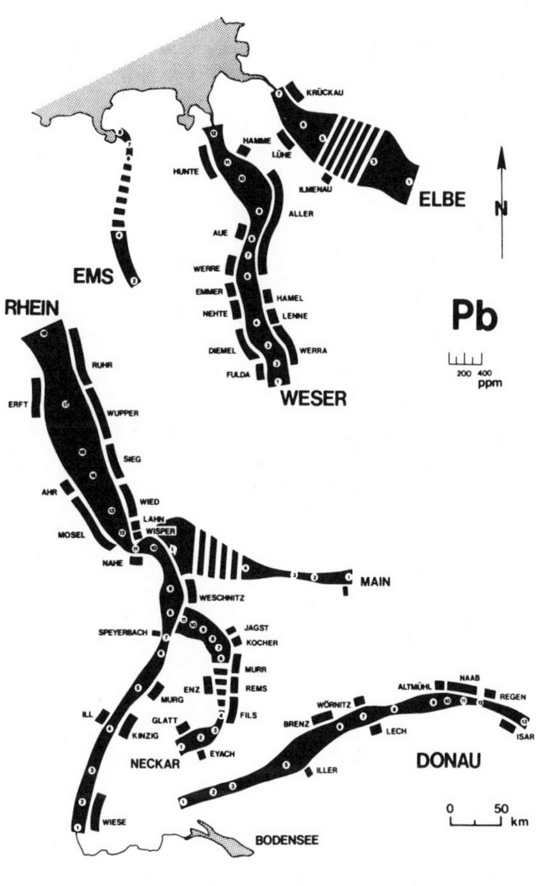

Fig. 3

- Lead concentration (in ppm) in the fraction < 2 μm the sediments of major rivers of the Federal Republic of Germany (from BANAT, FÖRSTNER a. MÜLLER 1972c).

- Blei-Konzentration (in ppm) in der Fraktion < 2 μm der Sedimente wichtiger Fließgewässer in der Bundesrepublik Deutschland (aus BANAT, FÖRSTNER u. MÜLLER 1972c).

An effect which can also be observed with copper, chromium, zinc and mercury is a distinct decrease of the lead content under marine influence in the estuaries. The same effect has been observed by DE GROOT et al. (1971) in the Rhine estuary. A possible explanation is the partial desorption of heavy metals by competing cations (esp. sodium) from sea water.

Variations of the *cadmium* contents are less significant. Only in the lower section of the Neckar river are the sediments highly polluted with cadmium.

The results of a detailed reconnaissance survey are presented in the following chapter.

4.3. Geochemical Reconnaissance of the Sources of Heavy Metal Pollutions: Cadmium in the Neckar

Fig. 4 shows that cadmium concentrations are extremely high in the lower part of the Neckar. Detailed investigations revealed maximum concentration in the 0.2–0.6 μm-fraction of the total < 2 μm fraction. For a geochemical reconnaissance survey, the 0.2–0.6 μm fraction was therefore chosen as the basis of comparison.

Table 3

- Average, minimum and maximum heavy metal concentrations (in ppm) in the < 2 μm fraction of river sediments from the Federal Republic of Germany. For comparison average values for shales (after TUREKIAN and WEDEPOHL, 1961). The enrichment factor indicates the mean enrichment of each element compared with the composition of argillaceous rocks (geochemical background).

- Durchschnitts-, niedrigste und höchste Schwermetallgehalte (in ppm) in der < 2 μm-Fraktion der Sedimente von Fließgewässern im Bereich der Bundesrepublik Deutschland. Zum Vergleich Durchschnittswerte in tonigen Sedimentgesteinen (nach TUREKIAN und WEDEPOHL, 1961). Der Anreicherungsfaktior gibt an, wie hoch die Anreicherung in der Tonfraktion der Fluß-Sedimente gegenüber dem Durchschnittswert für tonige Sedimentgesteine ist.

	Cd			Hg			Co			Pb			Cu			Ni			Cr			Zn		
	min	ϕ	max	min	ϕ	max	min	ϕ	max	min	ϕ	max	min	ϕ	max	min	ϕ	max	min	ϕ	max	min	ϕ	max
Rhine	3	9	23	1.2	6.3	17.8	17	26	43	83	251	566	42	192	408	50	164	416	33	330	1195	250	903	2061
Main	6	12	19	0.9	5.0	13.6	40	51	65	50	218	650	30	208	475	77	128	266	100	211	441	100	810	2100
Neckar	5	37	88	0.3	1.1	2.2	35	55	64	60	221	305	25	203	335	83	190	136	90	382	780	120	999	2100
Danube	7	14	29	0.6	1.5	3.0	32	47	62	81	156	312	50	232	500	96	125	165	56	187	581	250	699	1162
Ems	7	10	17	2.0	4.4	11.0	44	54	68	70	112	175	20	55	114	77	104	123	101	134	206	200	642	1408
Weser	6	14	23	1.0	2.3	7.0	41	57	70	150	241	405	25	115	210	72	98	148	100	281	860	400	1572	3100
Elbe	15	21	25	3.9	7.6	14.2	40	51	60	150	430	712	80	161	220	110	126	137	110	175	242	800	1425	2125
average Shale	0.3			0.4			19			20			45			68			90			95		
Concentration factor	50			10			2			10			4			2			3			10		

Element	1	2	3	4	5	6	7
Cd	Neckar	Elbe	Weser	Danube	Main	Ems	Rhine
Hg	Elbe	Rhine	Main	Ems	Weser	Danube	Neckar
Pb	Elbe	Rhine	Weser	Neckar	Main	Danube	Ems
Zn	Weser	Elbe	Neckar	Rhine	Main	Danube	Ems
Cu	Danube	Main	Neckar	Rhine	Elbe	Weser	Ems
Cr	Neckar	Rhine	Weser	Main	Danube	Elbe	Ems
Ni	Neckar	Rhine	Main	Elbe	Danube	Ems	Weser
Co	Weser	Neckar	Ems	Elbe	Main	Danube	Rhine

Table 4

- Order of succession of rivers according to their average content of heavy metals in the < 2 μm sediment fraction. 1 = highest, 7 = lowest average concentration.

- Reihenfolge der Flüsse nach dem mittleren Schwermetallgehalt der Fraktion < 2 μm der Sedimente. 1 = höchster, 7 = niedrigster Durchschnittsgehalt.

Between Heidelberg and Stuttgart (Fig. 5) a series of 22 samples was taken. From stations 1 to 10, the cadmium concentration increases more or less steadily from about 100 ppm to about 400 pm. Between stations 10 and 11, the concentration falls drastically to values less than 30 ppm, remaining at this level for the further stations 12–22.

These cadmium contents in the *sediments* of the Neckar led to the suspicion of a direct intoxication of the river water. Analyses of water samples, taken at distances of 1–2 km during July and October 1972, confirmed our fears. The maximum value at the second survey was 220 ppb (= 0.22 mg/l) cadmium at the inflow of the Enz river close to station 11.

Although this very high concentration was diluted by the water of the Neckar (having Cd-contents of less than 1 ppb), elevated Cd-concentrations could be found downstream until the Neckar flows into the Rhine. Between the mouth of the Enz and Heilbronn, a distance of about 20 km, the Cd-concentrations were distinctly higher at some localities than the maximum permissible limit of 10 ppb (World Health Organization, U.S. Public Health Survice) for drinking water purposes.

Further geochemical reconnaissance in the Enz river resulted in the discovery of the main source of the cadmium only some hundreds of meters upstream: a factory producing cadmium pigments.

In February 1973, the emission of cadmium was stopped by the opening of a treatment plant.

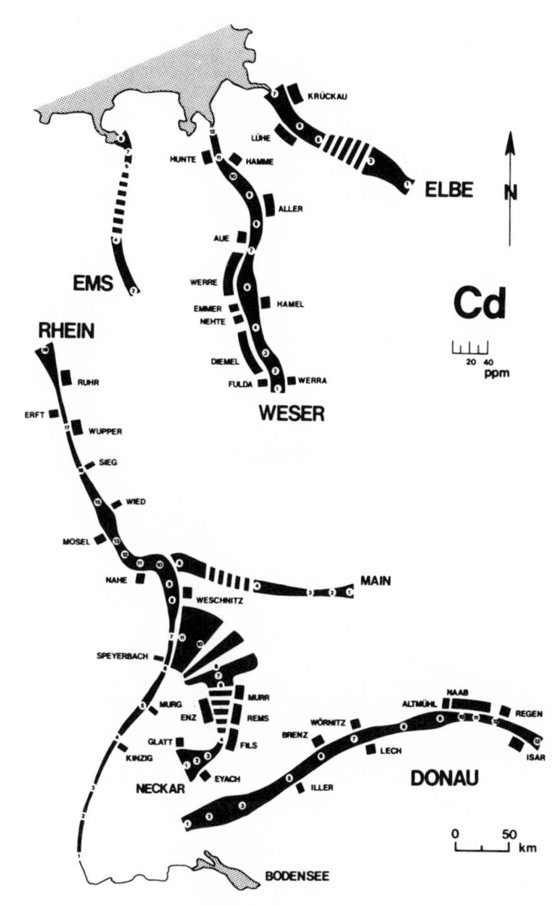

Fig. 4

- Cadmium concentration (in ppm) in the fraction $< 2\ \mu m$ of the sediments of major rivers of the Federal Republic of Germany (from BANAT, FÖRSTNER a. MÜLLER 1972c).

- Cadmium-Konzentration (in ppm) in der Fraktion $< 2\ \mu m$ der Sedimente wichtiger Fließgewässer in der Bundesrepublik Deutschland (aus BANAT, FÖRSTNER u. MÜLLER 1972c).

5. Provenance of Heavy Metals in Sediments

Heavy metal accumulations in sediments may have various sources. They are either directly induced by sewage or indirectly brought into the rivers from atmospheric emission, precipitation and by surface runoff.

As the most important source of heavy metals, industrial sewage must first be regarded. Table 5 contains a compilation of heavy metals found in major industries (after DEAN et al., 1972). This compilation shows that most of the heavy metals considered in this report have a multipurpose function.

A further important source are atmospheric emissions which above all contain cadmium (from melting ores and scrap), nickel and zinc (from combustion of coal) and lead (from gasoline additives).

An additional source has been recognized only in recent times: zinc, lead and copper as corrosion products from domestic and municipal pipe systems (HELLMANN 1972).

In the field of agriculture, mercury, being a constituent of fungicides, has to be mentioned. The usage of mercury containing fungicides, however, is decreasing steadily.

From this brief consideration, it becomes clear that in some cases difficulty arises in attributing a certain metal found in a contaminated sediment to a distinct source. A local emittent of a definite heavy metal, however, should be found by applying a geochemical reconnaissance survey.

6. Significance of Heavy Metals in Sediments

The critical question arising from the result of heavy metal studies in aquatic sediments is: Can, and to which extent, heavy metals in sediments affect the ecosystem?

As to our present knowledged, the heavy metals found in sediments present no *direct danger* as long as they are tightly bound to the sediment. The *potential danger*, however, lies in the possiblility that, under certain circumstances, (p.ex. changes in the redox potential within the sediment) a dissolution or desorption might lead to a release of metals into the water.

As long as only very little is known about the different states of bonding of heavy metals in a sediment (absorbed, complexed, organically or inorganically bound etc.), it is difficult to predict the behavior of a certain heavy metal under changed conditions.

On the other hand, the suspended load and the sediments of a river have the important function of buffering higher metal concentrations of the water, particularly by adsorption or precipitation.

Whether similar processes are also effective during the artificial enrichment of groundwater by bank filtration of river water, a procedure which will gain increasing importance in the next few years, has been investigated recently. Although the results should not be generalized and require further confirmation from other localities, they give evidence of the retaining capacity for heavy metals in the bank sediments through which the river water flows.

Fig. 5 summarizes the observations from 12 monthly investigations of a bank filtration test area at the Neckar near Heilbronn during 1972.

By passing throug the bank sediment, the various heavy metals show a different behaviour: iron and manganese are enriched whereas all other metals are more or less depleted. The highest reduction is obtained with cadmium

	Cd	Cr	Cu	Fe	Hg	Mn	Pb	Ni	Sn	Zn
Pulp, papermills, paperboard, building paper, board mills		x	x		x		x	x		x
Organic chemicals, petrochemicals	x	x		x	x		x		x	x
Alkalis, chlorine, inorganic chemicals	x	x		x	x		x		x	x
Fertilizers	x	x	x	x	x	x	x	x		x
Petroleum refining	x	x	x	x			x	x		x
Basic steel works foundries	x	x	x	x	x		x	x	x	x
Basic non-ferrous metals-works, foundries	x	x	x		x		x			x
Motor vehicles, aircraft-plating, finishing	x	x	x		x			x		
Flat glass, cement, asbestos products, etc.		x								
Textile mill products		x								
Leather tanning, finishing		x								
Steam generation power plants		x								x

Table 5

- Heavy metals found in major industries (after DEAN *et al.* 1972).
- Verwendung von Schwermetallen in wichtigen Industriezweigen (nach DEAN *et al.*, 1972).

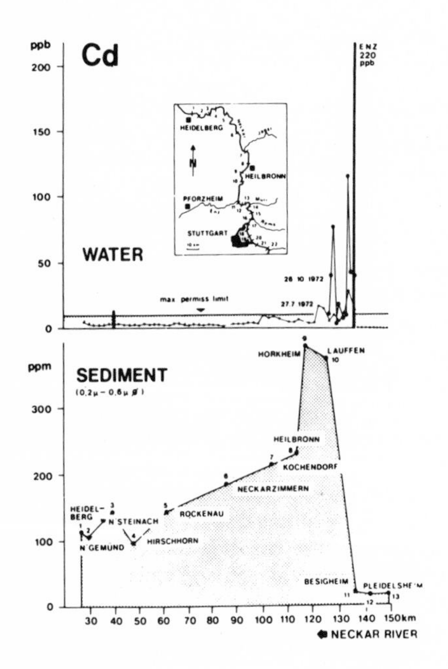

Fig. 5

- Cadmium pollution in sediments and water of the lower Neckar region.
- Cadmium-Verunreinigung von Sedimenten und Wasser des Gebietes des unteren Neckars.

(more than 95 %) whereas lead is reduced by only about 50 %. Since the retaining capacity of a bank filtration sediment is limited, a further increase of the heavy metal contents of the river water – possibly caused by an accident or by release of heavy metals from river sediments – might affect the quality of the drinking water.

As already mentioned, the heavy metal concentration of the river sediments decreases considerably as approaching the marine environment. It is to be assumed that his depletion already occurs in a state where the future sediment particles are still in suspension. The released heavy metals become an additional component of the sea water and in their dissolved state may easily enter the food chain.

Finally, the economic value of the heavy metals, bound to the suspended matter and dissolved in the water, which are each year lost to the sea, should be mentioned: In the Rhine alone, a total of 35 000 tons of zinc, 4 000 tons of copper and lead, 500 tons of cadmium and 150 tons of mercury are discharged into the North Sea (FÖRSTNER & MÜLLER 1973). The tonnage of zinc, mercury and cadmium corresponds to 0.8 %, 1.5 % and 3 % respectively, of the annual world production of these metals.

Since the natural resources of most heavy metals are very limited, the questions arises, how long can we afford the extravagance of wasting vital materials in this way. It should therefore be expected that economical rather than (eco-) logical aspects will lead to a reduction of the environmental pollution by heavy metals.

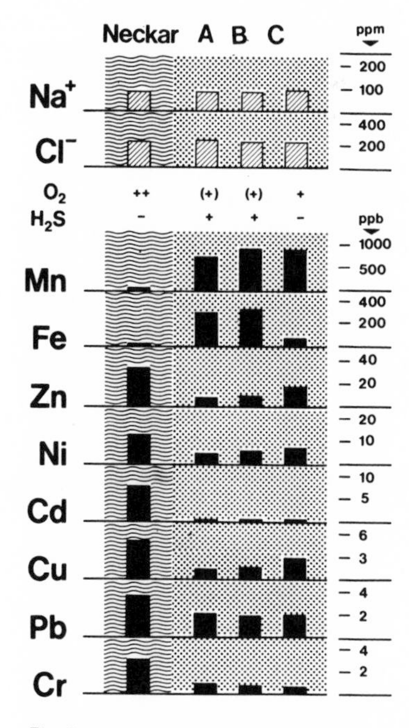

Fig. 6

- Behaviour of heavy metals during bank filtration in a test area near Heilbronn. The distance between the Neckar and well A is 40 m. The same distance also exists between wells A–B and B–C.

- Das Verhalten von Schwermetallen bei der Uferfiltration in einer Test-Strecke bei Heilbronn. Der Abstand zwischen dem Neckar und Brunnen A, wie auch zwischen den Brunnen A–B und B–C, beträgt jeweils 40 m.

References

BANAT, K., U. FÖRSTNER and G. Müller (1972a): *Schwermetall-Anreicherungen in den Sedimenten wichtiger Flüsse im Bereich der Bundesrepublik Deutschland – eine Bestandsaufnahme.* Unpubl. Report, 230 p., Heidelberg.

BANAT, K., U. FÖRSTNER and G. MÜLLER (1972b): Schwermetalle in den Sedimenten des Rheins. *Umschau in Wiss. und Techn.*, **72**, 192–193.

BANAT, K., U. FÖRSTNER and G. MÜLLER (1972c): Schwermetalle in Sedimenten von Donau, Rhein, Ems, Weser und Elbe im Bereich der Bundesrepublik Deutschland. *Naturwiss.*, **59**, 525–528.

BOWEN, H. J. M. (1966): *Trace elements in biochemistry.* Academic press, London and New York 235 pp.

DEAN, J. G., F. L. BOSQUI and V. H. LANOUETTE (1972): Removing heavy metals from waste water. *Environm. Science and Technology*, **6**, 518–522.

DE GROOT, A. J., J. J. M. DE GOEIJ and C. ZENGERS (1971): Contents and behaviour of mercury as compared with other heavy metals in sediments from rivers Rhine and Ems. *Geologie en Mijnbouw*, **50**, 393–398.

FÖRSTNER, U. and G. MÜLLER (1973): *Schwermetall-Anreicherungen in Binnengewässern.* Springer Verlag Heidelberg-Berlin-New York, in print.

HELLMANN, H. (1972): Herkunft der Sinkstoffablagerungen in Gewässern. *Deutsche Gewässerkunkl. Mitt.*, **16**, 137–141.

SAMES, C.-W. (1971): *Die Zukunft der Metalle.* Suhrkamp Verlag, Frankfurt/Main 240 pp.

WEBB, J. S. (1971): Regional Geochemical Reconnaissance in Medical Geography. *Geol. Soc. Amer. Man.*, **123**, 31–42.

ZÜLLIG, H. (1956): Sedimente als Ausdruck des Zustandes eines Gewässers. *Schweiz. Z. Hydrologie*, **18**, 7–143.

Acknowledgements

We wish to express our sincere thanks to the Deutsche Forschungsgemeinschaft and to the Ministry of the Interior of the Federal Republic of Germany for kindly supporting our investigations.

23

Reprinted from *U.S. Geol. Surv. Circ. 601-F,* 1-10 (1970)

Hydrologic Implications of Solid-Waste Disposal

By William J. Schneider

ABSTRACT

The disposal of more than 1,400 million pounds of solid wastes in the United States each day is a major problem. This disposal in turn often leads to serious health, esthetic, and environmental problems. Among these is the pollution of vital ground-water resources.

Of the six principal methods of solid-waste disposal in general use today, four methods—open dumps, sanitary landfill, incineration, and onsite disposal—carry an inherent potential for pollution of water resources. Seepage of rainwater through the wastes leaches undesirable constituents which reach the ground water in the area. This leachate is generally both biologically and chemically contaminated.

The extent of the pollution from this leachate is largely dependent upon the geologic environment in which the solid wastes are deposited. Pollution potential is highest in permeable areas with a shallow water table where the wastes are in direct contact with the ground water. In a relatively impermeable area, the pollution is generally confined locally to the vicinity of the waste-disposal site.

Site selection for disposal of solid wastes must be based on adequate water-resources information if pollutional potential is to be minimized. This will require regional as well as localized data on the water resources of the area. Only through such an approach can adequate protection be afforded to the environment in general and the water resources in particular.

INTRODUCTION

The disposal of solid-waste material—principally garbage and rubbish—is primarily an urban problem. However, unlike liquid waste disposal of sewage and industrial effluents, the problem has received only limited recognition. It is common practice in many metropolitan areas to overlook or ignore the consequences of waste-disposal programs. The full scope of the problem, though, cannot be ignored.

The urban population of the United States is now producing an estimated 1,400 million pounds of solid wastes each day. Disposal of these wastes is a major problem of all cities. In many instances, seemingly endless streams of trucks and railroad cars haul these wastes long distances—as much as hundreds of miles—to disposal sites. Based on a volume estimate of 5.7 cubic yards per ton of waste, this refuse is sufficient to cover more than 400 acres of land per day to a depth of 10 feet. Local governments spend an estimated $3 billion each year on collection and disposal, a sum exceeded in local budgets only by expenditures for schools and roads.

The disposal of these solid wastes poses many problems to local government agencies. Unfortunately, the problem is handled by many governments on the basis of expediency without due regard to environmental considerations. Garbage and rubbish are collected, hauled minimum distances commensurate with public acceptance, and dumped. Occasionally, the waste is either burned or mixed with soil to provide landfill. As long as the procedure removes the refuse and as long as the disposal site is not a health hazard and does not offend esthetic values too greatly, the operation is considered successful. Overlooked or even ignored is the effect of the disposal on the total environment, including the water resources of the area. Although the disposal of solid wastes can create many serious health, esthetic, and environmental problems, only the hydrologic implications—the effect upon water resources—are considered in this report.

TYPES OF SOLID WASTES

Our urban society generates many types of solid wastes. Each may exert a different influence on the

TABLE 1.—*Classification of refuse materials*

[Adapted from American Public Works Association (1966)]

Kind of refuse	Composition	Source
Garbage	Wastes from preparation, cooking, and serving of food; market wastes; wastes from handling, storage, and sale of produce.	Households, restaurants, institutions, stores, and markets.
Rubbish	Combustible: paper, cartons, boxes, barrels, wood, excelsior, tree branches, yard trimmings, wood furniture, bedding, and dunnage. Noncombustible: metals, tin cans, metal furniture, dirt, glass, crockery, and minerals.	Do.
Ashes	Residue from fires used for cooking and heating and from onsite incineration.	Do.
Trash from streets	Sweepings, dirt, leaves, catch-basin dirt, and contents of litter receptacles.	Streets, sidewalks, alleys, vacant lots.
Dead animals	Cats, dogs, horses, and cows	Do.
Abandoned vehicles	Unwanted cars and trucks left on public property	Do.
Demolition wastes	Lumber, pipes, brick, masonry, and other construction materials from razed buildings and other structures.	Demolition sites to be used for new buildings, renewal projects, and expressways.
Construction wastes	Scrap lumber, pipe, and other construction materials	New construction and remodeling.

water resources of an area. In order to understand the effect of each type, it is necessary to identify the various types as to the principal constitutents. Table 1 lists the various categories and sources of refuse material primarily generated by urban activities. Not included are wastes from industries and processing plants; hazardous, pathological, or radioactive wastes from institutions and industries; solids and sludge from sewage-treatment plants; and other special types of solid wastes. These items usually pose special handling problems and are usually not a part of normal municipal solid-waste-disposal programs. The following descriptions of the categories of solid wastes are abbreviated from descriptions by the American Public Works Association (1966).

Waste refers to useless, unused, unwanted, or discarded materials including solids, liquids, and gases.

Refuse refers to solid wastes which can be classified in several different ways. One of the most useful classifications is based on the kinds of material: garbage, rubbish, ashes, street refuse, dead animals, abandoned automobiles, industrial wastes, demolition wastes, construction wastes, sewage solids, and hazardous and special wastes.

Garbage is the animal and vegetable waste resulting from the handling, preparation, and cooking of foods. It is composed largely of putrescible organic matter and its natural moisture. It originates primarily in home kitchens, stores, markets, restaurants, and other places where food is stored, prepared, or served.

Rubbish consists of both combustible and noncombustible solid wastes from homes, stores, and institutions. Combustible rubbish is the organic component of refuse and consists of a wide variety of matter that includes paper, rags, cartons, boxes, wood, furniture, bedding, rubber, plastics, leather, tree branches, and lawn trimmings. Noncombustible rubbish is the inorganic component of refuse and consists of tin cans, heavy metal, mineral matter, glass, crockery, metal furniture, and similar materials.

Ashes are the residue from wood, coke, coal, and other combustible materials burned in homes, stores, institutions, and other establishments for heating, cooking, and disposing of other combustible materials.

Street refuse is material picked up by manual and mechanical sweeping of streets and sidewalks and is the debris from public litter receptacles. It includes paper, dirt, leaves, and other similar materials.

Dead animals are those that die naturally or from disease or are accidentally killed. Not included in this category are condemned animals or parts of animals from slaughterhouses which are normally considered as industrial waste matter.

Abandoned vehicles include passenger automobiles, trucks, and trailers that are no longer useful and have been left on city streets and in other public places.

METHODS OF SOLID-WASTE DISPOSAL

The disposal of these solid wastes generated by our urban environment is generally accomplished by one or more of six methods. All are currently in use to one degree or another in various parts of the United States. To a large extent, the method of waste disposal in any particular area depends upon local conditions and, to some extent, upon public attitude. In many areas several methods are employed. Each has its unique relation to the water resources of the area. The six general methods of solid waste disposal are:

1. Open dumps.
2. Sanitary landfill.
3. Incineration.
4. Onsite disposal.
5. Feeding of garbage to swine.
6. Composting.

Open dumps.—Open dumps are by far the oldest and most prevalent method of disposing of solid wastes. In a recent survey, 371 cities out of 1,118 surveyed stated that this method was emphasized within their jurisdictions. In many cases, the dump sites are located indiscriminately wherever land can be obtained for this purpose. Practices at open dumps differ. In some dumps, the refuse is periodically leveled and compacted; in other dumps the refuse is piled as high as equipment will permit. At some sites, the solid wastes are ignited and allowed to burn to reduce volume. In general, though, little effort is expended to prevent the nuisance and health hazards that frequently accompany open dumps.

Sanitary landfill.—As early as 1904, garbage was buried to provide landfill. Although in subsequent years, the practice was used by many cities, the technique of sanitary landfill as we know it today did not emerge until the late 1930's. By 1945, almost 100 cities were using the practice, and by 1960 more than 1,400 cities were disposing of their solid wastes by this method.

Sanitary landfill consists of alternate layers of compacted refuse and soil. Each day the refuse is deposited, compacted, and covered with a layer of soil. Two types of sanitary landfill are common: area landfill on essentially flat land sites, and depression landfill in natural or manmade ravines, gulleys, or pits. Depth of the landfill depends largely on local conditions, types of equipment, availability of land, and other such factors, but it commonly ranges from about 7 feet to as much as 40 feet as practiced by New York City.

In normal operation, the refuse is deposited and compacted and covered with a minimum of 6 inches of compacted soil at the end of each working period or more frequently, depending upon the depth of refuse compacted. Normally about a 1:4 cover ratio is satisfactory; that is, 1 foot of soil cover for each 4-foot layer of compacted refuse. Ratios as high as 1:8, however, have been used. The final cover is at least 2 feet of compacted soil to prevent the problems associated with open dumps.

Incineration.—Incineration is the process of reducing combustible wastes to inert residue by burning at high temperatures of about 1,700° to 1,800°F. At these temperatures all combustible materials are consumed, leaving a residue of ash and noncombustibles having a volume of 5 to 25 percent of the original volume.

Although incineration greatly reduces the volume and changes the material to inorganic matter, the problem of disposal is still present. Much of the residue is hauled to disposal sites or is used for landfill, although the land required for disposal of the residue is about one-third to one-half of that required for sanitary landfill. Some cities require that combustible materials be separated from noncombustibles prior to collection, while others use magnetic devices to extract ferrous metal for salvage.

The combination of urban growth, increasing per-capita output of refuse, and the rising costs of land for sanitary landfills has stimulated the use of incineration for solid-waste disposal. Today, there are an estimated 600 central-incinerator plants in the United States with a total capacity of about 150,000 tons per day.

Onsite disposal.—With the increasing rate of production of solid wastes in the urban environment, there is a growing trend toward handling this waste in the home, apartment, and institution. Onsite disposal has become increasingly popular during the past decade as a way of minimizing the waste problem at its source. Most widely used devices for onsite disposal are incinerators and garbage grinders.

Onsite incineration is used widely in apartment houses and institutions. The incinerators do, however, require constant attention to insure proper operation and complete combustion. Domestic incinerators for use in individual homes are not a major factor in solid-waste disposal, nor are they likely to be a major factor in the near future. Maintenance and operating problems are usually considerable.

Garbage grinders, on the other hand, are becoming increasingly prevalent in homes for disposal of kitchen food wastes. It is estimated that more than a million grinders are now in home use. The grinders are installed in the waste pipe from the kitchen sink; food wastes are simply scraped into the grinder, the grinder is started, and the water turned on. The garbage is ground and flushed into the sanitary-sewer system. In some local communities, garbage grinders have been installed in every residence as required by local ordinance.

Swine feed.—The feeding of garbage to swine has been an accepted way of disposing of the garbage part of solid wastes from urban areas for quite some time. Even as late as 1960, this method was employed in 110 American cities out of 1,118 cities surveyed on their solid-waste-disposal practices. In addition to the municipal practices of using garbage for swine feed, many cities and municipalities permit private haulers to service restaurants and institutions to collect garbage for swine feed. The feeding of raw garbage led to a wide-spread virus disease in the middle 1950's, which affected more than 400,000 swine. As a result, all States now require that garbage be cooked before feeding to destroy contaminating bacteria and viruses. However, according to the American Public Works Association (1966), more than 10,000 tons of food wastes—about 25 percent of the total quantity of garbage produced—is still used daily in the United States as swine feed.

Composting.—Composting is the biochemical decomposition of organic materials to a humuslike material. As practiced for solid-waste disposal, it is the rapid but partial decomposition of the moist, solid-organic matter by aerobic organisms under controlled conditions. The end product is useful as a soil conditioner and fertilizer. The process is normally carried out in mechanical digesters.

Although popular in Europe and Asia where intensive farming creates a demand for the compost, the method is not used widely in the United States at this time. Composting of solid-organic wastes is not practiced on a full-scale basis in any large city today. Although there are several pilot plants in operation, it does not seem likely that composting will be a major method of solid-waste disposal.

The selection of one or more of these methods of solid-waste disposal by a municipality depends largely on the character of the municipality. Geographic location, climate, standard of living, population distribution, and public attitudes play important roles in the selection. In general, the natural resources and environmental factors have been given only small recognition in this selection. Only recently has there been a considerable upsurge of scientific interest in the effects of solid-waste disposal on our water resources.

HYDROLOGIC IMPLICATIONS

TYPES OF POLLUTION

The disposition of solid wastes in open areas carries with it an inherent potential for pollution of water resources, regardless of the manner of disposal or the composition of the waste material. Of the six principal methods of solid-waste disposal, only swine feeding and composting offer no direct possibility of pollution of water resources from the waste material itself. Quite the contrary: properly composted garbage is a soil conditioner that improves the permeability of the soil and may actually assist in improving the quality of water that percolates through it. Although the cooked garbage that is fed to swine does not directly contribute to pollution of water resources, the manure from the feedlots may cause serious problems if not managed properly.

The type of pollution that may arise is directly related to the type of refuse and the manner of disposal. Leachates from open dumps and sanitary landfill usually contain both biological and chemical constituents. Organic matter, decomposing under aerobic conditions, produces carbon dioxide which combines with the leaching water to form carbonic acid. This, in turn, acts upon metals in the refuse and upon calcareous materials in the soil and rocks, resulting in increasing hardness of the water. Under aerobic conditions, bacterial action decomposes organic refuse, releasing ammonia, which is ultimately oxidized to form nitrate. In both landfills and open dumps, where decomposition is accomplished by bacterial action, the leachate has a high biochemical oxygen demand (BOD).

Table 2 indicates the magnitude of the constituents leached from solid wastes under various conditions. These data were compiled by Hughes (1967) from various sources.

TABLE 2.—*Percentages of materials leached from refuse and ash, based on weight of refuse as received*

[Adapted from Hughes (1967)]

Material leached	Percentage leached under given conditions *					
	1	2	3	4	5	6
Permanganate value 30 min	0.039					
Do 4 hr	.060	0.037				
Chloride	.105	.127		0.11	0.087	
Ammonia nitrogen	.055	.037		.036		
Biochemical oxygen demand	.515	.249		1.27		
Organic carbon	.285	.163				
Sulfate	.130	.084		.011	.22	0.30
Sulfide	.011					
Albumin nitrogen	.005					
Alkalinity (as $CaCO_3$)				0.39	0.042	
Calcium				.08	.021	2.57
Magnesium				.015	.014	.24
Sodium			0.260	.075	.078	.29
Potassium			.135	.09	.049	.38
Total iron				.01		
Inorganic phosphate				.0007		
Nitrate					.0025	
Organic nitrogen	.0075	.0072		.016		

* Conditions of leaching:
1. Analyses of leachate from domestic refuse deposited in standing water.
2. Analyses of leachate from domestic refuse deposited in unsaturated environment and leached only by natural precipitation.
3. Material leached in laboratory before and after ignition.
4. Domestic refuse leached by water in a test bin.
5. Leaching of incinerator ash in a test bin by water.
6. Leaching of incinerator ash in a test bin by acid.

RELATION TO HYDROLOGIC REGIMEN

That part of the hydrologic regimen associated with pollution from solid-waste disposal begins with precipitation reaching the land surface and ends with the water reaching streams from either overland or subsurface flow. The manner in which this precipitation moves through this part of the cycle determines whether or not the water resource will become polluted.

Precipitation on the refuse-disposal site will either infiltrate the refuse or run off as overland flow. In open dumps, there is little likelihood of direct runoff unless the refuse is highly compacted. In sanitary landfills, the rate of infiltration is governed by the permeability and infiltration capacity of the soil used as cover for the refuse. A part of the water entering the refuse percolates downward to the soil zone and eventually to the water table. If the water table is above the bottom of the refuse deposit, the percolating water travels only vertically through the refuse to the water table. During the vertical-percolation process the water leaches both organic and inorganic constituents from the refuse.

Upon reaching the water table, the leachate becomes part of and moves with the ground-water flow system. As part of this flow system, the leachate may move laterally in the direction of the water-table slope to a point of discharge at the land surface. In general, the slope of the water table is in the same direction as the slope of the land. The generalized movement of leachate in this part of the hydrologic cycle is shown in figure 1.

There are several well-documented cases of pollution caused by leachates from solid-waste-disposal sites, especially those compiled by the California Water Pollution Control Board (1961). Most of these studies, however, were able to determine only that the pollution originated from solid-waste-disposal sites; few, if any, data on the gross magnitude of the pollution and its fate in the hydrologic cycle are available.

One well-documented case is that of pollution from about 650,000 cubic yards of refuse deposited in a garbage dump near Krefield, Germany, over a 15-year period in the early 1900's. High salt concentrations and hardness were detected in ground water about a mile downgradient from the site within 10 years of operation. Concentrations up to 260 mg/l (milligrams per liter) of chloride and a hardness of 900 mg/l were measured—an increase of more

FIGURE 1.—Generalized movement of leachate through the land phase of the hydrologic cycle.

than sixfold in chloride concentration and fourfold in hardness. The pattern of pumping of wells in the area precludes detailed understanding of the course of the pollution in the ground water, but wells near the dumping site were still contaminated 18 years later.

In Schirrhof, Germany, ashes and refuse dumped into an empty pit extending below the water table resulted in contamination of wells about 2,000 feet downstream. The contamination occurred 15 years after the dump was covered; measures of hardness up to 1,150 mg/l were recorded as compared with 200 mg/l prior to the contamination.

In Surrey County, England, household refuse dumped into gravel pits polluted the ground water in the vicinity. Refuse was dumped directly into the 20-foot-deep pits where water depth averaged about 12 feet. Maximum rate of dumping was about 100,000 tons per year over a 6-year period, and this occurred during the latter part of the period of use (1954–60). Limited observations on water quality extending less than a year after the closing of the pits showed chloride concentrations ranging from 800 mg/l at the dump site, through 290 mg/l in downgradient adjacent gravel pits, to 70 mg/l in pits 3,500 feet away. Organic and bacterial pollution were detected within half a mile of the dumping sites, but not beyond. Because of the limited study period and the slow travel of the pollutants, the maximum extent of pollution was not determined.

More recently, a study was made of the ground-water quality associated with four sanitary landfill sites in northeastern Illinois (Hughes and others, 1969). At the DuPage County site, total solids of more than 12,500 mg/l and chloride contents of more than 2,250 mg/l were measured in samples collected about 20 feet below land surface under the fill. These were by far the highest concentrations measured at any of the four sites. In general, total solids ranged from 2,000–3,000 mg/l under the fill to as low as 223 mg/l adjacent to the fill.

HYDROLOGIC CONTROLS

The movement of leachate from a waste-disposal site is governed by the physical environment. Where the wastes are above the water table, both chemical and biological contaminants in the leachate move vertically through the zone of aeration at a rate dependent in part upon the properties of the soils. The chemical contaminants, being in solution, generally tend to travel faster than biological contaminants. Sandy or silty soils especially retard particulate biological contaminants and often filter them from the percolating leachate. The chemical contaminants, however, may be carried by the leachate water to the water table where they enter the ground-water flow system and move according to the hydraulics of that system. Thus, the potential for pollution in the hydrologic system depends upon the mobility of the contaminant, its accessibility to the ground-water reservoir, and the hydraulic characteristics of that reservoir.

The character and strength of the leachate are dependent in part upon the length of time that infiltrated water is in contact with the refuse and the amount of infiltrated water. Thus, in areas of high rainfall the pollution potential is greater than in less humid areas. In semiarid areas there may be little or no pollution potential because all infiltrated water is either absorbed by the refuse or is held as soil moisture and is ultimately evaporated. In areas of shallow water table, where refuse is in constant contact with the ground water, leaching is a continual process producing maximum potential for ground-water pollution.

The ability of the leachate to seep from the refuse to the ground-water reservoir is another factor in the degree of pollution of an aquifer. Permeable soils permit rapid movement; although some filtering of biological contamination may take place, the chemical contamination is generally free to move rapidly under the influence of gravity to the water table. Less permeable soils, such as clays, retard the movement of the leachate, and often restrict the leachate to the local vicinity of the refuse. Under such conditions, pollution is frequently limited to the local shallow ground-water reservoir and contamination of deeper lying aquifers is negligible.

Leachate that does reach the water table and enters an aquifer is then subject to the hydraulic characteristics of the aquifier. Because the configuration of the water table generally reflects the configuration of the land surface, the leachate flows downgradient under the influence of gravity from upland areas to stream valleys, where it discharges as base flow to the stream systems. The rate of flow is dependent upon the permeability of the rock material of the aquifer and on the slope of the water table. In flat areas or areas of gentle relief, minor local topographic variations may have no effect on the configuration of the water table, and movement of ground water may be uniform over large areas.

In some places dipping confined aquifers crop out in upland areas and thus are exposed to recharge. Contaminants entering the aquifer in these areas move downgradient into the confined parts of the aquifer. Although there is usually some minor leakage to confining beds above and below the aquifer, the contaminants in general will be confined to the particular aquifer, and water-supply wells tapping that aquifer will thus be subject to contamination to the extent that the contaminants are able to move from the outcrop to the wells.

Optimum conditions for pollution of the ground-water reservoir exist where the water table is at or near land surface, subjecting the solid waste to continual direct contact with the water. Such conditions commonly exist where abandoned quarries that penetrate the ground-water reservoir are used as refuse-disposal sites. The continual contact of the water with the refuse produces a strong leachate highly contaminated both biologically and chemically. Under hydrogeologic conditions of permeable materials and steep hydraulic gradients, the leachate may move rapidly through the ground-water system and pollute extensive areas. The hydrologic effects of solid-waste disposal in four geologic environments are shown in figure 2.

Figure 2A illustrates a waste-disposal site in a permeable environment. The waste is shown in contact with the ground water in a permeable sand-and-gravel aquifer underlain by confining beds of relatively impermeable shale. In this case, the potential for pollution is high because conditions of both high infiltration and direct contact between wastes and ground water exist. Because of the permeability of the aquifer, the contaminants move downgradient with the water in the aquifer and are diffused and diluted during this movement. In areas where the water table is below the bottom of the waste material, the degree of contamination is lessened because the wastes are no longer in direct contact with the ground water. In this case, leachate from the wastes moves vertically through the zone of aeration to the water table. It then enters the ground water and moves downgradient as in the case of a shallow water table.

Figure 2B illustrates a waste-disposal site in a relatively impermeable environment. In humid areas, the water table may be near land surface, and the disposed waste may or may not be in direct contact with the ground water. In the illustration, ground water is shown confined to the underlying limestone aquifer. The relative impermeability of the overburden prevents significant infiltration of the rainfall; consequently there is only minor leaching of contaminants from the wastes. Pollution is confined locally to the vicinity of the waste-disposal site; movement in all directions is inhibited by the inability of the water to

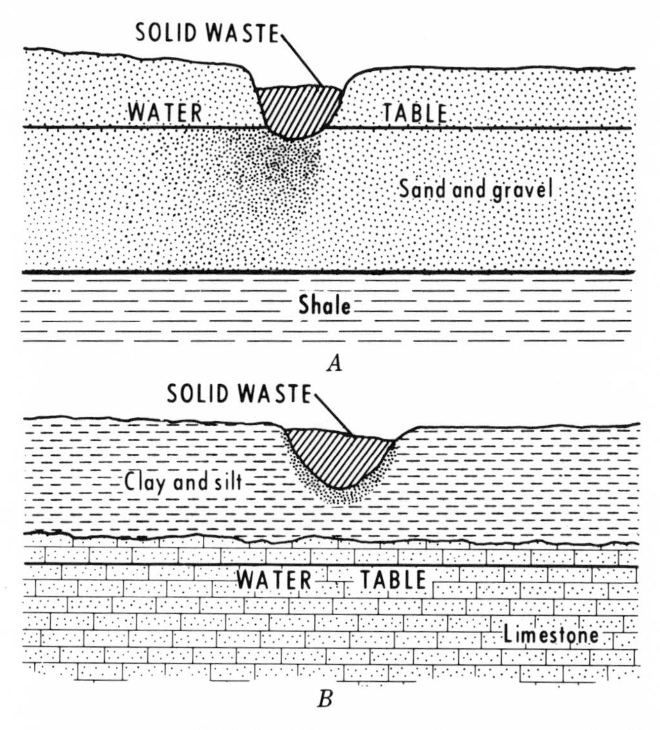

FIGURE 2 (above and right).—Effects on ground-water resources of solid-waste disposal at a site (*A*) in a permeable environment, (*B*) in a relatively impermeable environment, (*C*) underlain by a fractured-rock aquifer, and (*D*) underlain by a dipping-rock aquifer. Leachate shown in red.

move through the tight soils. If significant amounts of rainfall penetrate the wastes, a local perched water table may develop in the vicinity of the fill, and that water will likely be highly contaminated, both chemically and bacteriologically.

Figure 2C illustrates a waste-disposal site above a fractured-rock aquifer. The position of the water table in the overburden relative to the waste-disposal site is dependent upon the amount of infiltration and the geometry of the ground-water flow system. The water table shown here is below the body of waste. In this case, the potential for pollution is not high because of limited vertical movement of the leachate to the water table. However, the contaminants that reach the fractured-rock zone may move more readily in the general direction of the ground-water flow. Dispersion of the contaminants is limited because the flow is confined to the fracture zones. A thin, highly permeable overburden with a shallow water table (similar to that shown in figure 2A) overlying the fractured rock would provide an ideal condition for widespread ground-water pollution.

Figure 2D illustrates a waste-disposal site in a geologic setting in which dipping aquifers are overlain by permeable sands and gravels. In this illustration, the waste-disposal site is shown directly above a permeable limestone aquifer. Here leachate from the landfill travels through the sand and enters the limestone aquifer as recharge. Again, the strength of the leachate depends in part upon whether the water table is in direct contact with the waste. Leachate will move downgradient with the ground-water flow in both the sands and gravels and the limestone, as shown in the illustration. If the waste-disposal site were located above the less permeable shale, most of the leachate would move downgradient through the sand and gravel, with very little penetrating the relatively impermeable shale as recharge. However, in its downgradient movement, it would enter any other permeable formations as recharge.

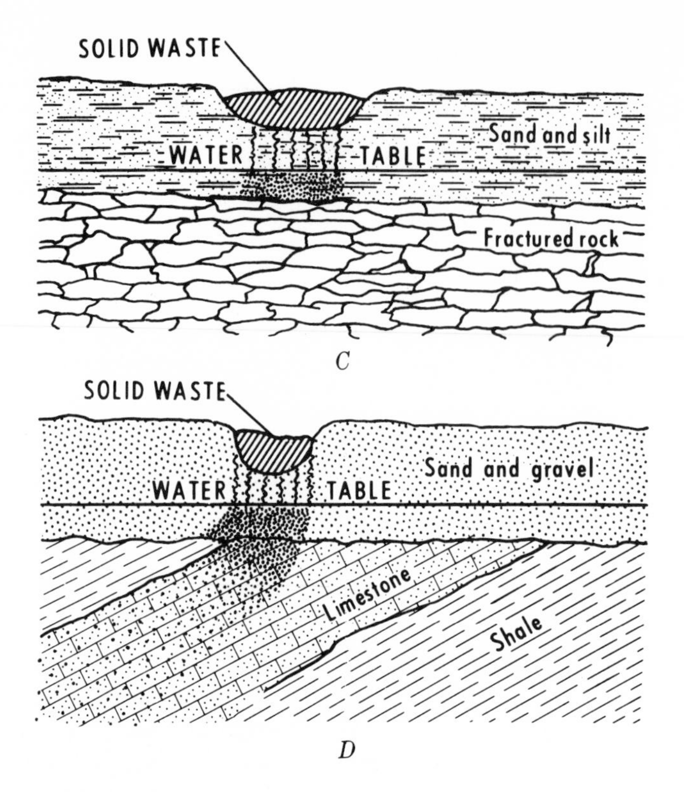

C

D

A high pollution potential exists also where waste-disposal sites are located on flood plains adjacent to streams. Water-table levels generally are near land surface in flood-plain areas, especially during the usual period of high water in winter and spring throughout much of the humid areas of the United States. In such environments the water may have contact with the refuse for extended periods, giving rise to concentrated leachate. The contaminated water moves through the flood-plain deposits and discharges into the stream during low-flow periods when the bulk of the streamflow is from ground-water discharge. The degree of pollution of the stream depends upon the concentration of the leachate, the amount of leachate entering the stream, and the available streamflow for dilution.

HYDROLOGIC CONSIDERATIONS IN SITE SELECTION

It is obvious that our current national policy of pollution abatement and protection of our natural environment requires full consideration of the water resources in selection of sites for solid-waste disposal. To date, with few exceptions, these considerations have been on a local scale, dealing primarily with the hydrological characteristics of the immediate site.

The American Society of Civil Engineers in a manual on sanitary landfill (American Society of Civil Engineers, 1959) discussed site selection from a hydrologic standpoint as follows:

> In choosing a site for the location of a sanitary landfill, consideration must be given to underground and surface water supplies. The danger of polluting water supplies should not be overlooked.

The report states further that:

> Sufficient surface drainage should be provided to assure minimum runoff to and into the fill. Also, surface drainage should prevent quantities of water from causing erosion or washing of the fill * * * Although some apprehension has been expressed about the underground water supply pollution of sanitary landfills, there has been little, if any, experience to indicate that a properly located sanitary landfill will give rise to underground pollution problems. It is axiomatic, of course, that when a waste material is disposed of on land, the proximity of water supplies, both underground and surface, should be considered * * * Also, special attention should be given areas having rock strata near the surface of the ground. For example, limestone strata may have solution channels or crevices through which pollution contamination

may travel. Sanitary landfills should not be located on rock strata without studing the hazards involved. In any case, refuse must not be placed in mines or similar places where resulting seepage or leachate may be carried to water-bearing strata or wells * * * In summary, under certain geological conditions, there is a real potential danger of chemical and bacteriological pollution of ground water by sanitary landfills. Therefore, it is necessary that competent engineering advice be sought in determining the location of a sanitary landfill.

Consideration of hydrology in site selection is required by law in several States. Section 19–13–B24a of the Connecticut Public Health Code requires that:

> No refuse shall be deposited in such manner that refuse or leaching from it shall cause or contribute to pollution or contamination of any ground or surface water on neighboring properties. No refuse shall be deposited within 50 feet of the high water mark of a watercourse or on land where it may be carried into an adjacent watercourse by surface or storm water except in accordance with Section 25–24 of the General Statutes which require approval of the Water Resources Commission.

The rules and regulations of the Illinois Department of Public Health require that:

> The surface contour of the area shall be such that surface runoff will not flow into or through the operational or completed fill area. Grading, diking, terracing, diversion ditches, or tilling may be approved when practical. Areas having high ground water tables may be restricted to landfill operations which will maintain a safe vertical distance between deposited refuse and the maximum water table elevation. Any operation which proposes to deposit refuse within or near the maximum water table elevation shall include corrective or preventive measures which will prevent contamination of the ground-water stratum. Monitoring facilities may be required.

Other States have similar regulations.

A common denominator in these sets of recommendations is the general concern for the onsite pollutional aspect. This is characteristic of most current approaches, especially from the engineering and legislative viewpoints. Another characteristic is the restrictive approach to the problem. Hydrologic conditions are documented under which disposal of solid wastes is either discouraged or prohibited. In general, the pollutional problem is treated more in local than in regional context. These are, of course, important considerations and should be followed in any site selection. In fact, even stronger guidelines are desirable, to the extent of requiring detailed knowledge of the extent and movement of potential pollution at any site before the site is activated.

The water resource, however, must be considered also as a regional resource, not just a localized factor. As such, it should be considered in a regional concept in its relation to solid-waste disposal. This, of course, requires that adequate regional information on the water resource is available. Given such information, the planner can weigh all available alternatives and insure that the final site selection is compatible with comprehensive regional planning goals and environmental protection. The Northwestern Illinois Planning Commission followed this comprehensive approach in developing its recommendations on refuse-disposal needs and practices in northeastern Illinois (Sheaffer and other, 1963).

It is, of course, quite possible that, in the comprehensive approach, some otherwise optimum sites for solid-waste disposal may be only marginally acceptable from a hydrologic viewpoint. Under such conditions detailed information on the hydrology should be obtained and detailed evaluations made of the impact of the potential waste disposal before the site is put into use; the actual impact should then be monitored during and after use. In general, although such studies are desirable for all solid-waste-disposal sites, they are essential where geologic, hydrologic, or other data indicate a possibility of undersirable pollutional effects.

The problem of solid-waste disposal is one of the most serious problems of urban areas. The ever-increasing emphasis on protection and preservation of natural resources though regional planning is evident today. The implementation of these commitments and goals can insure adequate protection of vital water resources from pollution by disposal of solid wastes.

SELECTED REFERENCES

American Public Works Association, 1966, Municipal refuse disposal: Chicago, Ill., Public Adm. Service, 528 p.

American Society of Civil Engineers, Committee on Sanitary Landfill Practices, 1959, Sanitary landfill: Am. Soc. Civil Engineers Eng. Practices Manual 39, 61 p.

Anderson, J. R., and Dornbush, J. N., 1967, Influence of sanitary landfill on ground water quality: Am. Water Works Assoc. Jour., April 1967, p. 457–470.

California Water Pollution Control Board, 1961, Effects of refuse dumps on ground water quality: California Water Pollution Control Board Pub. 24, 107 p.

Hughes, G. M., 1967, Selection of refuse disposal sites in northeastern Illinois: Illinois Geol. Soc. Environmental Geology Note 17, 18 p.

Hughes, G. M., Landon, R. A., and Farvolden, R. N., 1969, Hydrogeology of solid waste disposal sites in northeastern Illinois: Washington, D.C., U.S. Public Health Service, 137 p.

Sheaffer, J. R., von Boehm, B., and Hackett, J. E., 1963, Refuse disposal needs and practices in northeastern Illinois: Chicago, Ill., Northeastern Illinois Metropolitan Area Planning Comm. Tech. Rept. 3, 72 p.

24

Reprinted from *24th Intern. Geol. Congr. Proc.*, 13, 44–54 (1972)

Land Subsidence and Population Growth

NIKOLA P. PROKOPOVICH,
U.S.A.

ABSTRACT

Genetically, subsidence can be divided into exogenic and endogenic. Exogenic subsidence, barely known at the turn of the century, has become an important world-wide geologic process. It is commonly directly related to human activities such as: underground mining; extraction of crude oil and natural gases; overdraft of confined and free ground waters; irrigation and other types of wetting of certain semiarid and arid sediments; build-up of ground waters; drainage of terrains underlain by peaty sediments, etc. These activities usually cause compaction, due to either an increase of actual or effective stresses or removal of rock support. The surficial expression of compaction is land subsidence. Each case of subsidence is a direct result of the geologic past of the area. From the environmental standpoint, subsidence is usually an undesirable alteration of natural conditions and might be regarded as a form of pollution. Subsidence directly or indirectly leads to multi-million-dollar property losses and even to losses of human life.

Subsidence is most common in highly developed, densely populated areas. Its geographic-historic spread reflects an increasing population and growth of technological "know-how". It can be expected to accelerate with the advent of developing nations, particularly with future large-scale reclamation and resource developments of arid regions, which are unavoidable because of steadily increasing demands for food and natural resources by a rapidly growing population.

The association of subsidence and large irrigation canals, for example in California, U.S.A., is not accidental. The canals are designed to deliver water to water-deficient areas, which frequently are vulnerable to subsidence caused by overdraft of ground waters. The probability of subsidence in such areas should be anticipated in project planning, design and construction. Systematic leveling nets and comprehensive historic records on ground water or mining should help in early detection of subsidence.

INTRODUCTION

THE TERM "LAND SUBSIDENCE," which was rarely mentioned 40-50 years ago, has become more and more common, particularly after World War II. Numerous papers describing case histories and origin of subsidence are being published all over the world. In the United States, the Water Resources Division of the Geological Survey created a Special Project Office for subsidence study in California; a subcommittee on Land Subsidence has been formed within the Conservation and Disaster Prevention Committee of the Resources Council of the Prime Minister's Office in Japan; an International Symposium on Land Subsidence was held in Tokyo, Japan, in 1969 (UNESCO, 1969 and 1969A), etc.

Are these events coincidental? Was land subsidence not recognized in the past? Or is there a worldwide increase of subsidence with our old dependable "terra firma" suddenly becoming "terra infirma"? If so — why? Is subsidence an independent process or does it reflect other nongeologic changes on the earth? This paper discusses these questions, based on 20 years of studies of land subsidence by the U. S. Bureau of Reclamation, particularly in the State of California.

LAND SUBSIDENCE — —DEFINITION AND CLASSIFICATION

Natural processes are usually so complicated and interlocked that their genetic classifications are not an easy task. For the purpose of this paper, land subsidence is defined as "more or less vertical settling of the land surface." Such processes as erosion, deflation, excavations and others which involve mechanical or chemical removal of the original surface are not considered as subsidence. The definition, however, is only a working approximation. For example, landslides or soil creep are not considered to be land subsidence, whereas collapsing of caves or old mining works are recognized forms of subsidence. Wind erosion and oxidation-alteration of peat are commonly considered as subsidence (Stephens, 1969; Weir, 1950) though they are related to a partial removal of original land surfaces.

Land subsidence can be caused by numerous processes (Allen, 1969). Various forms of subsidence can be subdivided into two major genetic groups. (1) *Endogenic subsidence,* which is related to processes that originate within the planet such as epeirogeny, faulting and continental drift. With this meaning, the term subsidence was frequently used in the past (Lyell, 1853). Subsidence and resulting transgressions of the seas are generally a slow, regional process and are extremely important factors of geologic history. (2) *Exogenic subsidence,* which is related to processes originating at or near the earth's surface, including human activity such as mining, drainage, and withdrawal of ground water, oil or natural gas. This form of subsidence is generally more local and has become increasingly important.

Classification of some forms of subsidence is difficult; marine transgressions result from a form of endogenic subsidence, but some of them, related to the melting of Quaternary ice shields, are of exogenic origin. On the other hand, tectonic movements may be needed to create conditions permitting subsequent exogenic subsidence by withdrawal of confined ground waters.

This paper deals exclusively with various forms of exogenic subsidence. In general, they are surficial expressions of compaction of unconsolidated sediments caused by an increase of actual or effective stresses on compacting strata, or of collapse caused by the removal of rock support. Genetic subdivision of this type of subsidence is a classification of factors causing changes of stress or removal of support. In each case the susceptibility to subsidence is a direct result of the geologic past of the area.

DISTRIBUTION OF LAND SUBSIDENCE

The following text is not a complete worldwide review of land subsidence, but illustrates its widespread occurrence.

"Thermal Karst"

Subsidence by melting of buried ice blocks is common in areas of retreating permafrost and is caused by a removal of support. Temporary seasonal sinks are also frequent on arctic shores. The process was particularly common at the end of Pleistocene glaciations. Its numerous traces are scattered on flatlands of North America, Europe and Asia. At the present it is geographically more restricted. A few present sinks could be related to human activities such as forest fires, road construction, etc.

Karst Sinkholes

Sinks created by the solution of limestone, dolomite, gypsum and rock salt are genetically somewhat related to "Thermal Karst". Geological requirements for the formation of these features are the presence of soluble rock types and the availability of solvents and conditions for their circulation. Most such subsidence is not directly related to human activities, but in some cases (USBR, 1962) construction of dams can accelerate natural processes, particularly collapse of overburden into pre-existing cavities (Fig. 1 A).

Subsidence Due to Mining

Well described mining subsidence (Voight and Pariseau, 1970) is caused by removal of rock support at depth. It is common in old mining areas such as Appalachia and Boise, Idaho, in U.S.A., coal mining areas in England, the Ruhr Basin

FIGURE 1 — Different types of subsidence.
A: Karst sink developed during filling of the Anchor Reservoir on Owl Creek, Wyoming, U.S.A. Collapse into a pre-existing cavity in dolomite occurred in 1961 and was an acceleration of a natural process. "X" marks a bulldozer tractor. Courtesy of K. D. Hahn.
B: Subsidence sink made by an underground nuclear explosion. Courtesy of U.S.A.E.C.
C: Subsidence in peat. Old barn built at the ground level on piles became "elevated" due to compaction of peat within the piling depth. Holland Tract, Sacramento - San Joaquin River Delta, California, U.S.A., 1959. Courtesy of State of California, Department of Water Resources.
D: Subsidence due to withdrawal of crude oil and natural gas. Pier No. 1, Long Beach, Los Angeles Metropolitan area, California, U.S.A. Water level in the harbor is above the "dry land" protected by walls. Retaining walls under construction; existing walls are almost overflooded. February 1958. U.S. Navy Photograph.

in Germany and many others where it causes variable amounts of property loss and damage to buildings, roads and fields. Damage in densely populated areas with high land values is particularly costly. Deterioration of surface drainage is common (Wohlrab, 1969). Subsidence may affect conditions of river flow (Tison, 1969). Growing demands for mineral resources and expanded mining have resulted in a worldwide increase of this type of subsidence. Sinks related to underground nuclear explosions (Fig. 1B) are a special form of such subsidence.

Alteration of Peat and Subaqueous Sediments

A good illustration of peat subsidence is the Sacramento - San Joaquin Delta, California, U.S.A. (Fig. 1C). Slow decomposition and alteration of mud and peaty sediments in tidal swamps under natural anaerobic conditions were originally balanced by accumulation of new sediments. Consequently, there were no notable changes of elevation and deltaic islands were located at or just above sea level. Reclamation and cultivation of islands, construction of protective levees, plowing, grass- and peat-fires and other forms of "civilization" have caused a severe subsidence, particularly in the interior, more peaty, portions of islands (Weir, 1950). At present, most of the Delta is submerged below sea level. Some areas have elevations of minus 4.5 m and less. Breaks of levees cause periodic flooding of islands and require costly pumpage. Two islands, Franks Tract and Big Break, became permanently flooded.

Particularly devastating peat subsidence occurs in subtropical Florida (Stephens, 1969). The process is also common in the formerly glaciated Midwest, where it occurs even in downtown Minneapolis, Minnesota. It proceeds at a greater rate in a warm climate favorable for bacterial action. Large subsiding areas occur in peat districts in England, Germany, Poland and Russia.

An excellent example of subsidence related to dehydration-alteration of reclaimed marine lands is the Netherlands (Bennema *et al.*, 1954).

Subsidence Due to Withdrawal of Oil and Gas

Subsidence related to withdrawal of hydrocarbons could be expected in any productive oil or gas field. A good, widely publicized, example is the subsidence at Long Beach - Terminal Island, in the highly industrialized and populated Los Angeles Metropolitan Area (Mayuga and Allen, 1966). The subsidence center is located in the Port of Long Beach (Fig. 1D). Total maximum subsidence is of the order of 8.8 m, with horizontal displacement up to 3 m. The phenomenon has caused about $100 million damages in repair, maintenance and protective works. The damage has affected the Port and City of Long Beach, industrial facilities and the Naval shipyard, buildings, streets, railroad tracks, pipelines, wharves, bridges and hundreds of producing oil wells. Slow creep was accompanied by several small earthquakes. Following State legislation in 1958, repressurizing by injection of water was adopted and subsidence was successfully arrested. There are some theories that the failure of the Baldwin Hill Reservoir in December 1963 in Los Angeles was also related to subsidence (Hamilton and Meehan, 1971).

Similar subsidence was reported at Lake Maracaibo, Venezuela, where a 2-mile-long protective concrete wall was constructed to prevent flooding, and in the Goose Creek oil field, Texas, U.S.A., which became flooded by San Jacinto Bay (Poland and Davis, 1969).

Subsidence Due to Withdrawal of Ground Water

Water is the most important natural resource. Development of underground waters started in prehistoric time, but became particularly intensified with the growth of the energy and population growth. Land subsidence due to withdrawal of water, therefore, became one of the more common forms of subsidence (Poland and Davis, 1969). It could be caused by decrease of piezometric head and/or a lowering of unconfined water tables. In the first case, the increase of effective stress amounts to 1 ton/m^2 for each meter of piezometric decline and is applied below and in the aquiclude. A classic example of such subsidence is Mexico City, Mexico, where the total subsidence locally exceeded 7 m. (Marsal and Sainz-Ortiz, 1956).

In the case of overdraft of unconfined ground waters, an increase of intergranular pressure occurs which equals the sum of the decrease in buoyancy plus weight of water held by specific retention in sediments; stress increase is about 0.8 ton/m^2 for each meter of water table decline. This occurs only in the saturated portion of the aquifer system.

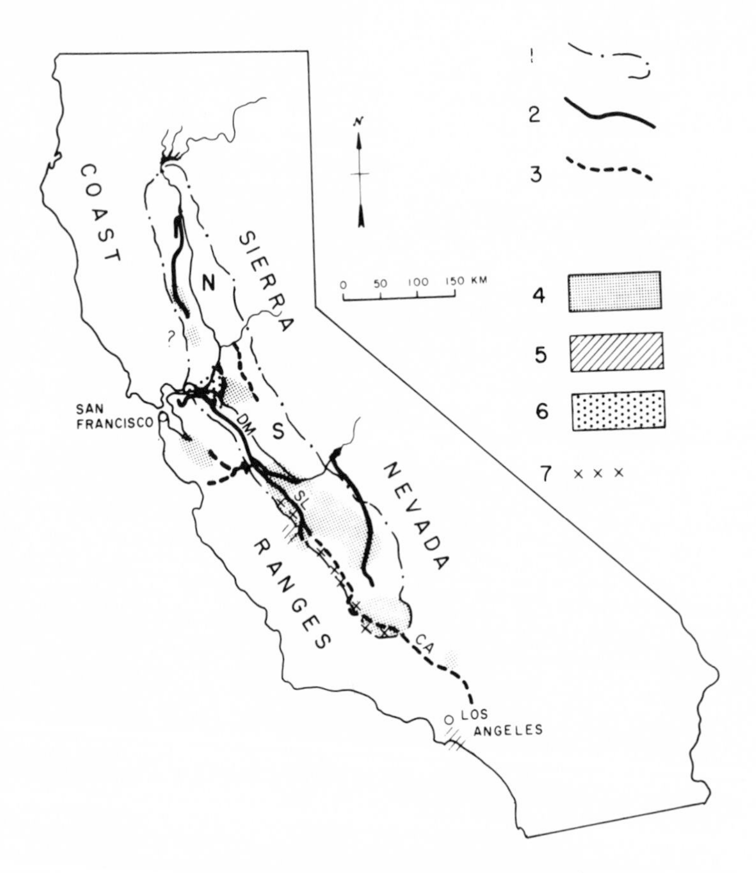

FIGURE 2 — Land subsidence in California, U.S.A.
1: Outlines of the Central Valley, which is composed of the Sacramento Valley (N) and the San Joaquin Valley (S).
2 & 3: Major canals; Completed (2) and proposed or under construction (3); DM = Delta-Mendota Canal, SL = San Luis Canal and CA = California Aqueduct.
Subsidence caused by: 4: Ground-water overdraft; 5: Oil and gas withdrawal; 6: Alteration of peaty deposits; 7: Hydrocompaction.

Historic-geologic requirements for these types of subsidence commonly are not fully recognized. Compaction (and subsidence) are possible only if sediments are not preconsolidated by the weight of overlying deposits. Unbalanced conditions could be created, for example, by folding of a confined aquifer during sedimentation. The folding creates an additional piezometric head which partially compensates for the weight of the newly deposited sediments. Such tectonic conditions probably existed in the San Joaquin Valley (Fig. 2), California, U.S.A. Land subsidence in this area has been caused mostly by pressure decline in a deep confined aquifer system used mainly for irrigation (Poland and Davis, 1969). The historic decline of piezometric levels along the alignment of the recently completed San Luis Canal on the western side of the valley was about 140 m. (Prokopovich, 1969, 1969A). Cumulative amounts of subsidence caused by the decline approached 7.5 m in some places (Fig. 3). The existence of subsidence was not recognized during the construction of the neighboring Delta-Mendota Canal. Consequently, lining, bridges, pipe crossings and other structures became partly or completely flooded (Prokopovich, 1969A) (Fig. 4AB) due to subsidence. Additional freeboard was built into the recently completed San Luis Canal to compensate for future subsidence (Prokopovich, 1969).

Similar subsidence occurs also on the east side of the valley in the vicinity of the Friant-Kern Canal (Lofgren and Klausing, 1969), on the southern end of the valley, along the California Aqueduct and in the Santa Clara Valley (Roll, 1967). Minor, genetically similar, subsidence occurs also near the Sacramento - San Joaquin Delta, near Los Angeles in the Mohave Desert and in the Sacramento Valley (Fig. 2).

Land subsidence due to withdrawal of water is common in other arid areas in the U.S.A. near Houston, Texas; in Arizona; near Denver, Colorado; and in Las Vegas, Nevada. In some places severe soil cracking is associated with the subsidence.

This type of subsidence is not restricted to arid lands; subsidence was recorded in Savannah, Georgia and in London, England. The Po Delta in Italy is a good example of subsidence due to pumpage of methane-bearing ground waters

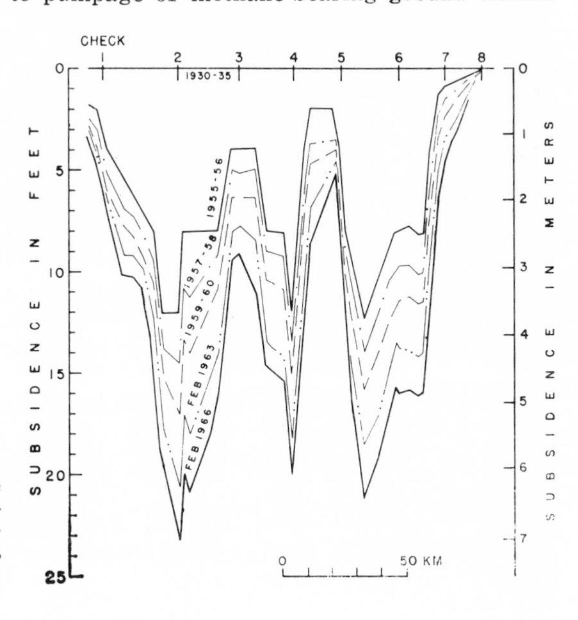

FIGURE 3 — Land subsidence along the San Luis Canal alignment prior to the Canal's completion—Western San Joaquin Valley, California, U.S.A.

since 1933. Over $30 million was spent on repairs and protective works in the area during a five-year period in the 1950's.

Japan is the best example of subsidence in a humid climate (UNESCO, 1969, 1969A). Japan is essentially limited to four islands totaling 369,662 km^2 and housing 99 million population. Flat alluviated lands comprise only 28 per cent of the country and underlie the main agricultural and industrial districts. Land subsidence has been recorded in 27 areas and occurs in each major alluviated region. Numerous, frequently leveled bench marks have been installed (approximately 1 bench mark per 0.1 square km). The devastating effects of subsidence are intensified by tidal waves related to earthquakes and typhoons. The combination of

FIGURE 4 — Subsidence due to withdrawal of confined ground waters.
A and B: Delta-Mendota Canal in west-central San Joaquin Valley, California, U.S.A.
A: Normal bridge over concrete-lined canal in non-subsiding area; about 1 m clearance.
B: Similar rehabilitated bridge in area with about 2 m of post-construction subsidence; original concrete lining and bottom of the bridge are submerged; protective walls at and along the bridge; raised road bed.
C: Subsidence in Tokyo, Japan. Densely populated industrial area "Koto Ward" submerged below sea level. Water in the Tatekawa Cr. above surrounding lands protected by walls. September 1969.
D: Subsidence in downtown Osaka, Japan. Tosabori Canal near the Osaka Royal Hotel has water level above the surrounding streets, protected by walls. Minor subsidence due to compaction of deltaic alluvium within the wall foundation separated steps from the street pavement.

flooded streets in densely populated areas, people stranded in floodwaters, and back-flushed waste disposal systems is disastrous.

The large extent and magnitude of subsidence in Tokyo, with a metropolitan population of 18 million, is a good example of subsidence in a deltaic region (ancient name of Tokyo — Edo — means an estuary). Subsidence was fully recognized only after the 1923 earthquake flood. Originally only flood plains were affected, but since 1957 subsidence has been recorded also on terraces. Total amount of subsidence since 1892 locally exceeds 4 meters. Subsidence rates decreased during and after World War II, reflecting war mobilization efforts and destruction. The remarkable post-war recovery and increase pumpage resulted in a significant increase of subsidence; large densely populated areas became submerged below sea level (Fig. 4C). Flood shelters, about 30 km of levees and over 20 emergency flood gates were constructed. In case of a tidal wave, gates are immediately closed to prevent flooding and stream water is pumped into the ocean by pumping plants. Ground waters occur at a shallow depth and cause flooding and swamping which require a complicated drainage system with a pump lift. In order to retard subsidence, local rigid control of industrial water and of domestic ground water pumpage were introduced in 1961 and 1963.

Osaka, the second largest city of Japan and a major industrial-commercial metropolis with a port, is also located on deltaic sediments. Recorded amounts of subsidence after 1935 locally exceed 2.8 meters (Fig. 4D).

Numerous subsided old bridges obstruct navigation in rivers and sloughs. Some basements are permanently flooded and the city has suffered several disastrous floods. Strict control of use of ground water for industrial and domestic purposes, including air conditioners, was introduced in 1960, and led to reduced amounts of subsidence. A program to enlarge existing dikes to the height of 6.6 meters and to construct additional flood gates was initiated in 1965. Existing features amount to 80 pumping stations, 190 kilometers of protective embankments, 40 flood-preventing locks and 500 flood-preventing gates.

The city of Niigata is located on the deltaic coastal plain. The subsidence is caused by pumpage of methane-bearing ground waters and exceeds 1.5 m. To improve the conditions, severe restrictions on gas extraction were introduced.

Frightening examples of subsidence caused by lowering of unconfined ground water occurred in South Africa (Bezuidenhout and Enslin, 1969) where drainage for gold mining resulted in slow subsidence and spectacular soil collapses into buried Karst sinks (Fig. 5, A and B.)

Hydrocompaction

An entirely different type of subsidence is hydrocompaction or "shallow subsidence." The first term reflects the nature of the process — spontaneous rapid collapse, slumping and cracking (Fig. 5C), which occur after wetting of certain dry, low-density, unconsolidated sediments. Hydrocompaction due to agricultural irrigation is common in several isolated areas totaling about 200 square miles in the western and southern San Joaquin Valley, California (Lofgren, 1969). Total amounts of hydrocompaction locally exceed 4.5 m. It causes severe damage to canals (Fig. 5D), irrigation ditches, wells, roads, buildings, pipelines and fields. In order to prevent such damage, about $4 million was spent for preconstruction consolidation by flooding of two reaches of the giant San Luis Canal (Prokopovich, 1969). Hydrocompaction is known in many semiarid areas in Arizona, Colorado, Kansas, Montana, Nebraska and Wyoming, U.S.A. (Lofgren, 1969), and in Eastern Europe and Asia (Drashevska, 1962).

FIGURE 5 — Different types of subsidence.
A and B: Sinks in originally flat land caused by dewatering for gold mining. Collapsing occurred into large pre-existing, buried Karst cavities in Transvaal dolomite. Republic of South Africa. Photographs courtesy of Dr. C. A. Bezuidenhout.
A. Destruction of the West Driefontein Gold Mine three-story crusher plant, in December, 1962. 29 lives were lost.
B. Sinkhole at Blyvooruitzicht, August, 1964. Five lives were lost.
C and D: Hydrocompaction of piedmont alluvium by wetting in the west-central San Joaquin Valley, California, U.S.A.
C. Large-scale cracks caused by natural flooding of dry alluvial sediments.
D. Damage to a test section of concrete-lined canal after water application.

DISCUSSION

From the environmental standpoint, subsidence is an alteration of natural conditions and, therefore, can be regarded as a form of pollution. In most cases, the changes have progressed in an undesirable direction such as development of swamps, flooding, offsetting of drainage and stream patterns, soil cracking and slumping to name just a few. The effects of subsidence on human structures and economy are even more serious. In many cases subsidence directly or indirectly leads to a multi-million-dollar expense in property losses and even to direct losses of human life.

With a few exceptions, such as Karst and permafrost subsidence which are more or less independent from human activities, most of exogenic subsidence is created or accelerated by humans. The first such subsidence occurred probably during the Stone Age as a minor cave-in when a growing family of Homo Sapiens became engaged in an enlargement of their underground shelter. The subsequent growth of population and increased knowledge of mineral resources, particularly underground mining, were reflected by subsidence near old mining communities. New types of subsidence were created by attempts to increase living and agricultural areas by reclamation of swamps and shallow marine areas.

The growth of affluent societies and unbelievable increase of technological "know how" (?) after World War II, multiplied by the population explosion, resulted in the present worldwide spread of subsidence. It is not a coincidence that the State of California, which is the most populated state in the United States, is so severely affected by subsidence. Subsidence is most common now in highly developed and densely populated areas (United States, Japan and Italy.) Even more populated, but less technologically developed, areas may be relatively immune to subsidence. The association of subsidence and large irrigation canals in California (Fig. 2) and elsewhere is not coincidental. The canals are designed to deliver water to water-deficient areas which frequently are vulnerable to subsidence related to overdraft of ground waters.

The present technological revolution and rapid progress of developing nations promises a notable future acceleration of subsidence, particularly in future irrigation and resource developments of arid regions which seem to be unavoidable to meet the demands of growing population. The possibility of pollution in over-sensitive arid conditions is much greater than in humid areas. Developing nations should by all means critically review both the achievements and failures of developed areas.

The probability of an already existing or potential future subsidence should be anticipated in all project planning, design and construction. Systematic leveling nets and comprehensive scientific, historic records on ground water or mining should help the delineation of hazardous areas and in planning corrective actions.

REFERENCES

Allen, A. S., 1969. Geologic setting of subsidence. Reviews in Engineering Geology, Vol. 2, p. 305-342, Geol. Soc. Am., Boulder, Colorado.

Bennema, J., Geuze, E. C. W. A., Smits, H., and Wiggers A. J., 1954. Soil Compaction in relation To Quaternary movements of sea level and subsidence of the land, especially in the Netherlands. Geol. Mijnbouw, New Ser. 16, No. 6, p. 173-178.

Bezuidenhout, C. A., and Enslin, J. F., 1969. Surface subsidence in the dolomitic areas of the Far West Rand, Transvaal, Republic of South Africa. p. 482-495. See UNESCO 1969A.

Drashevska, L., 1962. Review of Recent USSR publications in selected fields of engineering soil science. Reviews in Engineering Geology, 1, p. 197-221. Geol. Soc. Am.

Hamilton, D. H., and Meehan, R. L., 1971. Ground rupture in the Baldwin Hills. Science, 172, No. 3981, p. 333-344, Am. Assoc. Adv. Sci.

Lofgren, B. E., 1969. Land subsidence due to the application of water. Reviews in Engineering Geology, 2, p. 271-303, Geol. Soc. Am. Boulder, Colorado.

Lofgren, B. E., and Klausing, R. L., 1969. Land subsidence due to groundwater withdrawal; Tulare-Wasco Area, California. U.S. Geol. Surv. Prof. Pap. 437-B, p. B-1 to B-103.

Lyell, Charles 1853. Principles of geology or the modern changes of the Earth and its inhabitants. D. Appleton and Co., New York, p. 834.

Marsal, R. J., and Sainz-Oritz, I., 1956. Short description of the sinking of Mexico City. Bol. Geol. Mexicana, p. 11.

Mayuga, M. N., and Allen, D. R., 1966. Long Beach subsidence. Engineering Geology in Southern California. Spec. Publ. of Assoc. of Eng. Geol. p. 281-285, Glendale, California.

Poland, J. F., and Davis, G. H., 1969. Land subsidence due to withdrawal of fluids, p. 187-269. *In* Reviews in Engineering Geology. II. Geol. Soc. of Am. Boulder, Colorado.

Prokopovich, N., 1969. Prediction of future subsidence along Delta-Mendota and San Luis Canals, Western San Joaquin Valley, California. See UNESCO 1969A, p. 600-610.

Prokopovich, N., 1969A. Land subsidence along Delta-Mendota Canal, California. Rock Mechanics, p. 134-144.

Roll, J., 1967. Effect of subsidence on well fields. J. of Am. Waterworks Assoc., 59, No. 1, p. 80-88.

Stephens, John C., 1969. Peat and muck drainage problems. J. of the Irrigation and Drainage Division. Proc. of the Am. Soc. Civil Eng., 95, No. IR2, p. 285-305.

Stephens, John C., and Speir, William H., 1969. Subsidence of organic soils in U.S.A. See UNESCO 1969A, p. 523-534.

Tison, L. J., 1969. Influence des affaissements sur l'hydrologie tant de surface que souterraine. See UNESCO 1969, p. 242-249.

UNESCO, 1969. Land Subsidence. Publication No. 88 AIHS. (Proc. Int. Symp. on Land Subsidence, Tokyo, Japan), 1, p. 324.

UNESCO, 1969A. Land Subsidence. Publication No. 88 AIHS. (Proc. Int. Symp. on Land Subsidence, okyo, Japan), 2, p. 325-361.

USBR, 1962. Technical record of design and construction. Anchor Dam, constructed 1957-1961. Missouri River Basin Project, Wyoming. Denver, Colorado. Published by GPO, p. 148.

Voight, B., and Pariseau, W., 1970. State of predictive art in subsidence engineering. J. Soil Mech. and Foundation Div., Proc. Am. Soc. of Civil Eng., SM 2, p. 721-750.

Weir, Walter W., (1950). Subsidence of peat lands of the Sacramento - San Joauin Delta, California. Hilgardia (California Agricultural Experiment Station, Berkeley, California), 20, No. 3, p. 37-56.

Wohlrob, B., 1969. Effect of land subsidence caused by mining to the ground water and remedial measures. See UNESCO 1969A, p. 502-512.

25

Reprinted from *Geol. Surv. Alabama Circ. 83,* 1, 19-37 (1973)

SINKHOLE PROBLEM ALONG PROPOSED ROUTE OF INTERSTATE HIGHWAY 459 NEAR GREENWOOD, ALABAMA

J. G. Newton, C. W. Copeland, and W. L. Scarbrough

ABSTRACT

Sinkholes in and adjacent to the proposed right-of-way of Interstate Highway 459 near the community of Greenwood in Bessemer, Alabama, pose costly construction and potential maintenance and safety problems. More than 150 sinkholes, depressions, and related features have formed in or near the right-of-way. Their occurrence began about 1950 and continued through March 1972. The development of sinkholes is restricted to the area underlain by the Tuscumbia Limestone, lowermost beds of the overlying Floyd Shale, and in the immediate vicinity of faults. Complex geologic and hydrologic conditions were defined by 253 auger holes, 16 core holes, a shallow refraction seismic survey, and the use of multispectral photography and thermal infrared imagery.

A general lowering of the water table during the early 1950's or the preceding decade resulting from large withdrawals of ground water from wells and mines compounded with a prolonged drought during the 1950's makes the area prone to the development of sinkholes. Openings along faults and a fold provide hydraulic connection between aquifers at the surface and mines at depths exceeding 1,000 feet. Cessation of pumpage from wells and mines has resulted in conditions favorable to the recovery of the water table. Available information indicates that the water table could recover to its pre-1950 altitude as early as the summer of 1973 but the recovery may well extend over a much longer period of time.

Sinkholes occur where cavities develop in residual or alluvial deposits overlying openings in limestone. The downward migration of the deposits into openings in the underlying limestone and the formation and collapse of the cavities are caused or accelerated by a decline in the water table that results in a) an increase in the amplitude of water-table fluctuations, b) the increased movement of surface water through unconsolidated deposits into openings in bedrock in areas where recharge had previously been rejected, c) an increase in the velocity of movement of ground water, and d) loss of support to unconsolidated deposits overlying openings in bedrock. A recovery of the water table will probably result in a cessation of or drastic decrease in sinkhole development.

[*Editor's Note:* Material has been omitted at this point. Table 3, mentioned on p. 19, has also been omitted because of its length.]

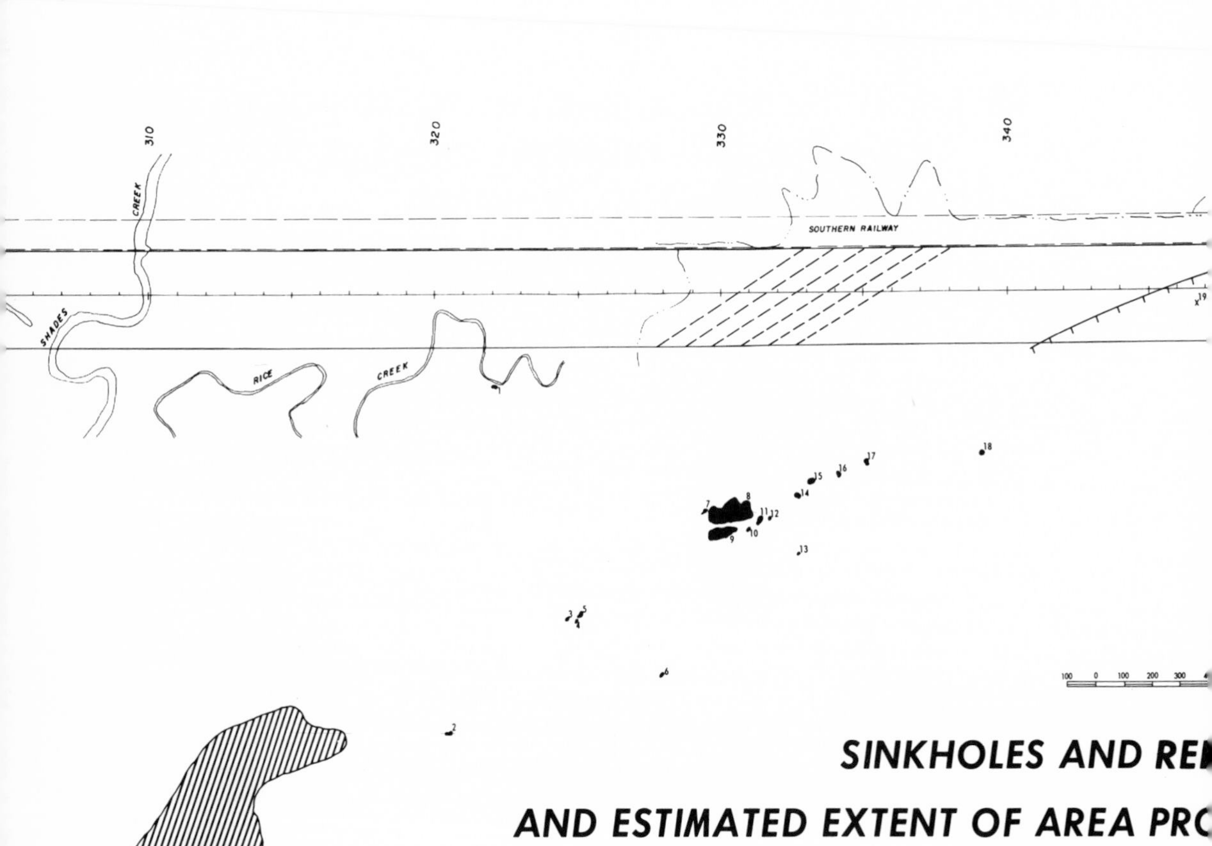

310
320
330
340
CREEK
SHADES
RICE
CREEK
SOUTHERN RAILWAY
1
2
3
4
5
6
7
8
9
10
11
12
13
14
15
16
17
18
19
100 0 100 200 300
SINKHOLES AND
AND ESTIMATED EXTENT OF AREA

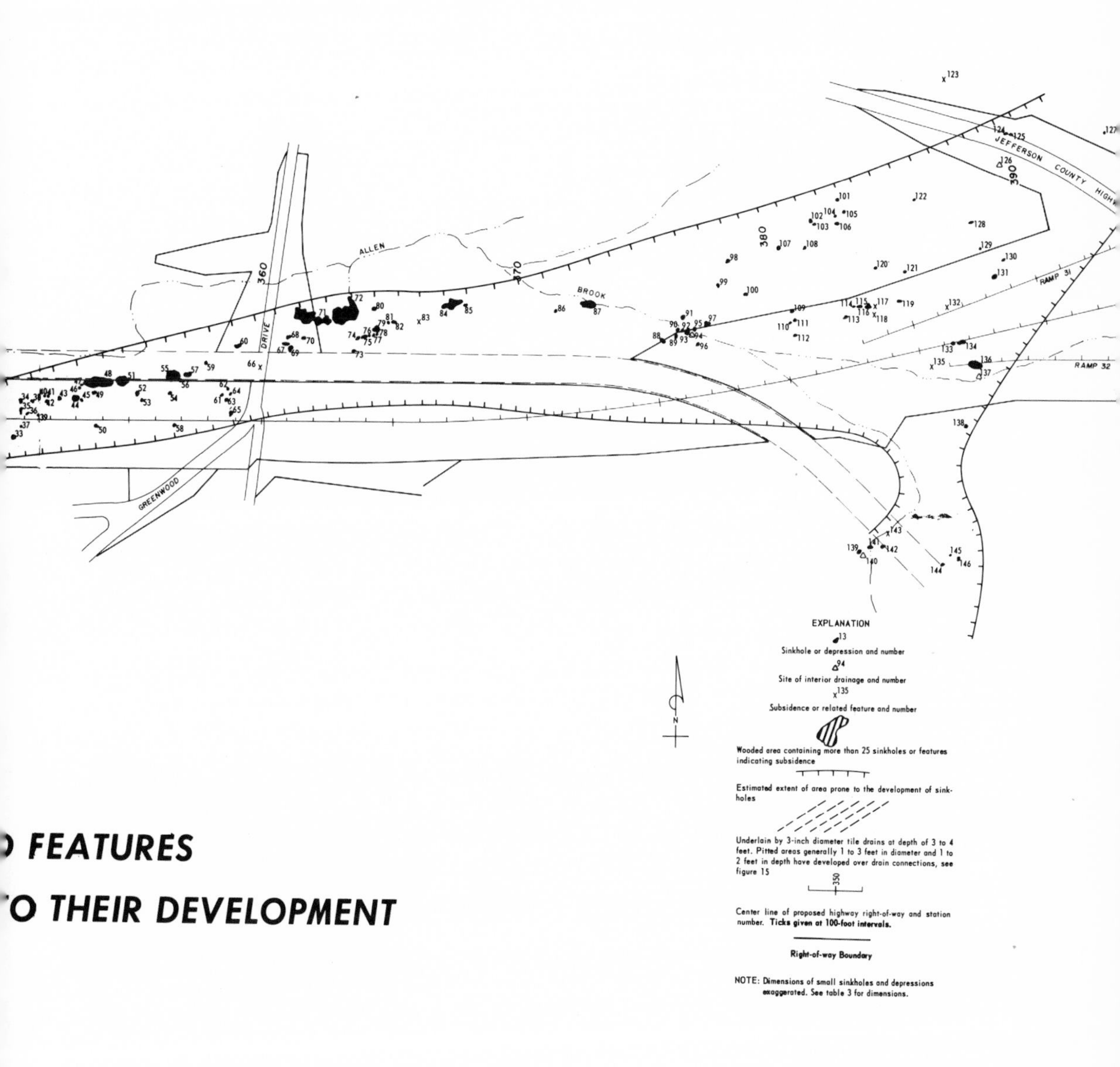

) FEATURES

'O THEIR DEVELOPMENT

OCCURRENCE OF SINKHOLES IN AREA OF STUDY

Sinkholes commonly occur in many areas underlain by carbonate rocks. Their occurrence in the Greenwood area and in other parts of Alabama poses many problems related to property damage, construction, and pollution. More than 150 sinkholes, depressions, or other related features inventoried in or near the proposed right-of-way are shown on plate 2. Their dimensions and other pertinent information are given in table 3. A few of the depressions inventoried may represent test pits excavated many years ago to locate limestone, however, most have resulted from subsidence that often precedes collapse.

HISTORY

The period during which sinkholes have formed in the area of study exceeds 20 years. The lack of recorded information necessitates relying on reports of events long past. Available information indicates that the occurrence of sinkholes started about 1950 with a collapse beneath a barn and tree (sinkhole 144 on pl. 2). Nearby sinkhole 141 reportedly formed in a road during the same year. Sinkholes 143 and 127 (pl. 2) formed in the mid-1950's beneath a railroad trestle and in the bottom of a pond resulting in its drying up. Sinkhole 9 formed beneath a major gas pipeline about 1956.

Sinkhole activity occurred continuously through the 1960's and during 1970. This activity was observed in open fields and near structures by numerous land owners. The same activity, based on collapses inventoried, was occurring in nearby wooded areas.

The development of sinkholes and related features continued unabatedly during fieldwork for this investigation from December 1971 to March 1972. During this period, four or more collapses occurred and subsidence was noted in at least three areas. A collapse (sinkhole 139 on pl. 2) beneath a bridged culvert that occurred on January 26, 1972, is shown in figure 10.

CAUSES AND DEVELOPMENT

Sinkholes result from a variety of causes, but the occurrence of numerous sinkholes in a restricted area is commonly associated with a substantial lowering of the water table which may be natural, man-created, or a combination of both. Excellent examples of both have occurred in Alabama (Newton and Hyde, 1971, p. 17).

Figure 10.—Collapse under bridged culvert.

A lowering of the water table resulting from withdrawal of ground water makes a part of the area of study prone to the development of sinkholes. The decline in the water table results in (1) a loss of support to the roof of cavities in bedrock that were previously filled with water and to residual clay or other unconsolidated deposits overlying openings in bedrock, (2) an increase in the velocity of movement of ground water, (3) an increase in the amplitude of water-table fluctuations particularly at lows where the levels are below those of previous record, and (4) the movement of water from the land surface to openings in underlying bedrock where recharge had previously been rejected because the openings were filled with water.

The removal of support following a decline in the water table can result in an immediate collapse of the roofs of openings in bedrock or can cause a downward migration of clay or other unconsolidated material into openings in underlying rocks. This migration into underlying openings results in the development of cavities in the unconsolidated deposits between the top of bedrock and the land surface. Information obtained from the study indicates that all sinkholes in the area resulted from the creation and collapse of cavities in unconsolidated deposits rather than from collapses in bedrock. Cavities in unconsolidated sediments overlying openings in carbonate rocks have been described, explored, or photographed in Africa, Pennsylvania, and Alabama (Donaldson, 1963; Jennings and others, 1965; Foose, 1967; and Newton and Hyde, 1971).

The principal forces resulting from the decline in the water table that cause or accelerate the creation and enlargement of cavities in residual clay or other unconsolidated sediments are recognizable in the area of study. Precipitation, with prevailing conditions, results in fluctuations of the water table that cause repeated movement of the water through openings in bedrock against overlying clay or other unconsolidated sediments. This causes a repeated addition and subtraction of support to the sediments and repeated saturation and drying. This process might be best termed "erosion from below" because it results in the creation of cavities, their enlargement, and eventual collapse. This relation of fluctuations in the water table to the contact between unconsolidated clay and the underlying bedrock is shown in hydrographs in figures 6 and 7.

The inducement of recharge through clay or other unconsolidated sediments to openings in bedrock results in the discharge of Rice Creek and Allen Brook into the subsurface when the water table is below the bottom of the stream channels. This results in the migration or transport of unconsolidated sediments overlying bedrock into underlying openings and the creation of cavities over bedrock that enlarge with time and eventually collapse.

The velocity of water moving in the subsurface is sufficient to transport unconsolidated sediments to points of discharge (faults or other openings) or to points of storage in openings in rocks at greater depths. Without this velocity, unconsolidated clay entering openings would eventually result in blockage and a cessation of growth of cavities. The continued growth of existing sinkholes shows that the water velocity is sufficient to carry away those sediments entering swallow holes. The velocity of water may also cause cavities in unconsolidated sediments filling fractures or other openings in bedrock where the water comes in contact with the sediments through other openings.

Sinkhole 74 (pl. 2 and table 3), which resulted from a collapse during December 24-27, 1971, is shown in figure 11. The top of

Figure 11.—Sinkhole resulting from collapse.

limestone, based on exposures in nearby sinkholes, is at a depth of 15 to 20 feet. The depth of the collapse shown in figure 11 is 14 feet. The cavity in residual clay that resulted in the collapse was formed by the fluctuation of the water table and the movement of water from the surface to the underlying opening in bedrock. The magnitude of water-table fluctuations in the cavity prior to its collapse is indicated by comparing the photograph of the dry sinkhole in figure 11 taken during a period of little precipitation to a photograph of the same sinkhole (fig. 12) taken after a significant rain. Fluctuations of the water level in this sinkhole and others nearby, including sinkhole 72 (fig. 7), correspond with those measured in nearby observation wells. The movement of water from the surface to the underlying opening in bedrock and its relation to the collapse is indicated by the number of small holes at the surface in the vicinity and their convergence into the collapsed cavity. A close examination of the cavity walls shows numerous circular holes connected with the land surface. Erosion around

Figure 12.—Water level in sinkhole after precipitation.

these holes at the surface forming significant depressions (fig. 13) where soil has been transported into the openings shows that considerable movement of surface water to the cavity has occurred. Uncollapsed overhanging parts of the roof of the cavity (fig. 11) indicate it's recent occurrence.

SIZE

Sinkholes in the area of study are generally circular, however, a substantial number are elongated or irregular in shape. Most of the elongated sinkholes are oriented parallel to the strike of underlying strata and have formed from the coalescence of two or more circular collapses. Sinkholes inventoried generally range from 3 to 75 feet in width, 3 to 144 feet in length, and 3 to 30 feet in depth to the top of bedrock. The average sinkhole is about 13 feet wide, 20 feet long, and 7 feet deep. The average length of individual collapses is considerably less than the average given because the longest sinkholes inventoried resulted from the coalescence of several collapses and the discharging of streams and floods into them. The average depth of individual collapses would

Figure 13.–Depression formed where soil has been eroded into subsurface.

be slightly greater than 7 feet because many of the collapses were partially filled prior to the present study.

PRESENT AND FUTURE ACTIVITY

The development of sinkholes in the area of study, based on collapses occurring during the investigation and observation of water-table fluctuations and water loss along streams can be described as "active and continuous." Available information indicates that the activity will cease or decrease drastically after the water table in the Tuscumbia Limestone has recovered to an altitude similar to its position prior to 1950. This altitude in the proposed right-of-way would be similar to or slightly above the altitude of the base of the channel in Allen Brook. The altitude of the base of the channel in the area of study generally ranges from 500 to 508 feet. Previous studies indicate that a cessation of or drastic decrease in the development of sinkholes results from a recovery of the water table (Foose, 1953, p. 644; Newton and Hyde, 1971, p. 20). The water table, based on the reported rate of recovery in the nearby mine, could recover to its pre-1950 level as early as the summer of 1973. However, available information indicates that the recovery will probably extend over a much longer period of time.

AREA OF POTENTIAL DEVELOPMENT

The geology and hydrology of the area of study governs the development of sinkholes. After more than 20 years of intensive sinkhole development, the boundaries of the area prone to their development have been fairly well defined by nature. The area in which sinkholes occur is confined almost entirely to the outcrop of the Tuscumbia Limestone, to the lower 20 feet of Floyd Shale overlying the Tuscumbia, and to the immediate vicinity of faults. Sinkhole 138 (pl. 2) is the one exception to these findings. It is located about 60 feet from a major fault in an area underlain by 40 feet of shale. It undoubtedly formed where weathered shale and overlying clay migrated downward into fractures or other openings associated with the nearby fault.

The estimated boundaries of the area prone to the development of sinkholes are shown on plate 2. Areas not prone to their development, based on available findings, are (1) those underlain by more than 40 feet of shale, (2) those underlain by the Fort Payne Chert above the basin of Allen Brook, and (3) those located 100 or more feet from faults where the underlying shale exceeds 40 feet in thickness.

SUPPORTING TOOLS

This investigation had the benefit of most major tools available to the geologist and hydrologist to aid in evaluating a complex problem. These tools included augering, core drilling, multispectral photography, thermal infrared imagery, and geophysics. All tools were beneficial although augering and drilling provided the foundation of interpretable quantitative geologic and hydrologic data.

Pre-project planning included the provision for early acquisition of multispectral photography, thermal infrared imagery, and geophysical determinations to provide a basis for test drilling to assure the economic acquisition of the maximum amount of interpretable data in the least amount of time. Unfortunately, uncontrollable delays reversed the planned orderly acquisition of data. The quantity of drilling and augering necessary to obtain results desired, due to the restricted project period, increased because of the absence of planning tools available. Even though their acquisition was made later than anticipated, these tools still provided a basis for major interpretations and verified earlier findings. A summary of their utility during this investigation will provide a foundation for evaluating their potential value in future projects.

MULTISPECTRAL PHOTOGRAPHY AND THERMAL INFRARED IMAGERY

Multispectral photography and thermal infrared imagery were obtained to aid in locating areas of water loss in streams, geologic structures, and to evaluate their utility in locating uncollapsed cavities in unconsolidated deposits overlying bedrock. An overflight altitude of 1,600 feet above mean terrain was selected to provide ground resolution of features or objects as small as 5 square feet in area. Four 70-millimeter cameras simultaneously exposed the following film types: infrared color, color Ektachrome, black and white infrared, and aerocolor. The photography was flown on January 18, 1972. Nighttime thermal infrared imagery was flown on January 18, 1972 and daytime imagery was flown on March 14, 1972.

All photography and the daytime thermal infrared imagery contributed substantially to the project. The interpretability of the nighttime imagery was limited because of weather conditions unfavorable for thermal mapping that normally occur during January. The imagery was obtained with a mercury doped cadmium-telluride detector with a temperature sensitivity of 0.5 C (centigrade) or better and a spectral region of 8.0 to 14 microns.

The color infrared and color Ektachrome defined existing and prior vegetative stress that resulted from subsidence and interior drainage through openings in unconsolidated sediments at the land surface. This definition aided in locating areas in which conditions are favorable for the formation of cavities in residual clay and in which they are present. A photograph (aerocolor) of an area of general subsidence in the proposed right-of-way between stations 346+00 and 350+00 (fig. 14) shows vegetative stress resulting

Figure 14.–Vegetative stress near sinkholes between stations 346+00 and 350+00. (Photograph by Environmental Systems Corporation.)

from subsidence or interior drainage. A cavity in unconsolidated clay was augered (sinkhole feature 19 on pl. 2 and table 3) beneath one of the sites where the stress is visible. All photography defined vegetative stress over tile drains and subsidence over couplings joining the drains that were buried beneath the land surface more than 40 years ago. A photograph in aerocolor shows (fig. 15) the subsidence and stress over drains located between stations 328+00 and 338+30 in the proposed right-of-way (pl. 2). This aided in separating minor subsidence occurring in unconsolidated deposits over shale from nearby subsidence over limestone.

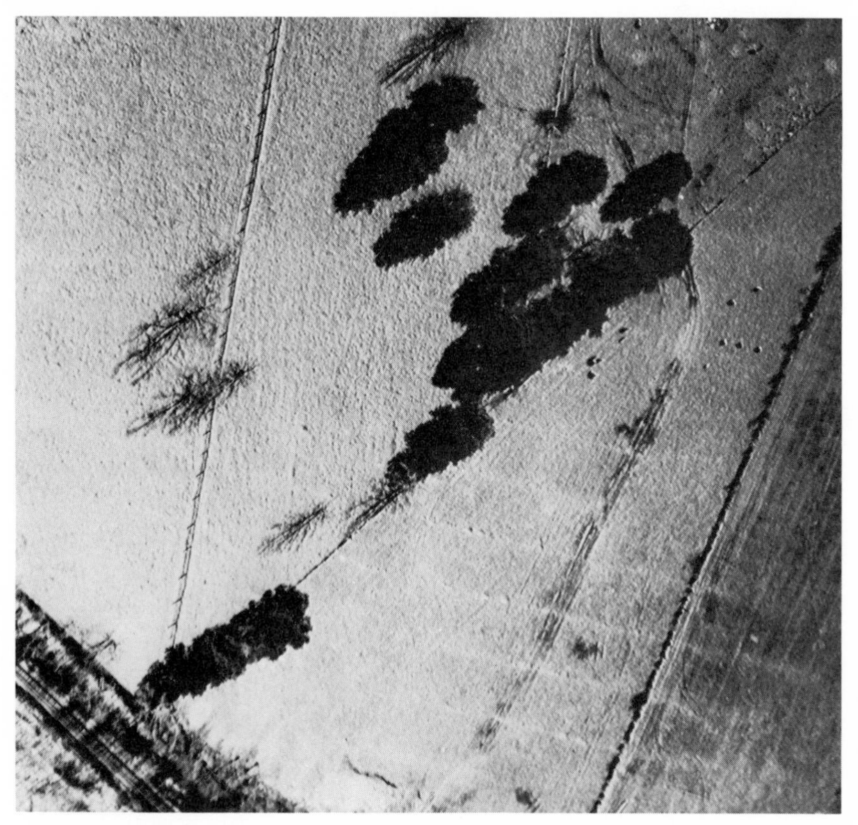

Figure 15.—Vegetative stress and subsidence over buried drains. (Photograph by Environmental Systems Corporation.)

Infrared black and white photography obtained to enhance water and land boundaries minimizes to some degree the obscuring effect of foliage in wooded areas. Figure 16 shows prominent lineaments formed by the channel of Allen Brook trending along a fault and associated highly inclined strata, a stream discharging into a sinkhole, and water-filled sinkholes.

Daytime thermal infrared imagery flown in March 1972 shows or indicates faults mapped during the project and points where water loss occurs. Standard analog processing (fig. 17) shows water loss along Allen Brook and its tributaries where they discharge into streambeds and sinkholes. Special processing of the

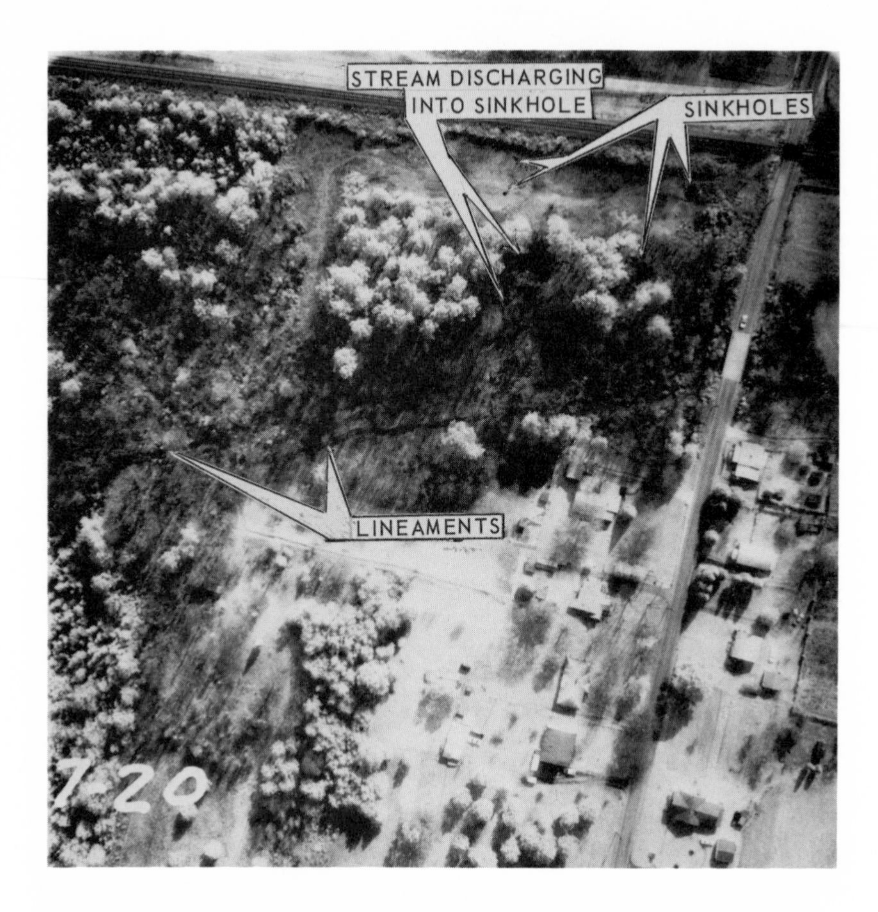

Figure 16.–Black and white infrared showing lineaments, water loss, and sinkholes. (Photograph by Environmental Systems Corporation.)

magnetic tape to enhance interfaces between adjacent temperature levels resulted in the recognition of lineaments not readily apparent on photography. The process enhancing the interfaces is referred to as "contouring." Lineaments located along the Dickey Springs Fault and previously unmapped fault crossing the proposed right-of-way near station 388+50 (fig. 3) are shown on figure 18. The easternmost lineament shown on figure 18 is apparently developed on an unmapped extension of a fault located east of the area of study (Simpson, 1965, pl. 1).

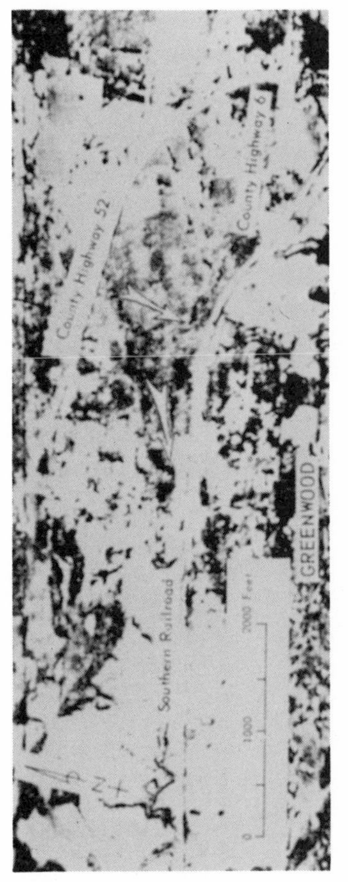

Figure 17.–Thermal infrared imagery showing water loss. (Imagery by Environmental Systems Corporation.)

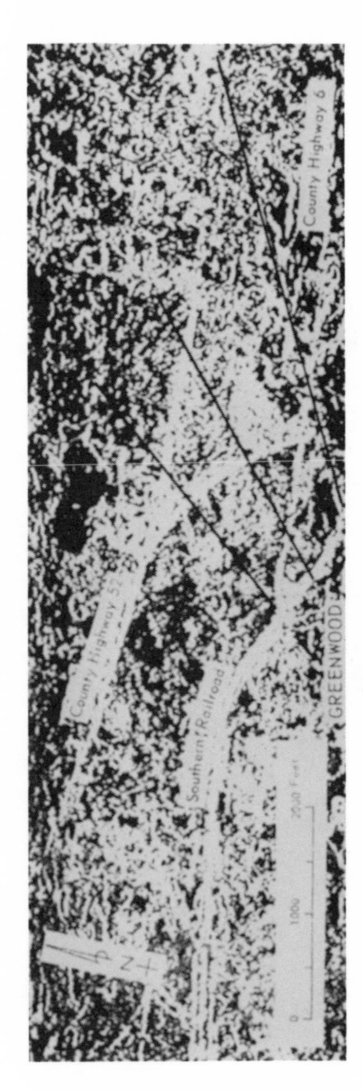

Figure 18.–Thermal infrared imagery showing locations of lineaments indicating faults. (Imagery by Environmental Systems Corporation.)

Most major geologic and hydrologic features in the area were located prior to the acquisition of photography and thermal mapping. The availability of these tools during early stages of the project would have resulted in less test drilling and "leg work." This, in turn, would have resulted in an earlier completion date and economic gain. The locating of geologic and hydrologic features that cause or are related to the development of sinkholes shows the potential value of applying multispectral photography and thermal infrared imagery in the evaluation of proposed highway corridors.

GEOPHYSICAL INVESTIGATION

A 14-trace portable refraction seismograph (fig. 19) with

Figure 19.–Portable refraction seismograph. (Photograph by T. V. Stone.)

timing, shot instant, and 12 signal traces was used in the investigation. The unit has a blaster synchronized with the recording mechanism. Twelve geophones detect the arrival of shock waves created when an explosive charge is detonated. Electrical impulses created by these geophones are transmitted through a connection cable to the recording unit where they are amplified and recorded on a film (fig. 20) from a moving optical system.

Seismic exploration was accomplished by laying out a line of geophones connected to the recording instrument and detonating a charge to transmit energy through the soil and rock layers to the geophones. The arrival time of energy to each geophone is dependent upon the physical characteristics of the underlying rock units. The highway right-of-way was investigated to a depth of approximately 80 feet by placing the 12 geophones 20 feet apart along a line 240 feet in length from the shotpoint to the twelfth geophone. This layout of shotpoint and geophones is referred to as a profile and a series of profiles make up a traverse. Explosive charges generally consisted of less than one stick of 40 percent gel dynamite. The charges were set in small diameter holes 3 to 4 feet deep and detonated with non-delay seismic blasting caps.

A nearly continuous traverse of the center line of the proposed highway was made from station 393+40 westward to station 343+50. The depth to bedrock along the traverse as determined by seismic methods is included on plate 1. The area between stations 360+00 and 370+00 was not investigated because homes are located near the center line.

The area west of station 343+50, underlain by alluvial and residual deposits and the Floyd Shale was reconnoitered with random, discontinuous profiles between stations 345+50 and 275+00 to detect possible near surface occurrences of Tuscumbia Limestone.

In the vicinity of the faults and sinkholes between stations 343+50 and 360+00 and from stations 370+00 to 393+40, fan shooting techniques (fig. 21) were used to explore the area between the center line and the outermost edge of each planned traffic lane. The fan shots were used to aid in determining the trend of fault zones and fractures, and in locating features related to the possible development of sinkholes. The principle involved is that arrival times of shock waves reaching the geophones will be delayed if the shock waves have passed through fractured or cavernous rock.

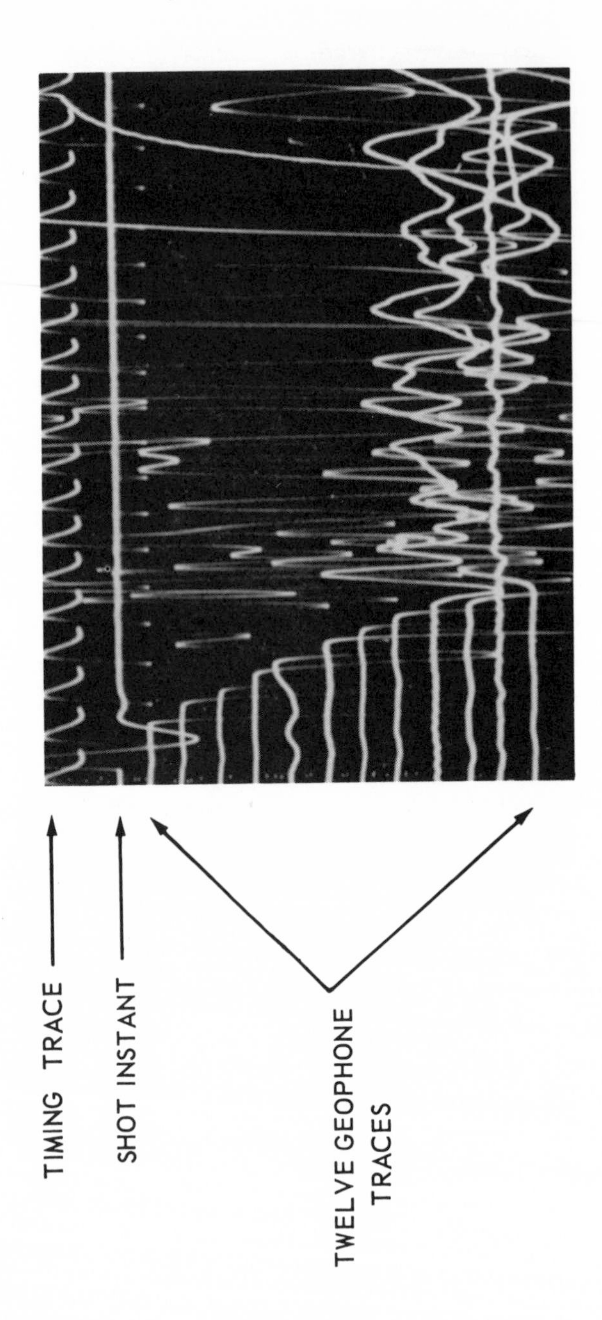

Figure 20.—Seismic record.

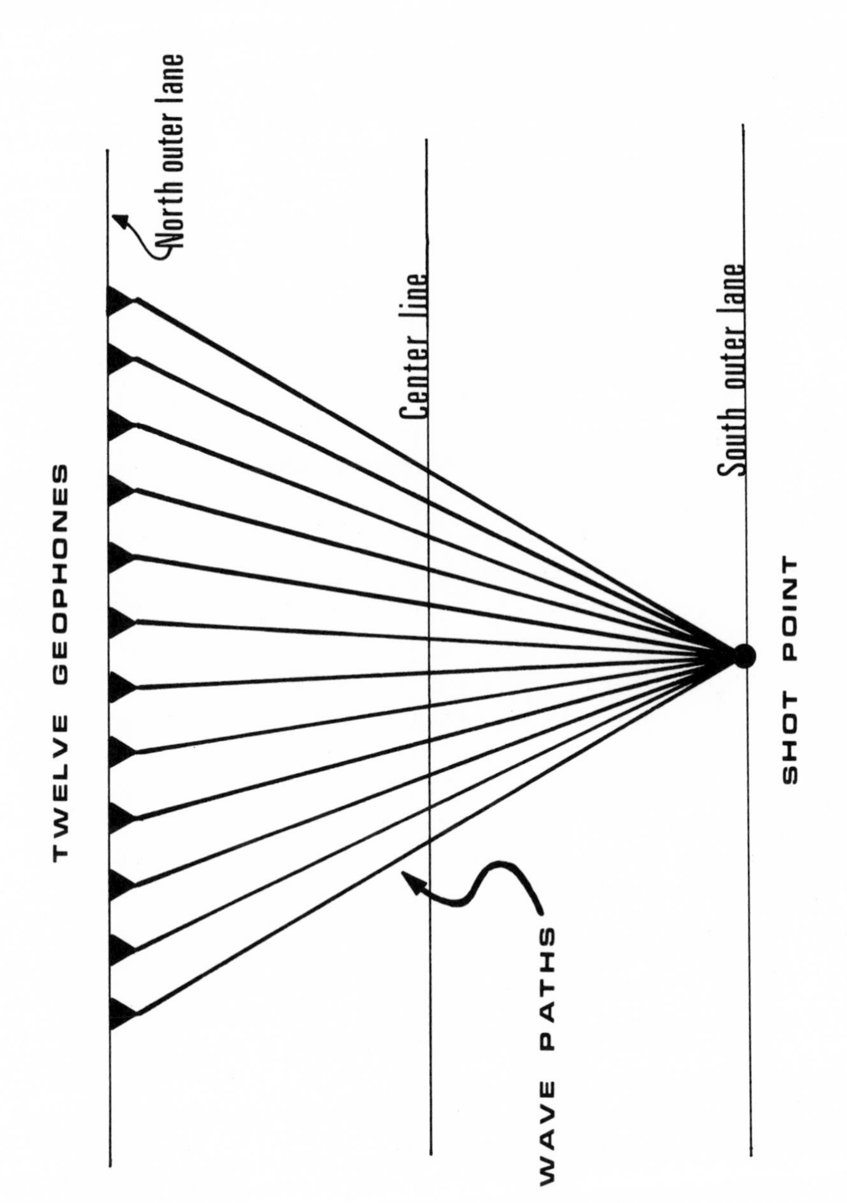

Figure 21.—Fan shot technique.

Variations in velocity detected by individual geophones establishes the approximate location of anomalous features. The fan shooting was arranged with the geophones spaced 50 feet apart along the outer edge of the traffic lane. This allowed coverage of an area 550 feet long. The shotpoint was placed 120 feet away on the outer edge of the other traffic lane in the center of the geophone spread. The locations of the shotpoints and geophones were then reversed to allow overall coverage of the roadway.

The location and trend of faults mapped by augering and drilling were similar to those of subsurface anomalies determined by the seismic survey. The depth to the top of limestone determined by the survey also corresponded with those determined by augering and drilling. The depths to bedrock recorded by the seismic equipment in areas underlain by Floyd Shale east of station 388+00 are about 10 feet deeper than those determined by drilling. The depth to bedrock determined by geophysical methods coincided with the depth to thin beds of limestone within the shale, whereas the auger tops represent the uppermost recognizable shale. The velocity contrasts between the residual deposits overlying the Floyd Shale apparently are not greatly different from the velocities of the slightly weathered shale occurring at the contact.

Most major geologic and hydrologic features in the area were located prior to acquisition of seismic information. The availability of geophysical data in early phases of the project would have facilitated test drilling and resulting interpretations. Seismic surveys can be utilized effectively in locating geologic conditions favorable to the development of sinkholes and in defining subsurface conditions in an area of active subsidence.

CONCLUSIONS

1. A general lowering of the water table due to years of ground-water withdrawals has resulted in the development of sinkholes. The decline in the water table probably began about 1950 but it may have begun at an earlier date.

2. The area prone to the development of sinkholes in the proposed right-of-way is located between stations 341+00 and 395+00 where, a) the Tuscumbia Limestone and lower 20 feet of the overlying Floyd Shale underlie the land surface or, b) the land

surface is underlain by a fault and 40 or less feet of Floyd Shale.

3. Sinkholes occur where cavities develop in residual or alluvial deposits overlying openings in limestone. The downward migration of these deposits into the openings and the formation and collapse of cavities are caused or accelerated by a decline in the water table that results in a) an increase in the amplitude of water-table fluctuations, b) the movement of water through unconsolidated deposits into openings in underlying bedrock where recharge had previously been rejected, c) an increase in the velocity of movement of ground water, and d) loss of support to unconsolidated deposits overlying openings in bedrock.

4. The development of sinkholes and subsidence in the area of study was active from December 1971 through March 1972. This development started about 1950.

5. Cessation of ground-water withdrawals at Greenwood in the mid-1960's and cessation of mine dewatering in 1970 has resulted in conditions favorable to a recovery of the water table. The major part of the recovery could occur as early as the summer of 1973 but total recovery may well extend over a longer period of time.

6. A recovery of the water table will probably result in a cessation of or drastic decrease in sinkhole development.

7. Grading and construction in areas underlain by limestone will create new connections between the land surface and openings in underlying bedrock that will alter natural drainage. These actions can result in the development of new sinkholes.

8. The construction of impermeable drainage facilities in the proposed right-of-way to allow rapid runoff of precipitation will tend to retard the development of sinkholes.

SELECTED REFERENCES

Butts, Charles, 1927, Description of the Bessemer and Vandiver quadrangles (Alabama): U.S. Geol. Survey Geol. Atlas, Folio 221.

Dix, C. H., 1939, Refraction and reflection of seismic waves, pt. 1, Fundamentals: Geophysics, v. IV, no. 2, p. 81-102.

Dobrin, M. B., 1960, Introduction to geophysical prospecting: New York, McGraw-Hill Book Co., Inc., 2nd ed., 446 p.

Donaldson, G. W., 1963, Sinkholes and subsidence caused by subsurface erosion: Regional Conference for Africa on Soil Mechanics and Foundation Engineering, 3rd, Salisbury, Southern Rhodesia 1963 Proc., C. 3, p. 123-125.

Foose, Richard M., 1953, Ground-water behavior in the Hershey Valley, Pennsylvania: Geol. Soc. America Bull. 64, p. 623-645.

________1967, Sinkhole formation by ground water withdrawal: Far West Rand, South Africa: Science, v. 157, p. 1045-1048.

Heiland, C. A., 1940, Geophysical exploration: New York, Prentiss-Hall, Inc., 1013 p.

Jennings, J. E., 1966, Building on dolomites in the Transvaal: The Civil Engineer in South Africa, v. 8, no. 2, p. 41-62.

Jennings, J. E., Brink, A. B. A., Louw, A., and Gowan, G. D., 1965, Sinkholes and subsidences in the Transvaal dolomites in South Africa: Internat. Conf. Soil Mechanics, 6th, 1965, Proc., p. 51-54.

Johnston, W. D., Jr., 1933, Ground water in the Paleozoic rocks of northern Alabama: Alabama Geol. Survey Spec. Rept. 16, 441 p.

Joiner, T. J., Warman, J. C., Scarbrough, W. L., and Moore, D. B., 1967, Geophysical prospecting for ground water in the Piedmont area, Alabama: Alabama Geol. Survey Circ. 42, 48 p.

Joiner, T. J., and Scarbrough, W. L., 1969, Hydrology of limestone terranes, geophysical investigations: Alabama Geol. Survey Bull. 94, pt. D, 43 p.

Meinzer, O. E., 1942, Hydrology, v. 9 of Physics of the earth: McGraw-Hill, 712 p.

Nettleton, L. L., 1940, Geophysical prospecting for oil: New York, McGraw-Hill Book Co., Inc., 444 p.

Newton, J. G., and Hyde, L. W., 1971, Sinkhole problem in and near Roberts Industrial Subdivision, Birmingham, Alabama—a reconnaissance: Alabama Geol. Survey Circ. 68, 42 p.

Powell, W. J., and LaMoreaux, P. E., 1969, A problem of subsidence in a limestone terrane at Columbiana, Alabama: Alabama Geol. Survey Circ. 56, 30 p.

Robinson, W. H., Ivey, J. B., and Billingsley, G. A., 1953, Water supply of the Birmingham area, Alabama: U.S. Geol. Survey Circ. 254, 53 p.

Simpson, T. A., 1965, Geologic and hydrologic studies in the Birmingham red-iron-ore district, Alabama: U.S. Geol. Survey Prof. Paper 473-C, 47 p.

Part VII

DATA GATHERING AND PRESENTATION

In this last part we concentrate on the problems of the relevancy of environmental geologic data to the needs of users. This topic has appeared recurrently in the present volume, and rightly so, since environmental geology is first and foremost an applied science. The responsibility of those who engage in some aspect of environmental geology commonly ends in communicating some type of information to persons other than geologists.

Most geologic data have been and are being gathered for the benefit of geologists. It should be feasible to convert these data into a form suitable for the nongeologist. This is the hope held out by those geologists who speak of converting basic geologic maps into derivative maps. To accomplish this result, the basic maps must contain something more than the historical-stratigraphic information. One may conclude that the premise of the basic mapping must contain an element of concern with the ensuing derivative map, so that the bridge from the first to the second is based on more than inference or guesswork.

As an example, we can proceed from a typical areal geologic map, with stratigraphic units described directly or in accompanying legends with their sometimes confusing lithologic groupings. Converting this map to an engineering geologic map seems simple enough: omit the stratigraphic terms and mark the map units with lithologic names. But any geologist knows that such a one-for-one exchange is fraught with conveyance of misinformation. Put another way, stratigraphic units are often a mixture (e.g., a rhythmic sand–shale formation). Perhaps nothing more can be done about it, but that is the problem to be solved in the first place by the geologist intending to produce an engineering geologic map.

The broader problem of land-use planning increases the need for thought about appropriate data gathering, so that the ultimate map will serve an intended purpose. The inputs must include all elements of the environment, or at least all that are significant; ideally, they should be determined in advance of the data-gathering effort.

The problem was examined by military geologists and their associates under the pressure of wartime need for prompt response. Extraordinary achievements in both data gathering and presentation came to light after the war. In respect to presentation in map form, the sophistication of the products of some organizations has not been duplicated since that time. These complex but ingenious cartographic solutions are reported to have been well received by military users. At the other end of spectrum, some organizations purposely produced extremely simple maps that left the user with no basis for making his own interpretations.

In conclusion, data gathering and presentation represent one of the most important aspects of environmental geology, which should be studied as carefully as the concept of environmental geology itself. On the whole, both concept and attention to purpose have been generally neglected. Thus, the thoughts expressed by the authors in Part I, Overview, and in Part VII, Data Gathering and Presentation, are the bookends that bind the entire rationale of environmental geology.

SUPPLEMENTAL READINGS

Algermissen, S. T. 1969. *Seismic Risk Studies in the United States.* 4th World Conf. Earthquake Eng. (Santiago, Chile) Proc., **1**, p. 14–27. (Reprinted by the U.S. National Oceanic and Atmospheric Administration, Washington, D.C., n.d.)

Anderson, J. R., E. E. Hardy, and J. T. Roach. 1972. *A Land-Use Classification System for Use with Remote-Sensor Data.* U.S. Geol. Surv. Circ. 671, 16 p.

Beckett, P. H. T. 1962. Punched Cards for Terrain Intelligence. *Roy. Engr. J.*, **76**, p. 185–194.

Branagan, D. F. 1972. *Geological Data for the City Engineer: A Comparison of Five Australian Cities.* 24th Intern. Geol. Congr. Proc., Sec. 13, p. 3–12.

Cluff, L. S., and B. A. Bolt. 1969. Risk from Earthquakes in the Modern Urban Environment, with Special Emphasis on the San Francisco Bay Area. In: *Urban Geology in the San Francisco Bay Region* (Edward A. Danehy, ed.), p. 25–64. Assoc. Eng. Geologists, San Francisco Bay Section, Spec. Publ.

Cratchley, C. R. 1972. *Engineering Geology in Urban Planning with an Example from the New City of Milton Keynes.* 24th Intern. Geol. Congr. Proc., Sec. 13, p. 13–22.

Frye, J. C. 1967. *Geological Information for Managing the Environment.* Illinois Geol. Surv. Environ. Geol. Notes 18, 12 p.

Gardner, M. E., and C. G. Johnson. 1971. Engineering Geologic Maps for Regional Planning. In: *Environmental Planning and Geology* (D. R. Nichols and C. C.

Campbell, eds.), p. 154-169. Washington, D.C., U.S. Geological Survey and U.S. Department of Housing and Urban Development (Proc. Symp. Eng. Geol. Urban Environ.)

Golodkovskaya, G. A., N. V. Kolomenskii, I. V. Popov, and M. V. Churinov. 1972. *Engineering Geological Mapping in the U.S.S.R.* 23rd Intern. Geol. Congr. Proc., Sec. 12, p. 57-64.

Grant, K. 1968. *A Terrain Evaluation System for Engineering.* Australia, Commonwealth Sci. Ind. Res. Organ., Div. Soil Mech., Tech. Paper 2, 27 p.

Hackett, J. E., and M. R. McComas. 1969. *Geology for Planning in McHenry County, Illinois.* Illinois Geol. Surv. Circ. 438, 31 p.

Jancić, Milosav. 1962. Engineering-Geologic Maps. *Vesnik Zavoda za Geološka i Geofizička I Stražívanja,* Ser. B, no. 2, p. 17-31. (Text in English.) Also published in German: Ingenieurgeologische Karten. *Deut. Geol. Ges. Z.,* **114**(2), 1962, p. 327-336.

LaMoreaux, P. E., et al. 1971. *Environmental Geology and Hydrology: Madison County, Alabama—Meridianville Quadrangle.* Geol. Surv. Alabama Atlas Ser. 1, 72 p.

Larsen, J. I. 1973. *Geology for Planning in Lake County, Illinois.* Illinois Geol. Surv. Circ. 481, 43 p.

Leopold, L. B., F. E. Clarke, B. B. Hanshaw, and J. R. Balsley. 1971. *A Procedure for Evaluating Environmental Impact.* U.S. Geol. Surv. Circ. 645, 13 p., 1 plt.

McComas, M. R., K. C. Hinkley, and J. P. Kempton. 1969. *Coordinated Mapping of Geology and Soils for Land-Use Planning.* Illinois Geol. Surv. Environ. Geol. Notes, 29, 11 p.

Matula, Milan. Engineering Geologic Mapping in Urban Planning. In: *Environmental Planning and Geology* (D. R. Nichols and C. C. Campbell, eds.), p. 144-153. U.S. Geological Survey and U.S. Department of Housing and Urban Development (Proc. Symp. Eng. Geol. Urban Environ.)

Moser, P. H. 1972. *Environmental Geology Studies in Alabama.* 24th Intern. Geol. Congr. Proc., Sec. 13, p. 37-43.

Pessl, Fred, Jr., W. H. Langer, and R. H. Ryder. 1972. *Geologic and Hydrologic Maps for Land-Use Planning in the Connecticut Valley with Examples from the Folio of the Hartford North Quadrangle, Connecticut.* U.S. Geol. Surv. Circ. 674.

Philipp, Hans. 1923. Die Methoden der geologischen Aufnahme. In: *Handbuch der biologischen Arbeitsmethoden* (E. Abderhalden, ed.), p. 395-484. Munich, Urban & Schwarzenberg.

Quay, J. R. 1966. Use of Soil Surveys in Subdivision Design. In: *Soil Surveys and Land Use Planning* (L. J. Bartelli et al., eds.), p. 76-87. Madison, WI, Soil Science Society of America and American Society of Agronomy.

Schneider, W. J., D. A. Rickert, and A. M. Spieker. 1973. *Role of Water in Urban Planning and Management.* U.S. Geol. Surv. Circ. 601-H, 10 p., 1 plt.

Sonne, Erich. 1936. Geologische und militärgeologische Karten. *Preuss. Geol. Landesanstalt Jahrb. 1935,* **56**, p. 192-195, 7 plts.

Steinitz, Carl. 1970. Landscape Resource Analysis. *Landscape Architecture,* **60**(2), p. 101-105.

Webster, Richard, and P. H. T. Beckett. 1970. Terrain Classification and Evaluation Using Air Photography—A Review of Recent Work at Oxford. *Photogrammetria,* **26**(2/3), p. 51-75.

Wilson, H. E. 1972. *The Geological Map and the Civil Engineer.* 24th Intern. Geol. Congr. Proc., Sec. 13, p. 83-86.

Abstracts (1972–1974) from Geological Society of America, Abstracts with Programs

Baker, V. R., and J. A. Pendleton. 1973. Urban Geological Problems of Boulder, Colorado. **5**(6), p. 462.

Cleaves, E. T., and A. S. Godfrey. 1973. Geologic Constraint Maps for Planning Purposes in the Piedmont of Maryland. **5**(2), p. 148.

Easterbrook, D. J. 1972. Environmental Geology of Whatcom County, Washington—A Possible Model for Applied Environmental Geology. **4**(7), p. 495.

Garner, L. E. 1972. Environmental Geology of the Austin, Texas Area. **4**(7), p. 515.

Halbig, J. B., and G. P. Sick. 1974. Environmental Geology and Land-Use Planning in Avon Township, Livingston County, New York. **6**(1), p. 33–34.

Hanshaw, B. B. 1972. Proposed Integrated Program for Providing Resource and Land Information. **4**(7), p. 526–527.

Hughes, T. H., S. H. Stow, and P. E. LaMoreaux. 1973. Urban Land-Use Capability Studies by Derivative Mapping of Geological Data in Alabama. **5**(7), p. 677.

Reid, Reginald, and E. L. Montgomery. 1972. Geologic Hazards Pilot Study, Flagstaff, Arizona. **4**(7), p. 634.

Stow, Stephen, T. H. Hughes, and D. E. Bolin. 1973. Derivative Land-Use Capability Mapping for Urban Planning—Tuscaloosa County, Alabama. **5**(5), p. 439–440.

Tilmann, S. E., C. M. Iversen, and S. B. Upchurch. 1974. Computer-Linked Terrain Analysis. **6**(6), p. 550.

Troughton, Gaël, and D. R. Hoffman. 1974. "Geo-Alert" Maps as an Aid to Environmental Planning and Open Space Development. **6**(3), p. 268.

Twiss, R. H. 1972. Use of New-Generation Geologic Information in Regional Planning. **4**(7), p. 695.

Van Driel, J. N., R. C. Palmquist, and L. V. A. Sendlein. 1973. A Computer Technique for Rapid Generation of Interpretive Planning Maps. **5**(4), p. 360.

Zeizel, Arthur, 1972. Communication—Keystone Between Geological Science and Metropolitan Problem Solving. **4**(7), p. 713.

Editor's Comments on Papers 26 Through 30

Hugh B. Montgomery has published a concise, systematic examination of data gathering and presentation of environmental data for purposes of local development planning, reproduced here as Paper 26. The coverage of topics depends on the scope of the planning project. For overall planning, a comprehensive survey of environmental resources is recommended, which will assist planners who are interested in determining suitability and feasibility of development for different purposes.

The presentation is usually in map form. Thus, the translation of basic data to units that have a meaning for nonscientists is especially important. The map, like the data gathered, must be directed toward planning decisions.

The paper is a useful guide to the varieties of data that might be pertinent and their sources. The information on sources given here is entirely concerned with United States agencies, mainly federal governmental, which tends to be ephemeral because of the periodic reorganizations and reassignment of responsibilities, but the types of sources will remain the same.

Montgomery is a consulting geologist. His interest and involvement in environmental planning studies and methodology are noteworthy.

The purpose of including Paper 27 is to show the early development of an engineering geologic map, based on experience gained by military geologists during World War I in handling this type of presentation. The authors of this short paper point out that there were signs of interest in engineering geology well before that war. However, the influence of wartime work on the subsequent development of applied geology was definitely significant.

The mapping for the Danzig area was based on practical criteria, as shown by the explanation of the map units, and the engineering implications are stated for each unit.

Stremme was a noted soil scientist, whose name is linked with the preparation of the first soil map of Europe. Moldenhauer was the source of the wartime experience in the collaboration; he published a dissertation with a meaningful title, in translation, "Conversion of the Historical-Geologic Map of the Danzig Area to a Technical-Geologic Map."

Paper 28 is a discussion of the engineering geologic map prepared for an urban area in Missouri, and exemplifies the treatment of the problem of presentation developed by the Missouri Geological Survey. The authors state that "the basic purposes of the map and text (on the reverse side) are to aid: (1) the engineer in initial site investigation and design of structures, (2) suburban and county planners, and (3) the developer or landowner in early anticipation rather than late awareness of construction problems."

The basic areal geologic map in this case shows fourteen formations. In the derivative map, six engineering geologic units were established. In the example, there are several soil subunits.

The significance of this paper lies in the opportunity for direct comparison between the basic map and the derivative map; sections are displayed side by side, which demonstrate the noticeable differences.

Paper 29 deals with a study for the new Montreal International Airport. The area has been investigated by a wide range of specialists for the purpose of determining "the capabilities of the land, mineral, and fluid resources with respect to their single, multiple, and sequential use."

In the discussed geoscientific study, a six-month period was involved from the gathering of data, especially those developed from field testing and laboratory analyses, to the completion of a computer map. The establishment of a data bank was a basic factor in moving the project ahead rapidly. The versatility of the data bank is illustrated by the maps that could be extracted. In addition to serving the purposes of the immediate project, the accumulated data can be used for other types of regional planning problems.

St-Onge is a research scientist with the Geological Survey of Canada; Scott is chief of the Terrain Sciences Division of the same organization. St-Onge is also a professor in the Department of Geography, University of Ottawa.

Our final paper exemplifies a type of data presentation that differs from what we have encountered previously. The author is concerned here with the contamination potential of waste-disposal sites in loose granular materials, and proposes a system that "is especially suitable for a quick initial appraisal of sites where geologic and hydrologic data are scarce." The system is based on an evaluation of the effect of five environmental factors on contaminants released near a site. The potential is rated by a point-count system, which assigns ranges of probability. At the same time, there is an estimate of the accuracy of prediction.

Such a system might also be adapted to establish evaluations of various other kinds of potential benefits and hazards resulting from some identified human action in a specified environmental setting. It would supplement the graphic presentation of data recommended in other discussions for transfer of information for practical purposes. One weakness of the majority of maps is the absence of a rating scheme. Most land-use and engineering geologic maps are based on units that have stated physical characteristics, but these are given in qualitative terms. An obstacle to quantified presentations may be the scale of many maps, which militates against introducing meaningful numbers on the face of the map. It is possible, as we see in an example, to introduce figures into units—admittedly not ratings directly, because the map area was relatively small and the scale was not suitable. Another significant lack in virtually all maps is acknowledgment to the user that the data are not all of equal reliability, and that a derivative map is usually composed of a patchwork of determinations (material from existing maps and other sources) tied together with strands of extrapolation.

26

Reprinted from *Appalachia,* 3(3), 1-11 (1969)

ENVIRONMENTAL ANALYSIS IN LOCAL DEVELOPMENT PLANNING

Hugh B. Montgomery

"We need help in fighting the disease that now threatens our planet—a sort of cirrhosis of the environment."
Conservationist David Brower

As the nation approaches a 300-million population and a trillion-dollar economy and as our strip cities, industrial complexes and communities grow, the time remaining to prevent the destruction of our natural environment and erosion of our resources needed to support life is becoming dangerously short.

As each man-made change occurs, it has effects on the environment which we, through shortsightedness, frequently fail to anticipate. And since our physical resources are limited and interdependent, the misuse of each resource alters the availability or usefulness of the others.

The nation is rapidly coming to realize that if action is not taken *now* on the national, regional and local levels to preserve our environment, we will soon lack the environment and resources necessary to sustain our civilization and perhaps human life itself.

Where will more and more people live, work and play? It may soon be true that in many metropolitan areas the nearest open space for a picnic within 25 miles of the downtown area will be the city dump.

Where will we dispose of the waste products of our increasing population and industrialized society?

The conservation of wildlife, wilderness and recreation land and the pollution of our water, land and air have finally become matters of public concern.

In Appalachia, where the mountainous environment has historically been an inspiration to poets and musicians but an impediment to planners, we have a unique opportunity to demonstrate the reciprocal relationship between resource conservation and economic progress.

Although many of Appalachia's environmental and resource problems are regional in character, it is often only at the local level that the planned use of our natural resources can be effective.

It has long been recognized in the region that resources analysis is an important step in certain types of local planning. In determining the location of an industrial plant or industrial park, for example, analyses of water supply and topography have become an important part of standard planning procedure.

Local planners are now becoming aware, however, that their efforts are more effective if a systematic analysis of the total environment is undertaken. In fact, many have discovered that only by incorporating environmental planning in the overall economic and social planning can development be successful. (See *Appalachia,* June-July 1969, pp. 32-35).

These local planners are seeking answers to such questions as:

- ☐ Why is environmental analysis important, and why should it be carried out during the very early stages of local development planning?
- ☐ What are the various aspects of the environment that should be inventoried and assessed in overall planning?
- ☐ What specific steps must be taken to incorporate environmental factors in any plan for local economic development?
- ☐ What resources exist or are needed to provide local planners with necessary information on environmental factors?

IMPORTANCE OF EARLY ANALYSIS AND PLANNING

On a Sunday in mid-September, a square block area of Scranton, Pennsylvania, pivoted slightly and settled a few inches toward the collapsed tunnels of an abandoned coal mine 90 feet below, leaving front doors that would not open and two-inch cracks in kitchen walls.

In other areas, automobiles have fallen into gaping holes in city streets because of the collapse of subsurface mines.

The Federal Bureau of Mines estimates that 2 million surface acres have already experienced mine subsidence, and that nearly a million more will sink by the year 2,000.

Item: In March 1964, an earthquake in Anchorage, Alaska, caused 14 percent of the city to slide into the Pacific, killing nine people and causing property damage valued at more than $300 million. The land- and mud-slide was caused by the fact that the city was built on clay subsoil which broke away when subjected to the tremors of the earthquake.

Item: In Columbiana, Alabama, in August and September of 1968, major cracks appeared in the concrete filter-plant water reservoir; the water storage tank leaned southeast, out of plumb by 13½ inches; commercial buildings and walls and sidewalks in a federal housing project were

Improper drainage under a parking lot caused two landslides in two years at a shopping plaza east of Pittsburgh.

damaged; water lines were broken and the tracks of the Louisville and Nashville Railroad were severed. The cause of the damage was a particular form of subsidence which is common in areas underlain by limestone. The flow of ground water erodes a subsoil layer which lies between the limestone and the topsoil, causing surface depression, tension cracks, and sinkholes.

Item: In May of 1967, rock slides closed off Route 28 in Pennsylvania's O'Hara Township. Some of the boulders were as large as buses, and three days were required to clear the blocked road of rocks and debris. The force of the slide, which extended for 100 feet, also damaged tracks of the Pennsylvania Railroad's main lines through the upper Allegheny Valley.

These are only a few examples of cases where public and private projects have been figuratively—and in some cases, literally—undermined by the lack of thorough environmental analysis.

Less spectacular, but equally significant, are areas where the interrelationship between the environment and man's activity offer opportunities for improvement of economic and social life.

- ☐ Knowledge of the type and location of various mineral resources permits hospitals and medical centers to plan staff capable of treating and undertaking research on the health problems associated with mineral extraction. The Environmental Health Centers established by the Public Health Service have done pioneer work in this field.
- ☐ Limestone terrains with open solution channels provide both wells for water supply and adjacent wells for waste disposal. The work of engineers planning water and sewerage systems is facilitated by resource surveys showing where this type of terrain exists.
- ☐ If soil is relatively impermeable, use of certain human waste disposal systems will result in malodorous and unhygienic conditions. Planners of public and private housing should avoid this unhealthy pitfall.
- ☐ If highways are built through highly erodable land, erosion control is essential to conserve neighboring soil and streams.
- ☐ If highways are built over land which contains valuable mineral resources, right-of-way becomes expensive, and the resources are unavailable for economic development.
- ☐ If landscapes are scarred, archaeological treasures inundated or virgin land and water defiled, the damage is irreparable and the loss unmeasurable, both in aesthetic and commercial terms.
- ☐ Local areas may suffer loss of potential mineral land tax revenue because precise information is lacking as to the location and extraction potential of mineral resources within their taxing jurisdictions.
- ☐ Foresighted resource analysis and environmental planning can help the local planner make wise investment decisions and facilitate his application for various types of funds which are available to finance activities in his area.

The planning activities of Washington County, Pennsylvania, furnish an excellent example of these benefits. Traditionally a coal-mining economy, the county was urged to consider making large-scale investments in reservoirs which would serve as the center of future recreational activities. Before making the decision, this county obtained professional advice on mine drainage, surface subsidence and the value of minerals in the area that would be affected by the reservoirs. As a result of these studies, it was decided that while construction of some of the reservoirs was feasible, others should not be built because of the negative overall effects of mining on the area. Time was saved, and unwise decisions were avoided.

At another time, when federal agencies were inventorying mine drainage sources in part of the same county, county officials hired their own personnel, arranged to have them trained by federal scientists and took the initiative in completing an inventory of the entire county. When mine drainage pollution control funds became available from the state government, Washington County was ready to file an immediate application, complete with all necessary detailed backup information.

The importance of environmental planning in connection with specific projects is being recognized more frequently by governmental bodies at all levels. For example, in 1968 Congress added to the federal highway legislation a clause requiring that in submitting plans for a federally aided highway project a state highway department must certify that it has considered the *"economic and social effects of the location of the project, its impact on the environment and its consistency with the goals and objectives of such urban planning as has been promulgated by the community."* The additional words reflect the growing concern of lawmakers that our environment be conserved and the use of our natural resources planned.

There is also a growth of interest in environmental planning at the state and local level. The June-July issue of this journal included an article describing the first environmental geology investigation which has been completed and will soon be published by Alabama's State Geological Survey. Stanley Munsey, director of the Muscle Shoals Development District, the area covered by the study, says that his district will find the study "invaluable." He stated that the industrial development committees of the district will use the information in brochures on industrial sites. The district staff expects to rely on it heavily in determining where facilities, highways and other investments can best go—in fact, in all future land-use planning. It will also furnish up-to-date geological information which technical and vocational schools will incorporate into courses of study.

ASPECTS OF ENVIRONMENT IMPORTANT TO PLANNING

Environmental resources are of two basic types: *biological* or living (plants and animals) and *physical* (air, water and land). Ecology is the science which explains the nature of the living resources in terms of the physical environment within which they are found and then describes the effects that the biological and physical resources have on each other. In this broad sense every element of the environment is important in creating a plan for economic development and understanding its long-term effects.

But in developmental planning some aspects of the environment are more important than others. Those aspects of the *physical* and geological environment that affect other economic and social development projects most are:

☐ AIR. Man can live only a few seconds without this essential resource. Yet we are polluting it with: exhausts from motor vehicles, airplanes and ships; smoke and gases from residential heating; wastes from industrial plants (especially steel, oil refining and coal-fired steam power); waste from mining; and certain construction materials, such as silts and sands, which are light and easily blown about, and which are frequently improperly stored. Public Health Service figures show that 72 million tons of carbon monoxide, 26 million tons of sulfur dioxide and 12 million tons of uncaptured airborne particulate matter were dumped into U.S. air in 1965. This pollution causes corrosion of materials, human illness—much of it serious—and personal discomfort.

☐ WATER. Water comes from two sources—streams and underground flows.

Streams. The quality of stream water is affected by two major factors: the biological systems which inhabit the streams (whose tenants range in size from bacteria to mammals as large as beaver) and the chemical processes which take place in the stream water. Any human activity which affects either of these factors will determine whether the water will be usable and at what expense. Prime examples are the

ENVIRONMENTAL PLANNING IN THE ALLEGHENY PLATEAU

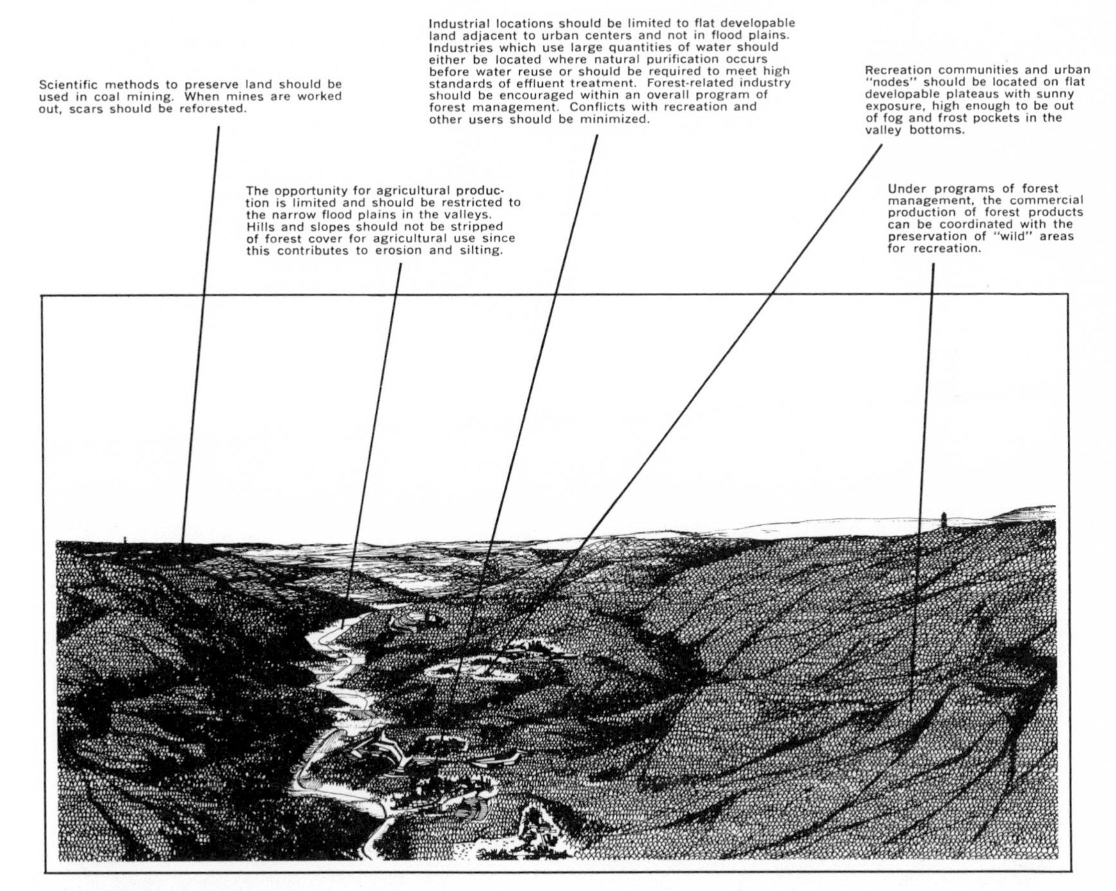

The schematic depiction of Maryland's Georges Creek Valley (above) portrays the various land-use patterns of the area—coal mining, agriculture, industry, recreation and timbering.

In a study prepared by the American Institute of Architects and students from the University of Pennsylvania's Department of Landscape Architecture and Regional Planning, the Potomac River basin was divided into four physiographic regions. The Allegheny Plateau Area, which is located where the Potomac begins its journey eastward to the sea, is epitomized by the Georges Creek Valley, a 64-square-mile area south of Frostburg, Maryland. The terrain is rugged. Extensive coal reserves have been largely depleted, leaving a legacy of acid drainage, land subsidence and erosion from spoil banks. Ample precipitation makes the forests moderately productive; with wise management they can be used both as a source of timber and as recreational development. The primary objectives of planning, according to the study, should be the reduction of acid effluent during mining operations, rehabilitation of spoil banks, reforestation of cutover forest lands and further development of the area's recreation potential.

discharge of human and industrial waste. Stream water *quantity* is also a matter of public concern. Planning can assure that future water supplies will be sufficient to meet total needs, and streams with unique qualities (for example, the "white water" rapids which attract adventurous tourists) can be preserved for special-priority uses. *Flow* of streams can also be controlled to prevent damaging torrents and floods.

Underground flow. Underground water performs two important functions: when it comes to the surface, it is a major source of supply, sustaining the flow and quality of streams; if it remains underground, it influences the performance of soils, rock strata and mineral resources. Underground water is "recharged"—and may be purified—by surface water which seeps down through the soil, subsoil and rock. If these upper layers are contaminated, however, the underground water is also contaminated by the recharging process. In order to maintain needed supplies of high-quality underground water, its use must be carefully planned, and there must be control of waste disposal which takes place in the subsurface rock strata and on the surface areas where recharge of underground water occurs.

□ LAND: SOIL AND TOPOGRAPHY. Soil is used for farming; it is frequently a mineral resource (as in the case of sand, gravel and clays); and it plays an important role in the construction of buildings and highways and in the operation of many systems of waste disposal. Since soils vary greatly in composition, their responses to water and gravity also vary. These variations have important effects on the productive capacity of the soil and on its ability to support heavy loads, serve as a medium for waste disposal or hold its shape and slope after excavation. Variations in topography—hills and valleys, plateaus and ridges—also play an important role in economic development. When we do not know about these differences or fail to recognize their importance in the planning process, the result is irreversible erosion, smelly waste disposal plants, highways that fall apart and buildings that sink.

□ LAND: ROCK. Although we use rock strata in the deeper subsurface layers of the land less frequently than we do the soil, these layers are of critical importance. They are a source of raw materials, including water; they serve as disposal sites; and they give the land "backbone" so it can support heavy structures on the surface. Improper use of this resource results in jeopardized water supplies, ineffective disposal systems and damaged buildings and highways.

□ LAND: MINERALS. The man-minerals relationship has written many of the most important pages in history —gold in our American west, diamonds in South Africa, uranium in Canada, oil in the Middle East and coal in Appalachia. However, the methods used to extract and purify these minerals have frequently degraded our water supplies, contaminated our land, polluted our air, caused subsidence of our land surfaces—and wastefully depleted our supplies of the minerals themselves. Long-range planning of mineral resource extraction can avoid these problems.

One example of this kind of planning is the multiple land use cycle developed in certain extractive industries where the minerals are found on or near the

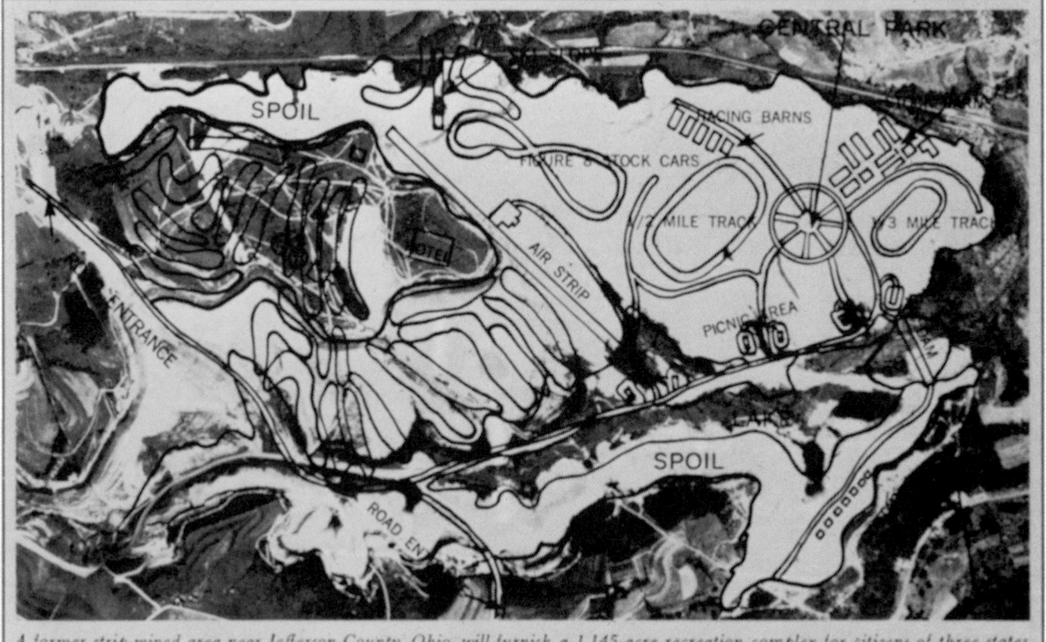

A former strip-mined area near Jefferson County, Ohio, will furnish a 1,145-acre recreation complex for citizens of three states.

surface and excavation sites can be filled and then used for other purposes. For example, the phosphate industry in Florida and the sand and gravel industries in New Jersey have transformed excavation sites into artificial lakes, which could be used for recreational purposes. In Western Maryland, strip coal mining areas are being filled with city refuse and will be planted over with trees and grass. In the future, these areas may be used as open-space recreational sites.

☐ LAND: HISTORICAL, AESTHETIC AND ARCHAEOLOGICAL SITES

Certain pieces of land are "where the action was" in our history. Others are superbly scenic, ecologically unique or geologically unusual. They are finite in number and unreproducible, and should be identified and preserved as part of our visual and inspirational heritage.

Watershed planning has assured flood prevention and municipal water supply for Cameron City, West Virginia.

STEPS TO INCORPORATE ENVIRONMENTAL FACTORS INTO LOCAL DEVELOPMENT PLANNING

All of the environmental elements described above should be considered by local planners in determining what kinds of development are feasible or preferable in an area.

Information about the natural resources or geologic makeup of an area may be useful in determining overall development goals or objectives or deciding upon priorities among projects. Once these are determined, however, several steps should be taken to assure that information about the environment is incorporated into project planning. Each planner should:

1. Determine what environmental information is essential to plan a project.
2. Survey how much information is readily available from existing sources and obtain it.
3. Organize efforts to obtain the remainder of the required information through consultation with professionals or appeals to local, state and federal agencies for expertise, or funds to hire experts.
4. Incorporate this environmental information into plans and project proposals.

Most information about our physical environment is usually presented in map form. Table 1 is designed to help the planner make a decision as to which maps would be most valuable in relation to a specific problem area. The left-hand column shows six major categories of planning assistance decisions: construction (which contains several important subcategories), recreation, mining, cleaning up the environment, preservation of prime agricultural land and preservation of the heritage. The right-hand column of the table shows the environmental maps which are most useful to planners, arranged according to the major categories of physical resources. The middle column performs a selection, indicating which maps would be of greatest value for each specific type of planning decision.

For example, if a planner is considering the construction of an industrial park, the maps which would be most relevant are those indicated by the colored lines. They are:

(1) Zones showing ability of areas to tolerate air pollution resulting from new industrial activity

(5) Size of area required for septic tanks or sewage disposal facilities

(6) Liquid waste acceptance capacity of soil and cost per 1,000 gallons

(7) Volume of solid waste which can be accommodated per acre and cost per acre of disposal

(8) Ground water level and yield

(9) Soils by origin

(10) Soil Slide zones

(11) Areas requiring intense rock blasting

(12) Rockfall and landslide zones

(13) Capacity of rock to accept liquid and solid waste

If, in addition, the area under consideration is one where mining has taken place in the past, maps (14) through (17) would be valuable:

(14) Cost per acre for reclamation

(15) Subsidence zones

(16) Underground mine fire areas

(17) Cost per cubic yard or acre for refuse treatment or for quenching or burial of mine fires.

Most maps listed in this table have been specifically designed to meet the needs of development planners. The drawing on page 8 illustrates schematically how such maps are prepared. Raw basic data are collected and measured, then organized and presented graphically. The basic data maps are then interpreted and translated into terms which have practical meaning to the planner. In the example shown on this drawing, geologic data on the various types of rock found in an area are ultimately translated into maps which show how much it would cost to pump 1,000 gallons of water in various parts of that area.

Following the sequence shown in this schematic presentation, Table 2 shows in detail the steps required to prepare the 18 maps which are valuable in making specific planning decisions—the same 18 maps which were discussed earlier in this article and listed in Table 1.

The left-hand column of Table 2 shows for each resource category the basic environmental data required; the next column lists the basic data maps which can then be prepared. The third and fourth columns show the data interpreted and translated into the final planning maps.

Sources of information for each of the four steps are shown at the bottom of each column.

In order to demonstrate the use of Table 2, let us return to the example of the planner who is interested in developing an industrial park. We found that six maps which would be relevant to his problem (numbers 5, 6, 7, 8, 9 and 10) were in the resource category of SOIL. Table 2 shows that in order to construct these maps the planner would need:

1. RAW BASIC DATA describing the location, distribution, composition, thickness and types of soil plus the history of origin of the area and its elevation above sea level at various points. From these data would be prepared

2. BASIC DATA MAPS: maps of soil, contour and land form.

As indicated in the bottom portion of the table, basic data can be obtained from various state and federal departments and commissions. For most areas, the soil map is available from the State Department of Agriculture and/or the U.S. Soil Conservation Service. Contour maps may be obtained from the State or U.S. Geological Surveys or the U.S. Army Map Service, and land form maps from the Geological Surveys or the Aeronautical Chart Information Service.

3. INTERPRETED DATA MAPS would then be prepared showing the thickness, fertility and engineering qualities (such as permeability) of the soil, the degree of land slope and the location of flood plains. This information would then be translated into:

4. MAPS USED BY PLANNERS, in this case maps (5) through (10), dealing with such practical concerns as the size of area required for a sewage disposal facility, cost per 1,000 gallons for disposal of liquid waste, cost per acre for disposal of solid waste, the volumes of both liquid and solid waste which can be accommodated in different areas, and information showing which areas should be avoided because of the danger of soil slides.

TABLE 1: ENVIRONMENTAL MAPS VALUABLE IN LOCAL PLANNING

TYPES OF DEVELOPMENT	*MAPS NEEDED*
CONSTRUCTION	All maps listed in the next column would be valuable in planning construction. Map 4 and maps 14 through 17 would be particularly important in areas where mining and agriculture are or have been significant activities. Map 18 would be relevant only in areas where special sites are located. Listed below are the maps which would be most useful for each category of construction:
Industrial parks and manufacturing plants	[1,] [5, 6, 7, 8, 9, 10,] [11, 12, 13]
Public buildings	1, 5, 8, 9, 10, 11, 12
Public utilities Water systems	2, 3, 8, 9, 10, 11, 12
Waste disposal facilities	5, 6, 7, 8, 9, 10, 13, 17
Electric power plants	2, 3, 9, 10, 11, 12
Transportation facilities, including highways	8, 9, 10, 11, 12, 15, 16
Housing developments	5, 6, 7, 8, 9, 10
DEVELOPMENT OF RECREATIONAL FACILITIES	Map 18 is of critical importance in this category. If construction of a facility is required, the maps listed above would be valuable. If the recreation facility is to be developed on an area where there has been mining activity, maps 14 through 17 are particularly useful. Maps 2 and 8 are of basic importance in development of any recreational facility.
MINING DEVELOPMENT, CONTROL AND RESTORATION	All maps would be valuable in this type of planning. Maps 14 through 17 would be particularly important in areas where there has been mining activity. Maps useful in the two categories below are:
Determination of commercial extraction sites	14, 15, 16, 17
Mine waste disposal	6, 7, 8, 9, 10, 13, 17
CLEANING UP THE ENVIRONMENT (Air, land and water pollution control)	1, 5, 6, 7, 8, 9, 10, 13, 14, 15, 16, 17
PRESERVATION OF PRIME AGRICULTURAL LAND	2, 3, 4, 6, 8, 9, 10, 14
PRESERVATION OF THE HERITAGE	2, 8, 18

TYPES OF MAPS USED IN PLANNING	
AIR	(1) Zones showing ability of areas to tolerate air pollution resulting from new activity
WATER	(2) Water yield per well or acre foot (3) Cost per 1,000 gallons of water
LAND: SOIL AND TOPOGRAPHY	(4) Agricultural products grown, including productivity per acre and dollar value of land (5) Size of area required for septic tanks or sewage disposal facilities (6) Liquid waste acceptance capacity and cost per 1,000 gallons (7) Volume of solid waste which can be accomodated per acre and cost per acre of disposal (8) Ground water level and yield (9) Soils by origin (10) Soil slide zones
LAND: ROCK	(11) Areas requiring intense rock blasting (12) Rockfall and landslide zones (13) Capacity to accept liquid and solid waste
LAND: MINERALS	(14) Cost per acre for reclamation (15) Subsidence zones (16) Underground mine fire areas (17) Cost per cubic yard or acre for refuse treatment or for quenching or burial of mine fires
SPECIAL SITES	(18) Location of historic, aesthetic and archaeological sites

The maps shown in the last two columns of Table 2 are not generally available to planners. They represent specialized analyses of basic data maps enriched by interpretation of other relevant engineering, technical and economic information, and are usually constructed by teams of natural resource scientists who can furnish expertise in geology, engineering and economic analysis.

How can a planner obtain the maps he needs? The answer to this question depends on the professional competence of his staff in the specialized fields described above. If the staff has the necessary training and experience, the basic geologic data can be obtained from the sources indicated at the bottom of Table 2. Other technical and engineering information can then be gathered and the final tables prepared.

If the planner's staff does not include experts capable of doing this type of analysis:

☐ Competent personnel can be added to the staff.

☐ Local, state and federal agencies can be stimulated to furnish the required expert assistance.

☐ Professional services can be purchased from nongovernmental sources.

The adjacent box lists the state agencies which may be able to furnish either technical assistance or funds to finance the purchase of such assistance. Federal programs of assistance (financial, technical, advisory and informational) listed in the *Catalog of Federal Domestic Assistance* published by the Office of Economic Opportunity are given on pages 10-11.

AGENCIES USEFUL IN ENVIRONMENTAL PLANNING

The local planning and development district in your area should be able to offer assistance and serve as the coordinator of environmental planning efforts. The following state agencies might be contacted for information or technical advice:

1) Geological Surveys
2) Mineral Resources Agencies
3) Soil Surveys
4) Water Resource Agencies
5) Environmental Health Divisions of Health Agencies
6) Environmental Conservation Practices Departments
7) Commerce Agencies
8) Public Utility Commissions
9) Wildlife Commissions
10) Forestry Departments
11) State Appalachian Offices

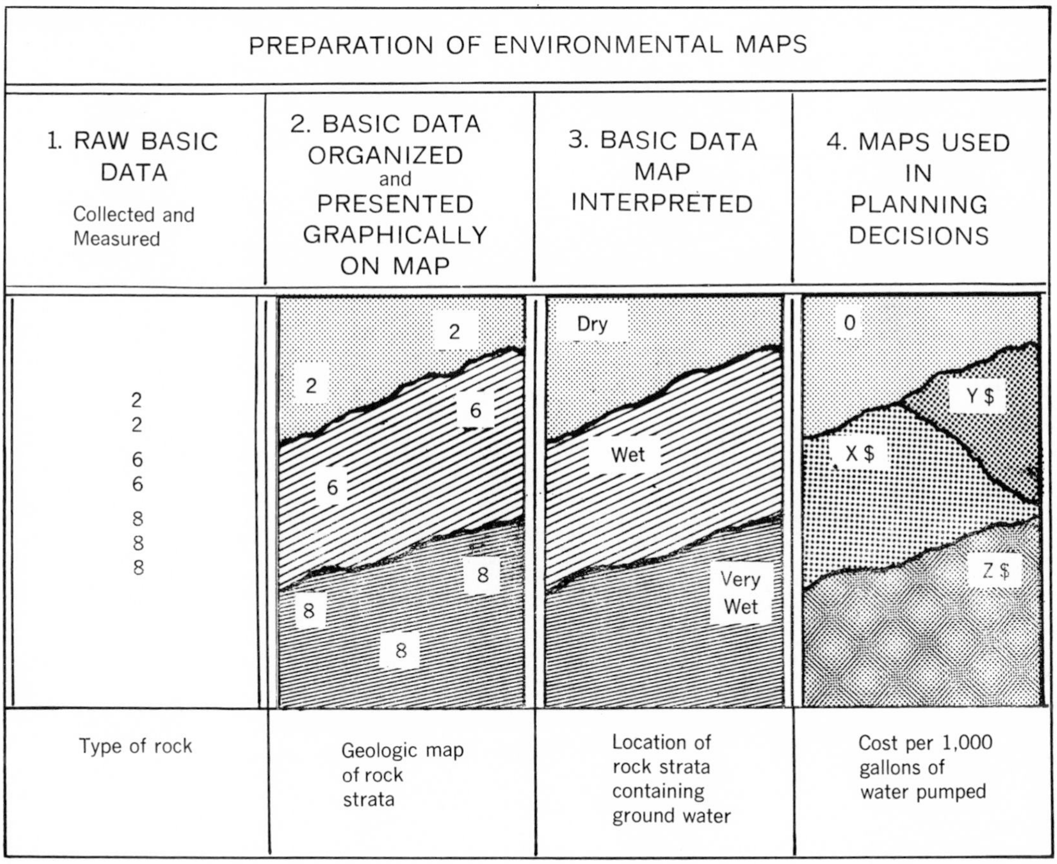

PREPARATION OF ENVIRONMENTAL MAPS

1. RAW BASIC DATA Collected and Measured	2. BASIC DATA ORGANIZED and PRESENTED GRAPHICALLY ON MAP	3. BASIC DATA MAP INTERPRETED	4. MAPS USED IN PLANNING DECISIONS
2 2 6 6 8 8 8	2 2 6 6 8 8 8	Dry Wet Very Wet	0 Y $ X $ Z $
Type of rock	Geologic map of rock strata	Location of rock strata containing ground water	Cost per 1,000 gallons of water pumped

TABLE 2: DATA AND MATERIALS NEEDED TO PREPARE ENVIRONMENTAL MAPS

	1. Basic Environmental Data	*2. Maps of Basic Data*	*3. Interpreted Basic Data Maps*	*4. Maps Used in Making Specific Planning Decisions*
AIR	Air pressures Moisture content Components Wind velocity	Airsheds map (a)	Patterns of pollutant distribution as affected by time of day and season	(1) Zones showing ability of areas to tolerate air pollution resulting from new industrial activity
WATER	Distribution Amount Quality	Groundwater geologic map and hydrologic atlas (b) Topographic map showing valleys and streams (c)	Water-supplying strata Groundwater recharge areas Groundwater quality Surface water flow	(2) Water yield per well or acre foot (3) Cost per 1,000 gallons of water
LAND: SOIL AND TOPOGRAPHY	Location Distribution Types Permeability Composition Thickness Elevation above sea level History of origin	Soil map (d) Topographic map (d) Land form maps (b) Orthophoto map (b)	Thickness Fertility Engineering qualities such as permeability Degree of slope Flood plains	(4) Agricultural products grown, including productivity and dollar value per acre (5) Size of area required for septic tanks or sewage disposal facilities (6) Liquid waste acceptance capacity and cost per 1,000 gallons (7) Volume of solid waste which can be accommodated per acre and cost per acre of disposal (8) Ground water level and yield (9) Soils by origin (10) Soil slide zones
LAND: ROCK	Distribution Thickness Composition Engineering qualities	Geophysical map (b) Geophysical map Orthophoto map (b)	Strata engineering qualities	(11) Areas requiring intense rock blasting (12) Rockfall and landslide zones (13) Capacity to accept liquid and solid waste
LAND: MINERALS	Location Distribution Amount Composition	Resource geologic map (b)	Mineral quality or quantity distribution Depth and type of overlying soil and rock Mineral areas Mine refuse areas	(14) Cost per acre for reclamation (15) Subsidence zones (16) Underground mine fire areas (17) Cost per cubic yard or acre for refuse treatment or for quenching or burial of mine fires
LAND: SPECIAL SITES	Location			(18) Location of historic, aesthetic and archaeological sites

Basic environmental data may be obtained from:
(a) Data sources listed in the next column.
(b) State and federal departments which carry on construction activities, such as agencies responsible for: water resource development, mines and minerals, highways, public buildings, public construction review bodies. For information as to whether and where these data can be obtained for a specific area, call or write the STATE GEOLOGICAL SURVEY (or its equivalent), the MAP INFORMATION OFFICE, U.S. Geological Survey, Room 1038, General Services Administration Building, Washington, D.C. 20242 or the GEOLOGIC INQUIRIES GROUP, U.S. Geological Survey, 801 19th St., N.W., Washington, D.C. 20242.
(c) Local and state public utilities commissions (for value and services information on water, gas and coal).

Types of Maps Published by Government Agencies, available at MAP INFORMATION OFFICE (address at left) includes addresses where maps may be ordered.
Maps in this column may be obtained from:
(a) Not generally available, but currently being developed for, some areas by State University Meteorological Schools and local Weather Bureau meteorological stations.
(b) May be available from State Geological Survey. General information on U.S. Geological Survey maps, publications and open file reports from GEOLOGIC INQUIRIES GROUP (address at left). Maps from WASHINGTON DISTRIBUTION SECTION, 1200 Eads Street, Arlington, Virginia 22202. Copies of reports containing maps from SUPERINTENDENT OF DOCUMENTS, Government Printing Office, Washington, D.C. 20402. Inquiries on water resources to: Mr. James Randolph, Inquiries Unit, WATER RESOURCE DIVISION, U.S. Geological Survey, Washington, D.C. 20242.
(c) Information from MAP INFORMATION OFFICE (address at left) and maps from WASHINGTON DISTRIBUTION SECTION (address in (b) above).
(d) Soil and topographic maps from sources in (b) above and from Commanding Officer, U.S. ARMY TOPOGRAPHIC COMMAND, Attention: 16230, Washington, D.C. 20315.
Aerial photographs usually available at State Departments of Commerce or Agriculture. Also write MAP INFORMATION OFFICE (address at left).

These maps are not generally available.
For some areas, they may be obtained from sources cited for basic data maps (see column at left).
If not available, they can be constructed by a professional geologist from basic data, available data maps and other relevant engineering, technical and economic information.

These maps, which are based on special analyses of basic environmental data, are not usually available, but can be constructed by physical resource teams using data and maps described in left-hand portion of this table.

TOOLS AVAILABLE TO IMPLEMENT LOCAL PLANS

This article has discussed the four steps that need to be taken to incorporate environmental factors into local development planning: determining what environmental information is essential; obtaining whatever is available from existing sources; organizing efforts to obtain the remainder; and incorporating all relevant environmental information into plans and project proposals. The tools used by the planner in these four steps are primarily technical—data, maps and professional expertise.

Once these four steps have been completed, the final and most important step remains—translating the plans and project proposals into reality. This step requires other types of tools, primarily administrative and institutional in nature.

☐ *Legislation.* Certain types of local legislation can be effectively used by the planner in implementing his programs: *zoning restrictions* for land, water and air use; *codes or standards of quality,* as in the case of construction; and *rules and regulations* (including a system of consistently imposed fines and penalties), which may be passed by local legislative bodies to control various types of activities. In addition, the planner can stimulate interest in state and federal legislation which will encourage environmental planning at all levels.

☐ *Scheduling.* One of the most important aspects of any plan is the timing of various activities which are scheduled to occur. Planners can use scheduling as a tool to insure that the physical environment is strengthened rather than depleted. For example, if there is need for clean water at a certain location on a stream, it may be more effective to schedule control of upstream industrial waste dumping *before*

FEDERAL PROGRAMS USEFUL IN ENVIRONMENTAL PLANNING

The *Catalog of Federal Domestic Assistance,* published in January 1969 by the Office of Economic Opportunity, lists programs useful in planning for environmental improvement. Several types of assistance are available:

☐ financial ☐ technical ☐ advisory
☐ informational (data, maps, publications)

A program described in the catalog may be directly related to a physical resource, as, for example, water supply and development studies or planning for water needs of the future. Other programs support the building and construction of facilities which capitalize on a resource or modify a problem, such as the restoration and preservation of an historic site or the building of a sewage treatment plant. Still other programs, such as training and education in the nature of environmental problems or in skills required for problem-solving, are not directly involved with manipulating the components of the environment but help make the climate more conducive to efficient management of physical resources.

Most programs listed are available to state and local governmental bodies of various types, although in some instances public or private educational or research institutions (and in a few cases individuals) are also eligible. Some programs require that applications must be channeled through established working relationships betwen a federal and a state agency.

In most instances the applicant initiates a program, with technical counsel from the federal agency involved. Although most programs which furnish technical counseling and advisory services are handled on a correspondence basis, a few provide for personal consultation.

If a planner learns of a program which would be beneficial to his overall plan but which he is not eligible to apply for or initiate, he may wish to encourage the appropriate body to take advantage of the program.

Programs of particular importance in the areas covered by this article are listed below, together with the page number in the catalog on which they are described.

AIR
- Air Pollution Control
  - air pollution, 76, 173, 175, 176
  - education, 173
  - environmental health, 173, 174, 175, 176
  - grants, 76, 173, 175
  - research, 483

LAND
- Erosion Control
  - Appalachia, 6, 341, 432
  - conservation, 55, 57, 58, 92, 93
  - fish and wildlife, 341
  - grants, 56
  - hydroelectric, 451
  - loans, 35
  - planning, 166, 333, 351, 422
  - reclamation, 6, 341, 345, 432
  - research, 1
  - resources, 56
  - soil and water conservation, 1, 3, 6, 57
  - transportation, 424
- Fish and Wildlife
  - conservation and development, 3, 6, 58, 341, 349
  - recreation, 58, 348
  - tideland studies, 74
- Geodetic Control Surveys
  - commerce, 74
  - maps, 356
  - research, 356
  - topographic surveys, 356
- Geology
  - maps, 356
  - research, 356
- Historic Preservation
  - buildings, 358
  - grants, 280, 283, 358
  - museum assistance, 518
  - new towns, 280
- Irrigation and Drainage, 35
- Land Reclamation
  - Appalachia, 6, 341, 432
  - fish and wildlife, 341
  - mining area, 341, 432
  - water resource projects, 345
- Minerals and Fuels
  - Appalachia, 341, 432
  - natural gas service, 453
  - solid wastes, 187, 188, 342
  - surveys, 356
- Open Space
  - community development, 281
  - community health, 183
  - grants, 183, 333, 351
  - highways, 421, 422
  - loans, 6, 38, 56
  - park planning, technical assistance, 357
  - planning, 166, 333, 351, 422
  - public health, 183
  - recreation, 56, 343, 524
  - regional economic development commissions, 507
  - urban affairs, 329
- Resource Development
  - conservation, 56
  - education, 43
  - grants, 56
  - maps, 356
  - mineral surveys, 356
  - research, 24, 46, 356
  - tideland studies, 74
- Soils and Surveys
  - agriculture, 1, 3, 6, 55, 56, 57
  - conservation, 55, 57
  - research, 1
- Topography
  - commerce, 74
  - maps, 356
  - topographic surveys, 356
- Watershed Development
  - agriculture, 58
  - Appalachia, 58
  - grants, 58, 351
  - planning, 333, 351
  - resources, 523, 524
  - TVA, 523, 524

WATER
- Fish and Wildlife
  - conservation and development, 3, 6, 58, 341, 349
  - hydroelectric, 451
  - marine harbor and water front services, 418
  - recreation, 58, 348
  - tideland studies, 74
  - water resources, 90
- Flood Prevention
  - agriculture, 58
  - conservation, 58, 93

constructing a water purification plant. After the source of pollution has been brought under control, the need for downstream purification can be reassessed and appropriate measures taken.

☐ *Budget and Finance.* Working with local legislative bodies, planners can determine where money needs to be spent and then raise part of the money through taxes, user fees and permit and license fees. These local fiscal tools can be used to reward development activities which conserve our environmental heritage and discourage those which do not. Many resource analysts believe that the use of our public resources, such as air and water, is a privilege which should be paid for, particularly when their use results in their degradation. They also feel that the money obtained from these payments should be used to restore the resources to their original condition. A local use tax on stream water, for example, can be an effective method of financing the purification of water which has been polluted during industrial use.

☐ *Organization of Compacts.* One of the most effective—and least used—tools available to local planners and legislatures is their ability to organize into compacts. Compacts are of various types. They may be intergovernmental (combining townships, counties, development districts or states) or interdisciplinary (bringing together experts from various professions). They may be formed to study problems, to propose plans or to carry out programs. They may deal with many resources or with one, as in the case of watershed compacts, which have been organized because state and local officials have realized the inefficiency and expense of handling water and sewage disposal on a community-by-community basis.

ADDENDUM BY HUGH B. MONTGOMERY

The author wishes to add the following comment to the material in paragraph 1, column 1, page 8:

Advanced environmental analysis for planning will need even more efficient literal and graphic shorthand for environmental communication and planning decision making. Under development is an approach that subjectively rates quantifiable components of the environment. These quantified indicators can then be combined into a number that conveys a level of quality at a specific time for a particular geographic area. The resulting combination or *index* can relate to a natural resource or a pollution characteristic or any combination of these characteristics individually or in any grouping appropriate to the decision-making process involved. For example, water quality could be characterized by a number of measurable or arbitrarily quantifiable components, such as acidity in the former and recreation utility in the latter. Each indicators could be combined and represented by a single index number, which would convey the quality of the component considered for the time and area for which the measured data are applicable. Another index might group two or more of the indicators to convey in a single number a more complex conception of the quality of the area. The index concept can be used to convey a third level of complexity in environmental analysis by combining characteristic indicators such as land and water resources and pollution. Finally, at a fourth level of complexity, environmental geology can merge even more intimately with environmental and development planning. This is done by merging indicators of the physical and natural environment with similar indicators of demographic and economic indicators. The possibilities of conveying by mapping and other communication means the complex present conditions and future alternatives for unified environmental and developmental concepts are infinite.

27

This abridged translation was prepared expressly for this Benchmark volume by Elisabeth W. Betz from "Ingenieurgeologische Baugrundkarte der Stadt Danzig," in Z. prakt. Geol., **29***(7), 97–100 (1921).*

ENGINEERING GEOLOGIC FOUNDATION MAP OF THE CITY OF DANZIG

Hermann Stremme and Erich Moldenhauer

Engineering geology, the scope and aims of which have been explained in fundamental publications by Karl Oebbeke (1904), Friedrich Rinne (1905), and Ernst Wochinger (1919), has in recent time developed in two directions: mineralogic-petrographic testing of rocks by J. Hirschwald and his associates, and cartographic representation for technical purposes based on general geologic data by military geologists.

It was military geology that first confronted numerous geologists with the problem of presenting practical geologic information to those untrained in geology.

For water investigations, geologists have prepared, in addition to the essential historic-geologic representation for scientists, a map free of stratigraphic data that shows only the movement of water in rocks.

For agricultural purposes, besides agronomic-geologic maps, geologists, such as J. Hazard, have produced purely agronomic maps without historic-geologic data.

In the case of military geologic maps, the suitability of terrain for excavation and tunneling, for structures for quartering of troops, for construction of camps, hospitals, roads, field railways, and canals, for effecting flooding, for obtaining construction materials, and for other practical uses was the motivation.

Hans Philipp (1920) and Julius Wilser (1920), both highly deserving for having organized the German military geologic effort (in World War I), published brief, but comprehensive, summaries of the results. Wilser provided examples of the maps that were prepared. Many articles on military geology appeared in German journals and newspapers during the war. A general review can be found in the *Comptes rendus des séances de l'Institut géol., Bucharest,* 1916.

Stimulated by the wartime work of Moldenhauer, the authors produced an engineering geologic map of Danzig in 1919, with a concept similar to that of military geologic maps. It has been used regularly by the municipal housing development agency and the Technical University.

The city of Danzig is generally characterized by a poor construction foundation. It is situated at a curve in the highly dissected, north–south oriented scarp of a diluvial plateau. The city proper is built on the alluvial flats of the Vistula valley. The southern and northwestern suburbs are expanding on dry terraces in front of the scarp; the western suburbs are in valleys and on top of the plateau.

By skillful analysis of available survey data, W. Geisler (1918) was able to trace the alluvial fan on which the Old City is built to arms of the Vistula and Mottlau rivers. For ships entering the Vistula from the Gulf of Danzig, the fan provided the first and, for some distance, only natural landing place (*Damm*) that is a direct dry connection to the plateau. The *Damm* is a very important traffic point, probably also in much earlier times when the great immigrations of Scandinavians occurred on the plateau margin and to the south.

The alluvial fan is penetrated by radial runnels and channels. The Mottlau cuts into it from the side. As is often the case with alluvial fans, a marshy depression (now only 1 to 2 meters above sea level) surrounds it. The new sections of the city occupy this depression. Artificial filling has taken place and is still in progress. From this area the Vistula flats, which were settled in recent centuries and are becoming more habitable, extend from south to east and then north.

As stated, this is a difficult area for construction, but it has been densely settled for a long time because of its importance as a transportation hub. Our task was to prepare the most precise foundation map based on technical considerations, at the scale of 1 : 10,000.

Among the rock and soil types in the North German Plain that have suitable bearing strength for construction, F. W. O. Schulze, F. Gerlach, L. von Willmann, and others identify gravel, medium- to coarse-grained sand, dry (not plastic water-loaded, and especially not muddy) clay, loam, and boulder clay in layers of 3 to 4 meters. The permissible load capacity amounts to about 5 to 8 kilograms per centimeter. Wet clay and loam, mixed with sand and gravel, gravelly sand, and silt are generally not regarded as capable of supporting loads. Also, humus soils, sapropel, mud, and man-made wastes are unsuitable.

Of special importance, in addition to the rock and soil types, is the position of the water table, into which one can build only with more or less costly construction processes, depending on the different kinds of groundwater (e.g., limestone, cement, and concrete are affected unfavorably by bog water).

Based on extensive comparisons made by F. Gerlach (1911), we arrived at units for our map, which was in color in the original, but is shown in samples here by patterns:

1. The load-carrying foundation is no more than 2 meters beneath the surface. The uppermost parts of the soil or rock form the sufficiently thick, dry, petrographically suitable base for construction, or contain a usable base beneath a less suitable cover not more than 2 meters deep. Since in north and east Germany, because of frost danger, the foundation must necessarily lie at least 1.5 meters beneath the surface, an extension to 2 meters represents no appreciable additional expense.

2. The load-carrying foundation lies 2 to 4 meters beneath the surface. This foundation depth is still appreciable. The basement must go below 2 meters, the usual situation with houses in which the cellar is used for living or heating space. It may be a problem in single-family homes, as a factor in increasing the construction cost considerably.

3. The load-carrying foundation lies between 4 and 6 meters. At this depth, the normal, simple foundation of structures terminates. For all structures one must now use artificial supports; however, for higher structures this is not a significant cost factor.

4. The load-carrying foundation lies between 6 and 10 meters below the surface. At this depth pilings and other support become essential. At a depth of more than 10 meters, the cost factor is a serious part of the total construction budget.

5. Investigation of the terrain shows that two subsurface levels for construction are present in parts of the area adjoining the Vistula and the Baltic Sea. Often the two load-carrying layers are separated by mud or sandy clay. The upper layers usually contain fine- to medium-grained sand, 4 to 7 meters thick, suitable for a foundation for light buildings. For heavy structures, foundations must usually be carried to the lower layer.

Figure 1 shows the different types of load-carrying capacities, as well as the rock and soil types that must be known for judging admissible compression and possible use as building material. The original map shows contour lines, which are omitted here. The southwestern corner of Danzig is represented. Loam and loamy sand indicate the diluvial plateau. The diluvial terrace has the best foundation of sand and gravel through a deeper section of which a filled fortification ditch passes.

Figure 2 shows a section of the Vistula alluvium upstream from Danzig, south of the new navigation channels. In the most suitable terrain for construction is a suburb, but there, too, the water table in filled areas is no more than 2 meters below the surface, generally less than 1 meter below sea level.

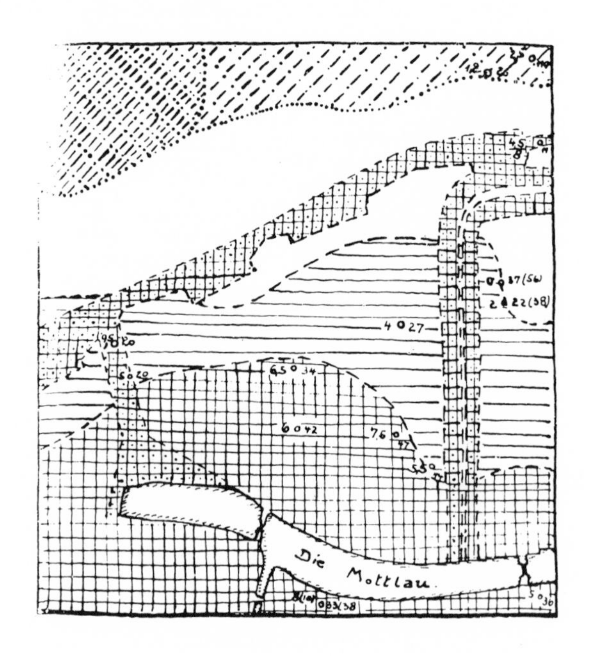

Figure 1 Foundation map of Danzig. Reproduced from *Z. prakt. Geol.*, 29(7), 99 (1921).

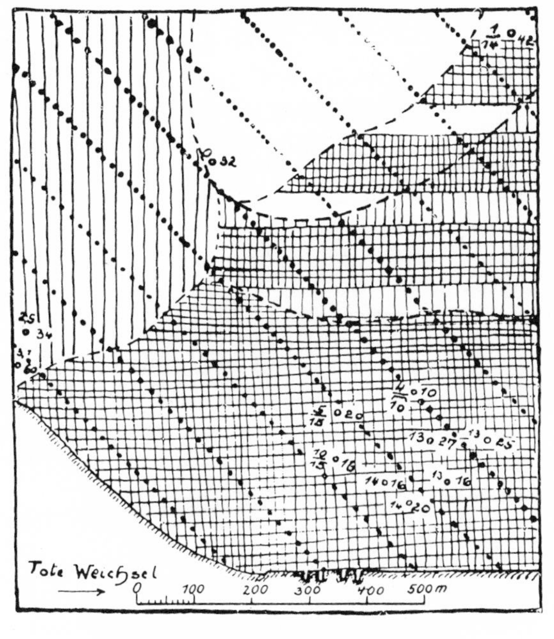

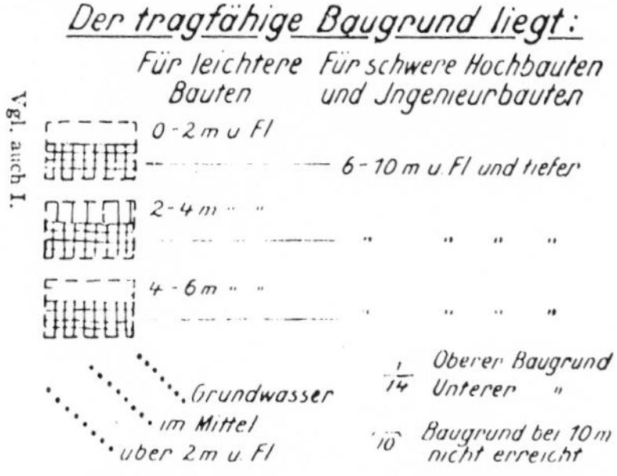

Figure 2 Foundation map of Danzig. Reproduced from *Z. prakt. Geol.*, 29(7), 99 (1921).

In both map sections, borings are marked by circles, and the depth of the load-carrying foundation and depth of the borehole are indicated. Such data are of great value along with the general determinations. Thus, in Figure 1 the end of the Mottlau is shown, where the construction of a skyscraper was planned. The borehole depth indicates that the load-carrying foundation lies 5 meters beneath the surface. Indeed, the borehole data were among the most important facts shown on the map.

REFERENCES

Anonymous. 1903, 1907, 1914. Ergebnisse von Bohrungen. Preuss. Geol. Landesanstalt, Mitt. a.d. Bohrachiv.

Geisler, W. 1918. Danzig, ein siedlungsgeographischer Versuch. Germany, Halle University, Dissertation.

Gerlach, F. 1911. Die elektrische Untergrundbahn der Stadt Schöneberg. Zeitschrift für Bauwesen, p. 20-22.

Moldenhauer, Erich. 1926. Die Baugrundkarte des danziger Stadtgebietes—Die Ausgestaltung der historisch-geologischen Karte des danziger Stadtgebieres zu einer technisch-geologischen. *Naturforschende, Ges. Danzig,* n.s. **17**(3) *Wiss. Abh.,* 95 p. (Technische Hochschule Danzig, Dissertation, 1919.)

Oebbeke, Karl. 1904. Die Stellung der Mineralogie und Geologie an den Technischen Hochschulen. Munich.

Philipp, Hans. 1920. Die Kriegsgeologie: Die technische Ausführung. In: Schwarte, Max (ed.), Dei Technik im Weltkriege. Berlin.

Rinne, Friedrich. 1905. Art und Ziel des Unterrichts in Mineralogie und Geologie an den Technischen Hochschulen. Deutsche Bauzeitung, nos. 36, 38,

Wilser, Julius. 1920. Angewandte Geologie im Feldzuge (Kriegsgeologie). Naturwissenschaften, vol. 8, p. 646-656.

Wochinger, Ernst. 1919. Beitrag zur Geschichte der Ingenieur-Geologie unter besonderer Berücksichtigung der Kriegsgeologie. Traunstein, 164 p. (Germany, Technische Hochschule München, Dissertation, 1917.)

28

Reprinted from *Assoc. Eng. Geol. Bull.*, 5(2), 109–121 (1968)

Missouri's Approach to Engineering Geology in Urban Areas

EDWIN E. LUTZEN
JAMES H. WILLIAMS

ABSTRACT

In 1966 the Missouri Geological Survey initiated a program of mapping engineering geology units in expanding urban areas such as St. Louis and Kansas City. After consulting with civil engineers and land use planners, it was decided to produce maps with text designed to illustrate and delineate engineering geology problems on $7\frac{1}{2}$-minute quadrangle bases.

The first completed map is the Maxville Quadrangle in Jefferson County. This county was chosen because of terrain complexity and suburban growth from St. Louis. The resultant map shows surface units whose boundaries may transgress geological and pedological boundaries. Major surface unit boundaries, based primarily on the engineering characteristics of the bedrock formations, outline engineering geology units. Overburden soils frequently exhibit variations in engineering characteristics that will influence urban development. Consequently, subunits were established for these variations. The field and laboratory criteria used in developing the map include basic soil mechanics, pedological, geophysical, geological, and groundwater data.

The projected series of maps will aid developers in site selection, private consultants in detailed studies, and city and county planners in developing zoning plans. Use of these maps will facilitate urban development by making possible the advance selection of exploration equipment and engineering procedures.

The Missouri Geological Survey has provided preliminary geologic information for public and private lake sites and other engineering structures since the early 1920's. Beginning in 1960, engineering

This paper was presented to the Annual Meeting of the Association of Engineering Geologists, October 1967, in Dallas, Texas. Edwin E. Lutzen is Engineering Geologist for the Missouri Geological Survey, Rolla, Missouri; James H. Williams is Chief of the Engineering Geology Section of the Missouri Geological Survey.

geologists have been employed fulltime in this capacity. As an illustration of the rapid increase in public awareness and need for this type of information, 32 requests pertaining to engineering geology were answered in 1960 while in 1966 the number was 428. These requests come from private individuals, state and federal agencies, and consulting geologists and engineers. The requests for information cover a variety of subjects, from building foundations to slope stability, but most of the inquiries concern sites for lakes and sewage treatment lagoons. As a rule the information supplied is a general geologic

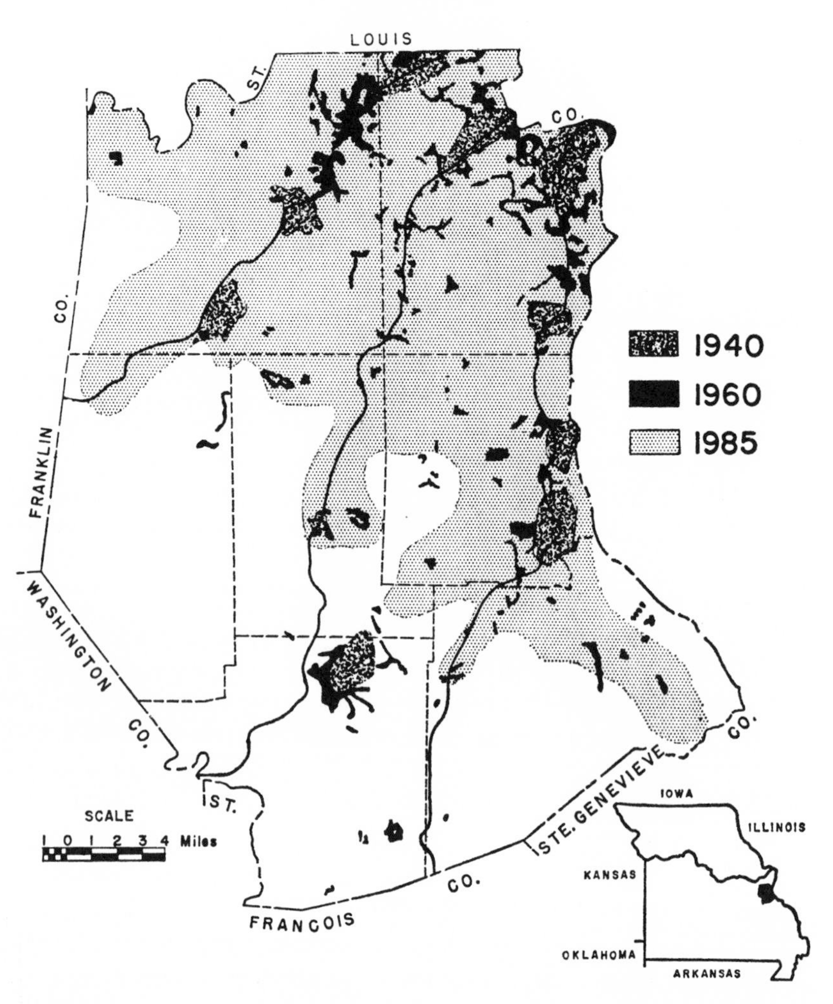

FIGURE 1. Urban expansion projected to 1985 in Jefferson County

appraisal of the location. Detailed geologic and soils studies, if needed, usually are performed by private consulting firms.

The Missouri Geological Survey began areal engineering geology studies in 1966 with the intent of providing data in published form. The studies were started southwest of St. Louis in adjoining Jefferson County where rapid urban expansion is outstripping construction of municipal water and sewerage facilities (Fig. 1). Moderate to rough topography, the complexity of surface materials, and many lakes for suburban recreation have resulted in a hodgepodge of poorly located sewage lagoons, leaking lakes, and housing developments on unstable slopes.

APPROACH TO THE PROBLEM

The Engineering Geology Section of the Missouri Geological Survey uses current field techniques to draw tentative conclusions about specific suburban problems such as sewage lagoons, lakes, and foundations. These conclusions are tempered by local groundwater conditions, soils mechanics data, and geomorphic history. Initially each site investigation in the state was approached as a separate problem until it was observed that persistent unit combinations of bedrock formations had similar characteristics. From this it was concluded that surface materials with similar major bedrock features that affect construction can be mapped as units. Further study indicated that the units could be broken down into subunits so that local variations such as soil changes, geomorphic features, groundwater, thickness, and permeability could also be mapped and described.

For explanatory purposes, "unit" could be analogous in rank to the geologist's term "formation," and "subunit" analogous to the term "member." Because bedrock similarities affecting construction are the criteria used to establish a unit, different rock formations of different ages could be mapped as one unit. Distinctive physical parameters of a unit are sufficiently broad so that a particular unit designation can be applied to an entire region.

The Maxville Quadrangle in Jefferson County, southwest of St. Louis, was the first area selected for engineering geology mapping. Field reconnaissance indicated that many of the units identified in individual site investigations throughout the state were also well developed and persistent in the Maxville Quadrangle. For ease and accuracy in presenting the areal features, $7\frac{1}{2}$-minute topographic (1:24,000) quadrangle maps were used as base maps on which to outline the units.

Descriptive text printed on the reverse side of the map presents in general terms the engineering geology of the quadrangle. Engineering characteristics of the units and subunits are described. Tabular presentation outlines basic soils mechanics data with ranges of numerical values rather than with specific values. Detailed information on groundwater is not presented because that data is available through a variety of publications of the Missouri Geological Survey.

The study of the engineering geology of the Maxville Quadrangle (Lutzen, 1968) showed that a map outlining major units of exposed materials, such as pinnacled bedrock, shales, sandstones, thick or thin residual soils, alluvium, and other materials, could be easily interpreted. To illustrate a use of the completed engineering geology map and text:

Sam Doe plans to build a lakefront subdivision using a lagoon for sewage disposal. He can determine from the engineering geology map what type of material is present in the area he has chosen. If he finds no obvious hazards, he can present his plans augmented with basic data from the published map to the county planner and consulting engineer. If the project appears initially feasible the consultant will be able to complete final feasibility studies with a minimum of preliminary time and expense. Of course, data from the engineering geology map could have provided sufficient information to reject Sam's plans in the initial feasibility phase. In this event the money that would have been spent on an investigation of an obviously unfeasible project could then be utilized in an area that appears more suitable for his development.

Thus, the basic purposes of the map and text are to aid: (1) the engineer in initial site investigations and design of structures, (2) suburban and county planners, and (3) the developer or landowner in early anticipation rather than late awareness of construction problems.

Current literature has no precedent in relating the nomenclature of soil and bedrock units to construction and urban planning. The Missouri Geological Survey, therefore, is employing Roman numerals for the major units and lower case letters for the subunits. Hence, Unit II in the Jefferson County area might be highly pinnacled Mississippian limestone with sinkholes and caves. In southeast Missouri Unit II, while developed on bedrock with similar engineering characteristics, might be Cambrian in age. Cherty residual soils on Mississippian limestones in Jefferson County and residual soils on Ordovician dolomites in the central Ozarks would both be assigned to Unit IV. These residual soils, locally 200 to 300 feet thick in the central Ozarks, would have subunits designating their separation from similar but thinner residual soils in Jefferson County. This classification method also provides that a rock formation and its variations in Jefferson County may be mapped as a different engineering unit in another area. For example Unit IV, consisting of cherty residual soils underlain by noncavernous Burlington-Keokuk limestone in Jefferson County, would not be assigned Unit IV classification in central Missouri where the Burlington-Keokuk limestone has large, active sinkholes and caves with a soil mantle of altered loess and glacial till.

The completed engineering geology map will have some resemblance to geology and soil maps, but there will be significant differences (Figs. 2 and 3). For example, the residual soil Subunit IVb in the Maxville Quadrangle crosses several cherty limestone formation boundaries. Unit III combines Mississippian and Ordovician formations because of shale slope stability problems.

It is helpful if detailed geologic and soil maps are available to assist in determining the boundaries of the engineering geology units. Agricultural soil maps are useful as they assist in determining the mode of soil origin and occasionally the underlying parent material. After sufficient reconnaissance to further establish the units, fieldwork is similar to customary geologic mapping. However, less time is required because the engineering geology units define obvious surface features. Other tools, such as aerial photography, topographic maps, and groundwater data, can be used as effectively as in geologic mapping. Because general construction needs are most important, the engineering geologist should be experienced in field geology; otherwise, he may exclude features important to engineering geology problems.

Representative sampling of different units is necessary to obtain soils mechanics data. Springs, caves, and streamflow (or lack of it), are noted. Successful or unsuccessful sewage lagoons, lakes, foundation problems, and examples of other engineering geology works are examined in the field. Drill hole data from the Missouri State Highway Department and well log data from the Missouri Geological Survey are studied to obtain residual soils thicknesses, groundwater data, and related information. Where there is little subsurface

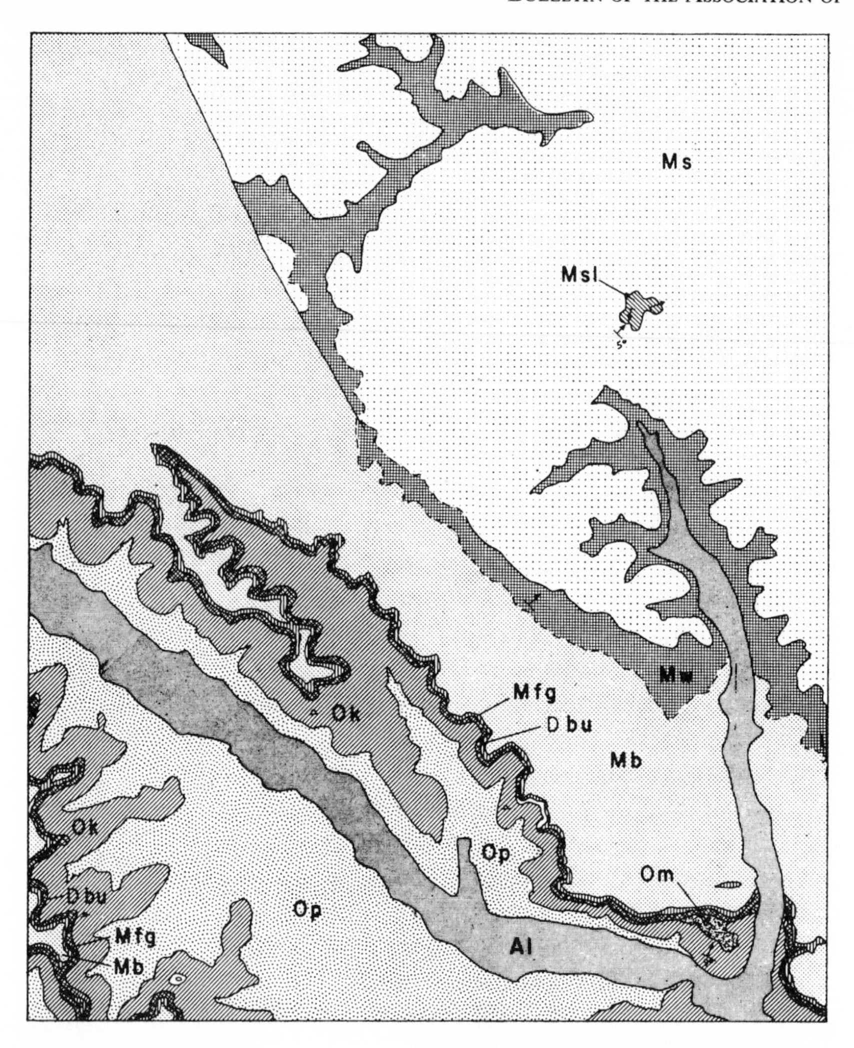

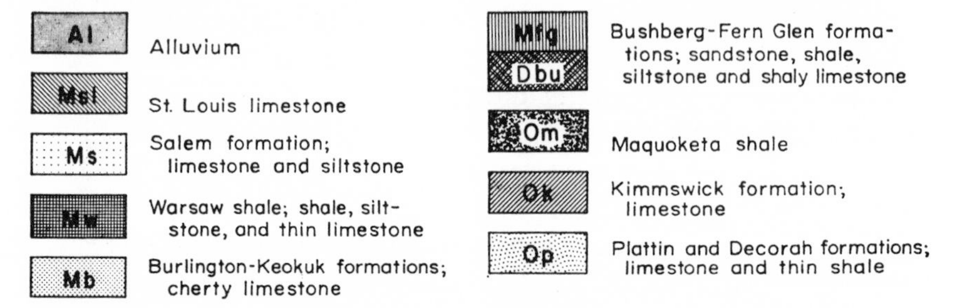

FIGURE 2. Areal geology map (based on preliminary geologic mapping by Stinchcomb, 1966)

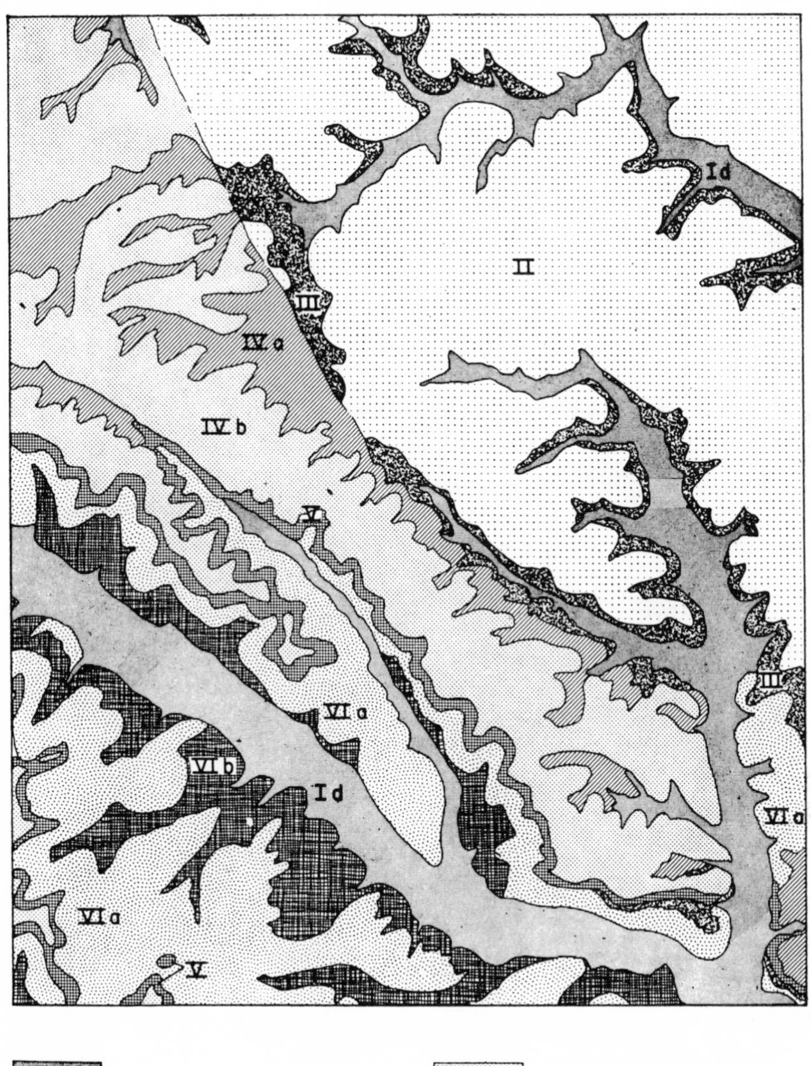

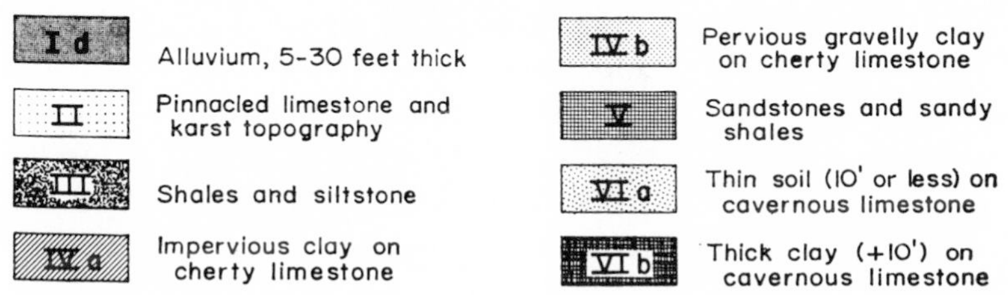

FIGURE 3. Engineering geology map of the area shown in Figure 2

information, a portable, 12-channel seismograph is utilized. Resistivity equipment may also be used in certain areas.

EXAMPLE—MAXVILLE QUADRANGLF

The Maxville Quadrangle is located on the northeastern flank of the Ozark uplift and is underlain by bedrock structure that is characterized by a northwest-southeast strike and a gentle regional dip northeastward into the Illinois basin. Complex geologic structures do not exist, but minor structural folds and faults are present. The area is tectonically stable.

Bedrock formations range from Ordovician to Mississippian in age. The oldest formations are Champlainian in age and consist of a sequence of medium to massively bedded limestones ranging in thickness from 250 to 300 feet. This sequence is overlain by a thin Upper Ordovician shale. The Ordovician sediments are overlain by a 15- to 20-foot section of Upper Devonian sandstones and limestones which in turn are overlain by Mississippian rocks (Osagean series) which consist of a 30-foot thick shaly limestone that grades upward into 100 to 150 feet of cherty limestone. The highest stratigraphic unit is of middle Mississippian age (Meramecian series) and is represented by approximately 200 feet of shale and limestone. Surficial units consist of eolian, colluvial, residual, and alluvial soils which have thicknesses varying from a few feet up to 90 feet. Pleistocene loess deposits are present on some ridges.

There are 14 geologic formations exposed within the Maxville Quadrangle (Stinchcomb, 1966). Six engineering geology units were established to represent the bedrock formations (Figs. 2 and 3). The following examples are taken from units that were mapped in the Maxville Quadrangle report. These selected examples, Units III, IV, and VI, are intended to show important criteria in establishing a meaningful engineering geology unit.

Unit III. The soils and bedrocks which are mapped as Unit III include soils derived from siltstones and shales of three rock formations: the Salem and Warsaw formations (Mississippian) and the Maquoketa formation (Ordovician). These formations are generally thin-bedded, silty, calcareous shale with local, thin limestone beds. The soils of this unit, especially in the more shaly areas, tend to creep on slopes greater than 30°. Creep may change to rotational slides

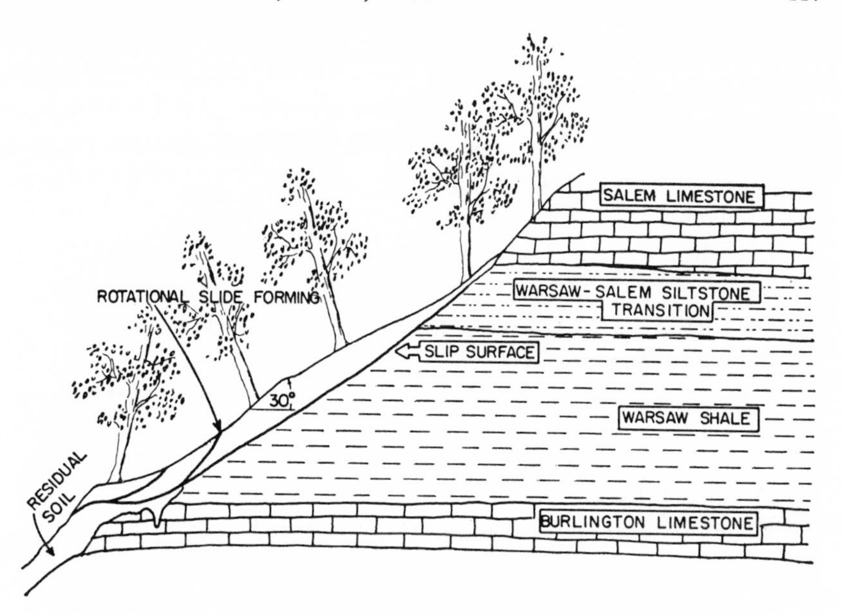

FIGURE 4. Creep slide with rotational slide forming at toe, typical of Unit III

near the toe of the slope (Fig. 4). The apparent natural stable slope is 30° or less. Swelling uplift pressures have been recorded in excess of 3000 psf as measured by FHA procedures. Ranges for the Atterberg limits are 35 to 54, liquid limit; 12 to 18, plastic index; and 16 to 21, shrinkage limit. For a soil that is composed predominantly of clay-sized particles, Unit III soil generally has a low moisture-density relationship with a maximum dry density ranging from 109 to 112 pcf at 12 to 18% moisture.

Unit IV. Unit IV in the Maxville Quadrangle is the Burlington and Keokuk formations, which are Mississippian limestones with varying amounts of nodular and bedded chert. These formations contain numerous enlarged vertical joints and horizontal bedding planes, and the rock surface is slightly pinnacled. The enlarged joints and bedding planes are commonly filled with loose, well-structured clay that has low to moderate permeability.

Field and laboratory work outlined two soil subunits in the Burlington-Keokuk outcrop area, but the subunit boundaries transgress the formational contact boundaries. The unit designation (IV) will carry over into other parts of the state where this type of bedrock exists. The engineering characteristics of the subunits will vary because of local changes in weathering processes that have affected the bedrock.

Subunit IVa is a light red to reddish-brown clay soil with small amounts of chert. The Atterberg indexes frequently fall within a few percentage points of the "A" line dividing the CH and CL clays on the Unified classification system. The natural permeabilities range from 0.8 to 0.08×10^{-4} cm/sec. The natural shear strength is generally sufficient for one tsf loading. This soil has excellent foundation characteristics as it is plastic and has favorable compaction qualities. When compacted, swelling pressures of the soil are generally less than 1500 psf. It is fairly stable soil maintaining apparent stability on natural slopes as steep as 55%. However, care should be taken to keep the soil in a near-natural moisture condition in a housing development, especially on steep slopes. Additional water from indiscriminant use of septic tanks and drain fields can substantially reduce soil stability. Water impoundments can be built with little or no difficulty in areas covered by IVa soils.

The IVb soil is generally a dark red to yellowish-red gravelly clay to clayey gravel. The fine portion of the soil, a well-structured fat clay (CH), comprises not more than 55% of the total volume of the soil. The gravel is composed of chert fragments from the parent bedrock. The chert gravel layers generally exhibit the bedding characteristics that were inherent in the bedrock.

Unit IVb makes a stable and durable foundation for routine building development as the strength of the soil frequently approaches that of a soft rock, such as poorly cemented sandstone. Seismic velocities up to 5,000 ft/sec have been measured in Unit IVb. There are some slopes that have remained stable with a vegetative cover at gradients greater than 60%. Unit IVb soil is poorly suited for water impoundments because of the nearly free-draining characteristics of the soil. Natural permeabilities have been measured as high as 8.5×10^{-4} cm/sec. Septic tanks and filter fields should not be considered in this unit and sewage treatment lagoons should be considered only when it is economical to treat the lagoons with a compacted soil pad (preferably borrow from Unit IVa soil) or another type of sealant.

Unit VI. The Kimmswick and Plattin formations of Ordovician age are crystalline limestones that have a distinctive pitted or honeycombed surface and are relatively free of chert. They can be readily separated by observation in the field, but the engineering characteristics of the two formations are so similar that they are mapped as one unit.

The Kimmswick and Plattin formations are generally separated by the Decorah formation which is a thin shale and shaly limestone less than 20 feet thick. It is impractical to map the Decorah as a separate unit at a 1:24,000 scale, but it is considered important in engineering geology studies. It could cause slides in rock cuts and has less permeability than the Kimmswick and Plattin formations.

Caverns of various sizes and shapes are quite common in both the Kimmswick and Plattin formations. The cavern alignment within the bedrock is joint controlled (Fig. 5). There are also scattered sinks and numerous springs.

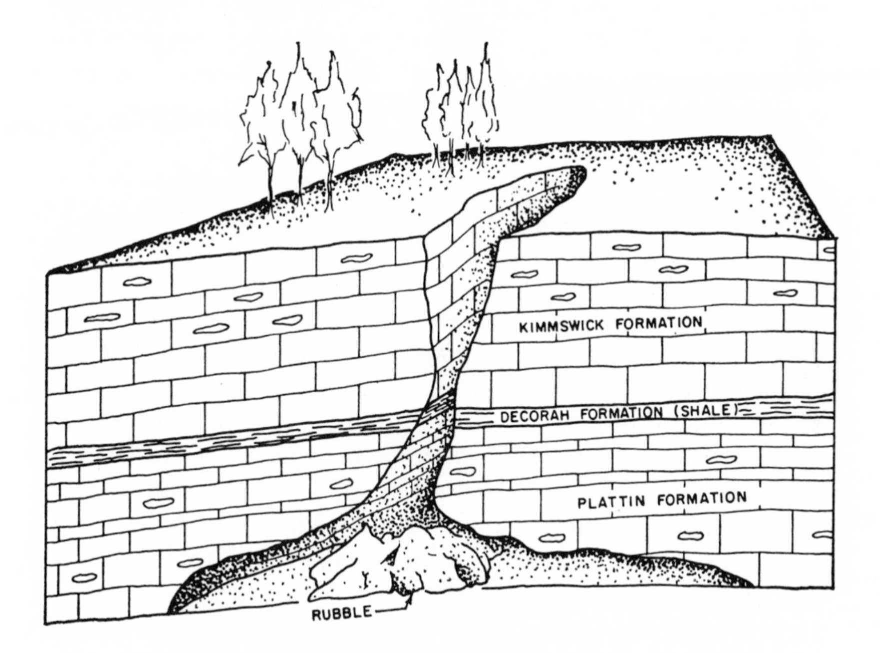

FIGURE 5. Typical cavern and sink formed at joint and bedding plane in Unit VI

One of the principal hazards in the Unit VI area is cavernous bedrock. Therefore, no major structure should be contemplated without drilling into the bedrock for at least 20 feet. Geophysical methods can be a great aid in locating possible cavernous areas.

Field and laboratory work demonstrated two distinct soils overlying these formations. These soils, not necessarily related to the underlying bedrock formation, are designated as Subunits VIa and VIb.

The soil of Subunit VIa is a firm, well-structured clay a few inches to 7 feet in thickness. Liquid limits of 45 are quite common, with a plastic index ranging from 15 to 24. Shrinkage limits of the soil range from 15 to 20. The moisture-density relationship is very erratic for clay, varying from 90 to 100 pcf dry density at 25 to 30% moisture. Swelling pressures as high as 2000 psf have been recorded. This soil when dried forms into hard, sand-sized cubes that are not readily broken down with common engineering practices. Compaction is a problem. The natural permeability of this soil has been measured as high as 5×10^{-4} cm/sec. Due to the high permeabilities, thinness of the soil, and the presence of caverns and enlarged joints in the bedrock, the Subunit VIa area is poorly suited for water retention structures. Septic tanks and fields should not be used because the septic discharge from the tanks and drain field is readily transmitted to shallow groundwater aquifers. Lagoons should only be considered when it is economically feasible to cover the floor and sides of the lagoon with not less than 18 inches of compacted clayey soils, possibly borrowed from Subunit VIb soil.

The soil of Subunit VIb ranges from 30 to 70 feet in thickness and is found on the more gentle slopes in the area of Unit VI. The liquid limits are generally less than 40 with a plastic index ranging from 15 to 20. The shrinkage limits are quite low, ranging from 8 to 13. The permeability of this soil is less than 2.8×10^{-4} cm/sec. With routine precautions and proper construction procedures, water retention ponds can be successfully constructed.

Stability of Subunit VIb is indicated by natural stable slopes with gradients up to 40%. Bearing capacities range from 0.6 to 1 tsf. With proper design and evaluation of the terrain, subdivisions and many varieties of commercial enterprises could be developed.

CONCLUSIONS

The engineering geology characteristics and the laboratory data of the units and subunits are presented as guidelines for land use planning and locating possible areas for construction. As an early phase in any development, routine engineering investigations should be conducted in the area, and very detailed engineering and geologic investigations are highly recommended in more hazardous areas. By knowing the particular characteristics of an area, a project can be successfully completed or rejected with minimum cost and effort because the exploration equipment and engineering requirements can be anticipated early in the project plans. The purpose of the Missouri Geological Survey engineering geology maps and texts is to alert the engineer, soils engineer, consulting geologist, land use planner, and owner to the physical characteristics of the terrain.

The techniques in engineering geology mapping used by the Missouri Geological Survey have proved useful in small sites where urban growth is intense and over wide areas for regional planning.

ACKNOWLEDGMENTS

The authors are indebted to various engineers who helped establish the criteria used in mapping the Maxville Quadrangle. Special thanks go to Eugene Brucker, Brucker and Thacker, St. Louis, Missouri; Bernard Browning, Browning Testing Laboratories, Fulton, Missouri; Tom Frye, University of Missouri, Rolla, Missouri; and Walter Eschbach, Jefferson County Planning and Zoning Commission, Hillsboro, Missouri. Staff members of the Missouri Geological Survey furnished field assistance and editorial advice.

REFERENCES

LUTZEN, E. E., "Engineering Geology of the Maxville Quadrangle, Jefferson County, Missouri," *Missouri Geological Survey*, 1968, Rolla, Missouri.

STINCHCOMB, B. L., "Geologic Map of the Maxville Quadrangle, Missouri," *Missouri Geological Survey*, unpublished manuscript files, Rolla, Missouri, 1966.

29

This article was modified from "Geoscience and Ste-Scholastique," by Denis A. St-Onge and John S. Scott, in Can. Geograph. Jour., *85(1), 232–237 (1972).*

GEOSCIENCE AND PLANNING THE FUTURE

Denis A. St-Onge, Marianne Kugler, and John S. Scott

Planning is no longer carried out solely to determine what is economically feasible; more and more the emphasis is on an attempt to define what is socially and culturally desirable. The value of a mineral or rock cannot be assessed as a separate entity from landscape; it is thus quite conceivable that mining or industrial development will be banned from an area because society has evaluated the landscape as more valuable. In planning the future we are not striving solely to ensure material comfort; we are also charting the role man will be expected to play in a given region. Hopefully this will take into account man's need for beauty and for communion with nature.

Technological capabilities, explosive rates of population growth, and the urban sprawl all militate for the incorporation of geoscientific data into the decision-making process to a far greater degree than was required in the past. This need implies a change in attitude on the part of the geoscientist who no longer can be satisfied with scientifically sound results published years after the original work was done. It is imperative now that he be deeply concerned with organizing his work in such a way that the results are available quickly and in a format that is understandable to the planners.

Detailed geomorphological and geological maps that describe the various components of a landscape have long played the role of information media between the specialist and the user. However, because of scientific terminology and complex legends, very few but geoscientists can understand these maps. Their generally high artistic quality does not make up for the years it takes to compile, draft, and print them. As a method of recording and transmitting urgently required geoscientific information these maps are inadequate.

The geoscientific study of the new Montreal International Airport at Ste-Scholastique, Québec, was an attempt to resolve some of the problems inherent to maintaining scientific excellence while attempting to make the results rapidly available to planners and engineers. Obviously, the construction of a large airport, with its complex of landing strips, control towers, warehouses, highways, and so on, will upset the delicate balance established between man and his environment. It is thus essential that scientists and planners cooperate to define a new equilibrium that will, as far as possible, respect both man and nature. To this end scientists from many disciplines have been studying the present ecosystem and the geological basement of Les Pays du Nord in order to plan well for the future. The disciplines represented cover a wide spectrum and the list appears nearly endless: group psychology, sociology, geography, geomorphology, geology, pedology, forestry, hydrology, zoology, and on and on. Of these, none were more important than the earth sciences, for optimum planning for regional development requires a thorough knowledge of the capabilities of the land, mineral, and fluid resources with respect to their single, multiple, and sequential use. The purpose of this paper is to describe the methods and some of the results of these geoscience studies, and to explain how it was possible, in less than 6 months, to progress from the first sample collected to a final, computer map.

In most regional planning, land-use capability, whether for agriculture, forestry, recreation, wildlife, industry, or urbanization, is commonly viewed in a two-dimensional or areal sense, with prime emphasis placed on the various aspects of the land surface. While such a view is justifiable from the standpoint of the inventory of land surface resources, the planning region is, in physical reality, a three-dimensional entity subject to continuing dynamic natural processes.

A geoscience approach to land-use planning must recognize this, and so the geoscience study of a region is directed toward obtaining the necessary data for producing a geological model to serve either as direct input into the planning or management process or to provide basic information for related studies. Figure 1 illustrates the major geoscience aspects of the region that relate to the character, form, and spatial distribution of geological materials, movement of fluids through these materials, and the rate and magnitude of surface and subsurface geological processes. Figure 2 illustrates the interrelation between geoscience activity and other studies.

Geoscience data in the form of qualitative descriptions and quantitative measurements are derived both from existing sources of information and from field studies. In the field, analyses of available surface exposures are supplemented by test borings, sampling, surface and borehole geophysical surveys, instrument measurements, and laboratory analyses of selected materials. Then comes the efficiency of the computer.

Modern geoscientific technology makes it possible to utilize computers not only for storage and treatment of data, but also for the production of maps and graphs as well as analogue models. A data bank thus becomes the central core of an information system that makes possible the continual revision and refinement of information. The principal elements of a geoscientific study are basic data on the physical characteristics of the bedrock, the unconsolidated deposits, and the subsurface fluids. Other aspects, needed for enlightened regional planning, must be supplied by specialists from other disciplines, such as geography, pedology, forestry, biology, and hydrology. In the Ste-Scholastique region, urgent need for the information required to avoid extensive damage to the ecosystem is forcefully demonstrated by the gigantic machinery that is rapidly reducing a series of picturesque, rolling farms to the crude, dull monotony of a landing strip.

In the case of Ste-Scholastique, an agreement between the Québec Department of Natural Resources (QDM), the Federal Department of Regional Expansion and Economics (DREE), and the Geological Survey of Canada (GSC) allowed for a distribution of tasks within a unified project. Funding was provided by DREE, hydrogeological studies and the financial administration of the project were the responsibility of QDM, and the GSC carried out several studies related to the geology of the area, which are described in more detail below.

To assemble a competent crew is obviously the first and most important necessity. Deep and shallow borings, the gathering of samples, laboratory analyses, and data compilation demand an array of talents that range from mechanics, carpentry,

GEOSCIENCE ASPECTS

1. Bedrock-lithology structure distribution
2. Configuration of bedrock surface
3. Surficial Deposits; character, distribution thickness
4. Geotechnical aspects engineering properties of materials, terrain capability
5. Geomorphology - landscape form processes.
6. Geological hazards - landslides, erosion susceptibility
7. Construction material potential bedrock & surficial materials
8. Hydrogeology-physical & chemical aspects of groundwater flow aquifer potential, fluid waste disposal
9. Seismic effects, bedrock & surficial materials

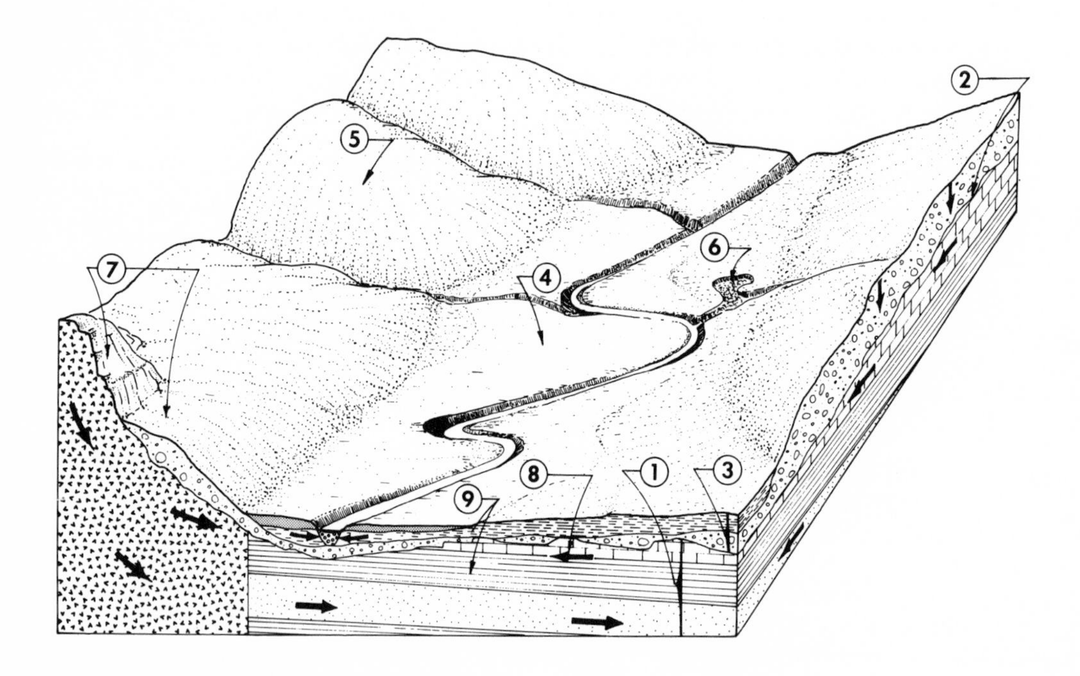

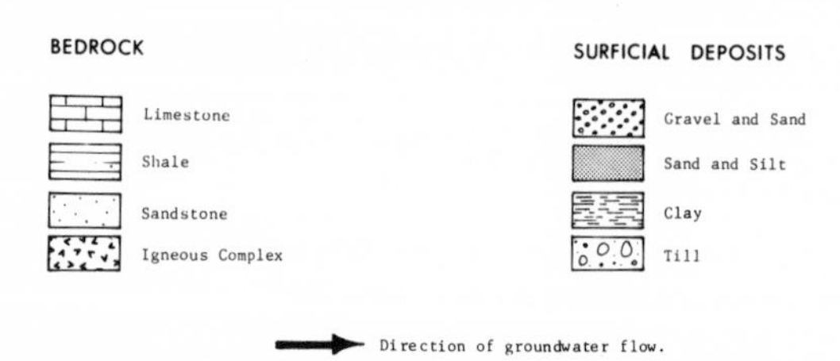

Figure 1 Study model for geoscience applied to regional mapping.

mathematics, and drafting to programming. Students, both male and female, from l'Ecole Polytechnique, the universities of Ottawa, Laval, Québec at Montréal, McGill, and Waterloo, finally met in Ste-Scholastique for what was to become an extraordinary adventure, with highlights ranging from mass hysteria to exuberant enthusiasm.

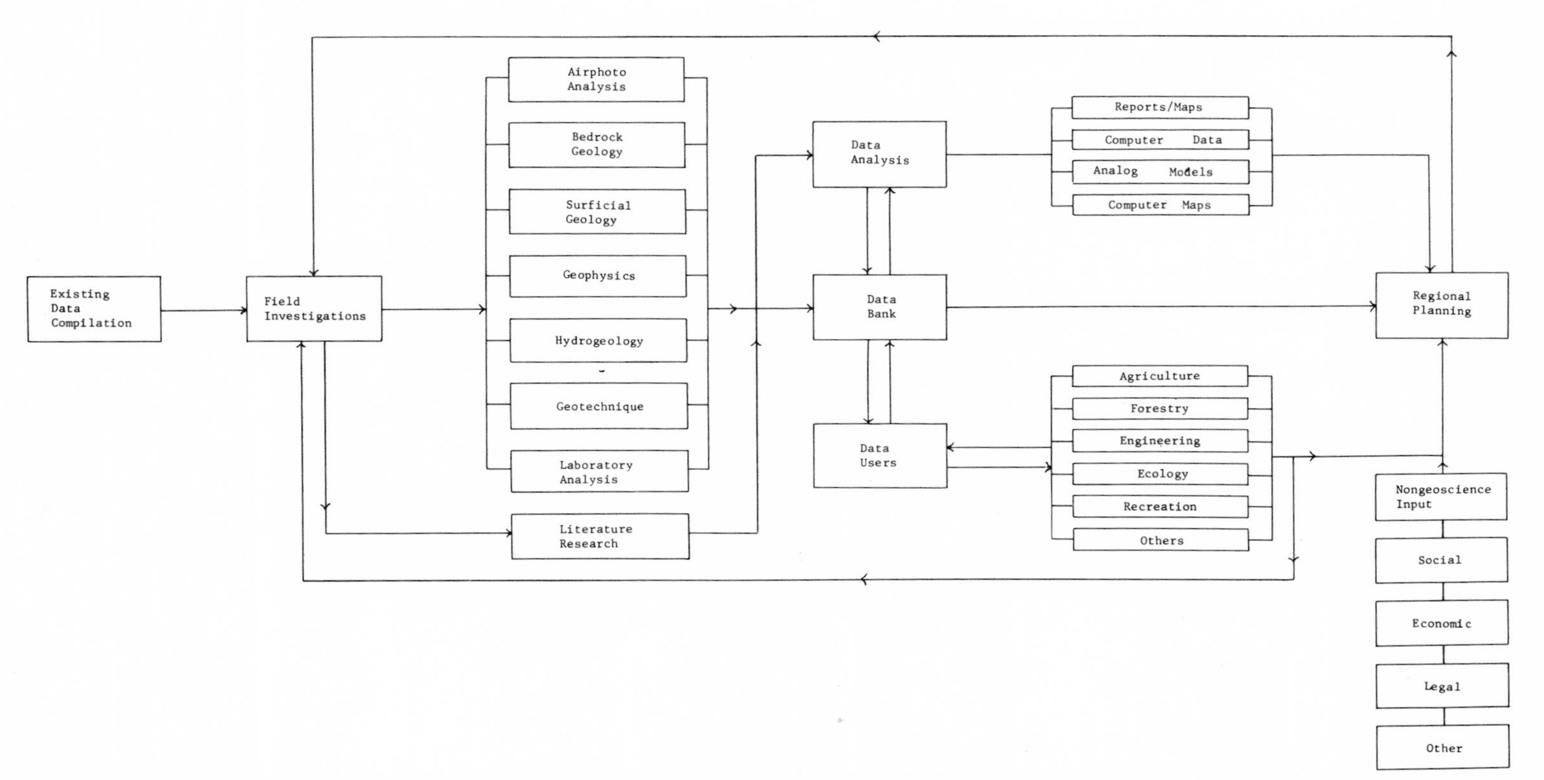

Figure 2 Interrelations between geoscience activities and other studies in a regional planning context.

All the work except for the computer treatment of data was done in the field. An old house in Ste-Scholastique took on new life as a storeroom, laboratory, and dormitory. A mobile laboratory parked in the back of the house served for the more delicate analyses, thus avoiding the effects of unstable floors on delicate balances.

Two series of samples were used to determine the physical and chemical properties of the material from the bedrock to the surface. During the summer of 1971, cores were obtained to depths of 50 to 75 centimeters in order to determine the properties of surface materials. The density of sampling was determined in part by local relief and in part by changes in the nature of the material. Figure 3 shows the sequence of analyses carried out on each sample. In conjunction with this work, two other groups determined the thickness of surface deposits by seismic methods. The raw seismic data were sent by cable to Montreal where they were processed by a computer and returned the next morning to a terminal in Ste-Scholastique as printouts.

Using this information, a program of deep sounding was established. A Meyhew 2000 rotary drill, belonging to the QDM, drilled 220 holes on an average of approximately one hole per 2.5 square kilometers. In each hole geophysical measurements were carried out to determine the porosity, density, salinity, and natural conductivity of the various materials. These logs and the cuttings brought up by the drilling fluid made it possible to determine the most propitious depths at which to take samples from the sides of the holes with a side-wall sampler. The samples were placed in plastic containers and then stored in the laboratory awaiting processing. The size of the samples obtained by this means was less than in the shallow holes, and so the sequence of analyses was somewhat different, as shown in Figure 4. The results of the various analyses were transcribed on a compilation sheet, which is the first step in forming a data bank (Fig. 5).

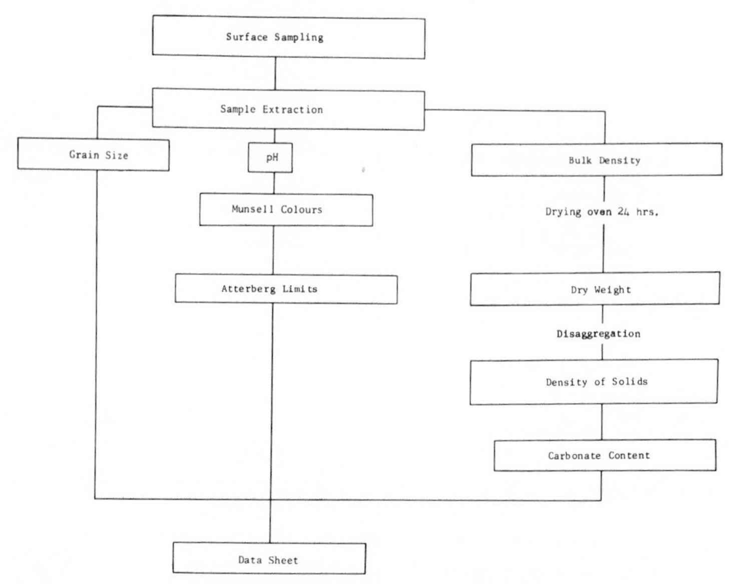

Figure 3 Flow chart depicting the sequence of laboratory analyses performed on surface samples collected at 50- to 75-cm depths.

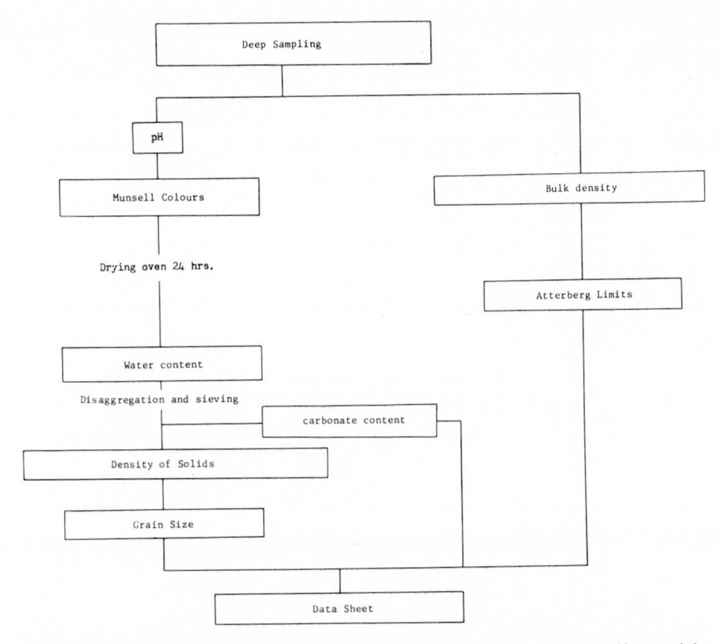

Figure 4 Sequence of laboratory analyses on samples collected in deep holes with the side-wall sampler. The sequence was modified from that shown in Figure 3 to allow for smaller samples collected at depth.

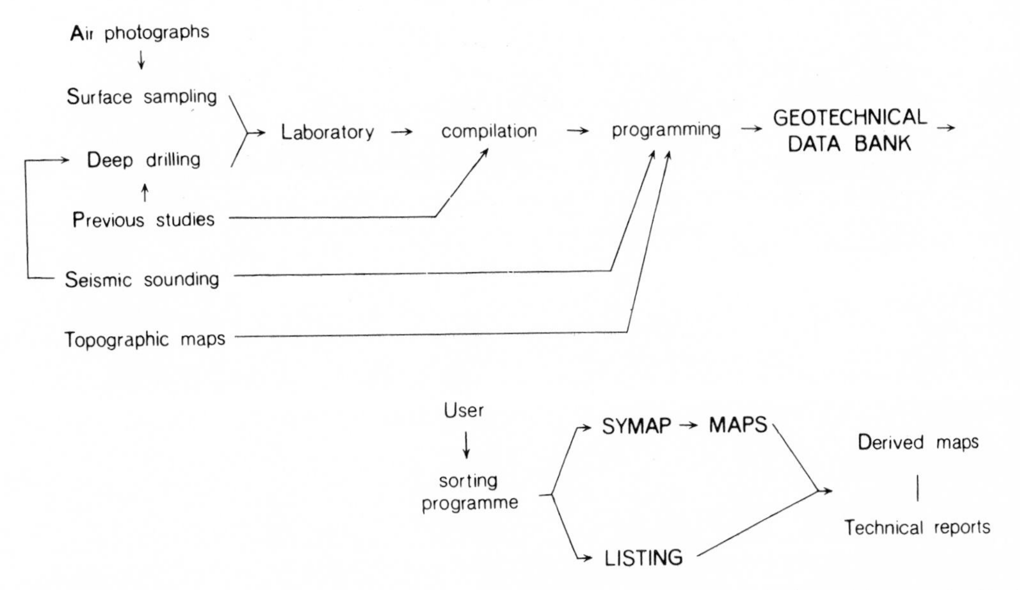

Figure 5 Flow chart showing sequence of events from the initial gathering of data to the final product.

To maximize use while keeping the programming fairly simple, the data bank was constructed as a series of homogeneous information blocks, with easily recognizable signals between each block (Fig. 6). This type of structure is easily expanded by the addition of data blocks to the information system. Five blocks were used in the initial Ste-Scholastique data bank:

A: Limits of the region for which information is required.
B: Results of analyses of surface samples.
C: Results of analyses of borehole samples.
D: Seismic results.
E: Digitized information: roads, rivers, railroads, villages, till, and bedrock outcrops.

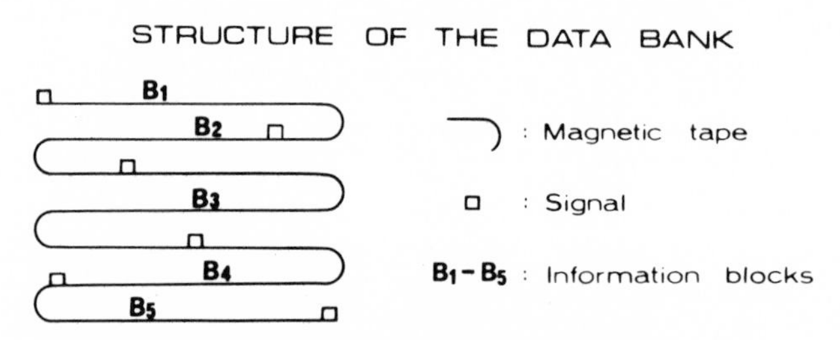

Figure 6 Data bank categorized in blocks to allow for easy expansion of the information system.

Depending on the user's requirements, information can be extracted from the data bank either in the form of printouts or in the form of maps produced from the program SYMAP V (Figs. 7 to 10). It would be possible to extract from this data bank tridimensional models that would depict either the relationship of materials to present landforms or the changes in thickness of various units.

From this mass of information special maps can be constructed to illustrate specific regional planning problems, for instance maps showing best locations for industrial zones.

The study in the Ste-Scholastique region has shown how geosciences can contribute significantly to long-term regional planning; it also indicates that change need not be equated with destruction and that sequential land use need not necessarily correspond to a gradual deterioration of our environment.

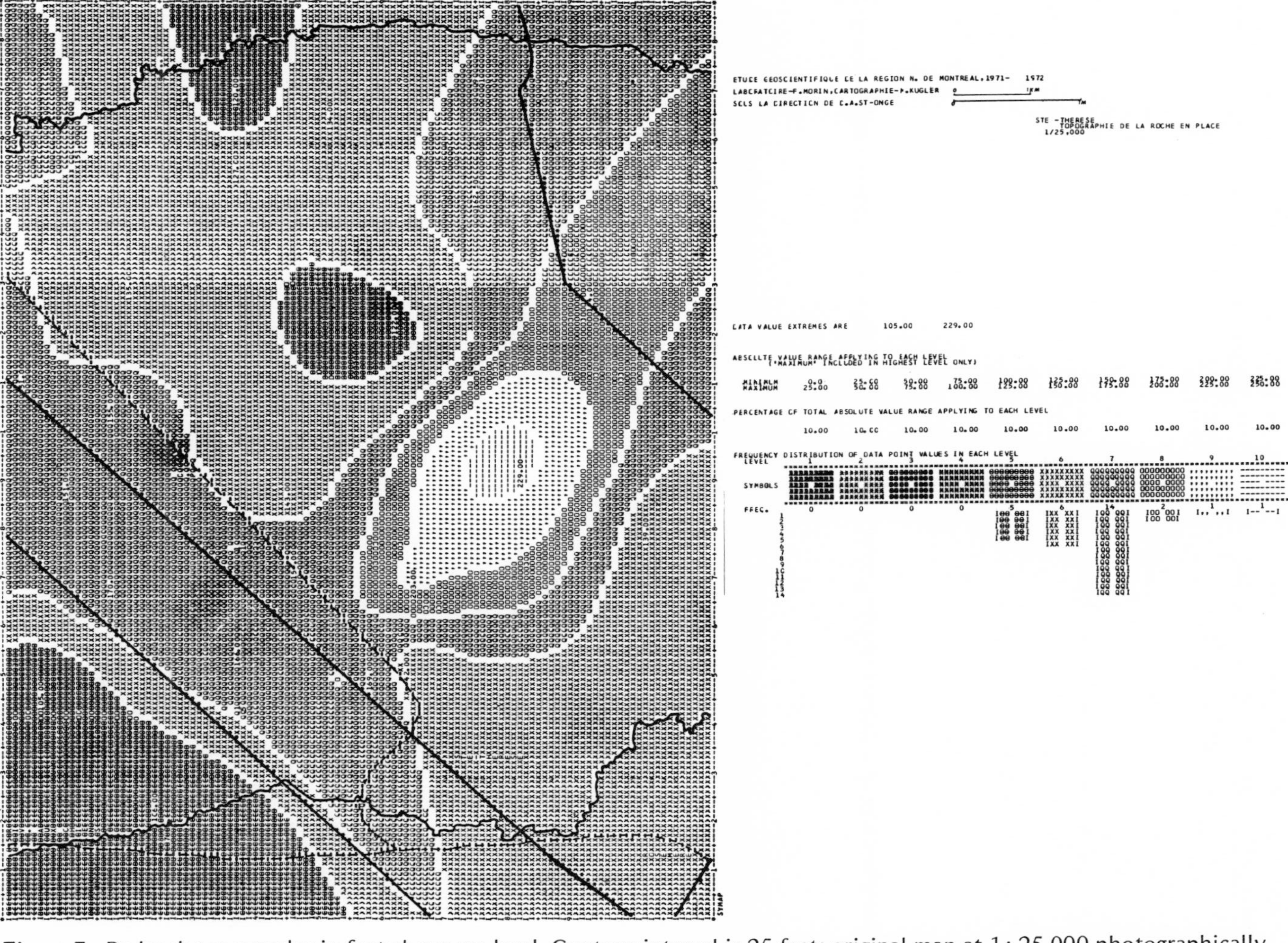

Figure 7 Bedrock topography in feet above sea level. Contour interval is 25 feet; original map at 1 : 25,000 photographically reduced.

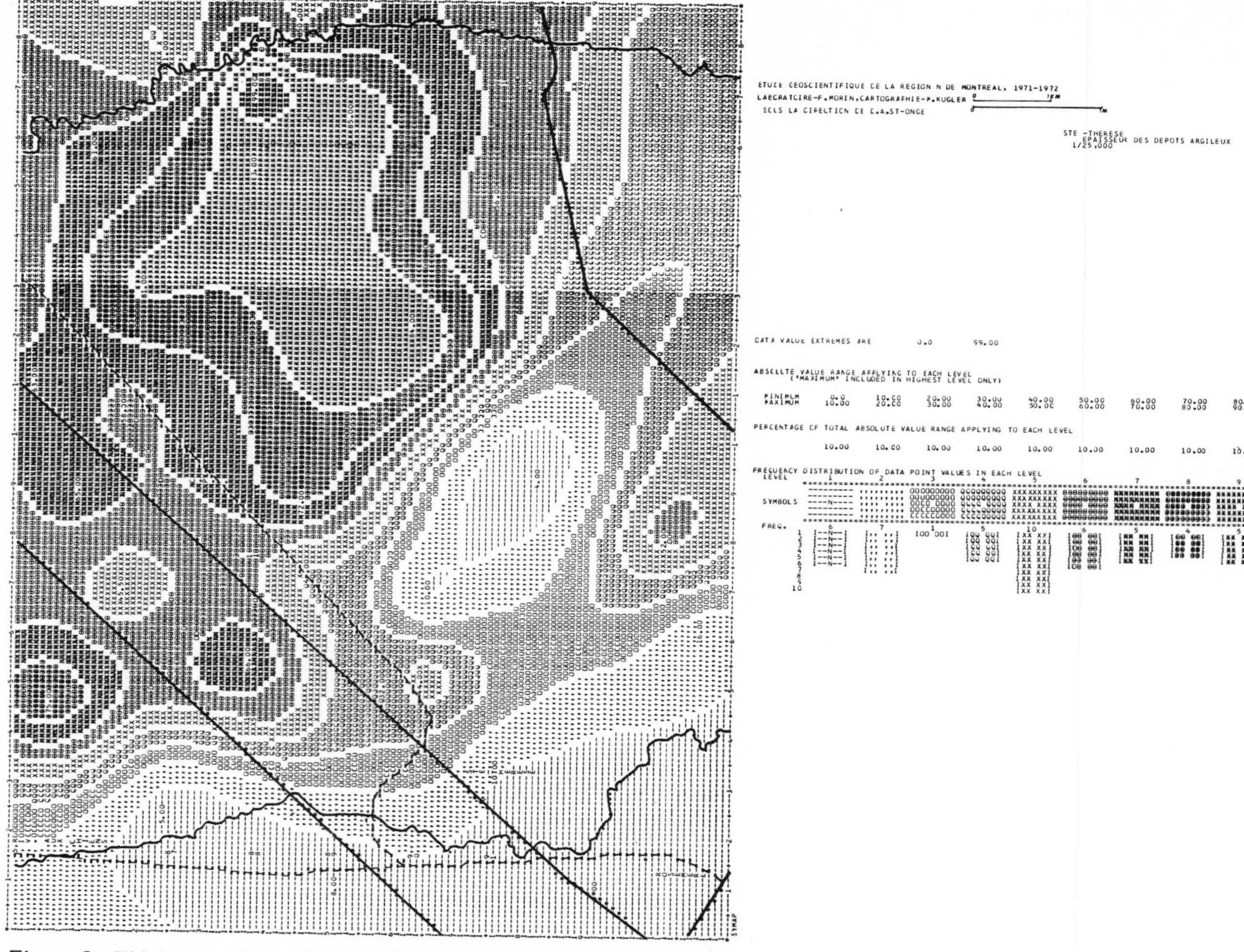

Figure 8 Thickness of clay deposits (in feet). This is a marine silty clay deposit of Champlain Sea age; slump and mud flows are common in this material; other maps have been produced to describe some of its geotechnical properties, i.e., Atterberg limits, plasticity index, porosity, percentage of material <4 micrometers.

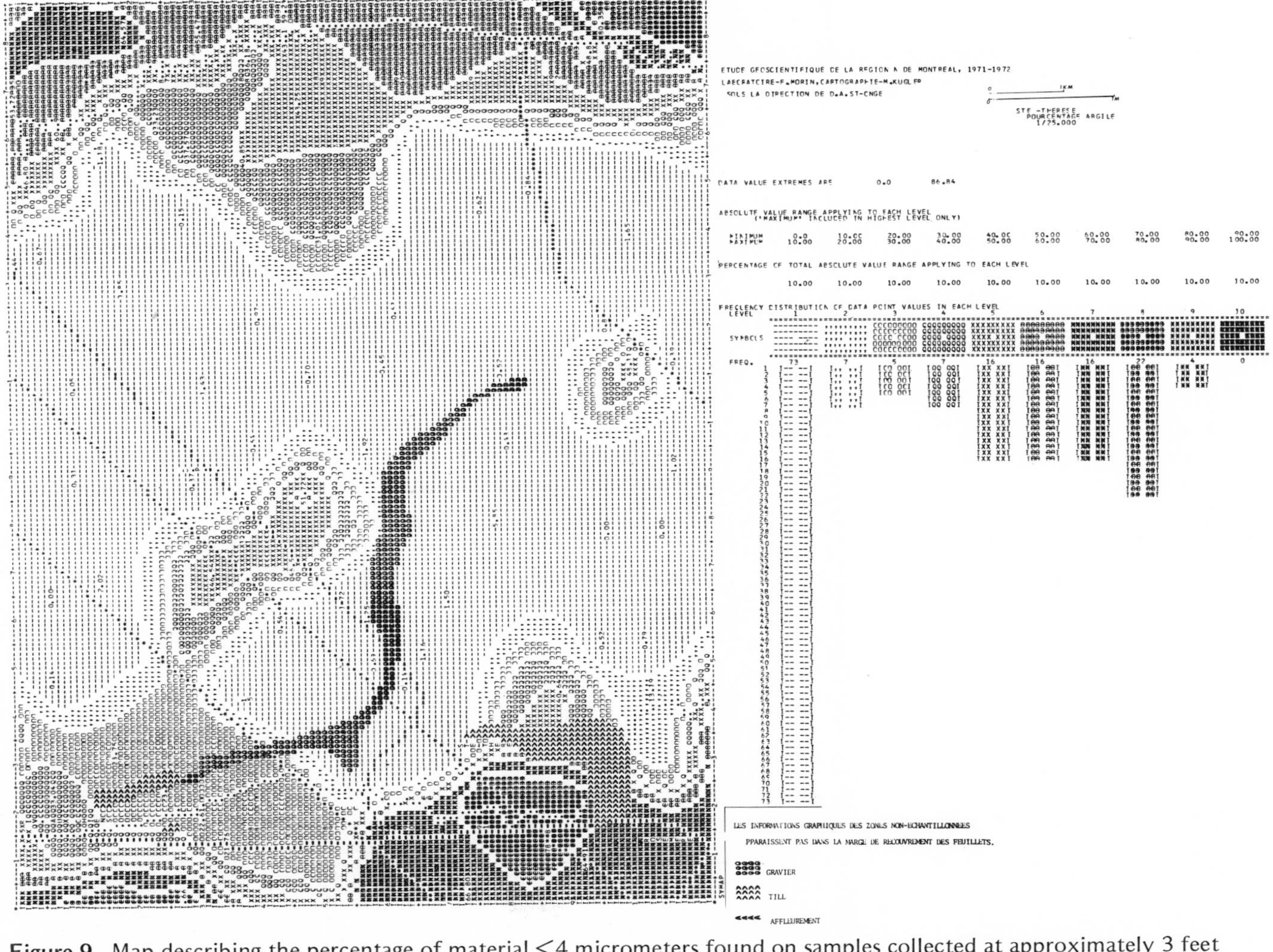

Figure 9 Map describing the percentage of material <4 micrometers found on samples collected at approximately 3 feet below ground surface. Esker gravel, till, and bedrock outcrops are shown on the map from digitized information. These areas are not included in the computed data used by the program to draw the isolines. The large area in the middle of the map with very low values (0 to 10 percent) is a sand plain.

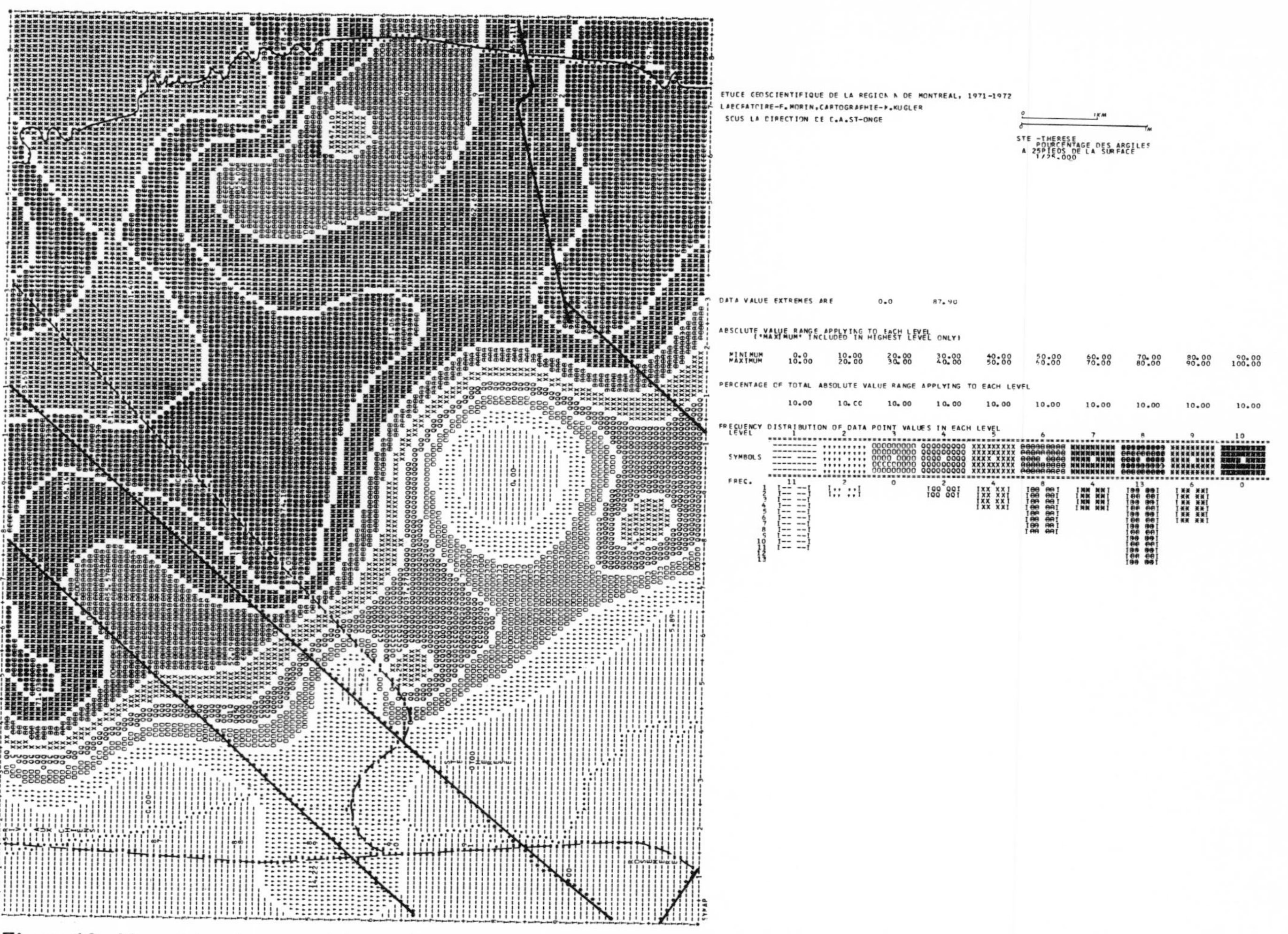

Figure 10 Map giving the same information as the map in Figure 9, but for a depth of 25 feet below ground surface. It thus describes one horizon of the Figure 8 map and indicates that the material underlying the sand shown in Figure 9 is rich in clay.

SELECTED BIBLIOGRAPHY

Carter, R. W. 1961. Magnitude and frequency of floods in suburban areas. Short papers in geology and hydrology sciences. U.S. Geol. Surv. Prof. Paper 424 B, p. B9–B11.

Clark, T. H. 1952. La région de Montréal (feuilles de Laval et de Lachute). Min. Rich. Natl. (Qué), Rappt. Géol. 46, 150 p.

——. 1972. Région de Montréal. Min. Rich. Natl. (Qué.) Rappt. Géol. 152, 244 p.

Gadd, N. R. 1960. Surficial geology of the Bécancourt map-area (Quebec). Geol. Surv. Can. Paper 59-8, 34 p.

——. 1971. Pleistocene geology of the Central St. Lawrence Lowland. Geol. Surv. Can. Mem. 359, 153 p.

Guy, H. P., and Jones, D. E., Jr. 1972. Urban sedimentation in perspective. J. Hydraulics Div. Amer. Soc. Civil Engr., p. 2099–2115.

Hagerstrand, T., and Kuklinsky, A. R. 1971. Information systems for regional development: a seminar. Lund Studies in Geography, ser. B, no. 37, 266 p.

Hoffman, H. J. 1972. Stratigraphie de la région de Montréal. Livret Guide, Intern. Geol. Congr. Montréal, 34 p.

International Geological Congress. 1972. Stockage, récupération et traitement des données géologiques par ordinateur. Sec. 16, Intern. Geol. Congr. Montréal, 218 p.

Karrow, P. F. 1961. The Champlain Sea and its sediments. *In* Soils in Canada, R. F. Legget (ed.), Roy. Soc. Can. Spec. Publ. 3, p. 97–108.

Laverdière, C., Bertrand, J., Carette, N., and Guimont, P. 1972. Les paysages physiques et leur origine morphogénétique. Ecol. Zone Aéroport Intern. Montréal Rappt. Prélim. 3, 102 p.; Centre Rech. Ecol. Montréal.

Leopold, L. B. 1968. Hydrology for urban land planning. U.S. Geol. Surv. Circ. 554, 18 p.

McHarg, I. L. 1969. Design with nature. Doubleday & Company, Inc., Garden City, N.Y., 198 p.

Marsan, A. 1972. Intégration des facteurs écologiques dans la planification. *In* Perspectives en écologie humaine, G. E. Bourgoignie (ed.), Chap. 2, Ed. Univ., Paris, p. 31–69.

Matula, M. 1971. Engineering geologic mapping and evaluation in urban planning. *In* "Environmental planning and geology," Proc. Symp. Engr. Geol. in the Urban Environ., San Francisco, 1969; U.S. Geological Survey, Department of the Interior and Office of Research and Technology, U.S. Department of Housing and Urban Development.

Ritchot, G. 1964. Problèmes géomorphologiques de la vallée du St-Laurent; 1ière partie: l'assiette rocheuse du paysage laurentien. *Rev. Géog. Montréal,* 23(1). p. 5–64.

St-Onge, D. A., and Scott, J. S. 1972. Geoscience and Ste-Scholastique. *Can. Geograph. Jour.,* **85**(1), p. 232–237.

Tomlinsson, A. F. (ed.). 1970. Environment information system. First symposium on geographical systems, UNESCO/Intern. Geogr. Union, Ottawa, 161 p.

Wilson, A. E. 1964. Geology of the Ottawa-St. Lawrence Lowland (Ontario and Quebec). Geol. Surv. Can. Mem. 241, 66 p.

30

Reprinted by permission of the Association from *Jour. Amer. Water Works Assoc.,* 56(8), 959-967, 969-970, 974 (1964)

System for Evaluation of Contamination Potential of Some Waste Disposal Sites

Harry E. LeGrand

A paper presented on Nov. 8, 1963, at the Chesapeake Section Meeting, Washington, D.C., by Harry E. LeGrand, Ground Water Hydrologist, Water Resources Div., USGS, Washington, D.C.

INTERSPERSED with water wells and springs in the United States are 23,000,000 septic tanks [1] and many other thousands of sites where a great variety of wastes are disposed of at the ground surface or slightly below it. These wastes become a part of the subsurface water system, and their soluble constituents move toward points of ground water discharge in stream valleys or water wells. In the majority of instances the concentration of contaminants is reduced to harmless levels before wells are reached.

Yet in an alarming number of places, contamination of ground water is a serious problem. As contaminated ground water spreads, it denies to man's use the storage space formerly occupied by uncontaminated water. Thus, the volume of potable water in underground storage is shrinking while the need for water increases. Other serious aspects of contamination are the danger it poses to health and the high costs of its elimination or prevention.

Contamination Potential

There is no conventional method for evaluating the contamination potential (susceptibility to contamination) of areas or sites where wastes are released to the ground. Few people are capable of making such evaluations; they have frequently had to be made by engineers with limited knowledge of the geologic and hydrologic environment or by geologists with limited knowledge of the biologic, chemical, and economic aspects of waste disposal. Guidelines established by health officials, such as a minimum distance between a well and a waste disposal site, are enforced, and certain aspects of geology are sometimes given consideration.

Where the consequences of contamination are high costs or severe risks to health, field investigations may be undertaken to predict and monitor the movement of contaminants in the ground. Not only is monitoring expensive, but its use must necessarily come after commitments for waste disposal or water supply practices have been made. Early and precise predictions of the behavior of wastes in the ground are almost impossible, because of the difficulty of getting the necessary precise knowledge of the hydrogeologic environment and of integrating such knowledge with the characteristics of the various types of wastes.

One develops a false sense of security if he pursues only that aspect of the problem for which he can get a precise value without realizing that the precision is wiped out when one must integrate the precise value with crude estimates of the values for other aspects. If progress is to be made, one must avoid this error, on the one hand, and a feeling of futility, on the other. It is with this attitude that the author has developed the present system of evaluation.

Evaluation System

The article describes a relatively simple system for evaluating the contamination potential of areas where wastes are released in loose granular earth at or near ground surface. The system is based on characterizing a "problem site" in terms of the probable effect of five environmental factors on contaminants released near the site. The system is especially suitable for a quick initial appraisal of sites where geologic and hydrologic data are scarce. Use of the system requires some knowledge of the kinds of wastes involved, the methods of disposal, the general behavior of contaminants in the ground, and the geologic factors that control the movement.

The system has practical use for contaminants that attenuate or decrease in potency in time or by oxidation, chemical or physical sorption, and dilution through dispersion. Such contaminants include sewage, detergents, viruses, and radioactive wastes. Some chemical wastes attenuate only by dilution; among these are chlorides, nitrates, and certain minor elements of considerable toxicity even in extremely small amounts. The method should not be used in evaluation of disposal sites for mixed wastes, such as those found in refuse dumps and sanitary land-fills, if the critical consideration is the movement of chemical wastes that attenuate slowly.

Limitations of System

The system is intended as a guide for sanitary engineers and others interested in water and waste management in the evaluation of sites for which there are few data. No system can be precise and foolproof under all conditions, and the use of numerical ratings in the system does not imply precision. No conflict should develop with the necessarily specific requirements of health officials; for example, health officials may specify that for a site to be acceptable the maximum height of the water table must be no closer than, say, 4 ft below the waste disposal point or that the minimum ground distance between a well and a waste disposal site be, say, 50 ft.

One can never hope to say with finality that a well site will always be favorable or that a waste disposal site will not lead to contamination of a nearby well. Factors that control the spread of contamination are interrelated; some are relatively fixed; others vary with time or are changed by man's activities. The possibility that contaminants will enter a given potable water supply and render it unacceptable depends on a series of contingencies; for example, an acceptable waste disposal site may become unacceptable if a well is developed a short distance away, but it may remain acceptable if the rate of disposal is reduced, the type of waste changed, or the pumping rate of the well is reduced enough to reverse the slope of the hydraulic gradient between the well and disposal site and thus prevent contamination.

Variety of Contaminants

The variety of types of waste materials makes the ground water contamination problem a very complex one. Disposal at a given site may become unacceptable if the rate of disposal is increased or if one of the wastes released does not attenuate in the ground. For example, a septic tank system that may have operated safely for years could become instantaneously unsafe if a mass of exotic, not easily attenuated, chemicals were to suddenly drain into the system. Only rarely is adequate monitoring of waste disposal provided. Thus, not only broadly applicable realistic standards for waste disposal practices but also readily adaptable provisional policies to impose on these standards are urgently needed.

Types of contaminants include sewage (which may contain detergents and viruses), oil field brines, pesticides, organic and inorganic industrial wastes, and complex wastes from refuse; radioactive wastes may also be included, but these are released only in a few places and under rigidly controlled conditions. Disposal is chiefly by surface dumping or through septic tanks, cesspools, impoundments, lagoons, and disposal wells. As most wastes are released at or slightly below ground surface, water table aquifers are especially vulnerable to contamination. The discussions in this article apply to water table aquifers rather than artesian aquifers.

Need for Standards

At present only crude trial and error methods are available for determining the limits of safe waste disposal practices. Excessive restraints may result in excessive costs; on the other hand, too few restraints may not afford sufficient protection of health. A real need exists to define the acceptable limits [2] at an early stage of planning, before commitments for water supply and waste disposal practices have been made and before damage has been done.

It is well understood that a contaminant in the ground moves in the same direction as the subsurface water and that wastes released near the ground surface may become undetectable near the points of disposal or may travel through the ground for great distances. Some people consider the natural environment to be so complex as to preclude prediction of the effects of percolation except through detailed field studies. The geologic and hydrologic complexities are evident, and some intensive field studies will always be needed; on the other hand, preliminary evaluation based on a minimum of site data must be made for planning purposes, and these data must often suffice for construction and operation purposes as well.

Water Table Aquifers

There are two kinds of water table aquifers, each characterized by a range of porosity and permeability peculiar to it. One is composed of loose, granular earth materials; the other, of dense rocks with linear openings. The loose granular materials include clays, silts, and sand accumulated as sedimentary deposits or as residual, weathered, and decay products of the underlying rocks; water and some contaminants move through the small interconnected pores in these materials. In dense, hard rocks the only interconnected megascopic pore spaces are fractures or other linear openings.

The method presented in this article for evaluating the contamination potential of a given site is based on the following three categories of site geology:

1. Unconsolidated granular materials extending 100 ft or more below ground surface (typical of coastal-plain formations and of alluvial deposits in semiarid regions)
2. Unconsolidated granular materials at the ground surface underlain at shallow depths by dense rocks with linear openings (referred to as the two-media site and typical of humid regions, especially where residual soils have developed on consolidated rocks and where the consolidated rocks are covered by glacial or flood plain deposits)
3. Dense rocks at the ground surface with the movement of fluids only through interconnecting joints or solution channels.

These distinctions in geologic conditions are important, because they represent distinctions in movement of waste in water and in retention of wastes by sorption in the rock materials. For example, the sorptive capacities of loose granular materials are much greater than those of fractured rocks, and the intergranular movement of water in loose materials contrasts greatly with the constricted movement of water in fractured rocks.

Point-Count System

The proposed system for evaluating sites is based on weighted values for five factors whose relative significance can be evaluated from measurements or estimates made at the sites. Each factor tends to have a direct or inverse relation with each of the other factors, as is shown in Fig. 1. The relation between any two factors is not necessarily mutually exclusive and may not be distinctive. The relations apply to natural conditions; for example, the distance and gradient factors of Fig. 1 pertain to the zone between a waste disposal site and a stream and not necessarily to the zone between a waste disposal and a well site. Quantitative studies have been drawn upon for preparation of the system, but the scheme is empirical and designed to give an approximate evaluation.

The system was developed by repeated trial and adjustment and has been checked on a variety of actual and hypothetical field conditions. The results were good for one-medium sites (disposal sites and wells both in loose granular materials) and fair to good for two-media sites (disposal sites in loose granular materials and wells in underlying rock). The system applies to the contamination of ground water but not to contamination of the ground surface resulting from the inability of wastes to infiltrate the ground or stay below the surface.

To apply the system to a given hydrogeologic situation, one scans the lower scale for each factor on the appropriate chart (Fig. 2 or 3), selects the approximate value for that factor in the given situation, and reads the point value directly above. For example, in Fig. 2, if the depth of the water table below the contaminant is more nearly 40 than 30 or 50 ft, 5 points should be given for this factor. The sum of the points for all five factors indicates the degree of probability that a site will become contaminated by bacteria or detergents. The contamination potential chart for radioactive material is shown in Fig. 4. The system may also be used for other types of contaminants but should be modified for contaminants that attenuate only by dilution.

The five factors are: water table, sorption, permeability, water table gradient, and distance to point of use.

Water Table

The water table is a determinable but fluctuating boundary between the zone of aeration and the underlying zone of saturation. It is the rest level of water in the ground but not necessarily the level at which water may first be struck in a well.

The chief factors that control the thickness of the zone of aeration, or the depth to the water table, are frequency and intensity of precipitation, topography, and permeability. The water table generally lies deeper in regions of scarce rainfall than in humid regions. The depth of the water table tends to change with surface topography, lying deeper beneath interstream areas, being closer to the land surface in lowlands, and coinciding with the surface of the perennial streams. The water table tends to lie closer to the ground surface in relatively impermeable materials, such as clays, than in relatively permeable materials such as coarse sands. In dense unfractured rock, the water table may be absent or discontinuous.

The thickness and nature of the zone of aeration are important considerations in the management of waste disposal in the ground. In most places loose granular materials occupy at least part of the zone of aeration, and waste products tend to be stationary except when leached or carried downward by precipitation or by waste seepage. The great reliance on the zone of aeration for natural contamination control stems chiefly from the degradation of some contaminants by oxidation or other processes, and the sorption of contaminants.[3]

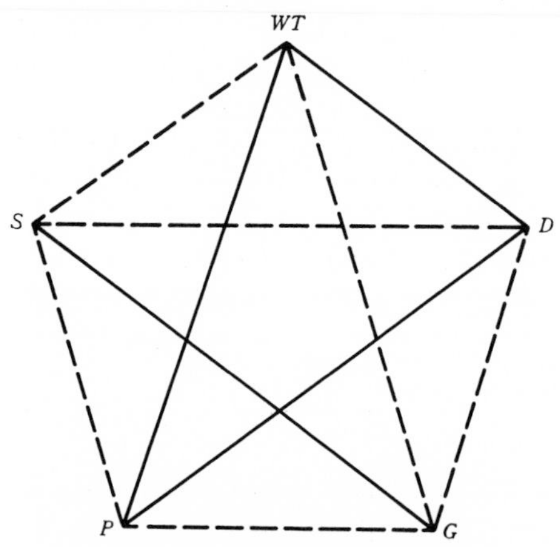

Fig. 1. Diagram of Environmental Factors

WT *stands for water table;* D, *distance;* G, *gradient;* P, *permeability; and* S, *sorption. The solid lines connect directly related factors; the dashed lines, inversely related factors. Relationships are not necessarily mutually exclusive and may be lacking or obscured because of overriding factors.*

The number of points given increase with depth of water table, but the increases are not in a simple arithmetic progression; for example, 5 points are given for a 40-ft depth and 10 for

1,000-ft depth. At many sites, contaminants released at the ground surface will be greatly reduced in intensity before reaching a 40-ft deep water table and may even be undetectable. The estimate of the depth to the water table should be the average position of sorption, and the extent to which sorption occurs is not easily determined. Much valuable research has been done on the sorptive capacity of various clays, especially for radioactive wastes, but it is difficult for research to keep pace with the increasing complexity

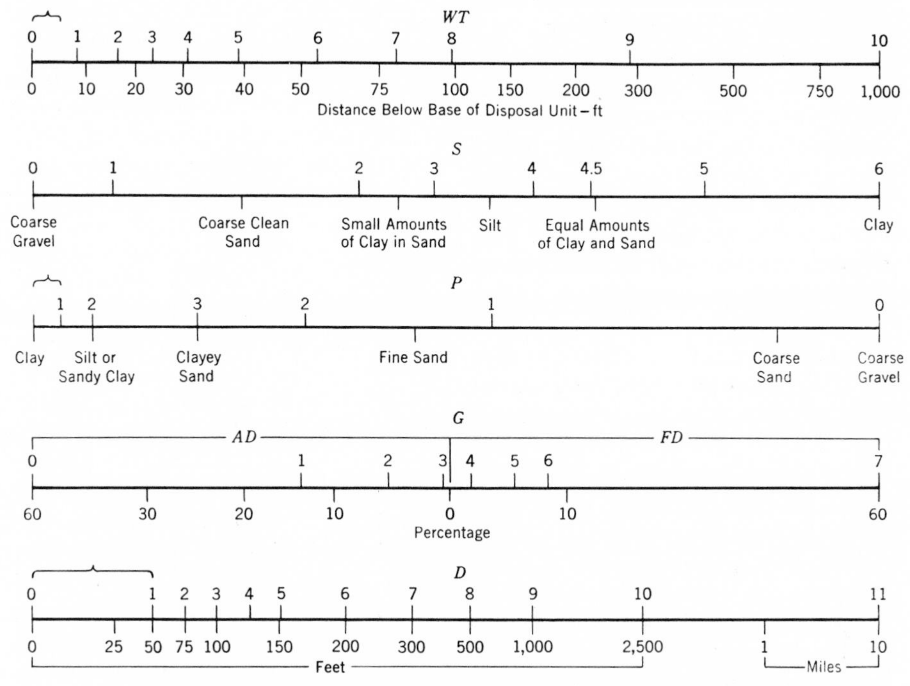

Fig. 2. Rating Chart for Sites in Loose Granular Materials

The scales for the various factors are labeled as follows: WT, *water table;* S, *sorption;* P, *permeability;* G, *gradient; and* D, *distance. On all scales the point values are indicated by the upper scale; the brackets indicate unacceptable ranges for any factor, except the two brackets on the gradient scale, one labeled* AD, *which is for an adverse direction of flow (toward point of water use), and one* FD, *which is for a favorable direction of flow.*

the highest water table, or, at least, the highest position of the water table 95 per cent of the time.

Sorption

Many contaminants are retained on earth material by chemical and physical

of waste products. Clays tend to have a greater sorptive capacity than sands; thus there is an inverse relation between sorption and permeability. Dense consolidated impermeable rocks and rocks permeable only in linear openings such as fractures tend to have poor sorptive capacities. An inverse relation also exists between sorption and the natural depth of the water table, at least in humid regions; for example, the water table tends to be nearer the ground surface in clays than in sands.

The sorption scale values suggested in Fig. 2–4 are qualitative and are not as precise as persons concerned with management of radioactive wastes would like. Six points are allotted if the contaminant moves entirely through clay from the point of contact with the ground to a point of water use. In many sites a contaminant may move through several types of materials; if so, averaging is necessary.

To get a good sorption value, it is necessary to express the types of earth materials through which the contaminant is likely to move as segments of equal length. For example, if 30 ft of clay underlies a septic tank 150 ft from a well drawing water from coarse clean sand, the total moving distance of the contaminant is at least 180 ft; thus, there are six segments of 30 ft each; in the evaluator's judgment, four of these 30-ft segments may be coarse sand ($1\frac{1}{2}$ points) and two may be clay (6 points), or an average of about 3 points.

The length of the column through which sorption occurs extends from the point of release of the waste, which is generally in the zone of aeration, to the point of water use. Although some clays have greater sorptive capacities than others for certain contaminants, no distinction is made between clays in this system; however, if the sorptive capacity of the earth material at a given site is known, the point count obtained from the chart may be adjusted.

Permeability

Although quantitative research is being done on the permeabilities of various earth materials, only qualitative distinctions are made in this method. The permeability of some coarse sands may be hundreds of times greater than that of some clays. In sands and clays the flow of water is by pervasive movement through pores separating mineral grains; in many consolidated rocks the permeability is determined by linear openings, such as joints. Great variations in degree and in type of permeability are found locally. Contrasts of permeability are especially sharp in the vertical direction, because there is a tendency for sedimentary rocks to alternate in depositional sequence and to be only gently inclined; also the interface between loose granular soil and underlying jointed rock is distinct and commonly horizontal. The changes in permeability of the underlying strata must be seriously considered, because water will flow readily through permeable zones and slowly through impermeable zones.

For the present system, distinctions are made only between the relatively poor permeability of clay, the moderate permeability of fine sand, and the good permeability of coarse sand and gravel. Owing to the interdependence of the five factors and to the distribution of permeability-related points in the other factors, permeability as a factor is assigned no more than 3 points. A sandy clay may be assigned 3 points, but a relatively tight and impermeable clay may be assigned only 1 point because it might not sorb the waste to an acceptable degree. A coarse sand may be assigned no points, and consolidated rock would receive no more than 1 point, whether it has fractures and solution openings or not.

Water Table Gradient

Both the direction and the rate of flow of ground water are important considerations in evaluating the possibilities of contamination at a specific site. Measurements of water levels in wells and the preparation of a water table map are major steps in solving or avoiding the more serious contamination problems. If a current water table map is not available or the cost of making one is not justified, synthetic premises may be used to construct a hypothetical water table map

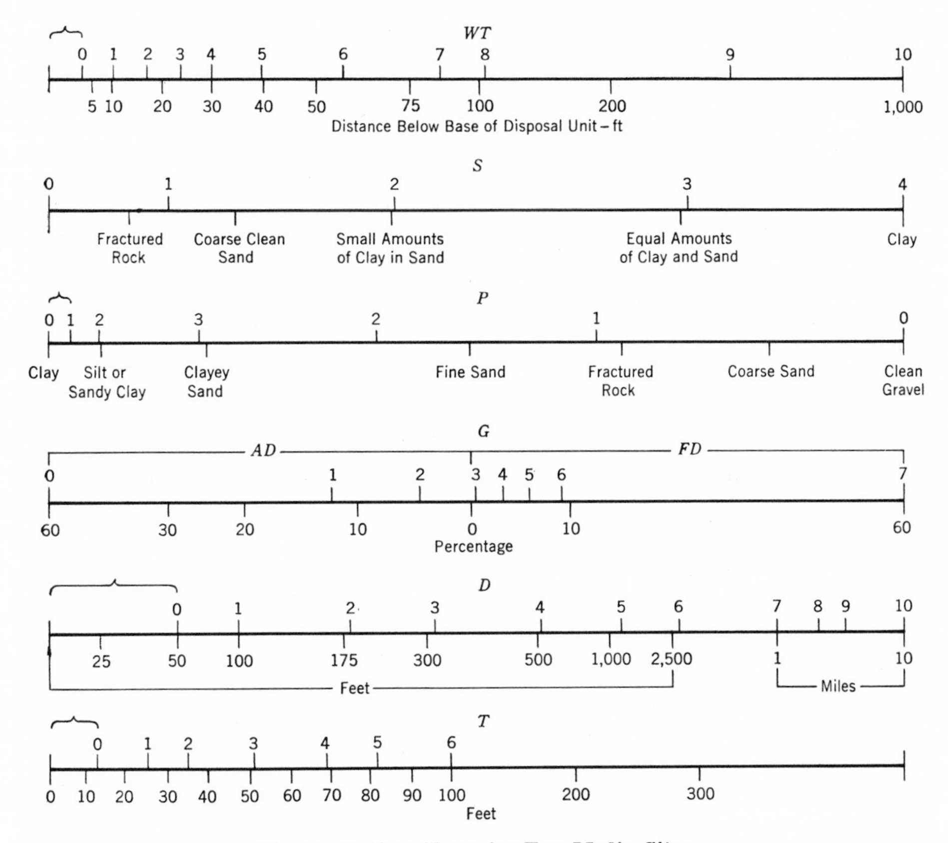

Fig. 3. Rating Chart for Two-Media Sites

The scales for the various factors are labeled as follows: WT, *water table;* S, *sorption;* P, *permeability;* G, *gradient;* D, *distance; and* T, *thickness of porous granular materials below disposal point. On all scales the point values are indicated by the upper scale; the brackets indicate unacceptable ranges for any factor, except the two brackets on the gradient scale, one labeled* AD, *which is for an adverse direction of flow (toward point of water use), and one* FD, *which is for a favorable direction of flow.*

or to visualize the general gradient of the water table and the general direction of water movement.

It is important to know whether a contaminant is moving toward or away from a water supply. Less than 3 points are allotted if the contaminant is moving toward the supply. The number of points if the flow is away from the source is arbitrary, 6 points being assigned for a water table gradient of about 10 per cent and 7 for a gradient of 60 per cent. It can be argued fairly that even a slight gradient away from supply is acceptable and that 7 points should be assigned for a gradient as low as 1 per cent. If the contaminated water is known to be moving in a favorable direction and if there is no likelihood that a water supply will be developed downgrade from the contaminated water, 7 points may be assigned regardless of the percentage of gradient.

The value for the water table gradient must be approximated or averaged, because the gradient steepens around pumping wells and in zones of natural discharge along streams. The points assigned for flow toward the supply are based on the assumption that all the contaminants will move toward a point of water use. Where contaminants move toward a specific well, there is generally another direction of movement of parts of the contaminants away from the well or spring (Fig. 5). Where these conditions are considered likely, one additional point may be given if not more than 75 per cent of the contaminants are moving toward the particular well or spring and two additional points may be given if not more than half of the contaminants are moving toward the well. The natural ground water flow is altered not only by the pumping of wells but also by the development of mounds on the water table beneath sites where liquid wastes are released. Situations *B* and *C* of Fig. 5 are adaptations from diagrams by da Costa and Bennett [4] that show the pattern of flow in the vicinity of a recharging and discharging pair of wells in an aquifer with areal parallel flow.

Although one's knowledge of the direction and rate of flow of ground water is likely to be much less complete than one would like, it is possible to make an effective evaluation of the site. The ranges of conditions near the ends of the scale for the water table gradient—that is in the ranges of 0–2 and of 5–7 points—are relatively easy to determine; conditions in the range of 1–4 points are difficult to evaluate, especially where sporadic pumping of a well may change the gradient and direction of contaminated water significantly for intermittent periods. One can only average the gradients in a general way and select the point value that seems proper.

For the purposes of the point system, the factor of water table gradient is modified for disposal of radioactive wastes (Fig. 4), because there is generally no intended ground water use downgradient, and because the distance to the zone of ground water discharge in a stream valley is an overriding consideration.

Distance to Point of Use

The chance of contamination decreases as the distance between the source of contamination and a point of water use increases. Some reasons for this are:

1. Dilution tends to increase with distance.

2. Sorption tends to be more complete with increase of distance.

3. Time of travel tends to increase with distance, thus decay or degradation is more complete.

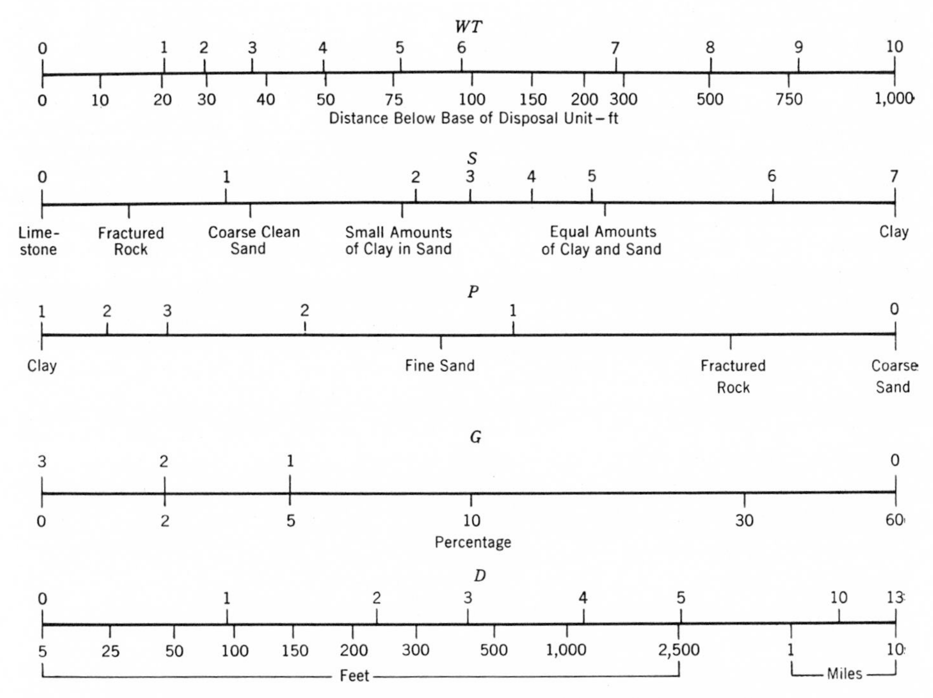

Fig. 4. Rating Chart for Radioactive-Waste Disposal Sites

The scales for the various factors are labeled as follows: WT, *water table;* S, *sorption;* P, *permeability;* G, *gradient; and* D, *distance. On all scales the point values are indicated by the upper scale. The chart is not applicable if more than 50 per cent of the distance (D) is through limestone or fractured rock.*

4. The water table gradient tends to decrease with distance, so that the velocity of flow decreases with distance from the disposal site.

The distance between a disposal point and a point of water use may be as little as 5 ft or as great as 10 mi. The greater distances are commonly in sparsely populated or relatively arid regions. The chart indicates straight-line distances and not necessarily the distance the fluid travels, which may be much greater. As attenuation is normally great near the source of contamination, point values are not distributed evenly with distance. For example, a distance of 200 ft has a point value of 6, and a distance of 1,000 ft has a point value of 9.

[*Editor's Note:* Certain tables and some text material have been omitted at this point.]

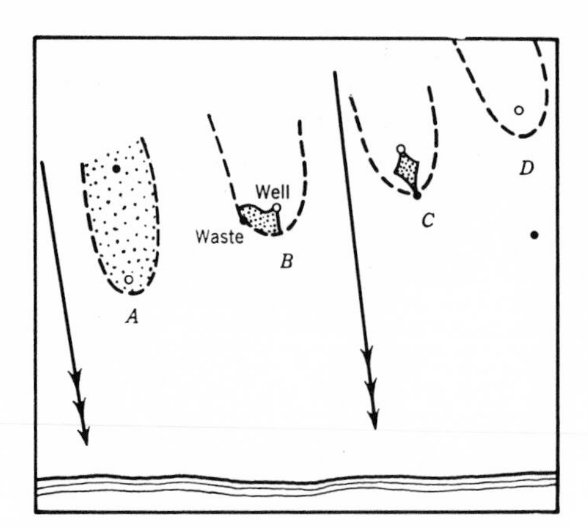

Fig. 5. Effect of Contamination Potential of Relative Positions of Well and Disposal Point

The arrows indicate the water table gradient; the dashed line, the area within which water may move toward well; and the speckled area, the zone of potential contamination. The open circle is for the well; the solid circle, the waste disposal point. Situation A *shows the disposal point directly upgradient from the well, with all waste moving toward the well; at* B, *the two points are aligned at right angles to the gradient, with some contaminated water (possibly) moving toward well;* C *shows disposal point downgradient from well, with some contamination moving toward well;* D, *same as* C *but beyond the influence of pumping.*

References

1. Environmental Health Problems. Report of the Committee to the Surgeon General. USPHS Bul. 908. US Govt. Printing Office, Washington, D.C. (1963).
2. STONE, R. V. The Way We Do It. Proc. Symp. Ground Water Contamination, USPHS. US Govt. Printing Office, Washington, D.C. (1961).
3. LEGRAND, H. E. Graphic Evaluation of Hydrogeologic Factors in the Management of Radioactive Wastes. Proc. 2nd AEC Working Meeting on Ground Disposal of Radioactive Wastes. TID7628. US Dept. Commerce, Washington, D.C. (1962).
4. DE COSTA, J. A. & BENNETT, R. R. The Pattern of Flow in the Vicinity of a Recharging and Discharging Pair of Wells in an Aquifer Having Areal Parallel Flow. *Intern. Assn. Com. Subterranean Waters, Intern. Union Geodesy Geophys.* (1961).

AUTHOR CITATION INDEX

SUBJECT INDEX

About the Editor

FREDERICK BETZ, JR., was associated for many years with the U.S. Geological Survey. He has also worked in various aspects of geoscience and related subjects with several other governmental agencies and private organizations, including the Space and Environmental Sciences Program of Texas Instruments, Inc., and the Coastal Plains Center for Marine Development Services as its Executive Director. He is now engaged in independent consulting.

Long before the term environmental geology became known, Dr. Betz was attracted to problems that demanded the participation of geology with other disciplines in examining the implications of human use of environment. Extensive experience in Europe provided him with added perception of the historical and geographic scope of environmental concern on the part of geoscientists. He is also known for his active involvement in the study and solution of problems of communication that arise from the necessity of conveying geologic information to specialists in other disciplines and to concerned laymen.

Dr. Betz received an A.B. from Columbia University in 1934 and a Ph.D. in geology from Princeton University in 1938.